NATIONAL INNOVATION AND THE ACADEMIC RESEARCH ENTERPRISE

NATIONAL INNOVATION AND THE ACADEMIC RESEARCH ENTERPRISE

Public Policy in Global Perspective

Edited by
DAVID D. DILL
and
FRANS A. VAN VUGHT

The Johns Hopkins University Press

Baltimore

© 2010 The Johns Hopkins University Press
All rights reserved. Published 2010
Printed in the United States of America on acid-free paper
2 4 6 8 9 7 5 3 1

The Johns Hopkins University Press
2715 North Charles Street
Baltimore, Maryland 21218-4363
www.press.jhu.edu

Library of Congress Cataloging-in-Publication Data

Dill, David D., 1940–
National innovation and the academic research enterprise : public policy in global perspective /
edited by David D. Dill and Frans A. van Vught.
p.   cm.
Includes bibliographical references and index.
ISBN-13: 978-0-8018-9374-2 (hardcover : alk. paper)
ISBN-10: 0-8018-9374-7 (hardcover : alk. paper)
1. Education, Higher—Economic aspects—OECD countries. 2. Education, Higher—Effect
of technological innovations on—OECD countries. 3. Technological innovations—Government
policy—OECD countries. 4. Higher education and state—OECD countries. I. Vught, Frans van.
II. Title.
LC67.68.O43 D55 2009
338.4′3378—dc22   2009002887

A catalog record for this book is available from the British Library.

*Special discounts are available for bulk purchases of this book. For more information,
please contact Special Sales at 410-516-6936 or specialsales@press.jhu.edu.*

The Johns Hopkins University Press uses environmentally friendly book materials, including recycled
text paper that is composed of at least 30 percent post-consumer waste, whenever possible. All of our
book papers are acid-free, and our jackets and covers are printed on paper with recycled content.

For
Caitlin and Matthew
Jasper and Lonneke

CONTENTS

As global forces transform the basis of economic development, policymakers in the mature economies have focused increasingly on promoting innovation and technical change as principal means of sustaining international competitiveness. In the leading developed countries, these national innovation policies have begun to shape and supersede traditional science and technology policies and are now wielding a significant influence on the university sector. Among the participant countries in the Organisation for Economic Co-operation and Development (OECD), the Academic Research Enterprise (ARE), which performs a substantial and growing portion of national research and development, particularly basic research, is the essential source of research human capital and is a primary channel by which new knowledge contributes to social and economic betterment. Consequently, governments in OECD countries have come to believe that higher education is an engine of economic development and that the effective steering of the ARE is a critical means of improving national innovation.

This book is the first systematic, comparative analysis of national innovation policies and their impact upon the ARE of the leading developed nations. The new concern with the relationship of the ARE to national systems of innovation poses questions for policymakers and researchers that have yet to be effectively addressed. The book therefore draws upon empirical research to assess current national reforms, the policy instruments that are being used to implement them, and the intended and unintended impacts they appear to have on the ARE.

We are deeply indebted to the contributions of our international colleagues who joined with us in this ongoing scholarly effort, which involved a number of meetings and paper revisions. The project began at a Consortium of Higher Education Researchers (CHER) annual meeting in Jyväskylä, Finland, proceeded through a special seminar among the authors in Seville, Spain, and continued via much e-mail correspondence as well as discussions at professional meetings in Dublin, Ireland, and Pavia, Italy. During the journey we lost our good friend and colleague Maurice Kogan, but we are honored that Maurice's contribution to Mary Henkel's paper on the United Kingdom is one of his last publications in the field.

We want to express our appreciation to the Ford Foundation for its generous support of our project, and most especially to Jorge Balan, who had the wisdom to see the value of the topic. Ramon Marimon kindly provided us with visiting professorships at the Max Weber Program of the European University Institute while we were completing the book. We also wish to convey our thanks to Asta Crowe for her typically fine administrative support of the project, to Jennifer Miller for her significant writing assistance on several of the chapters, to Brian Stucky and Karin Reese for technical assistance with the figures in the text, and to Karen DeVivo and Lois Crum for their very skillful and professional copyediting of a complicated text. We are also indebted to Ashleigh McKown of the Johns Hopkins University Press for her unstinting support and assistance in bringing our book to fruition.

NATIONAL INNOVATION AND THE ACADEMIC RESEARCH ENTERPRISE

CHAPTER ONE

Introduction

DAVID D. DILL AND FRANS A. VAN VUGHT

Over the past two decades, higher education policy in the developed countries has undergone unparalleled transformation. Much of the change has been motivated by an increased appreciation of the influential role that human capital now plays in the new global economy. As Ralf Dahrendorf (1995) perceptively observed, higher education has now become the primary determinant of an individual's "life chances." Consequently, all developed countries have adopted a higher education policy of "massification" (Trow 2000), rapidly expanding first- and second-level degree enrollments in their higher education systems. National debates about higher education reform have therefore been dominated by concerns with the educational function of the university. Correspondingly, the rapidly growing research on higher education policy has focused attention on the issues of university access, the efficient financing of university education, and the means of assuring the quality of academic provision (see, e.g., OECD 2006).

Over the same period, however, there has been a growing appreciation among policymakers in the developed countries of the way in which international forces have altered the basis of economic development (Soete 2006). In the global market, natural resources are no longer a key factor in economic growth. In many developed countries, there is an observable trend toward de-industrialization, an international outsourcing of increasing numbers of traditional industries as well as routine service activities, and a corresponding government concern about how to promote innovation and technical change as principal means of sustaining international competitiveness. This new policy orientation has been informed by the emerging research on National Innovation Systems (NIS), and in the leading developed countries, national innovation policies have begun to shape and supersede traditional science and technology policies (Balzat 2006; Nelson 1993; OECD 2005).

The emerging literature on National Innovation Systems highlights the contribution that academic research and research training make to industrial innovation (Mowery and Sampat 2004). Among the member countries of the Organisation for Economic Co-operation and Development (OECD), the academic research enterprise (ARE) performs a substantial and growing portion of national R&D (research and development), particularly basic research. The ARE is the essential source of research human capital and is a primary channel by which new knowledge contributes to social and economic betterment. Consequently, governments in many countries have come to believe that higher education is an engine of economic development and that the effective steering of the ARE is a critical means of achieving economic competitiveness (Laredo and Mustar 2001).

The new focus on the relationship of the ARE to national systems of innovation poses questions for public policy that have yet to be effectively answered. This volume intends to respond to these questions. In this context, we touch upon diverse issues such as the increasing political priority of innovation policies, the relationship between universities and industrial innovation, the concentration or dispersion of academic research, the nature of intellectual property rights, the appropriateness of immigration policies, and other issues that have not previously been considered components of traditional research and higher education policies. But our main focus is on national innovation policies, on the policy instruments that are being used, and on the impacts they appear to have on the ARE.

In the sections that follow, we further define what we mean by the ARE, outline the differences in orientation between the national science policies that previously guided university research activities and the emerging national innovation policies now influencing higher education, and introduce the sample of countries chosen for our analyses. The concluding section is a brief overview of the chapters that follow.

The Nature of the Academic Research Enterprise

The focus of our inquiry is the ARE, the academic sector's proportion of a nation's research system. The national research systems of the developed countries are composed of differentiated sectors: the R&D laboratories of profit-making industry and private nonprofits, government research institutes, and the research carried out in the higher education sector. Industrial laboratories, government research institutes, and higher education institutions play a significant role in the R&D activities of the OECD countries (table 1.1), but the relative size and influence of the sectors vary

TABLE I.I.
Percentage of Gross Expenditure on R&D (GERD), by sector, 2006 or latest available year

	Business enterprises	Government	Higher education	Private nonprofit
Australia	54.1	16.0	26.8	2.7
Canada	52.4	8.8	38.4	0.4
Finland	71.3	9.3	18.7	0.6
Germany	69.6	13.9	16.5	—[a]
Japan	76.4	8.3	13.4	1.9
Netherlands	55.3	13.8	28.1	0.7
United Kingdom	61.6	10.6	25.6	2.2
United States	70.3	11.1	14.3	4.2
EU27	62.6	13.8	22.5	1.1
OECD	68.0	11.8	17.6	2.6

Source: OECD 2007a.
[a] Included elsewhere.

among countries and are influenced by the design of public policy. The size of the private and nonprofit R&D sector is directly influenced by government contracts, particularly for military projects, and may be positively stimulated by R&D tax incentives. The size of the government-operated R&D sector is, by definition, a function of public policy. The academic sector is the leading performer of basic research and is also highly dependent upon national government for its financial support. The academic sector's proportion of national R&D performed varies significantly among the mature economies, averaging 17.6% for the overall OECD in 2005. The proportion of national R&D performed by the higher education sector, measured by Gross Expenditure on R&D (GERD), has increased among the OECD nations, from 16.3% in 1995 to 17.6% in 2005 (OECD 2007a).

The research activity of the academic sector is often categorized by the terms *university, university research,* and *research university,* but we have chosen instead to use the term *academic research enterprise* (ARE) in our analyses. This term was applied to an international study of research policies in the industrialized nations conducted under the auspices of the US National Science Foundation (NSF) in the early 1990s (Government-University-Industry Research Roundtable 1990) and has since been utilized as a means of clearly focusing on the contributions of the academic sector to national systems of research.

The massification and globalization of higher education have compromised the traditional terms used to categorize the research activity of the higher education sector. The title *university* is legally protected in many countries and is customarily reserved for institutions that conduct basic research and scholarship and also provide research doctoral education. But this legal constraint on the university title does not exist in the United States, which is generally acknowledged as having the largest number of academic institutions performing internationally recognized research and where some of those institutions do not use the university title. Furthermore, many countries, such as Australia, Finland, Germany, the Netherlands, and the United Kingdom, expanded their systems of higher education in the years following World War II by developing so-called nonuniversity institutions (e.g., the polytechnic sector), which combine an emphasis on technical and professional education with applied research. In some cases the research activities of these institutions may have an economic and social impact on society that is at least as great as that of the research activity of institutions in the same country traditionally named universities. In any event, over time the growing political influence of these nonuniversity academic institutions has permitted them to directly challenge and, in a number of countries, to eliminate the titular distinction between the university and the nonuniversity sector, thereby diluting the supposedly close association of the term *university* with significant academic research. Finally, in many European countries where the university title is controlled by the state, government funding policies supposedly assured a common "gold standard" of academic research and doctoral education among all universities. However, recent assessments and international rankings based upon research productivity and impacts (EC 2003; Institute of Higher Education 2007) have revealed substantial variation in the output and significance of research among the universities in the developed nations. In sum, the variation in academic research activity among higher education institutions in many countries is likely greater than common understandings of the terms *university* and *university research* would imply.

Furthermore, empirical studies of national innovation performance suggest that a focus on academic research and scholarly output alone is not a sufficient guide for contemporary public policy. Recent research has revealed that measures of research doctoral production and of active linkages between industry and universities, such as business support for university research, are significant additional predictors of national innovation and technical change (Balzat 2006; OECD 2005; Schmoch, Rammer, and Legler 2006). This empirical research reinforces the theoretical framework underlying the concept of National Innovation Systems (NIS), which emphasizes the influence that the supply of highly skilled human capital and the means of disseminating academic knowledge have on economic and social development in the industrialized nations.

The quality and the productivity of research doctoral training are critical, not only because research doctorates are an essential input to academic and industrial research, but also because mobile doctoral graduates are an important means of communicating new theoretical insights and emergent research methods to the larger society (Cohen, Nelson, and Walsh 2002). Furthermore, while the necessity of linking research and teaching, as advocated by Wilhelm von Humboldt in nineteenth-century Germany, may be debatable in institutions engaged primarily in first- and second-level higher education, the case for linking research with teaching is indisputable at the level of PhD training (Clark 1995). The joint production of research and doctoral education is the feature that most clearly distinguishes the research activity of institutions of higher education from industrial research laboratories and government research institutes. Therefore, public policies addressing academic research must necessarily be concerned with the interrelated system of research doctoral training.

Finally, the empirical research on NIS confirms that there is substantial variation across the developed countries in the impact of academic research on industrial innovation (Balzat 2006; OECD 2005; Schmoch, Rammer, and Legler 2006). At the heart of the NIS framework is the assumption that new knowledge must be effectively disseminated and absorbed if innovation and economic growth are to be produced. Therefore, interaction and cooperation among universities and businesses—that is, effective knowledge transfer—are essential conditions for socially beneficial academic research.

As a result, our conception of the ARE not only includes the customary activities of academic research and scholarship, but it also encompasses the related activities of research doctoral education and the means by which the outcomes of academic research are transferred to society.

The Evolution of Science Policy

The framework conditions and public policies governing academic research in the OECD nations experienced substantial reform in the closing decades of the twentieth century and are still being adjusted in the first decade of the twenty-first (Geuna, Salter, and Steinmuller 2003). The reforms have reflected changing expectations among policymakers about the function of higher education research, particularly its contribution to innovation, technical change, and economic development. These changes can be illustrated by tracing the evolution of science policy following World War II and its eventual absorption into national innovation policies (Lundvall and Borrás 2004).

The research and doctoral training function of higher education first came to fruition during the early part of the nineteenth century in Germany. In subsequent decades the research function of universities was adopted by other countries in Europe, North America, and eventually Asia. Prior to World War II, regional and national governments supported academic research and doctoral education primarily for historical and cultural reasons. The prevailing belief in the university as a bastion of culture and basic research, independent of markets and governments, tended to discourage closer connections between academic research and industrial activity in many countries. Still, even in this early period, there were important examples of the potential relationship between academic research, innovation, and economic development.

The first science-based industry emerged in nineteenth-century Germany in the field of synthetic dyestuffs, as a result of the existing strong university research system (Murmann 2003). Both the universities and the German government were openly supportive of the application of new discoveries in organic chemistry to industry. The actions taken by the German *Länder* to increase the public moneys supporting chemistry at German universities, in addition to the introduction of new laws to protect patent rights in private businesses, were the principal reasons German industry led the world in dyestuffs and later in organic chemical products up until World War II. In contrast, this belief in close cooperation between universities and industry was largely incompatible with the institutional framework governing universities in Britain, Germany's leading industrial rival; as a consequence, England's fine chemicals industry declined throughout the twentieth century.

Similar cooperation between universities and industry developed in agriculture in a number of countries during the nineteenth century (Lundvall and Borrás 2004). The US federal policy on land-grant universities and agricultural extension stations, initiated in the 1860s, is a frequently cited successful instance of the positive role that academic research and thoughtful knowledge transfer have played in innovation and economic development. Another example is the Danish state policy establishing the Agricultural University in Copenhagen in 1856 and the Agricultural Research Station in 1883, which helped to diffuse good dairy practices. These actions facilitated the development of a strong Danish dairy industry in the late nineteenth century.

The important contributions of academic researchers to military innovation and technology during World War II altered the prevailing public perception of the research function of universities. As outlined in Vannevar Bush's (1945) report in the United States, *Science: The Endless Frontier,* the experience of the war suggested that significant investments in academic research could benefit not only national security but also health and economic growth. Bush's report outlined the influential "linear" or "science-push" model of innovation. That model assumed that if government invested significant sums in university basic research, then applied research, technical developments, innovation, and benefits in the form of wealth, health, and national security would follow automatically. After World War II, many developed countries adopted the linear model, implementing new national science policies with the expectation that larger investments in academic research would stimulate economic development and welfare. While in subsequent years the rapidly growing expenditures on academic research were also reinforced by the pressures of the arms and space races between the United States and the USSR, public support of academic research was increasingly justified by the science-push model of innovation articulated in Bush's famous report.

Consistent with the linear model, the major issues for science policy in the post-war years were assuring sufficient resources for academic research, determining how to apportion them wisely, and developing decision processes to inform these judgments (Lundvall and Borrás 2004). In most OECD countries the principal policy instruments for funding university research continued to be undesignated grants to state institutions (i.e., general university funds or GUF) supplemented by peer-reviewed research grants awarded by discipline-oriented research councils (the "dual funding" model). In contrast, the policy instrument adopted for most federal academic research funding in the United States was a competitive, peer-reviewed grant system open to all institutions of higher education, public and private. In many countries, government funding decisions regarding new fields of academic research were often informed by foresight exercises carried out by members of the academic community. These collected science policy instruments purposely limited the ability of government to steer research in the university sector, provided substantial autonomy to academics in the selection of topics for research, and—in the developed countries—assured constant growth in academic research funding.

By the close of the 1980s, the traditional science policies based upon the linear model came under serious question (Martin 2003). In the United States, the end of the cold war lessened the need for research investments for national security, and in most developed countries the rising costs associated with education and the aging population—particularly mass higher education, health care, and social welfare—led to constraints on public expenditures for research. Growing concerns about public accountability inspired the development of new instruments for evaluating and assessing the outputs of academic research (Geuna and Martin 2003).

A number of influential analyses attacked the underlying assumptions of the linear model and called for a new conceptualization of the relationships between universities and industry (Etzkowitz and Leydesdorff 1997; Gibbons et al. 1994). The Mode 2 thesis (Gibbons et al. 1994), for example, suggested that the nature and locus of knowledge production was undergoing fundamental change. Mode 1 represented the existing science-push model, in which knowledge was produced by autonomous academic researchers working within their disciplines, primarily in universities, with limited accountability to society. In this mode academic research had little connection to societal needs and the research produced was often not taken up by users. In contrast, the newly emerging Mode 2 research was multidisciplinary in character, was carried out in nonuniversity as well as university settings, and was characterized by interinstitutional collaboration. In Mode 2 research, societal needs are influential from the outset in the definition of research problems, and accountability to funders is explicit.

The Mode 2 analysis captured some of the visible changes in university and industry relationships occurring in the last decades of the twentieth century. With the slowing of government funds for academic research, many universities were now aggressively seeking "third-stream" funding from industry to support their research activities. The new concept of an "entrepreneurial university" (Clark 1998), which closely cooperates with industry in its research activities, became an influential strategic guide to university reform in the more competitive international academic-research environment.

However, the strong version of the Mode 2 argument, which suggested it was a brand new form of knowledge production, is not supported by the empirical evidence (Martin 2003; Pavitt 2001). Mode 2–type interdisciplinary, collaborative, and socially useful research has always been a part of the university. Academic science has always been a dynamic system in which new interdisciplinary fields of research, such as biochemistry and cognitive science, have continually emerged from existing disciplines. Cross-institutional collaborations were an integral part of university research in the nineteenth century, as noted in the previously described examples of cooperation between university researchers and agricultural extension services in Denmark and the United States. Socially useful research has been carried out in the engineering departments of universities in Germany and Japan since the nineteenth century, and the eminent technical universities of Germany, Switzerland, the Netherlands, and the United States have long effectively combined research on national needs with distinguished fundamental research (Martin 2003). A more defensible interpretation of the Mode 2 thesis is that universities may now be returning to the more balanced combination of applied and basic knowledge production that characterized higher education prior to World War II. The post–World War II euphoria with basic science and the strong belief in the linear model had strengthened the voices of those within universities who advocated a more pure and autonomous form of academic research. As a consequence, during the postwar period, research to meet the needs of society and the economy often came to be perceived as a less central, if not deviant, form of knowledge production.

While the Mode 2 thesis raised useful questions about the science-push perspective, the limitations of the linear model that served as the basis for national science policies in the last half of the twentieth century were more clearly revealed by the new research on the economics of innovation that emerged in the 1980s. This new NIS perspective emphasizes the interactive character of the generation of ideas, scientific research, and the development and introduction of new products and processes.[1] The NIS approach takes an explicit policy orientation, and its framework has been promoted by influential international organizations such as the OECD, the World Bank, and the European Commission (EC) (Balzat 2006). As a result, the NIS perspective is informing the national policies of many developed nations and thereby altering their traditional science and higher education policies. In this book we focus on these national innovation policies. We describe and analyze these policies and explore their consequences for the ARE in the various countries.

The NIS Perspective and the Academic Research Enterprise

The development of a standard methodology for collecting and using statistics about research and development over the past fifty years by the OECD (i.e., the Frascati Manual of the 1960s and the Oslo Manual of the 1990s) helped economists to achieve a better understanding of the positive relationships between innovation, international competitiveness, and economic growth (Soete 2006). Researchers subsequently discovered that the key to international competitiveness among the developed countries was national "factors that influence the development, diffusion, and use of innovation" (Edquist 1997, 14), that is, the NIS. The new economic research revealed industrial innovation to be decidedly nonlinear. Instead, innovation has proved to be an interactive, reciprocal process involving different actors and organizations (Nelson 1993). From the outset, academic institutions were identified as playing a critical role in the NIS, and the evidence suggests that, if anything, their influence has grown over time (Mowery and Sampat 2004). However, the NIS research emphasized that while the tangible outputs of academic research—publications and patents—remain important, equally significant to successful innovation is the contribution of highly skilled human capital in the form of new science and engineering (S&E) and research doctoral graduates (Cohen, Nelson, and Walsh 2002). Most importantly, and in sharp contrast to the linear assumptions of the science-push model, the NIS perspective stresses the role of linkages among the various actors and organizations that participate in the overall innovation process (Edquist 1997; Nelson 1993). These linkages include not only formal knowledge transfer arrangements between universities and industry, such as science parks and joint university-industry research ventures, but also soft linkages—the many channels of communication by which knowledge is exchanged. Finally, a distinguishing difference between the NIS perspective and traditional science policy is the NIS approach's emphasis on institutional framework conditions: the governance processes, regulations, incentives, and underlying beliefs that shape innovative behavior (Balzat 2006).

The NIS perspective thereby reflects the increasing emphasis within the social sciences on the role institutions play in ensuring, defining, and steering market and nonmarket interactions (Eggertsson 2005). The research on NIS emphasizes how institutional framework conditions vary across nations and are "path dependent" (Liebowitz and Margolis 1995) in that they are shaped by the distinctive history and political forces within each country. A similar emphasis on historical forces has also characterized studies of national higher education systems where the differing governance traditions of market competition (in the United States), state supervision (in Europe), and professional control (in the Westminster countries, i.e., Australia, New Zealand, Canada, and the United Kingdom) have long influenced academic behavior (Clark 1983). Institutional structures are known to change over time, but slowly (North 1990), and there is the potential danger of "lock-in," in which cultural beliefs and political forces sustain existing institutions to the point where they are no longer efficient for the parent society (Liebowitz and Margolis 1995). While the internationalization of market forces and their common impact upon developed countries would suggest that national studies are becoming outmoded, the path-dependent character of innovation systems is one reason national borders still matter. The observed variance in NIS performance across countries motivates national-level studies seeking generalizable policy reforms and instruments that may prove to be more effective in the new, more common, international environment. At the same time, the design of appropriate national innovation policies requires insight into the institutional characteristics of the relevant country.

This same general point is true regarding attempts to reform national AREs and is well illustrated by the systematic effort to develop a common research area in Europe (EC 2007). The European Commission has articulated a comprehensive strategic approach to reforming the framework conditions for innovation of its member nations, including new indicators for the innovative capacity of small and medium-sized enterprises, a new central agency for funding basic research, and capacity-building initiatives to restructure academic institutions. But these reforms must be implemented in countries with historically different higher education systems, which inevitably are resistant to change. Similar implementation challenges confront any country attempting to apply an NIS approach to innovation policy.

Over the past twenty years, the NIS perspective clearly has influenced national reforms in science and higher education policy in the developed nations (Laredo and Mustar 2001; Lundvall and Borrás 2004; Rammer 2006). One version of the NIS perspective aims at promoting innovation within the existing institutional context of higher education through national and state-level incentive programs for basic research in fields deemed critical to future industrial innovation, such as biotechnology, information and communication technology, medical technology, nanotechnology, new materials, and environmental technologies. The more systemic and laissez-faire version of the perspective focuses on changing the framework conditions of higher education institutions to promote innovation. This latter approach includes changes in higher education governance processes and legal frameworks; the development of new yardsticks for the evaluation of academic research activity; and the adoption of new incentives to promote the transfer of academic research to society, an issue not traditionally considered part of science and higher education policy. Examples of this approach include changes in the laws governing intellectual property rights and academic labor markets; the introduction of competitive market forces into higher education systems; the transformation of institutional financing of research (GUF) into competitive research funding; the deregulation of university management; the evaluation of academic research ex post facto, utilizing new performance indicators; and novel initiatives to strengthen and reform research doctoral education as well as various incentive schemes designed to encourage more effective university-industry linkages.

As these reforms indicate, the leading OECD nations are implementing new policy instruments in their efforts to improve the effectiveness of their AREs. Policy instruments are "the tools of government" (Hood 1983), the mechanisms that governmental actors employ in their policy strategies to try to "produce" certain outcomes. Without these instruments, governmental policies would be no more than abstract ideals or fantasies. Policy tools can be categorized as legal, financial, and information policy instruments (Van Vught 1994). The reforms associated with the NIS perspective seem to suggest a significant shift in the types of policy instruments now being applied to steer the ARE in the leading nations. The specific nature and impact of these new instruments therefore deserve scrutiny.

In addition, as noted in the above discussion of national framework conditions, innovation and technical change are influenced by both formal and informal institutions. University behavior is similarly shaped by governance processes and regulations established by the state as well as by underlying academic beliefs (Clark 1983). Beliefs, to adopt the metaphor of Max Weber, often act like "railway switchmen," setting the tracks along which interests propel actions (noted in Clark 1995, 256). The preceding discussion about national science policies in the developed countries since World War II, for example, suggests how academic beliefs, such as the science-push model, the necessary unity of teaching and research, and the essential incompatibility of basic and socially useful research, significantly influenced the evolution of national ARE structures as well as academic research behavior. The NIS perspective and its proposed reforms clearly challenge a number of these underlying academic beliefs (Martin 2003).

Not surprisingly, the NIS perspective has provoked controversy within the academic community and raised questions among policymakers regarding the empirical evidence supporting the advocated reforms and policy instruments. This study focuses on issues related to these discussions. More specifically, we address the following questions:

- What is the influence of national innovation policies on ARE reforms among the leading OECD countries?
- What are the policy roles of the various governments at supranational, national, and state or province levels with respect to the ARE?
- What are the major policy instruments used to steer the ARE by these various policy actors?
- What are the effects of the use of these policy instruments, and what do these effects imply for the ARE in their respective contexts?

THE STUDY SAMPLE

Our study includes analyses of the ARE policies in Australia, Canada, Finland, Germany, Japan, the Netherlands, the United Kingdom, and the United States. In addition, to gain a deeper understanding of the complex interplay among different levels of public policy and their possible impacts on the ARE, we have included analyses of the relevant supranational policies of the European Union as well as of US state-level policies in California and Pennsylvania. Our focus on the developed nations is purposeful. Empirical evidence suggests that investments in innovation and higher education, particularly postgraduate education, are more likely to enhance growth for countries closer to the technological frontier (Aghion et al. 2005). Therefore, a careful investigation of the relationship between national innovation policies and the ARE in the advanced nations is most likely to yield useful insights.

The eight selected nations rank among the top eighteen countries in the world in the performance of their NIS as assessed by the European Commission (Hollanders and Arundel 2006). The eight countries together also carried out over 72% of the total Higher Education R&D (HERD) performed in the OECD nations in 2005 (OECD 2007a). Some general insights into the relationship between the NIS and the ARE can be deduced by a comparison of relevant indicators among the eight countries (table 1.2).

Public discussion of national innovation is often overly focused on the single indicator of Gross Expenditure on R&D (GERD) as a percentage of GDP. This indicator provided the highly visible target of 3% by 2010 for the EU countries in the context of their Lisbon Strategy, intended to make the EU the most dynamic knowledge-based economy in the world. However, as noted, by comparing the indicators in table 1.2, overall GERD is strongly influenced by private investment in R&D. While industrial investments in R&D can be positively affected by national tax policies, research on the NIS has identified a number of national funding indicators other than GERD that better predict innovation performance and also more directly influence knowledge generation by the ARE (Balzat 2006; OECD 2005; Schmoch, Rammer, and Legler 2006). At the national funding level, these include HERD as a percentage of GDP and the percentage of HERD financed by industry. An examination of these indicators in table 1.2 reveals, for example, that the very visible and large US investment in HERD is more clearly understood as a reflection of the size of its economy rather than as a unique national policy commitment to academic research. Finally, while comparative analysis of higher education systems often emphasizes the more entrepreneurial orientation of US universities and their close ties to industry, the United States actually trails many other developed countries in the proportion of academic R&D supported by business.

The mounting importance of science, technology, and innovation for economic growth has led to increased national investments in knowledge. The OECD suggests that national expenditures on R&D, software, and higher education can serve as proxies for knowledge investment (OECD 2007b). Among all OECD countries (table 1.2), this investment averaged 4.91% of GDP, with the United States investing the most (6.56%), followed by Finland (5.92%) and Japan (5.33%). The OECD suggests that the United States and Japan appear to be moving more rapidly to a knowledge-based economy than the EU (3.62%), because their respective expenditures in knowledge as a percentage of GDP have grown more rapidly than the average for the EU since 1994.

As noted, the literature centered on NIS performance has shown that the proportion of awarded S&E degrees, the proportion of awarded S&E research doctorates, the number of academic research publications, and their relative impact

TABLE 1.2.
Comparative data on ARE among selected OECD countries, 2006 or latest available year

	Australia	Canada	Finland	Germany	Japan	Netherlands	UK	US	EU	OECD
GERD as % of GDP	1.78	1.97	3.45	2.51	3.33	1.73	1.78	2.62	1.74	2.25
Government-financed GERD as % of GDP	0.70	0.66	0.87	0.70	0.56	0.64	0.58	0.77	0.61	0.66
HERD as % of GDP	0.48	0.75	0.65	0.41	0.45	0.49	0.45	0.37	0.39	0.40
HERD (in $M, total)	3,950	7,993	1,075	10,443	17,525	2,552	8,999	45,831	52,235	136,585
% of HERD financed by industry	5.7	8.3	6.6	14.1	2.8	6.8	4.6	4.9	6.5	6.1
Investment in knowledge as % of GDP	3.94	4.45	5.92	3.90	5.33	3.75	3.50	6.56	3.62	4.91
S&E degrees as % of total new degrees	21.4	19.4	29.9	29.8	25.0	16.1	27.9	14.7	23.4	21.2

New S&E PhDs as % of relevant age cohort	0.6	0.3	0.7	0.7	0.3	0.5	0.9	0.3	0.6	0.5
Scientific articles per million pop.	791.2	783.2	997.9	536.9	470.3	830.6	810.8	725.6	573.2	440.5
Relative prominence of cited scientific literature[a]	0.71	0.85	0.83	0.82	0.58	0.97	0.86	1.03	0.74	0.76
Triadic patent families per million pop.[b]	20.2	25.4	50.3	76.0	119.3	72.6	26.4	55.2	32.4	43.9

Sources: OECD 2007a; 2007b.

Note: GERD = Gross Expenditure on Research & Development. HERD = Higher Education Research & Development. S&E = Science & Engineering.

[a] A country's share of cited literature adjusted for its share of published literature. An index greater (less) than 1.00 would indicate that the country is cited relatively more (less) frequently than is indicated by the country's share of scientific literature.

[b] A triadic patent family is a set of patents taken at the European Patent Office, the Japanese Patent Office, and the US Patent and Trademark Office that shares one or more priorities.

are consistently significant indicators of knowledge development within the ARE (Balzat 2006; OECD 2005; Schmoch, Rammer, and Legler 2006). As emphasized in our definition of the ARE, the provision of both skilled human capital and new academic research is a critical function of higher education in the developed economies. In addition, patenting activity, as measured by the number of times the same invention qualified for a patent in the European, Japanese, and US patent offices ("triadic patent families"), is also an important indicator of national innovation and technical change. The relationship between basic academic research and patenting activity is complex and clearly not linear (Meyer 2000). Patenting behavior is most strongly influenced by industrial expenditures on R&D (OECD 2005), a relationship that is obvious in the case of Japan (table 1.2). However, research on patent citations (Narin, Hamilton, and Olivastro 1997) suggests an increasing intracountry connection between basic science research papers and applied technology, especially in high-technology fields. This raises the important question of the extent to which the low level of patenting observed in some countries with high levels of scientific output is indicative of weaknesses in the related national policies on the ARE (Martin 2003).

In sum, and as outlined above, our conception of the ARE as encompassing academic research, research doctoral training, and academic knowledge transfer clearly reflects this emerging evidence on the role and influence of higher education in the NIS.

ORGANIZATION OF THE BOOK

The chapters that follow present case analyses of public policies affecting the Academic Research Enterprise. We begin with discussions of Australia, Canada, and Japan. Next we examine the leading nations in Europe, first by way of a discussion of relevant EU policies and then by looking at Finland, Germany, the Netherlands, and the UK. We then analyze the US ARE, beginning with relevant federal-level policies and moving to examinations of state-level policies in Pennsylvania and California. We conclude with an assessment of what we have learned from the collected studies.

As V. Lynn Meek, Leo Goedegebuure, and Jeannet van der Lee note in their chapter, Australia is the only one of our sampled countries in which total government investment in R&D exceeds private investment. Australia has consequently adopted an aggressive reform program for its ARE, implementing a number of policies designed to achieve increased productivity in academic research through greater concentration and selectivity. General university funds were centrally reallocated to the Australian Research Council for competitive award, and a portion of the traditional university grant has been distributed by a series of performance-based funding schemes using institutional indicators, such as success in attracting research grants, research student productivity, and publications. Research training support is now similarly performance-based. Both the federal and the state governments have implemented priorities for academic research funding, emphasizing strategic technical fields. Australia has also financed its Cooperative Research Centres program to encourage greater research collaboration among industries, research organizations, universities, and government agencies. Characteristic of the "policy borrowing" clearly visible among national innovation systems, Australia is now implementing a Research Quality Framework closely modeled on the UK Research Assessment Exercise (RAE) experience. Meek, Goedegebuure, and Van der Lee provide some evidence that these policies have increased university attention to the management of research and research training and have increased productivity in publications, but the evidence for improved knowledge transfer and technical innovation is less clear. Australian universities do, however, exhibit a shift in emphasis from teaching to research as well as significant decreases in expenditures for basic research in mathematics, the basic sciences, the humanities, and the social sciences.

Donald Fisher and Kjell Rubenson's chapter examines ARE policy in Canada. Canada's history of "soft federalism" had led to a fragmented national science and technology policy in which provincial governments retained primary responsibility for higher education with the assistance of unconditional fund transfers from the federal government. Fisher and Rubenson review how this federal-provincial balance radically shifted as the federal government adopted a more formal science and technology policy increasingly informed by a national innovation agenda. The relatively low levels of R&D performed by government and industry have motivated the Canadian federal government to invest a larger proportion of GDP in academic research than any of our other sample OECD countries. During the past decade, the federal government also adopted a series of policies that substantially altered the Canadian framework conditions for the ARE. These included the reorganization of the existing research funding councils, the creation of the Networks of Centres of Excellence (NCE), and the introduction of new initiatives directly funding the universities for research chairs, research infrastructure, and indirect costs. As Fisher and Rubenson note, the NCE program helped increase industrial support for academic research and played a key role in redirecting the culture of Canadian universities toward commercially oriented science. A number of the examined federal instruments had innovative designs. The Canadian Foundation for Innovation (CFI) and the Canadian Graduate Fellowship (CGF) program were established as public foundations to buffer them from changes in government and government policy. By funding only 40% of each CFI program, the federal government was able to induce provincial and private support for the foundation's programs.

Akira Arimoto's chapter reviews the rapidly changing framework conditions of the ARE in Japan. While Japan is frequently placed among the global innovation leaders and is considered a country from which others could learn to improve their performance, this assessment is based primarily on industrial indicators rather than on measures of the ARE (Hollanders and Arundel 2006). Therefore, as Arimoto explains, the Japanese central government is intervening with major new policies designed to strengthen the ARE, a system still evolving from the substantial restructuring that occurred after World War II. Arimoto notes how the government's *Jutenka* policy has successfully restructured academic research activity by shifting university research from a primarily undergraduate to a graduate orientation. Graduate schools, nominally implemented following World War II, are now more fully realized as independent, internal university organizations with supplementary funding. This shift has been accompanied by a series of new, government-research-funding initiatives that have adopted a process of competition-based allocation utilizing third-party evaluation and by changes to the legal status of the national universities to national corporations with enhanced autonomy. Arimoto also notes that while Japanese universities' tradition of knowledge transfer has been minimal, recent government initiatives to reform intellectual property rights and to permit professors greater latitude for involvement in the private sector have been associated with a dramatic increase in the number of university-based venture start-ups. Arimoto reviews the impacts of these new policies, including the increasing differentiation observed among academic institutions as well as among subject fields and faculty members within the universities.

Frans A. van Vught's chapter examines the EU supranational policies intended to make research, higher education, and innovation in the member countries of the EU more competitive with North America and Asia. The early EU Framework Programmes were the central policy instruments for steering research and technology, but Van Vught suggests that these mechanisms were only partly effective because the collective research system of the member states suffered from fragmentation and duplication of effort and resources. In 2000 the European Commission (EC) proposed the development of a European Research Area (ERA). This goal was adopted by the EU member countries as part of the Lisbon Strategy and has been actively pursued by the EC through the development of a common EU research policy. The sixth and seventh Framework Programmes provided financial incentives for networking among centers of research excellence in the EU and for greater research cooperation among governments, the universities, and industry. These latter initiatives included the technology platforms and joint technology initiatives intended to foster technical innovation. The ERA-NET Scheme stimulates the development of a network among national research councils as a means of better coordinating national research programs across the EU, and the creation of the European Research Council provides substantial new competitive funding for basic research. Van Vught notes that these financial instruments have been augmented by EU initiatives designed to promote greater researcher mobility among institutions, sectors, and countries as well as by voluntary guidelines intended to clarify intellectual property rights and patenting processes within the EU. The several ministerial agreements that have accompanied the Bologna process, designed to make academic degree standards and quality assurance standards more comparable throughout Europe, also have dramatically altered the environment of EU higher education, encouraging the reform and strengthening of research doctoral education in the university sector. Van Vught assesses the impacts of these policy approaches on the EU ARE in the multi-echelon system of the EU. In addition he explores the EC's use of policy instruments, including its creative application of information instruments through the "open method of coordination."

Finland is frequently cited as a leader in national innovation and is one of the few EU nations to exceed the Lisbon Strategy goal of expending 3% of GDP on R&D. Seppo Hölttä's chapter reviews the national policies designed to make the ARE an integral part of the Finnish innovation system after a deep recession in the early 1990s. While many countries are experimenting with incremental governance reforms to better coordinate the many government ministries and agencies involved in their national innovation systems, Finland early on instituted the Science and Technology Policy Council, chaired by the prime minister, to develop policy on higher education research, researcher training, industry, and technology. Finland also rapidly shifted from its long tradition of direct state supervision of higher education to an innovative policy instrument of state steering through results. The government now develops formal performance contracts with each university based upon measurable outputs, such as the numbers of PhDs awarded, publication rates, the ability to attract external research funding, and the designation of research centers of excellence. In return, the university sector has received substantially increased academic and financial autonomy and is to be awarded a more independent status under law. Knowledge transfer has been encouraged though the competitive grants of TEKES, the National Agency for Technology and Innovation, which has a budget almost double that of the Academy of Finland. Hölttä's analysis reveals why Finland is one of the best examples in the world of national government policies' stimulating regional innovation.

Jürgen Enders's chapter on Germany reveals a country confronting the complexities of a federal system with some similarities to the systems of Australia, Canada, and the United States, but with a uniquely structured ARE. Public research institutes in Germany are an important component of its R&D system but traditionally have been independent of both the government and the universities. Enders suggests that the growth of these institutes, along with the teaching demands accompanying the massification of German higher education, substantially weakened the research capacity of the traditional university sector. As in other EU nations, the German federal government has pursued deregulatory reforms to rebuild the ARE, delegating greater autonomy to the states (*Länder*), which, in turn, have adopted different regulatory regimes. The national government has also introduced greater competition into the ARE via the Excellence Initiative, an instrument intended to strengthen the international standing of the university system. As Enders notes, this initiative has encouraged mergers among several universities and proximate public research institutes, thereby increasing the critical research mass of the universities. Federal financial support for technology transfer through so-called lead projects has appeared to reinforce the already existing close ties between established industries and universities rather than assisting the development of emerging technologies. Germany has long been among the leading nations in PhD production because of attractive opportunities for doctoral graduates in the labor market, but PhD study has been poorly organized and graduates' specializations have not always reflected national needs. A federal program supporting graduate colleges and, as part of the Excellence Initiative, new graduate centers has improved the efficiency of research training. Enders notes that these new policies appear to have improved the overall productivity of the German ARE, but deeply ingrained beliefs regarding the unity of teaching and research, "egalitarian homogeneity" among the universities, and the ongoing tensions between the federal and state governments have limited the impacts of these reforms.

Ben Jongbloed's chapter on the Netherlands examines a country that has traditionally adopted a "hands-off" regulatory approach to the ARE. Academic research funding continues to follow a dual funding model in which the majority of support is awarded on a lump-sum basis to institutions. But over recent decades, the national government has implemented thoughtfully designed research funding instruments that stress research performance, quality, and relevance. The Dutch ARE continues to exhibit high levels of productivity and quality, as indicated by conventional publication measures, but innovation indicators suggest a "valorization gap" in which research knowledge has not led to equivalent technological and economic development. Recent governance changes have therefore permitted the

National Research Council to play a more strategic role in ARE policy. They have also introduced initiatives that concentrate funding in strategic fields of research, provide innovation vouchers designed to encourage new technology start-ups, and implement a "smart mix" strategy designed to reward entrepreneurial scientists and encourage greater university-industry collaboration. The Netherlands was also an early leader in developing research schools, interuniversity networks of academic staff and research students, as a means of creating more internationally competitive research doctoral programs. Reviewing the impact of these systematically designed financial incentives, Jongbloed suggests ironically that the largest positive impact on the ARE may have been derived from a nonfinancial instrument—a program of research assessments in which the quality, productivity, relevance, and viability of each institution's research program was reviewed by an external team of academic peers.

The distinctive institutional framework of UK universities—once characterized as the "private management of public money"—has influenced the higher education policies of the other Westminster countries and served as an ideal for many nations. Mary Henkel and Maurice Kogan's chapter reviews the far-reaching changes in UK science policy over the past two decades and their effects upon the ARE. Growing concern that the UK had failed to maximize the economic benefits of its strong science base led successive governments to adopt a much more directive national policy framework for the ARE, featuring strong mechanisms of coordination, regulation, and accountability. Henkel and Kogan systematically explore the efforts to restructure the work of the university research councils to better steer academic research; the Foresight programs that were designed to identify future priorities for science funding as well as to improve research collaboration among universities, government, and industry; the "third-sector" funding of knowledge transfer activities; and the policy goal of full-costing academic research. The most influential policy for change, however, has been the Research Assessment Exercise (RAE). While the RAE thus far appears to have reflected academic values, the policy has clearly increased the selectivity and concentration of research in the university sector. Henkel and Kogan examine the impacts of the new framework conditions on economic innovation, the overall UK ARE system, the universities, and individual academics.

The massive scale—representing 34% of all OECD national expenditures for higher education R&D in 2005 (OECD 2007a)—and complex federal and state structure of the US ARE make it difficult to capture in a single chapter. Given the significant international attention focused on US ARE policy, we have included separate but interrelated chapters on US federal policy and on policies in two of the leading states in academic science and technology: Pennsylvania and California.

David D. Dill's chapter on US national policy underscores the central role federal financing plays in academic R&D. While from a comparative perspective, US national ARE policy is judged to be poorly coordinated, federal research policy is distinctive in its use of mission-oriented agencies to award funds, its process of competitive allocation to both public and private universities, its reliance on merit-based peer review of research projects, and its use of indirect cost reimbursement. These distinctive framework conditions have helped to concentrate academic research in universities with a critical mass of talent and also have provided incentives for innovative interdisciplinary research. While research doctoral education is largely unregulated and less dependent upon federal funds, federal financing has had a greater influence on doctoral and postdoctoral education in the sciences and engineering. Federal regulations redefining intellectual property rights in the 1980s and federal subsidies for university-industry cooperative research have helped deepen and broaden long-standing university efforts to support knowledge transfer. Dill's analysis, however, suggests that in the new global competition for academic research, the traditional strengths of US federal policy are increasingly offset by discovered weaknesses. Recent econometric studies suggest declining research productivity in the US ARE. Also, the strong incentives for exclusionary licensing of academic knowledge may compromise the public benefits of "open science." Finally, the heavy reliance on programmatic research funding as the primary means of financing both academic R&D and research doctoral education appears to provide inadequate support for basic research in core scientific disciplines as well as insufficient incentives for the development of domestic human capital in critical fields of doctoral education.

Roger L. Geiger's chapter on Pennsylvania includes an insightful overview of ARE policies at the state level in the United States. Given the previously described US tradition of "science federalism," in which the federal government is primarily responsible for supporting basic research, state ARE policies have customarily focused on endogenous growth. Geiger usefully categorizes state policies as traditional "downstream" strategies, which focus on university support for local industries; more contemporary "cluster" strategies, which attempt to foster innovation by linking universities and start-up firms through enhanced technology transfer activities; and emerging "upstream" strategies, in which states invest in building academic quality and capacity in strategic, science-based fields. His analysis of Pennsylvania explores the implementation of the unique downstream Ben Franklin Partnerships policy as well as the Pennsylvania nanotechnology cluster initiative. Geiger's chapter suggests that Pennsylvania, like other US states, lacks the institutional framework to effectively insulate state ARE policy from politics, resulting in the "scatteration" of state-financed academic research activity. Similarly, Pennsylvania lacks mechanisms to objectively assess and evaluate research proposals and to implement science and technology initiatives. As a result, the state's investments in the ARE may be an inefficient approach to technical innovation. In contrast, Geiger's analysis suggests that the state's laissez-faire regulatory approach to state-supported higher education has proved to be a particularly effective ARE policy in the competitive US context. For example, while Penn State University, like other US state-supported universities, has seen its state allocations decline over the last decades, the university's independence from state regulation, including tuition restraints, has permitted it to increase its net revenue per student over the same period and to strengthen its international standing in research.

William M. Zumeta's chapter examines California, the leading US state in both industrial and academic R&D. With its distinctive Master Plan for Higher Education and an economy that is equivalent to the fifth-largest in the world, California's policies for the ARE have long attracted international interest. California is similar to Pennsylvania, but dissimilar from many other US states, in that the academic reputation of its university sector has often been attributed to the universities' constitutional autonomy from state government policies; this independence extends to personnel, contracting, budgeting, and tuition fee regulations. California's ARE policy, however, differs from Pennsylvania's in emphasizing a more upstream research funding strategy. While the university system also has thus far compensated for declining state revenues with tuition increases, Zumeta examines the difficulties this financing strategy creates for the universities' ability to compete successfully for the best research doctoral students. California has also provided substantial basic research funding for its ARE, but it has been focused in scientific fields associated with politically popular health issues, such as HIV-AIDS, tobacco-related diseases, cancer, and stem cell research. Zumeta's analysis of the new, highly visible California Institute for Regenerative Medicine illustrates the strains that state-based research policy creates for the effective conduct of academic basic research, including state efforts to recoup funding by asserting property rights over research discoveries. His conclusion that California shows little evidence of the institutional structures and "honest brokers" necessary to effectively steer the state's ARE is similar to Geiger's evaluation of Pennsylvania.

In the concluding chapter, David D. Dill and Frans A. van Vught review the implications of the collected case studies for the design of future public policies for the ARE. They examine the potential policy roles of governmental actors at supranational, national, and subnational levels, and the general aspects of their respective policies. They summarize the noted impacts of the primary policy instruments—

financial, regulatory, and information-based—currently applied by the leading industrial nations. They then explore two particular issues that emerge from the case studies: the relationship between university autonomy and regulation, and new forms of governance. They end by suggesting a needed new policy perspective—the mutual learning strategy.

NOTE

1.   The NIS perspective is presented under various names, such as the evolutionary approach (Nelson and Winter 1977), the technological paradigm (Dosi 1982), the technological innovation systems approach (Carlsson 2002), and the concept of sectoral systems of innovation (Malerba 2002).In this study, we have adopted the NIS perspective inspired by authors like Freeman (1982) and Dosi (1984) and further developed by Lundvall (1992), Nelson (1993), and Edquist (1997).

REFERENCES

Aghion, P., L. Boustan, C. Hoxby, and J. Vandenbussche. (2005) Exploiting States' Mistakes to Identify the Causal Impact of Higher Education on Growth. Working Paper, Harvard University Department of Economics, www.economics.harvard.edu/faculty/aghion/papers_aghion.

Balzat, M. (2006) *An Economic Analysis of Innovation: Extending the Concept of National Innovation Systems*. Cheltenham, UK: Edward Elgar.

Bush, V. (1945) *Science: The Endless Frontier*. Washington, DC: Government Printing Office.

Carlsson, B. (ed.). (2002) *Technological Systems in the Bio Industries: An International Study*. Boston: Kluwer Academic.

Clark, B. R. (1983) *The Higher Education System: Academic Organization in Cross-National Perspective*. Berkeley: University of California Press.

———. (1995) *Places of Inquiry: Research and Advanced Education in Modern Universities*. Berkeley: University of California Press.

———. (1998) *Creating Entrepreneurial Universities: Organizational Pathways of Transformation*. Oxford: IAU Press.

Cohen, W. M., R. R. Nelson, and J. P. Walsh. (2002) Links and Impacts: The Influence of Public Research on Industrial R&D. *Management Science* 48:1–23.

Dahrendorf, R. (1995) Whither Social Sciences? The 6th ESRC Annual Lecture, Economic and Social Research Council, Swindon, England, October 19.

Dosi, G. (1982) Technological Paradigms and Technological Trajectories: A Suggested Interpretation of Determinants and Directions of Technical Change. *Research Policy* 11:47–162.

———. (1984) *Technical Change and Industrial Transformation*. New York: St. Martin Press.

Edquist, C. (1997) *Systems of Innovation: Technologies, Institutions, and Organizations*. New York: Francis Pinter.

Eggertsson, T. (2005) *Imperfect Institutions*. Ann Arbor: University of Michigan Press.

Etzkowitz, H. and H. Leydesdorff (eds.). (1997) *Universities and the Global Knowledge Economy: A Triple Helix of University-Industry-Government Relations*. London: Francis Pinter.

European Commission (EC). (2003) *Third European Report on Science and Technology Indicators 2003: Towards a Knowledge-Based Economy*. Luxembourg: EC.

———. (2007) *The European Research Area: New Perspectives*. Brussels: EC, http://ec.europa .eu/research/era/pdf/era_gp_final_en.pdf.

Freeman, C. (1982) *The Economics of Industrial Innovation*. London: Francis Pinter.

Geuna, A. and B. R. Martin. (2003) University Research Evaluation and Funding: An International Comparison. *Minerva* 41:277–304.

Geuna, A., A. J. Salter, and W. E. Steinmuller (eds.). (2003) *Science and Innovation: Rethinking the Rationales for Funding and Governance*. Cheltenham, UK: Edward Elgar.

Gibbons, M., C. Limoges, H. Nowotny, S. Schwartzman, P. Scott, and M. Trow. (1994) *The New Production of Knowledge: The Dynamics of Science and Research in Contemporary Societies*. London: Sage.

Government-University-Industry-Research Roundtable. (1990) *The Academic Research Enterprise within the Industrialized Nations: Comparative Perspectives*. Washington, DC: National Academy Press.

Hollanders, H. and A. Arundel. (2006) *2006 Global Innovation Scoreboard (GIS) Report*. Maastricht, Netherlands: Maastricht Economic and Social Research and Training Centre on Innovation and Technology (MERIT), http://trendchart.cordis.lu/scoreboards/score board2006/pdf/eis_2006_global_innovation_report.pdf.

Hood, C. (1983) *The Tools of Government*. London: Macmillan.

Institute of Higher Education. (2007) *Academic Ranking of World Universities—2007*. Shanghai: Shanghai University, http://ed.sjtu.edu.cn/ranking.htm.

Laredo, P. and P. Mustar. (2001) *Research and Innovation Policies in the New Global Economy: An International Comparative Analysis*. Cheltenham, UK: Edward Elgar.

Liebowitz, S. J. and S. E. Margolis. (1995) Path Dependence, Lock-In, and History. *Journal of Law, Economics, and Organization* 11:204–226.

Lundvall, B-Å. (1992) *National Systems of Innovation: Towards a Theory of Innovation and Interactive Learning*. London: Francis Pinter.

Lundvall, B-Å. and S. Borrás. (2004) Science, Technology, and Innovation Policy. In J. Fagerberg, D. C. Mowery, and R. R. Nelson (eds.), *Oxford Handbook of Innovation*, 599–631. Oxford: Oxford University Press.

Malerba, F. (2002) Sectoral Systems of Innovation and Production. *Research Policy* 31:247–264.

Martin, B. R. (2003) The Changing Social Contract for Science and the Evolution of the University. In A. Geuna, A. J. Salter, and W. E. Steinmuller (eds.), *Science and Innovation: Rethinking the Rationales for Funding and Governance*, 7–29. Cheltenham, UK: Edward Elgar.

Meyer, M. (2000) Does Science Push Technology? Patents Citing Scientific Literature. *Research Policy* 29 (3): 409–434.

Mowery, D. C. and B. N. Sampat. (2004) Universities in National Innovation Systems. In J. Fagerberg, D. C. Mowery, and R. R. Nelson (eds.), *Oxford Handbook of Innovation*, 209–239. Oxford: Oxford University Press.

Murmann, P. (2003) *Knowledge and Competitive Advantage: The Coevolution of Firms, Technologies, and National Institutions.* Cambridge: Cambridge University Press.

Narin, F., K. S. Hamilton, and D. Olivastro. (1997) The Increasing Linkage between U.S. Technology and Public Science. *Research Policy* 26 (3): 317–330.

Nelson, R. (1993) *National Innovation Systems: A Comparative Analysis.* Oxford: Oxford University Press.

Nelson, R. and S. Winter. (1977) In Search of a Useful Theory of Innovation. *Research Policy* 6:36–76.

North, D. C. (1990) *Institutions, Institutional Change, and Economic Performance.* Cambridge: Cambridge University Press.

Organisation for Economic Co-operation and Development (OECD). (2005) *Governance of Innovation Systems.* Vol. 1, *Synthesis Report.* Paris: OECD.

———. (2006) *Education Policy Analysis: Focus on Higher Education.* Paris: OECD.

———. (2007a) *Main Science and Technology Indicators.* Paris: OECD.

———. (2007b). *OECD Science, Technology, and Industry Scoreboard 2007.* Paris: OECD.

Pavitt, K. (2001) Public Policies to Support Basic Research: What Can the Rest of the World Learn from US Theory and Practice? (And What They Should Not Learn). *Industrial and Corporate Change* 10 (3): 761–779.

Rammer, C. (2006) Trends in Innovation Policy: An International Comparison. In U. Schmoch, C. Rammer, and H. Legler (eds.), *National Systems of Innovation in Comparison: Structure and Performance Indicators for Knowledge Societies,* 265–286. Dordrecht, Netherlands: Springer.

Schmoch, U., C. Rammer, and H. Legler. (2006) *National Systems of Innovation in Comparison: Structure and Performance Indicators for Knowledge Societies.* Dordrecht, Netherlands: Springer.

Soete, L. (2006) Knowledge, Policy, and Innovation. In L. Earl and F. Gault (eds.), *National Innovation, Indicators, and Policy,* 198–218. Cheltenham, UK: Edward Elgar.

Trow, M. (2000) From Mass Higher Education to Universal Access: The American Advantage. *Minerva* 37 (4): 303–328.

Van Vught, F. A. (1994) Policy Models and Policy Instruments in Higher Education. In J. C. Smart (ed.), *Higher Education: Handbook of Theory and Research,* 10:88–126. New York: Agathon Press.

CHAPTER TWO

Australia

V. LYNN MEEK, LEO GOEDEGEBUURE, AND
JEANNET VAN DER LEE

This discussion of the Australian research enterprise starts with a general overview
of its structure and function. We then describe and analyze the national policy
framework in which Australian academic research takes place. As can be seen, Australia has had its fair share of reviews and government policy initiatives. This section
also discusses such issues as the research-industry interface, higher degree research
education, compliance with ethical policies governing all research on either humans
or animals, and other regulatory and quality assurance matters. We find that market discipline and the principles of concentration and selectivity have significantly
shaped the Australian academic enterprise in recent decades. However, this by no
means implies that burdensome bureaucratic regulation has disappeared from Australian higher education. Before providing concluding comments, we analyze in the
penultimate section of the chapter some of the consequences of the various policies
shaping academic research in Australia and suggest what may be major reform issues
for the future.

The Context of the Australian Academic Research Enterprise

Australia is a constitutional democracy consisting of a federation of six states and
two territories. In the Australian federal system, the powers of the commonwealth
are limited to areas deemed to be of national importance. As discussed in more detail below, just how far those areas extend is the subject of considerable debate.

Whereas in terms of landmass Australia is the sixth-largest country in the world—
approximately the same size as the continental United States—its population is only

slightly larger than that of the Netherlands. Most of the nation's approximately 21 million people (0.3% of the world population) are highly urbanized.

In recent years the Australian economy has grown faster than the economies of most other OECD (Organisation for Economic Co-operation and Development) countries, while maintaining low inflation and high employment. In May 2008 the rate of unemployment was approximately 4.3% of the workforce (Australian Bureau of Statistics 2008a). Since 1997 Australia has been a net exporter of R&D services, while education services have been an important export commodity for a somewhat longer period of time (ARC 2006, 4). "Since 1982, education services exports have grown at an average annual rate of around 14 per cent in volume terms, with their share in the value of total exports increasing from less than 1 per cent to almost 6 per cent in 2007," according to the Reserve Bank of Australia (2008, 1). In 2007 the education share of services exports was 26.1%, and it was 5.8% of total exports, with higher education being the most significant contributor. Of Australia's main exports, education services are ranked third, behind coal and iron ore, amounting to around $12.6 billion[1] in 2007 (Reserve Bank of Australia 2008).

Australia has a well-developed but comparatively small science base, with the majority of its R&D effort concentrated in the public sector. In comparison to the size of the nation, Australia's contribution to world science is impressive, particularly with respect to medical and health disciplines, biological sciences, and astronomy.

Over the past decade or so, Australia's investment in knowledge (defined by the OECD as including R&D, education and training, and software) as a percentage of Gross Domestic Product (GDP) has varied from a low of 3.7% in 1993 to a high of 4.12% in 2002 (ARC 2006, 9). This places Australia among the top 50% of OECD countries but below the OECD average of 5.2%. In the past three decades, gross expenditure on R&D has quadrupled, from $3.1 billion in 1976–77 to $15.8 billion in 2004–5 (Australian Bureau of Statistics 2008d). Total expenditure on R&D in higher education increased by about 25% between 2004 and 2006, from $4.3 billion to $5.4 billion (Australian Bureau of Statistics 2008b), driven primarily by a $460 million increase in spending on medical research (Matchett 2008).

For a number of historical and geographical reasons, the funding of R&D is more dependent upon the public purse in Australia than in most other developed countries (Davis and Tunny 2005; Productivity Commission 2007). In 2004 government-financed expenditure on R&D was 0.70% of GDP, compared to an OECD average of 0.68% (Productivity Commission 2007, 346).

In contrast, business expenditure on research and development (BERD) has traditionally been low compared to other OECD countries. For example, in 2005–6, BERD was $10.08 billion or 1.04% of GDP (Australian Bureau of Statistics 2008d), compared to 1.53% for the OECD (OECD 2007, 29). This is mainly because most of the large multinational corporations in Australia have their headquarters elsewhere and historically have conducted little of their R&D in Australia. Unlike the United States and the UK, Australia has very few private foundations for Australians to look to for research support (Wills 2001), and there is a much lower level of endowment support than some of the major US universities enjoy. Australian governments have instituted numerous policies and programs, such as tax concessions, in an attempt to boost business investment in research, as is discussed in more detail later in the chapter.

However, the historically low profile of business R&D in Australia appears to be moderating. From about the mid-1990s, BERD began to exceed that of government (ARC 2006, 15).

If we separate university expenditure on R&D from that of other government agencies, then the increase in business expenditure is shown to be even more prominent in recent years. According to Shanks and Zheng (2006, 29), "R&D activity in government agencies has increased, with expenditure growing in real terms at a rate of 2.0 percent a year since the 1970s. But, because that rate was slower than BERD, the government-agency share of GERD [excluding universities] declined from 48 per cent to 20 per cent." Of course, such historical comparisons are devilishly complex, particularly when we take into account the fact that business R&D may be subsidized by public funds, and research in universities and other government agencies may be supported by the private sector (Universities Australia 2008, 10).

Australia lags behind many other OECD countries in terms of Gross Domestic Expenditure on R&D (GERD) as a proportion of GDP. Australia's total expenditure in 2002–3 was 1.6% of GDP compared to an OECD average of 2.5%. By 2004–5, Australia's expenditure on R&D had increased to 1.76% of GDP, and the OECD average was 2.26% (Australian Bureau of Statistics 2008c). There have been calls from such bodies as the Australian Vice-Chancellors' Committee (AVCC), renamed Universities Australia as of 2007, for Australia to increase its investment in research to 2% of GDP by 2010 and 3% by 2020 (Universities Australia 2008, 10).

The traditionally relatively low level of investment in R&D from the private sector has meant that historically government has had to play a leading role in funding Australian science and innovation (table 2.1). The federal government channels its support for R&D through a variety of schemes and organizations, the two major ones being the Commonwealth Scientific and Industrial Research Organisation (CSIRO) and the nation's universities. The former receives about $700 million annually directly from the government (CSIRO 2007) and the latter $4.3 billion (Group of Eight 2008, 29). Of course, commonwealth support for universities is for

TABLE 2.1.

Gross Expenditure on R&D, by sector, Australia, in current $M

	1996–97	1998–99	2000–2001	2002–3	2004–5
Business	4,234.7	4,094.7	4,982.6	6,940.3	8,446.2
Government					
Commonwealth	1,266.6	1,179.4	1,404.8	1,531.3	1,573.4
State/territory	797.7	863.6	951.0	950.9	977.3
Total	2,064.3	2,043.0	2,355.8	2,482.2	2,550.7
Higher education[a]	2,307.6	2,555.1	2,789.8	3,429.6	4,282.8
Private nonprofit	185.8	225.3	289.0	359.5	493.2
Grand total	8,792.4	8,918.1	10,417.2	13,211.6	15,772.9

Source: Australian Bureau of Statistics 2008d.

[a] Data for the calendar year ending within the financial year.

teaching as well as for research. Some 27% of GERD is performed by the higher education sector, which is a bit above the European Union average of 22.5% but much higher than that of the United States (14.3%) or Japan (13.4%) (OECD 2007, 26). A greater proportion of Australia's R&D workforce is located in higher education than is the case for most OECD countries (OECD 2007; Universities Australia 2008, 6).

In terms of type of activity, experimental research and applied research consume the largest proportions of GERD, followed by strategic basic and basic research, both of which, as a proportion of GERD, have declined since 1990 (fig. 2.1). These decreases have led to calls for increased investment in basic research on the argument that "free riding on the rest of the world's research is not a realistic option" (Group of Eight 2008, 22; see also Universities Australia 2008), and to claims for recognition of the full costs of research.

As might be expected, the largest proportion of pure basic research and, to a lesser extent, strategic basic research takes place in the universities, while most of the experimental development takes place in the business sector. About one-third of applied research is carried out in universities, one-third in other government agencies (particularly CSIRO), and one-third in the business sector (table 2.2).

Illustrative of the debate that runs across government, industry, and higher education on the level of public and private expenditure on R&D is the way in which each presents the data so as to support particular points of view. Universities Australia tries to emphasize the relatively low levels of government and business expenditure and does this through continuous reference to comparative data (see Universities Australia 2008 for the latest example). The previous Liberal coalition government,

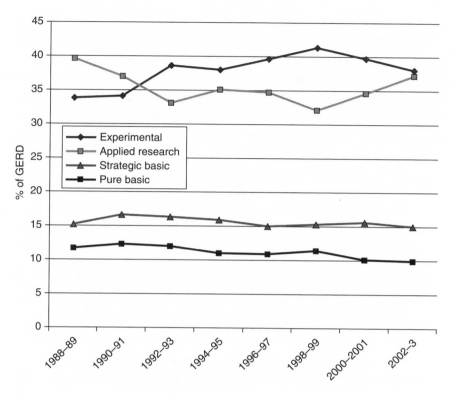

Fig. 2.1. Gross Domestic Expenditure on R&D, by type of activity. *Source:* ARC 2006, 11.

TABLE 2.2.
Gross Expenditure on R&D, by type of activity, 2003

Sector	Pure basic research	Strategic basic research	Applied research	Experimental development
Business	4%	18%	35%	85%
General government: commonwealth	8%	25%	16%	6%
General government: state	4%	7%	15%	2%
Higher education	79%	42%	32%	6%
Private nonprofit	5%	8%	3%	1%
Total ($M)	1,240	1,904	4,379	4,727

Source: AVCC 2005, table E.9.

basically on a day-to-day basis, referred to forecast expenditure under the Backing Australia's Ability agenda (a 2001 policy initiative to boost research funding over a ten-year period), while not providing much accessible data on actual expenditures. The new Labor government has launched two major reviews (discussed in a later section). And the business sector makes little reference to funding but stresses the need for practical and strategic relevance of public research. These points are revisited in the next sections of this chapter.

The Policy Framework
HISTORY

An exceptional feature of the Australian higher education sector is that the states have legislative control of higher education institutions, while financial responsibility rests with the commonwealth. Historically and constitutionally, all forms of education in Australia have been primarily a matter for the states. But in the years since the Second World War, there has been substantial and increasing federal intervention in higher education. Decisions by successive governments have significantly changed the Australian higher education landscape and have ensured that the federal government would dominate planning and funding of this sector.

In 2008 the nation's higher education sector consisted of thirty-seven public universities, some of them quite large, with enrollments in excess of 45,000 students; two small private universities; and a number of small, specialist institutions, both public and private. In 2005 an Australian branch of the American Carnegie Mellon University was established in Adelaide, as was a business development office of the British Cranfield University. Additionally, an agreement was recently signed for the 2009 establishment of a school of energy and resources in Adelaide by University College London (University World News 2008). Up to 2007 there were more than 150 non-self-accrediting higher education providers registered by the states and territories (DEEWR 2008a).

In 2007 Australia had nearly 1 million students enrolled in higher education courses, about one-quarter of whom were full-fee-paying overseas students. Up to this time, the defining characteristic of an Australian university was its strong endorsement of two principles: the unity of teaching and research, and the need for a broad, comprehensive curriculum. But the former Liberal coalition government actively challenged these principles.

From 2005 up to the end of 2007, successive federal education ministers called for the commonwealth to assume full legislative as well as financial control of higher education. They were motivated, in part, by the desire of the then federal government to introduce more fee-for-service private higher education providers. In July 2006 the then minister announced, under the banner of enhancing diversity, that she had achieved agreement with her state and territory counterparts to "provide greater choice for students to study at a variety of high quality higher education institutions." The minister said the new set of National Protocols for Higher Education Approval Processes would "allow . . . higher education providers to accredit their own courses, bypassing costly and time consuming reaccreditation processes run by State Governments" (Bishop 2006). Previously, only universities could accredit their own courses. The agreement also gave "specialist institutions . . . access to a university title" and reduced "research and higher degree teaching requirements for new universities in their first five years of establishment." This decision has the potential of transforming the Australian higher education landscape more than any other decision in the past decade. It remains to be seen whether the new Labor government will continue along a similar policy line with respect to this issue.

Since the early 1990s, the Australian higher education sector has experienced profound change, driven by, among other things, massification—the increase in student numbers that accelerated throughout the 1980s and 1990s. One of all governments' key strategies to cope with the rapid expansion of higher education has been to encourage institutions to diversify their funding base and to adopt market-like behavior. Australia is possibly the quintessential example of marketization and internationalization of higher education. Presently, the commonwealth provides only about 40% of the cost of higher education. The other main sources of funding are domestic and international student fees, followed by research grants, consultancies, investments, and so forth.

In most OECD countries, while private expenditure on higher education has risen more rapidly than public expenditure, public expenditure has expanded as well. Australia appears to be the exception (OECD 2006). Funding of Australian higher education increased during the period 1996–2005 (1996 was the year the former Liberal coalition government gained power) with respect to all sources of revenue. However, direct public funding from the commonwealth declined, as is illustrated in figure 2.2. The Higher Education Contribution Scheme (HECS)—tuition fees for Australian students collected through the tax system—was introduced in 1990. Funding constraints have been a major factor impacting the Australian research enterprise.

In the late 1980s, the then Labor government, which initiated extensive reforms of the higher education sector, explicitly stated that it was not prepared to fund growth entirely from the public purse. The former Liberal coalition government had gone even further in demanding that an increasing proportion of the financing

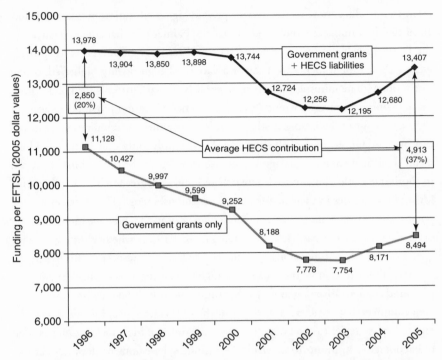

Fig. 2.2. Funding per government-supported university student. *Source:* Kniest 2007, 27. *Note:* HECS = Higher Education Contribution Scheme; EFTSL = Effective full-time student load.

of higher education should come from sources other than the public purse. In Australia, as elsewhere, the past two decades have seen the development of an approach to higher education steering that is quite different from what prevailed previously. This new approach is characterized by

- reductions in public expenditure;
- increased emphasis on efficiency of resource utilization;
- increased emphasis on performance measurement, particularly in terms of outcomes;
- increased emphasis on demonstrable contribution through both research and teaching to the economy of the nation; and
- the strengthening of institutional management, including research management, and of the policy and planning role of individual institutions.

Considerable pressure has been placed on Australian academic staff to become more competitive, productive, and accountable, while simultaneously being more entrepreneurial and innovative. This pressure applies as much to research as to teaching (Coates et al. 2008).

Table 2.3 gives a detailed breakdown of money specifically targeted for research in the higher education sector; the government component of full-time staff salaries (25%) nominally devoted to research is excluded. As the table illustrates, about two-thirds of university income specifically for research comes from government sources, and one-third comes from business and industry.

Table 2.4 gives a rough idea of expenditure on R&D by source of funds, type of research, type of expenditure, field of study, and discipline for the years 1988 to 2002. As mentioned above, in 1974 almost total funding for higher education was assumed by the commonwealth, and since then the funding and policy influence of state governments on higher education has been on the wane. This is reflected in the low percentage of state and local government expenditure (2% to 3%). However, it is worth noting that in recent years some states have been targeting university funding, particularly in the biotechnology fields, in the belief that such investment will strengthen the local economy—a clear commitment to the notion of the knowledge economy.

TABLE 2.3.
Research income, by source, 2003

Research income source	Total ($M)	% of total research income
Commonwealth competitive grants	628	42
Noncommonwealth competitive grants	22	2
Total national competitive grants	650	44
Local government	8	1
State government	144	10
Other commonwealth government	114	8
Total other public sector funding	267	18
Australian contracts	169	11
Australian grants	74	5
Donations, bequests and foundations	70	5
International funding	138	9
Syndicated R&D	2	0
Total industry and other funding	452	31
Commonwealth grants to CRC's	72	5
Nonuniversity participants	26	2
Third-party contributions	14	1
Total CRC funding	112	8
Total university research income	1,481	101
Total government grants	917	62

Source: AVCC 2005, table E.4.

Note: Percentages do not add to 100 because of rounding.

TABLE 2.4.
University expenditure on R&D, 1988–2002, in percentages

	1988	1990	1992	1996	1998	2000	2002
Source of funds							
General university funds				65	64	63	59
Commonwealth schemes				16	17	17	15
Other commonwealth government				7	7	6	12
Total commonwealth government	91	88	91	89	88	87	86
State and local government	2	3	2	2	3	3	3
Business enterprise	3	2	3	5	5	5	5
Other Australian	4	6	4	3	3	3	3
Overseas	1	1	1	1	2	2	3
Type of research							
Pure basic research	38	41	40	34	34	31	28
Strategic basic research	24	22	24	25	25	24	23
Applied research	31	31	30	35	35	38	41
Experimental development	7	6	6	6	6	8	8
Type of expenditure							
Staff	69	62	64	51	52	50	49
Other capital	8	10	11	6	6	6	5
Other current	20	22	21	42	41	42	42
Land and buildings	3	6	5	2	1	2	4
Field of study							
Medical and health sciences	17	19	19	21	23		
Biological sciences	14	12	12	12	12		
General engineering	14	7	7	7	7		
Other sciences and technologies	26	32	32	32	31		
Social sciences	18	22	22	19	19		
Humanities	12	9	10	8	8		
Research fields, courses and disciplines							
Medical and health science						24	25
Physical, chemical, math, and earth science						14	13
Biological science						12	12
Engineering and technology						11	11
Agricultural, vet and environmental science						7	7

	1988	1990	1992	1996	1998	2000	2002
Information, computing, and communication						4	4
Education						3	4
Economics, commerce, and management						6	6
Behavioral science						3	3
Humanities						3	3
Other						11	11
Total ($M)	1,077	1,351	1,695	2,308	2,600	2,775	3,430

Source: AVCC 2005, table E.7.

Notes: Classification of fields of study by disciplinary group changed between 1998 and 2000. Percentages may not add to 100 because of rounding.

The figures in table 2.4 also reconfirm the decline in university expenditure on pure basic research and the corresponding increase in funding applied research. We return to this point later in the analysis section. As might be expected, expenditure on medical and health sciences has increased over the years, as has expenditure on technologies.

Competitive Allocation of Research Funding. Since the early 1990s, federal governments of both political persuasions have encouraged competition among institutions, particularly with respect to research funding. With the intention of increasing competition over research funding even further, in June 1999 the then Liberal coalition government released a discussion paper on research and research training entitled *New Knowledge, New Opportunities* (Kemp 1999b). The paper identified several deficiencies in the existing framework that were considered to limit institutional capacity to respond to the challenges of the emerging knowledge economy. These included funding incentives that did not sufficiently encourage diversity and excellence; poor connections between university research and the national innovation system; too little concentration by institutions on areas of relative strength; inadequate preparation of research graduates for employment; and unacceptable wastage of resources, associated with low completion rates and long completion times for research graduates. A particular concern was with research training and the funding of PhD and research master's students (discussed further below).

The Howard Liberal coalition government released its policy statement on research and research training, *Knowledge and Innovation: A Policy Statement on Research and Research Training*, in December 1999 (Kemp 1999a). Major changes to the policy and funding framework for higher education research in Australia were identified. The principal ones were

- a strengthened Australian Research Council and an invigorated national competitive grants system;
- performance-based funding for research student places and research activity in universities, with transitional arrangements for regional institutions;
- the establishment of a broad quality verification framework supported by Research and Research Training Management Plans; and
- a collaborative research program to address the needs of rural and regional communities (Kemp 1999a, iv).

These changes were put into effect by two new performance-based block funding schemes intended to "reward those institutions that provide high-quality research training environments and support excellent and diverse research activities" (Kemp 1999a, 15).

The Institutional Grants Scheme and the Research Training Scheme. The Institutional Grants Scheme (IGS) supports the general fabric of institutions' research and research training activities (DEST 2005a). Funding under the IGS is allocated on the basis of a formula. The components and weightings are success in attracting research income from a diversity of sources (60%); success in attracting research students (30%); and the quality and output of research publications (10%) (DEST 2004b, 4).

Funding for research training is allocated on a performance-based formula through the Research Training Scheme (RTS). Institutions attract funded Higher Degree Research (HDR) places based on their performance through a formula comprising three elements: numbers of all research students completing their degree (50%); research income (40%); and a publications measure (10%) (DEST 2004d, 7).

Research Infrastructure Block Grants and Regional Protection Scheme. In 2003 the two schemes outlined above, together with the Research Infrastructure Block Grants, which are also allocated on the basis of institutions' relative success in attracting competitive research funds, accounted for close to $1 billion in total university funding, or about 11% of total funding (DEST 2005b). These schemes' share of overall research funding dropped from 63% in 2003 to 57% in 2008 due to the increase in Australian Research Council (ARC) grants (table 2.5). Mainly for political reasons, the Regional Protection Scheme was introduced to buffer regional universities from the most severe effects of research funding competition. An additional cap on funding redistribution exists in the sense that in any one year no institution can receive funding that is either decreased or increased over the previous year by more than 5%.

Investigator-driven research projects are supported mainly by the ARC (for non-

TABLE 2.5.
University research funding, by scheme, 2001–8, in $M

Research funding	2001	2002	2003	2004	2005	2006	2007	2008
Research training scheme	541	540	541	541	540	540	540	540
Institutional grants scheme	274	284	285	285	284	284	284	284
Research infrastructure block grants	86	119	140	160	179	192	192	192
Regional protection scheme		2	3	6	3	3	3	3
ARC grants	257	285	364	414	472	535	535	535
NHMRC grants	159	193	205	205	205	205	205	205
Other research programs	7	36	65	66	64	62	9	9
Total research funding	1,323	1,459	1,603	1,677	1,747	1,822	1,769	1,769

Source: AVCC 2005, table E.5.

   Note: Detail may not add to total because of rounding.

medical areas of research) and the National Health and Medical Research Council (NHMRC) (for medical-related projects). The ARC supports a variety of other programs as well, including key centers in specialized areas. Table 2.5 summarizes the amount of research income generated by the various competitive research funding schemes.

Research Priorities. Commonwealth changes to research funding have required Australian universities to rethink much of their approach to the management of research and research training. High on the agenda has been the need to identify priorities, concentrate research effort, and develop a set of performance indicators and sophisticated research management information systems (Wood and Meek 2002). The former government's intention to introduce a UK-style Research Assessment Exercise (see below) further exacerbated the situation, at least until very recently.

Coupled with the introduction of new research funding mechanisms has been government intervention in the setting of research priorities. At the beginning of 2002, the Howard Liberal coalition government announced, as a result of a "consultation" process that was far from transparent, that a proportion (33%) of the ARC's funding would be targeted for research in the following four priority areas: nano- and biomaterials, genome/phenome research, complex/intelligent systems, and photon science and technology.

In May 2002 the government instituted a review process to further set national research priorities for government-funded research in the areas of science and engineering. According to the government, the priorities "highlight[ed] research areas of particular importance to Australia's economy and society, where a whole-of-Government focus has the potential to improve research, and broaden policy outcomes" (DEST 2002a, 1). The priorities announced at the end of 2002 were

- an environmentally sustainable Australia;
- promoting and maintaining good health;
- frontier technologies for building and transforming Australian industries; and
- safeguarding Australia (DEST 2002b, 1).

These priorities subsume the ARC research priorities mentioned above. When the priority review process was first initiated, the intention was to follow the research priority-setting exercise in the sciences and engineering with one in the social sciences and humanities. But that did not happen. Rather, subgoals for each priority area were written in such a way that the social sciences and humanities could be incorporated. Nonetheless, while broad in scope, the priorities are "hard-science" oriented and mainly emphasize areas of immediate economic relevance. The research priorities are applicable across all commonwealth research agencies and funding bodies. ARC grant applications, for example, ask applicants to explain how their research meets national priorities and is of national benefit.

Concentration and selectivity remain the key issues in research. This means that universities have to identify strengths and make hard decisions about allocating resources to some areas and not to others. It is fairly obvious that those areas best able to commodify their intellectual wares are the ones that stand to gain the most from the new funding regime. It is also fairly obvious that these areas are not randomly distributed across the academy. And it is not just science and technology who are the winners, but also those subfields that can lay claim to short- to medium-term economic returns on their efforts. There is a danger that basic science will be further ignored and, in particular, that those disciplines traditionally associated with basic research, such as chemistry and physics, will go into further decline.

Under the new funding formula for research students, universities earn income not only through student load but also through rates of completion. This formula presents a particular difficulty for faculties in the humanities and social sciences that often have a large number of research students who traditionally study part-time, take considerable time to complete their degrees, and have low completion rates compared to other disciplines. While absorbing a large amount of initial RTS load allocation, such areas may lose their student load in the future if completion rates are outside the formula guidelines. In protecting its overall share of the national research student quota, a university may decide that some subjects in the arts, humanities, and social sciences are ones that it can ill afford (Wood and Meek 2002).

Either wittingly or unwittingly, management within universities is playing the research-concentration and priority-setting game, with the potential result of segmenting academic staff into research haves and teaching staff have-nots. Directing research funds and infrastructure to priority areas means that nonpriority areas will have fewer resources to conduct research. This appears to be exactly what was intended: "It seems timely to challenge the assumptions of the academic model of much of the past century, and validate alternative academic career paths. Some academics may choose to specialise in teaching, and become 'teaching-only' academics. Some academics may choose to specialise in research" (DEST 2002c, 48).

The intention is not merely to decouple research from teaching, but to simultaneously tie research more closely to the needs of industry and the economy, while reducing unit cost. In Australian universities, management in many institutions strongly promotes those areas of the enterprise that appear to turn a profit, while shedding investment in less lucrative activities, such as the humanities, ancient history, and some modern languages. Given the decline of public funding and rising student numbers in a highly competitive and volatile market, institutional leaders may well indeed argue that they have no other choice.

As can be seen from this review of research funding policies, at the sector level concentration and selectivity have become the order of the day. But these sector-wide funding principles, not surprisingly, have also been translated into principles adopted by individual institutions. Universities are going through a process of identifying and rewarding research-active staff, identifying their research priorities and areas of strength, recruiting staff who can bring with them external research funding, and channeling funding to areas where it is hoped there will be significant growth in research.

DOCTORAL EDUCATION POLICY

As with funding of research generally, support for research training is guided by principles of concentration and selectivity. As discussed above, the RTS funds are paid to institutions upon HDR student completions, based on the following parameters: high-cost and low-cost fields at a 2.35:1 ratio, with completion payments also weighted for the level of the degree at 2:1 for doctorate versus master's completions (Group of Eight 2008, 77). Actual levels of payment currently are the subject of debate within the framework of the two system reviews. The universities argue that these levels are based on out-of-date drivers: "The current split reflects former funding cost relativities from the early 1990s which themselves largely replicated cost relativity estimates made by the former Australian Universities Commission in the 1970s. . . . It is time to take a fresh look" (Group of Eight 2008, 77).

All domestic HDR students receive an HECS exemption scholarship; that is, while tuition fees are charged for HDR courses, as for undergraduate courses, they are paid by the government, not by the student (but only for three years, or two years in the case of research master's degrees). After that, the institution must decide whether it will allow the student to continue to study while it collects no revenue, or whether it will charge the student a fee or terminate candidature.

When the RTS was introduced, it was argued that it would be an effective mechanism for freezing out weaker institutions from graduate research training. But the scheme has had several unintended consequences: for example, some of the strongest research universities actually have lost HDR revenue under the application of the original funding formula (subsequently modified). Also, the process has been slowed by the restrictions mentioned above on how much research income can be won or lost in any one year. Nonetheless, the former government was adamant that there is room in Australia for only a few comprehensive research universities. Debate on this issue is continuing through the various higher education reviews initiated in 2008 by the newly elected Labor government.

While all domestic HDR students receive tuition fee exemption scholarships, a variety of other research training awards also contribute to living expenses and some research-related costs. The main ones are the Australian Postgraduate Awards (APA) and International Scholarships administered by the Department of Education, Science and Training (DEST), which has become the Department of Education, Employment and Workplace Relations (DEEWR). There is a set number of APA scholarships allocated to institutions on a competitive basis. Within the parameters of DEST guidelines, institutions allocate the scholarships to subject areas and to students on a competitive basis. Applicants for ARC projects involving an industrial partner may request one or more PhD scholarships as part of the grant. Table 2.6 provides the number of HDR awards and ARC-funded postdoctoral fellowships and senior researcher fellowships, which are awarded on a competitive basis. It is interesting to note that, while the DEST awards have remained steady over the years, the ARC industry linkage scholarships have tripled in number. In 2008 the newly elected Labor government announced that the total number of APA holders would be doubled by 2012.

The competitive allocation of research funds has made Australian universities highly conscious of the importance of their research profiles, some would say to the detriment of teaching. However, the increased emphasis on HDR students has strengthened the way in which these students are supervised in many institutions. Many universities have or are in the process of introducing a "learning contract" between supervisors and research students. They also have policies guaranteeing minimum facilities to research students and have identified suitable staff eligible to supervise HDR students.

TABLE 2.6.
DEST/ARC postgraduate research training awards (number of new awards),
1996–2003

	1996	2000	2001	2002	2003
Australian PG awards (with stipend)	1,550	1,550	1,550	1,550	1,550
Australian PG awards (industry)	150	334	350	397	461
International PG research scholarships	300	300	300	310	330
Research fellowships (postdoctoral and above)	100	100	100	165	166
Total	2,100	2,284	2,300	2,422	2,507

Source: AVCC 2005, table E.2.

KNOWLEDGE TRANSFER POLICY

Regarding linking publicly funded research to economic growth and development, DEST (2006b, 35) wrote: "A key focus of Australian policy developments over recent years has been on enhancing the capacity of universities to contribute more actively to innovation and knowledge transfer through the commercialisation of their intellectual property (IP). While it has been common for many years for universities to have commercialisation offices or companies, commercialisation of IP has until recently been seen as a secondary priority compared with the core teaching and research missions."

Since the late 1990s, a raft of government policies and programs have come into being, designed specifically to connect academic research to economic development, with a particular emphasis placed on the role of universities in innovation. These programs not only encourage collaborations nationally but also support academics in forming collaborative networks with industry internationally. In addition, a number of small programs provided by such entities as the learned academies and small private foundations support meetings and conferences, personnel exchange, and so on. There are also nationally funded programs for international exchanges and the formation of international networks that may explicitly involve partners from industry.

The university-industry liaison office has become an integral part of management at most Australian universities. Many universities have established their own commercial arm for marketing research innovations and creating spin-off companies and related ventures (with varying degrees of success). Liaison officers and related research managers have formed their own professional association, the Australasian Research Management Society.

The way Australian universities internally organize and manage knowledge transfer varies considerably. Some universities integrate commercialization with their overall research management structures, while others have established separate offices for commercialization purposes. Harman and Harman (2004, 163) identify four models of university technology transfer support structures: "specialist expertise located in university research offices; specialist university research commercialisation offices; university-owned companies providing research commercialisation support; and research commercialisation support provided by companies jointly owned by groups of universities, or through collaborative arrangements."

According to Harman and Harman (2004, 163), "support for research commercialisation via specialist expertise in a single university research office is the model adopted by most universities on their entry into serious efforts to file patents and commercialise inventions and discoveries." One advantage of this approach is that the research office provides a "one-stop-shop," both for the university's staff and for firms interested in engaging university researchers. However, with the growing complexity of managing commercial contracts, negotiating patents, and protecting intellectual property, there has been a tendency toward greater specialization of tasks.

As a result, some universities have established "technology transfer support by specialised university technology transfer offices" (Harman and Harman 2004, 164). These innovation offices usually operate in parallel with the normal research office charged with handling government-funded research grants.

The third, more complex and expensive approach is "technology transfer support by university-owned companies" (Harman and Harman 2004, 164). An advantage of this approach is that technology transfer activities are placed on a more professional, businesslike footing, and it allows for the recruitment of staff specialized in the area of research commercialization and intellectual property. However, such companies can be expensive to operate, take considerable time before any return on investment becomes apparent, and, to a considerable degree, remove technology transfer activities from the direct control and scrutiny of senior university management (Harman and Harman 2004). Australian and other universities have had variable experience with the creation of separate technology transfer companies, and the failure of some such ventures has been as notable as the success of others. One problem is that, in their initial establishment, projections of success and substantial wealth generation often far exceed what can be reasonably expected.

The fourth model identified by Harman and Harman (2004, 166) is "technology transfer supported by jointly owned university companies or by consortia." Increasingly, some universities are cooperating in raising venture capital, in providing seed funding for spin-off companies, in sharing expensive research facilities, and in promoting research. A particular type of consortium that has become entrenched in Australia is the Cooperative Research Centre (CRC).

Of the variety of government initiatives to encourage greater university-industry interaction, two stand out: the ARC linkage program and the just-mentioned CRCs.

As the ARC (2008c) notes, Linkage Projects aim to

- encourage and develop long-term strategic research alliances between higher education institutions and industry in order to apply advanced knowledge to problems or to provide opportunities to obtain national economic or social benefits;
- support collaborative research on issues of benefit to regional and rural communities;
- foster opportunities for postdoctoral researchers to pursue internationally competitive research in collaboration with industry, targeting those who have demonstrated a clear commitment to high quality research;
- provide industry-oriented research training to prepare high-caliber postgraduate research students; and
- produce a national pool of world-class researchers to meet the needs of Australian industry.

In terms of the strategic research alliance mentioned above, industry is required to make a contribution to the project.

In 1990 the Australian government established the CRC program in response to "perceived weaknesses in the institutional framework for Australia's R&D effort" (Howard Partners 2003, 4), with a vision for creating a "one-stop-shop" for innovation. The CRC program thus was designed to facilitate collaborative research partnerships "between industry, research organisations, education institutions and government agencies" (Howard Partners 2003, 3), bringing together researchers and research users. The emphasis of the program is on "collaboration between business and researchers to maximise the benefits of research through an enhanced process of utilisation, commercialisation and technology transfer. It also has a strong education component with a focus on producing graduates with skills relevant to industry needs" (DIISR 2008b, 1). There are currently fifty-eight CRCs operating across six broad sectors: manufacturing technology (9), information and communication technology (5), mining and energy (7), agriculture and rural-based manufacturing (16), environment (12), and medical science and technology (9) (DIISR 2008a).

To partner in a CRC, a group is required to provide both cash and in-kind support (Cooperative Research Centres Association 2005). The bulk of the cash contributions comes through the CRC program, while the bulk of the in-kind contributions comes from universities, totaling more than $11 billion since the start of the program (Universities Australia 2008).

POLICIES ASSURING THE ACCOUNTABILITY OF PUBLICLY
SUPPORTED RESEARCH UNIVERSITIES

Quality Assurance. The introduction of quality assurance mechanisms in Australia is more recent than in most OECD countries. Quality assurance was not put on the Australian political agenda until the early 1990s. The 1991 publication by the Hon. Peter Baldwin, the minister for higher education, *Higher Education: Quality and Diversity in the 1990s,* and the subsequent establishment of the Committee for Quality Assurance in Higher Education (CQAHE) in 1993 were the first real indicators of government concern in this area. The committee's guidelines clearly conveyed the principle that quality assurance should be based on a reward system tied to the achievement of excellence (Meek and Wood 1997).

The CQAHE operated for only three years. In 2000 the present approach to quality assurance was established in the form of the Australian Universities Quality Agency (AUQA). AUQA conducts regular rounds of higher education institution quality audits, but the agency has focused more on institutional policies and processes that support teaching than on research. Government, nonetheless, has taken a keen interest in the assessment of research quality, the most recent example being the failed attempt to introduce the Research Quality Framework (RQF).

Research Quality Framework. The notion of a new quality assessment system for research was introduced in 2004. It followed recommendations made in the *Review of Closer Collaboration between Universities and Major Publicly Funded Research Agencies* (DEST 2004e) and the *Evaluation of Knowledge and Innovation Reforms Consultation Report* (DEST 2004a) for a mechanism to more adequately measure both the quality and the impact of publicly funded research. As such, it clearly fits the more general public-sector reform agenda of "value for money" and the further introduction of competition in the higher education sector. While at the time the idea was launched, the program's actual scope and mechanisms were unclear, the Howard Liberal coalition government was outspoken that the RQF should

- be transparent to government and taxpayers so that they are better informed about the results of the public investment in research;
- ensure that all publicly funded research agencies and research providers are encouraged to focus on the quality and relevance of their research; and

- avoid a high cost of implementation and imposing a high administration burden on research providers (DEST 2004c, 1).

At great expense—despite the last point—to both individual institutions and the sector as a whole, the RQF was developed over the three-year period 2005 to 2007, with the intention that it would become operational in 2008. The scheme took on some of the characteristics of the UK Research Assessment Exercise (RAE) but with a much heavier emphasis on impact in addition to quality. The principles used to define impact resulted in a wave of criticism, particularly from the Group of Eight universities (a self-selecting group of research intensive universities), which feared that this would open the doors for low quality research.

In 2007 the Labor Party promised to abort the RQF if elected. This is exactly what happened following its election in late 2007. On December 21, 2007, the minister for innovation, industry, science and research, the Hon. Kim Carr, in canceling the RQF, stated that "the RQF is poorly designed, administratively expensive and relies on an 'impact' measure that is unverifiable and ill-defined." His press release went on to state that "the Rudd Government is committed to a new streamlined, internationally-recognised, research quality assurance process using metrics or other agreed quality measures appropriate to each research discipline" (Carr 2007).

However, the demise of the RQF does not mean that the evaluation of research has been taken off the Australian political agenda. In June 2008 a consultation paper was released describing a replacement framework—the Excellence in Research for Australia (ERA) Initiative (ARC 2008b). The ERA is to be developed by the ARC in conjunction with the Department of Innovation, Industry, Science and Research and administered by the ARC. The aims of the ERA Initiative are to

- identify excellence across the full spectrum of research activity;
- compare Australia's university research effort against international benchmarks;
- create incentives to improve the quality of research; and
- identify emerging research areas and opportunities for further development (ARC 2008b, 5).

The proposed approach will be to "evaluate research undertaken in higher education institutions using measures of research activity and intensity; indicators of research quality; and indicators of excellent applied research and translation of research outcomes" (ARC 2008b, 6).

So far, the response from the sector has been cautious but positive. Who will be winners and who will be losers will depend considerably on the details and the perceived outcomes.

Although the ERA could have a major impact on the structure and function of the higher education and research system, the key question will be how many resources can and will be freed up for reinvesting in the sector. No unambiguous signals of a major financial injection have yet been detected despite the government's commissioning even more reviews of the sector. The *Review of Australian Higher Education: Discussion Paper* released in June 2008 raises many pertinent questions but provides no indication as to preferred directions and policy options, and it has skirted the funding question (DEEWR 2008b). A green paper from the Review of the National Innovation System, chaired by Dr. Terry Cutler, was released in the second half of 2008, recommending full-cost funding of research.

The RQF was a costly failure, although institutions may have benefited from having to take a very close look at their research outputs. But, had the RQF survived, one could ask what "real" purpose it would achieve. If that purpose was to concentrate research funding on the highest performing institutions, it could be argued that this was already being achieved. The vast majority of research funding already flows to a strictly limited group of institutions (tables 2.7 and 2.8); the RQF probably would have merely reinforced the Matthew effect. This may also turn out to be the result of the ERA Initiative.

Allocation of other research income, such as ARC grants, demonstrates a similar degree of concentration of funding in the same select institutions (table 2.8).

Ethics. All Australian universities and other research organizations have animal research ethics committees and human research ethics committees. Every research project must be reviewed by a research ethics committee to determine if the proposed research adheres to general ethical principles and to national and state legislation and codes of practice. With respect to human research ethics, the general ethical principles are set out in the *National Statement on Ethical Conduct in Research Involving Humans* (recently reviewed; see AVCC 2006). The principles are integrity, respect for persons, beneficence, and justice. All university and other research organization ethics committees must function within the parameters of a clear and prescriptive compliance framework set forth in the National Statement.

A substantial and increasing body of literature in Australia and elsewhere documents the centrality of ethical conduct to the research enterprise. Through a variety of mechanisms, universities and other research organizations are being held more accountable for the ethical conduct of their researchers (Allen 2002, 2). It has been noted in much of the literature that as a result of increased external pressure to more closely approve and monitor the ethical conduct of researchers, many institutions have adopted a rigid, standardized, legalistic, and narrow interpretation of "proper" ethical research behavior. The literature also indicates that best practice in this area

TABLE 2.7.

Australian Research Infrastructure Block Grant and HDR scholarship allocations, 2006

	RTS grant	IGS grant	RIBG	APA grant Amount	APA grant Places	APA grant RPS	Total
Go8 universities							
Melbourne	$59,906,570	$33,985,085	$26,050,735	$10,560,034	178	0	$130,502,424
Sydney	56,985,729	30,523,273	23,727,428	8,901,571	150	0	120,138,001
Queensland	53,922,836	28,731,438	19,709,820	8,528,635	142	0	110,892,729
NSW	42,568,321	23,022,144	17,028,054	6,937,380	112	0	89,555,899
Monash	40,211,886	21,370,274	14,184,850	6,810,756	113	0	82,577,766
ANU	27,393,976	17,664,169	17,248,705	4,901,640	85	0	67,208,490
WA	29,579,096	16,821,078	14,806,590	4,737,369	80	0	65,944,133
Adelaide	29,757,358	16,371,441	14,247,813	4,381,363	73	0	64,757,975
Subtotal	340,325,772	188,488,902	147,003,995	55,758,748	933	0	731,577,417
% of grand total	60.49	63.65	73.54	59.88	59.77	0.00	63.35
All other institutions	222,318,228	107,624,098	52,904,005	37,362,252	628	3,086,000	423,294,583
Grand total	$562,644,000	$296,113,000	$199,908,000	$93,121,000	1,561	$3,086,000	$1,154,872,000

Source: DEST 2006a.

Note: RTS = Research Training Scheme; IGS = Institutional Grants Scheme; RIBG = Research Infrastructure Block Grant; APA = Australian Postgraduate Awards; RPS = Regional Protection Scheme.

TABLE 2.8.
ARC Discovery Grant and Linkage Grant allocations commencing in 2008

	ARC Discovery	ARC Linkage	Totals
Go8 Universities			
Melbourne	$38,004,295	$8,047,807	$46,052,102
Sydney	34,497,035	3,599,507	38,096,542
Queensland	28,724,683	8,085,401	36,810,084
NSW	26,004,779	11,410,443	37,415,222
Monash	27,659,169	4,397,096	32,056,265
ANU	30,827,792	2,264,528	33,092,320
WA	13,089,935	1,581,127	14,671,062
Adelaide	14,143,514	615,643	14,759,157
Subtotal	212,951,202	40,001,552	252,952,754
% of grand total	70.80	62.78	69.39
All other institutions	87,845,201	23,715,587	111,560,788
Grand total	$300,796,403	$63,717,139	$364,513,542

Source: ARC 2008a, 2008d.
   Note: ARC Linkage funding commenced in July 2008.

adopts quite the opposite approach (Allen 2002). Some Australian universities are attempting to apply a "lighter touch" to ethics approval, particularly for low-risk projects. Overall, however, a fairly rigid compliance culture is establishing itself in Australian higher education.

Assessment

It is quite difficult for several reasons to assess the impact and outcomes of policy on higher education and its research effort. Data is often sporadic, out of date, and difficult to obtain. But, more importantly, the effects of particular policies often take a considerable amount of time to appear.

   This problem is exacerbated when the implementation of the funding dimensions of particular policies is relegated to the relatively distant future—which is the case for many Australian policies. The stark effects of the Howard Liberal coalition government's 1996 financial cutbacks affecting higher education did not become blatantly apparent until the end of that decade. The impact of new money committed through Backing Australia's Ability is only starting to emerge. But while the analysis of higher education and research policies is necessarily complex, it is also vital, because of the importance of the sector to the nation's economic and social well-being. Given the caveats mentioned above, in this section we summarize what appear to be some of the key trends and issues in Australian research policy and effort.

The report *Mapping Australian Science and Innovation* lists a number of weaknesses of Australian science (see DEST 2003, 12, 14–17, 19, 20, 22, 24). These include

- Australia's scientific standing in the world may be at risk and, in general, Australian science and patented technology has limited visibility and impact on the development of world technologies.
- Business innovation involving R&D and development of new technology remains low by international standards. . . .
- Investment in the development of strategic Information and Communications Technology (ICT) capability is low, which may weaken the innovation base and the future competitiveness of the economy.
- Australia's commercialisation record . . . remains low compared to other countries and is uneven within and across different research sectors. Continuing barriers to commercialisation include lack of access to early stage capital, a shortage of management and entrepreneurial skills and lack of fully effective links between researchers and industry. . . .
- Challenges remain in fostering science and innovation collaboration and linkages, especially between publicly funded research providers and industry. . . .
- Australia's research infrastructure is under pressure in terms of investment and maintenance, and in leveraging access to international research infrastructure in an environment of increasing scale, costs and technical complexity. . . .
- The long-term sustainability of Australia's skills base in the enabling sciences is under pressure in some areas with declines in participation in most science subjects in Year 12 and in S&T subjects at the undergraduate level at university. . . .
- Availability of innovation skills and cultural attitudes towards innovation limit Australia's innovation potential. . . .
- While total gross expenditure on R&D as a proportion of GDP is now some 50 per cent higher than in 1981, Australia continues to rank towards the bottom of OECD countries in terms of R&D investment.
- Government support for business R&D is low by international standards, being less than half that of the leading OECD countries.

The single most fundamental issue facing Australian higher education in general and research specifically remains, not surprisingly, funding. But more is at issue than just money. At the heart of the problem is whether Australia is to have a publicly supported, publicly subsidized, or fully private higher education system. Some have argued that little will be achieved with respect to funding until government of whatever persuasion agrees to restore full supplementation of operating grants for academic salary increases. While the former government in its latest round of reforms committed some new money to the sector, most of it was absorbed by salary increases. Moreover, as indicated above, the main funding reform has been to shift even more of the burden to the student consumer. But student fees will not support an increased research effort. In fact, with an ever-worsening staff-to-student ratio, in some universities money earned through research effort is actually subsidizing teaching through payment of staff salaries.

Australian higher education faces fundamental structural and long-term funding issues. The longer research infrastructure is allowed to decline, the more difficult it becomes for the nation to recover its R&D standing relative to the rest of the world. A past president of the AVCC observed that "the pace of change in public investment in universities is such that if our universities get too far behind those in other countries we will not catch up" (Chubb 2000, 3). He also raised the concern that "Australia will become an importer of knowledge and an exporter of talent and that we will have too few educated personnel locally to add value to the efforts of others let alone enough to produce from our own." As we argued earlier, for a number of historical and structural reasons, the research effort is more dependent on public support in Australia than in most OECD nations. However, the ideological commitment of the Howard Liberal coalition government was to the market and to privatization. As yet, there is little indication that the Rudd Labor government will depart significantly from this approach to the steering of Australian higher education.

The election campaign in 2007 featured the introduction of an "education revolution." But, so far, the main initiatives have been the two national reviews mentioned above: the Review of Australian Higher Education (the Bradley Review) and the Review of the National Innovation System (the Cutler Review). Both reviews have broad remits. Announced in March 2008, the Bradley Review was to "examine and report on the future direction of the higher education sector, its fitness for purpose in meeting the needs of the Australian community and economy and the options for ongoing reform" (DEEWR 2008c, 1). "In particular, the Review Panel will examine the current state of the Australian higher education system against international best practice and assess whether the education system is capable of: contributing to the innovation and productivity gains required for long term economic development and growth; and ensuring that there is a broad-based tertiary education system producing professionals for both national and local labour market needs" (DEEWR 2008b, 76).

The Cutler Review, started in January 2008, was to "provide advice on the role of basic research . . . [and] consider the issues of technology transfer and adaptation" as well as that of "research training and business' access to researchers." Furthermore, "the full cost of research" will be examined and a "comprehensive review of existing policies and programs will be undertaken" (Carr 2008). Featured in both reviews are increased research collaboration between public and private sectors and skills development in relation to the demands of the knowledge economy.

A deep issue in Australian higher education research is the connection between teaching and research. On the one hand, no country can afford to fund all of its universities as if they were world-class, research-intensive institutions. On the other hand, there are those who argue that all university teaching must be informed by research. Moreover, each institution has its own special arguments as to why it should be recognized as a leading research university (whether or not the facts support such arguments). The collapse of the binary system of higher education in the early 1990s has exacerbated this problem. The introduction of new performance-based research funding measures mentioned above (RTS and IGS) is designed to concentrate research funding on the research performers. It is too early to tell whether the policies will have the desired effect, since a cap has been placed on how much funding individual universities can lose or gain through the application of the policies. But in the longer term, more radical policies may be necessary.

The Howard Liberal coalition government instituted a number of policies to boost business investment in R&D. *Mapping Australian Science and Innovation* pointed out that "Australia is the only country in which business funding of R&D as a percentage of GDP is lower than government funding of R&D as a percentage of GDP" (DEST 2003, 367). The review in a background paper also observed that a key OECD finding is that "rapid growth in research and development is largely driven by increases in business-performed research and development" (AVCC 2003, 10). Given the country's history of investment in R&D, it is probably necessary to attempt to increase the share coming from business and industry. But this should complement, not diminish, the investment from other sectors, particularly government.

With research policy strongly based on principles of concentration and selectivity, it is hardly surprising that the government would wish to set national research priority areas. The danger here, however, is that if funding becomes progressively concentrated in priority areas, innovation may be "straightjacketed." This is one of the dilemmas a small country with a limited science base faces. While the nation cannot adequately fund all kinds and aspects of modern research, it must maintain a broad enough science base to participate in advances in knowledge globally. According to the AVCC (2003, 22), "The key issue is plurality: as a nation we need to support a range of research, and do so by a number of different means. Allowing any single approach to dominate would inevitably result in a diminished overall research capacity and a weaker national innovation system. The impact of research prioritisation should be restructured to recognise this fact."

Another aspect of priority setting is the prominence given to science and engineering at the expense of the social sciences and humanities. Current priorities pay little more than lip service to the social sciences. Much of the present thinking is based on the assumption that worthwhile research means commercialization, and commercialization means science and technology. Again, a more balanced approach is necessary. The social sciences have much to add, particularly to the nation's social and cultural prosperity. They also have an important role to play as critic of the environmental and social consequences of scientifically and technologically driven development. However, with an increasing emphasis on commercialization, the university's role of "speaking truth to power" may be lost sight of. There is some evidence to suggest that this is a significant problem in the US higher education sector (Newman, Couturier, and Scurry 2004). Even the AVCC (2003, 12) agrees that "recent priority setting in research has underrated the contribution made by the social science and humanities" and asserts that "the research base must include the humanities and social sciences. The focus in recent years on Science, Engineering and Technology (SET) has marginalised areas of research and scholarship which play an important role in society, and which provide the vital critical and creative underpinnings of many other disciplines. The social sciences and humanities provide an understanding of ourselves and of the human and natural world around us, and work with the sciences towards resolving the full spectrum of problems and challenges which confront us."

Related to the issue of priority setting is the emphasis placed on pure basic research relative to applied and developmental research. Both government and institutional management have been very interested in the commercialization of research outcomes. This has resulted in a shift of funding over the years from pure basic research to applied research, as is depicted in table 2.9. The emphasis on applied research reflects the concern by both government and institutional leaders that research outcomes be commercialized, which in turn leads to the funding of the type of research most likely to achieve this result (table 2.10).

Nonetheless, if basic, "blue-sky" research is progressively diminished, the fountain of ideas and advances in knowledge that feed other forms of research and technological innovation may dry up as well (AVCC 2003, 7; Universities Australia 2008). This possibility is reflected in the submissions to the Cutler Review, where, as mentioned earlier, a case is made for reinvestment in basic research (Group of Eight 2008; Universities Australia 2008).

TABLE 2.9.
University expenditure on research and experimental development, by type of research activity, 1988–2002, in percentages

Type of research	1988	1990	1992	1996	1998	2000	2002
Pure basic research	38	41	40	34	34	31	28
Strategic basic research	24	22	24	25	25	24	23
Applied research	31	31	30	35	35	38	41
Experimental development	7	6	6	6	6	8	8
Total	100	100	100	100	100	101	100

Source: AVCC 2005, 8.
Note: Percentages may not add to 100 because of rounding.

Society remains the major research category with respect to the socioeconomic objective of research, partially because health research is classified under society. The category of economic development is steadily increasing, while the most alarming trend is the sharp decline in nonoriented research, or what used to be classified as "advancement of knowledge." Fields that fall into the nonoriented research category include mathematical sciences, physical sciences, chemical sciences, earth sciences, biological sciences, political science and public policy, studies in human society, and behavioral and cognitive sciences (AVCC 2002, 2).

Another big-ticket item on the policy agenda relates to the matter of institutional diversity. The revision of the higher education protocols mentioned above is one means of promoting diversity through encouraging the introduction of private providers, particularly at the teaching end of the scale. With respect to comprehensive research universities, the former minister went on the record to say that Australia needs only about twelve such institutions. In recent years, there has also been talk that Australia requires one or two "world class" universities, with various members of the Group of Eight putting up their hand to take on the task of becoming such an institution.

But, so far, the Australian policy environment has been poor with respect to promoting substantial institutional diversity and encouraging institutions to create niche markets. While various research policies, including the failed RQF, have been designed to promote competition for research funding, a result of that competition has not been differentiation despite the obvious concentration of research funding in a few universities. The evidence suggests that the competitive research funding environment tends to lead institutions to imitate one another rather than to consciously diversify and seek niche markets (Meek et al. 1996). Clearly, some institutions win much more than others in the competition. Nonetheless, all universities

TABLE 2.10.
Socioeconomic objective of research, by type of funds received: Percentage of total HERD for each socioeconomic objective

	Economic development			Society			Environment			Nonoriented research		
	1996	1998	2000	1996	1998	2000	1996	1998	2000	1996	1998	2000
All sources	21	23	29	25	27	40	7	7	6	46	42	25
Commonwealth national competitive grants	21	24	27	25	26	39	8	7	6	46	43	28
State and local government	21	27	31	51	47	48	10	9	12	18	17	10
Business	43	42	44	21	22	32	9	10	7	26	25	16
General university funds	18	22	28	24	27	40	7	7	5	50	44	27
Overseas funds	23	26	27	36	32	47	6	4	5	33	38	21

Source: AVCC 2002.

see the necessity to play the research funding game, to win not only money but also prestige.

Conclusion

A more sophisticated approach to the role of research in higher education and its contribution to national innovation and well-being is starting to emerge in Australia. But both Australian government policymakers and universities still have much to learn. In many respects, short-term economic gain is increasingly guiding investment in research, possibly to the detriment of the longer-term viability of the nation's research enterprise. In early 2007 the federal government's own Productivity Commission (2007, xxiii) concluded that "Universities' core role remains the provision of teaching and the generation of high quality, openly disseminated, basic research. Even where universities undertake research that has practical applications, it is the transfer, diffusion and utilisation of such knowledge and technology that matters in terms of community wellbeing. Commercialisation is just one way of achieving this. The policy framework for universities should encourage them to select the transfer pathway that maximises the overall community benefits, which will only sometimes favour commercialisation for financial gains."

The trend toward privatization of Australia's public universities, at least in terms of funding, has led to a view of higher education and its research functions largely in terms of private rather than public benefits. But that trend neither is inevitable nor has it proceeded uncontested. In terms of policy, it is mainly a question of balance between the public and the private. But understanding what is the "correct" balance requires a good deal more knowledge and research than is currently available. It also requires political will. Australia, in late 2007, ushered in a new government with educational reform high on its agenda. It is too early to reflect upon the newly elected Labor government's approach to research and innovation. But if the past is any guide, then it is safe to conclude that, while the direction of change may be uncertain, it is highly unlikely that the pace of change will abate.

NOTE

1. Throughout this chapter, all monetary figures refer to Australian dollars.

REFERENCES

Allen, A. L. (2002) *The New Ethics: A Guided Tour of the 21st Century Moral Landscape.* New York: Hyperion.

Australian Bureau of Statistics. (2008a) *6202.0—Labour Force, Australia, May 2008.* Canberra: ABS, www.abs.gov.au/ausstats/abs@.nsf/mf/6202.0#.

―――. (2008b) *8111.0—Research and Experimental Development, Higher Education Organisations, Australia, 2006.* Canberra: ABS, www.abs.gov.au/AUSSTATS/abs@.nsf/Latest products/8111.0Main%20Features32006?opendocument&tabname=Summary&prodno =8111.0&issue=2006&num=&view=.

―――. (2008c) *8112.0—Research and Experimental Development, All Sector Summary, Australia, 2004-05.* Canberra: ABS, www.abs.gov.au/AUSSTATS/abs@.nsf/Latestproducts/8112 .0Media%20Release2004-05?opendocument&tabname=Summary&prodno=8112.0& issue=2004-05&num=&view=.

―――. (2008d) *Research and Innovation, Research and Development (R&D).* Canberra: ABS, www.abs.gov.au/Ausstats/ABS@.nsf/7d12b0f6763c78caca257061001cc588/5916DED2FA D9589BCA2573D20010C477?opendocument.

Australian Research Council (ARC). (2006) *ARC Submission to the Productivity Commission Public Support for Science and Innovation.* Canberra: Australian Government, www.arc .gov.au/pdf/ARC_submission_PC_06.pdf.

―――. (2008a) *Discovery Projects Selection Report for Funding Commencing in 2008.* Canberra: Australian Government, www.arc.gov.au/ncgp/dp/DP08_SelRpt.htm#summary.

―――. (2008b) *Excellence in Research for Australia (ERA) Initiative: Consultation Paper.* Canberra: Commonwealth of Australia, www.arc.gov.au/pdf/ERA_ConsultationPaper.pdf.

―――. (2008c) *Linkage Projects.* Canberra: Australian Government, www.arc.gov.au/ncgp/ lp/lp_default.htm.

―――. (2008d) *Linkage Projects Round 2 Selection Report for Funding Commencing in July 2008.* Canberra: Australian Government, www.arc.gov.au/ncgp/lp/LP08_R2_Selection.htm.

Australian Vice-Chancellors' Committee (AVCC). (2002) *Positioning Australian Universities for 2020: An AVCC Policy Statement.* Canberra: AVCC, www.universitiesaustralia.edu.au/ documents/publications/policy/statements/2020.pdf.

―――. (2003) *Advancing Australia's Abilities: Foundations for the Future of Research in Australia.* Canberra: AVCC, www.avcc.edu.au/documents/publications/policy/submissions/ BAA2_AVCC_Statement_Dec03.pdf.

―――. (2005) *Key Statistics—Research.* Canberra: AVCC, www.avcc.edu.au/content.asp? page=/publications/stats/research.htm.

―――. (2006) *Review of the National Statement on Ethical Conduct in Research Involving Humans.* Canberra: AVCC, www.universitiesaustralia.edu.au/documents/publications/ policy/submissions/AVCC-Submission-Ethical-Conduct.pdf.

Baldwin, Hon. P. (1991) *Higher Education: Quality and Diversity in the 1990s: A Policy Statement by the Hon. Peter Baldwin, MOP, Minister for Higher Education and Employment Services.* Canberra: AGPS.

Bishop, J. (2006) *Greater Diversity in a High Quality Higher Education Market.* News release.

Canberra: Media Unit, DEST, July 7, www.dest.gov.au/Ministers/Media/Bishop/2006/07/B005070706.asp.

Carr, Hon. K. (2007) *Cancellation of Research Quality Framework Implementation.* Canberra: Australian Government, http://minister.innovation.gov.au/Carr/Pages/CANCELLATION OFRESEARCHQUALITYFRAMEWORKIMPLEMENTATION.aspx.

————. (2008) *Leadership in Higher Education for a New ERA.* Address to the Senior Staff Forum, Innovative Research Universities Australia, Flinders University, SA, February 6, http://minister.innovation.gov.au/carr/Pages/LEADERSHIPINHIGHEREDUCATION FORANEWERA.aspx.

Chubb, I. (2000) *Our Universities: Our Future—An AVCC Discussion Paper.* Canberra: AVCC, www.universitiesaustralia.edu.au/documents/publications/policy/statements/ouof _4p.pdf.

Coates, H., L. Goedegebuure, J. van der Lee, and V. L. Meek. (2008) *The Australian Academic Profession in 2007: A First Analysis of Results.* Armidale, NSW: Centre for Higher Education Management and Policy, University of New England.

Commonwealth Scientific and Industrial Research Organisation (CSIRO). (2007) *Record Funding Package Backs CSIRO's Future.* Clayton South, Victoria: CSIRO, www.csiro.au/news/FundingPackageForCSIRO.html.

Cooperative Research Centres Association. (2005) *About the CRC Programme.* Canberra: CRCA, www.crca.asn.au/.

Davis, G. and G. Tunny. (2005) International Comparisons of Research and Development. *Economic Roundup,* Spring, 63–82.

Department of Education, Employment and Workplace Relations (DEEWR). (2008a) *Higher Education Summary.* Canberra: Australian Government, www.dest.gov.au/sectors/higher_education/.

————. (2008b) *Review of Australian Higher Education: Discussion Paper.* Canberra: Australian Government, www.dest.gov.au/NR/rdonlyres/06C65431-8791-4816-ACB9-6F1FF9CA3042/22465/08_222_Review_AusHEd_Internals _100pp_FINAL_WEB.pdf.

————. (2008c) *Review of Australian Higher Education: Terms of Reference.* Canberra: Australian Government, www.dest.gov.au/NR/rdonlyres/E447DAE5-E64F-4302-A356-0ACE 0BD2A1C5/20706/HEReviewToR1.pdf.

Department of Education, Science and Training (DEST). (2002a) *The Framework for Setting National Research Priorities.* Canberra: Australian Government, www.dest.gov.au/sectors/research_sector/policies_issues_reviews/key_issues/national_research_priorities/backgrnd/framework.htm.

————. (2002b) *National Research Priorities.* Canberra: Australian Government, www.dest.gov .au/sectors/research_sector/policies_issues_reviews/key_issues/national_research_priorities/.

————. (2002c) *Striving for Quality: Learning, Teaching and Scholarship.* Canberra: Australian Government, www.backingaustraliasfuture.gov.au/pubs.htm#2.

————. (2003) *Mapping Australian Science and Innovation—Main Report.* Canberra: Australian Government, www.dest.gov.au/sectors/research_sector/publications_resources/profiles/documents/final_report_x3_pdf.htm.

————. (2004a) *Evaluation of Knowledge and Innovation Reforms Consultation Report.* Can-

berra: Australian Government, www.dest.gov.au/sectors/research_sector/policies_issues
_reviews/reviews/previous_reviews/government_response_to_research_reviews.htm.

———. (2004b) *Institutional Grants Scheme 2004 Guidelines.* Canberra: Australian Government, www.dest.gov.au/highered/research/igs.htm.

———. (2004c) *Research Quality Framework, Consultation Discussion Starter.* Canberra: Australian Government, www.dest.gov.au/sectors/research_sector/policies_issues_reviews/key_issues/research_quality_framework/documents/rqf_disc_paper_rtf.htm.

———. (2004d) *Research Training Scheme 2004 Guidelines.* Canberra: Australian Government, www.dest.gov.au/highered/research/rts.htm.

———. (2004e) *Review of Closer Collaboration between Universities and Major Publicly Funded Research Agencies.* Canberra: Australian Government, www.dest.gov.au/sectors/research_sector/policies_issues_reviews/reviews/previous_reviews/research_collaboration_review/documents/pub_x3_pdf.htm.

———. (2005a) *Institutional Grants Scheme.* Canberra: Australian Government, www.dest.gov.au/sectors/research_sector/programmes_funding/general_funding/operating_grants/institutional_grants_scheme.htm.

———. (2005b) *Research Quality Framework: Assessing the Quality and Impact of Research in Australia—Issues Paper.* Canberra: Australian Government, www.dest.gov.au/resqual/issues_paper.htm.

———. (2006a) *Allocation_Advice_080807.xls—Summary of 2006 Research Block Grant Amounts.* Canberra: Australian Government, www.dest.gov.au/sectors/research_sector/programmes_funding/general_funding/rbgrants/.

———. (2006b) *OECD Thematic Review of Tertiary Education: Country Background Report—Australia.* Canberra: Australian Government.

Department of Innovation, Industry, Science and Research (DIISR). (2008a) *About CRCs.* Canberra: Australian Government, www.crc.gov.au/Information/ShowInformation.aspx?Doc=about_CRCs&key=bulletin-board-information-about-crcs&Heading=About%20CRCs.

———. (2008b) *Cooperative Research Centres.* Canberra: Australian Government, www.crc.gov.au/.

Group of Eight. (2008) *Adding to Australia's Capacity: The Role of Research Universities in Innovation—A Submission from the Group of Eight to the Review of the National Innovation System, April.* Canberra: Go8.

Harman, G. and K. Harman. (2004) Governments and Universities as the Main Drivers of Enhanced Australian University Research Commercialisation Capability. *Journal of Higher Education Policy and Management* 26 (2): 153–169.

Howard Partners. (2003) *Evaluation of the CRC Programme 2003.* Canberra: Howard Partners, www.howardpartners.com.au/publications/crc-report.pdf.

Kemp, D. (1999a) *Knowledge and Innovation: A Policy Statement on Research and Research Training.* Canberra: Commonwealth of Australia, www.dest.gov.au/archive/highered/whitepaper/report.pdf.

———. (1999b) *New Knowledge, New Opportunities: A Discussion Paper on Higher Education Research and Research Training.* Canberra: Commonwealth of Australia, www.dest.gov.au/archive/highered/otherpub/greenpaper/index.htm.

Kniest, P. (2007) Who Is Being Selective in Their Choice of University Funding Data? *Advocate* 14 (1): 26–27.

Matchett, S. (2008) Research Funds Healthy. *Australian,* June 16, www.theaustralian.news.com.au/story/0,25197,23868383-12149,00.html.

Meek, V. L., L. Goedegebuure, O. Kivinen, and R. Rinne. (1996) *The Mockers and the Mocked: Comparative Perspectives on Differentiation, Convergence and Diversity in Higher Education.* Oxford: Pergamon.

Meek, V. L. and F. Q. Wood. (1997) *Higher Education Governance and Management—An Australian Study.* Canberra: AGPS.

Newman, F., L. Couturier, and J. Scurry. (2004) *The Academic Marketplace: Preserving Higher Education's Soul in the Age of Competition.* San Francisco: Jossey Bass.

Organisation for Economic Co-operation and Development (OECD). (2006) *Education at a Glance 2006: OECD Indicators 2006.* Paris: OECD.

———. (2007) *Main Science and Technology Indicators.* Paris: OECD.

Productivity Commission. (2007) *Public Support for Science and Innovation.* Melbourne: Productivity Commission, www.pc.gov.au/__data/assets/pdf_file/0016/37123/science.pdf.

Reserve Bank of Australia. (2008) *Australia's Exports of Education Services.* Sydney: Reserve Bank of Australia, www.rba.gov.au/PublicationsAndResearch/Bulletin/bu_jun08/aus_exports_education_services.html.

Shanks, S. and S. Zheng. (2006) Econometric Modelling of R&D and Australia's Productivity. Staff Working Paper, Productivity Commission, Melbourne, www.pc.gov.au/research/staffworkingpaper/?a=37183.

Universities Australia. (2008) *Universities Australia Submission to the Review of the National Innovation System, April.* Canberra: Universities Australia.

University World News. (2008) *UK-Australia: First British Campus Down Under.* London: University World News, www.universityworldnews.com/article.php?story=20080530080712633.

Wills, P. (2001) *The National Investment in Science, Research and Education.* Address to the National Press Club. Canberra: National Press Club of Australia, August 21.

Wood, F. Q. and V. L. Meek. (2002) Over-Reviewed and Under-Funded? The Evolving Policy Context of Australian Higher Education Research and Development. *Journal of Higher Education Policy and Management* 24 (1): 7–25.

CHAPTER THREE

Canada

DONALD FISHER AND KJELL RUBENSON

The contours and the content of the academic research enterprise in Canada have experienced major and at times dramatic changes over the past twenty years. As the university sector in Canada is almost entirely a public venture, it follows that these changes have necessarily been housed in shifts in government policy. Further, the constitutional division of powers between the two levels of government has meant that the provinces take responsibility for education and the federal government takes the major responsibility for research funding.

The federal government developed a science and technology policy for the first time in the late 1980s. The Conservative government at that time and the subsequent Liberal governments have taken a neoliberal stance. This stance has translated into policies that explicitly try to move universities closer to the market, that emphasize commercialism and utilitarianism, and that in general weaken the boundary separating the academy from industry. These policy shifts are set within and have contributed to the emergence of the knowledge society (Drucker 2002).

The stage was set in 1989 with the creation of the Networks of Centres of Excellence (NCE). The drama unfolded in a major way after 1996, as the Liberal federal government created new policy instruments and restructured the Medical Research Council (MRC) into the Canadian Institutes of Health Research (CIHR) as the means for distributing massive increases in research funding to the university sector. The main story here is that the federal government withdrew funds from unaccountable block transfers to the provinces for postsecondary education (PSE) and simultaneously increased funding for research. Throughout this period, federal and provincial governments across the political and ideological spectrum have subscribed to the belief that investments in R&D (research and development) will translate into economic development.

Overview of the Structure and Functions of the Academic Research Enterprise

Canadian federalism is characterized by a major paradox.[1] On the one hand is the constitutionally derived responsibility the provinces have for social welfare, health, and education. On the other hand is the federal responsibility for concerns of national interest, for equality of treatment and opportunity, for economic development, and for Indians and lands reserved for Indians. This paradox has led to a major line of tension in federal-provincial relations as each jurisdiction attempts to fulfill these responsibilities. The provinces to varying degrees have attempted to protect the constitutional division of powers by either blocking or accommodating federal interference. Quebec has played the most significant role in both protecting its own autonomy and, by extension, pushing the federal government to observe at least the relative autonomy of the other provinces. This major line of tension has been influenced through time by structural factors that are simultaneously national and global.

Federal PSE policy has gone through some significant and at times dramatic shifts. The overlap of responsibilities between the federal and provincial levels of government produces a continual struggle for recognition, credit, and increased accountability. Federal governments have used the powerful instrument of federal spending power[2] to intervene with the enormous weight of federal taxes as the means for channeling funds directly to federal priorities and as levers for realigning the behavior of provincial legislatures. What has emerged from the basic line of tension and the ensuing struggle is a patchwork of indirect and direct federal spending and an assortment of conditional and unconditional federal-provincial agreements governing grants and transfers. Commentators have variously labeled the federal-provincial PSE relationship as "soft federalism" (Jones 1996), "checkerboard federalism" (Bakvis 2002), and "collaborative federalism" (Cameron 2004; Noël 2002; Simeon and Robinson 2004).

Over the past two decades, the federal government has used its spending power to reduce indirect transfers to PSE and to channel more money into direct funding to universities for research, research chairs, research infrastructure, and the indirect costs of research. Federal administrations have become stronger and more dominant in the federal-provincial relationship since the mid-1990s. Yet, we should note that while federal ministerial portfolios cover health and social welfare, they do not cover PSE (Young and Levin 2002), except in the case of Indian and Inuit education.

Federal governments have always viewed PSE in relation to their own responsibilities for national well-being. These include supraprovincial issues, such as defense, foreign affairs, and the national economy, as well the imperative to ensure that all citizens enjoy similar rights and living standards regardless of their province or territory of residence. The federal case for involvement in PSE has repeatedly invoked the senior government's responsibility for national economic policy, including human resource development, and for the educational and occupational standards that ensure citizens' interprovincial mobility and equity.

The federal government's responsibility for economic development has led it to support university-based research. Through national research councils, institutes, and intramural funding as well as various intermediary bodies like the Canadian Foundation for Innovation (CFI), the federal government has become the largest source of support for university-based research. Consequently the federal government wields considerable influence over this aspect of PSE. Similarly, the federal government has a role in financing student PSE, in the past through the Canada Student Loans Program (CSLP) and currently by funding fellowships through the national granting councils, the Canada Millennium Scholarship Foundation (see CMSF 2001, 2003), and the Canadian Graduate Fellowship program.

The main sources of academic research funding are through a dual support system. Until recently, infrastructure has been covered primarily through block grants from the provinces, which in turn have been heavily dependent on transfers from the federal government. Project and program grants have come primarily from three national funding bodies. In some provinces, notably Quebec and Ontario, additional funds have been provided through a parallel set of granting structures. Since 1996 the federal government has created a series of R&D programs that cut directly across the infrastructure/project divide. Funds are disbursed primarily using three mechanisms: grants directly to faculty members for research projects, capital funding on a shared-cost basis for infrastructure projects, and grants directly to universities for indirect costs of research. The funding is disbursed by federal agencies on a competitive basis and awarded in accordance with federal criteria, which include merit and national interests.

As the federal government has decreased its commitment to providing unconditional transfers to the provinces for PSE over the past ten years, it has substantially increased the funds allocated for R&D (Doern and Sharaput 2000; Fisher et al. 2006). The underlying policy decisions are set within a science and technology policy that emerged from competing definitions of *science, utility,* and the *public good.* By promoting industry's access to publicly funded research, the science and technology policy recognizes that scientific research is simultaneously fundamental and useful. But the policy also skews the balance in favor of private interests and commercial science.

The Canadian academic research enterprise has three main components: government and quasi-government funding agencies, universities, and faculty and doctoral students. Currently, the federal government provides research funding for academic disciplines and fields through three national funding bodies: CIHR, the Natural Sciences and Engineering Research Council (NSERC), and the Social Sciences and Humanities Research Council (SSHRC). In 2002–3, federal spending on these three bodies amounted to approximately Can$1.6 billion[3] (Fisher et al. 2006). In 2004–5, the three bodies awarded approximately $1.42 billion in research grants (CAUT 2007).[4] In addition, the government has created a series of policy instruments to provide additional research and capital funding. The list includes the Networks of Centres of Excellence (NCE) program, the Canadian Foundation for Innovation (CFI), the Canadian Research Chair (CRC) program, and indirect costs of research funding. In 2006 the NCE was funding twenty-one networks for a total of $570 million in current and longer-term spending (NCE 2007).[5] By November 2006, the CFI had awarded a total of $2.5 billion for both research and infrastructure (CAUT 2007).[6] By March of that year, almost 1,700 of the projected 2,000 CRC chairs had been awarded, and the entire original fund of $900 million had been distributed (CRC 2006). Finally, the government in 2005–6 distributed $300 million in indirect cost payments to 140 PSE institutions, including all those that grant degrees. As noted in the introduction to this book (see table 1.1), in 2006 the proportion of the Gross Domestic Expenditure on Research and Development (GERD) carried out in the higher education sector stood at 38.4% (OECD 2007). This proportion is higher than for the other seven countries covered in this volume and twice the OECD (Organisation for Economic Co-operation and Development) average for this sector, which was 17.6%. Similarly, in 2006 Canada had the highest Higher Education Expenditures on Research and Development (HERD) as a percentage of GDP (at 0.75%) in our sample of countries (introduction, table 1.2). Yet GERD as a percentage of GDP in 2006 stood at 1.97%, which was below the OECD average of 2.25% (OECD 2007).

In 2004 the total GERD stood at just over $26 billion. Governments and higher education funded 23% and 16%, respectively, while business enterprise accounted for 49%. Of the total GERD, two provinces, Ontario and Quebec, accounted for $11.7 and $7.2 billion, respectively, or approximately 73% (Statistics Canada 2007). These are also the two provinces that have been the most committed to creating provincial research funding structures parallel to the federal ones. Some sense of this commitment can be inferred from GERD as a percentage of GDP in Quebec and Ontario, which in 2004 stood at 2.72% and 2.26%, respectively. These percentages are well above the Canadian average and either level with or above the OECD average (OECD 2007).

Canada has 83 degree-granting institutions, which employ approximately 34,000 full-time faculty members. Four of these institutions are very small, private institutions that play little or no part in the academic research enterprise. Of the remaining institutions, 36 are either university colleges, specialized colleges or institutes, or universities focusing mainly on undergraduate education. Again, while many of these institutions have research aspirations and expect their faculty to bring in grants and conduct research, as a group they account for a small portion of the total amount of R&D. The 43 universities that remain have research and doctoral study as part of their central mission. Yet even they are divided into two groups: 20 medical/doctoral universities that are the most research-intensive, and 23 comprehensive universities that are like the former group in every way except that they do not have a medical school. In 2004, 15[7] of the 43 universities referred to above accounted for 80.69% of the total university research income (table 3.1). The 8 universities with research income in excess of $200 million accounted for 60.44%. While the universities of Manitoba and Saskatchewan did make the list of 15, Dalhousie University was below the cut-off threshold of $100 million, and the next university from the Atlantic region (Memorial University) was well below the threshold in the $40-to-$50-million range. The next 8 universities accounted for another 10.56% of the total.

Of the 1,689 CRCs awarded (CRC 2006), 883 (52.3%) were awarded to the 8 universities discussed above that received more than $200 million in research income. When we extend the list to the top 15, then the number of awards goes up

TABLE 3.1.
University research income, Canada, 2004

Funding level ($M)	No. of universities	Total ($M)	%
> 200	8	3,051,557	60.44
100–199.999	7	1,022,528	20.25
50–100	8	533,010	10.56
10–49.999	15	321,888	6.37
5–9.999	10	77,784	1.54
1–4.999	13	37,339	0.74
< 1	22	5,132	0.10
Total	83	5,049,238	100.00

Sources: Statistics Canada, www.statcan.ca; adapted from CAUT 2006, table 5.6, p. 46.

to 1,181 (69.9%). The other 58 universities with CRCs accounted for 508 awards or 30.1% (table 3.2).

Both tables 3.1 and 3.2 are somewhat misleading in that the Université du Québec has nine campuses spread throughout the province, and each one reports separately. The total research income for all the campuses in 2004 was well in excess of $100 million. This added information serves to highlight the dominance of Ontario and Quebec in the academic research enterprise. These two provinces contain 23 of the 43 research-oriented universities and 10 of the group of 15 referred to above. In 2004–5, Ontario and Quebec accounted for 37% and 29%, respectively, of the federal funding coming from the three granting councils. In turn, British Columbia and Alberta accounted for 12% and 11%, respectively, which means that these four provinces received 89% of this funding (CAUT 2007).

In 2002–3 Canadian universities had a total enrollment of 32,000 in doctoral programs (Statistics Canada 2007) and awarded 3,861 doctorates (table 3.3). Males accounted for 58% of this total. The gender divide is still somewhat predictable. The disciplines awarding the majority of doctorates to women were education (67%); visual and performing arts, and communications technologies (59%); and social and behavioral sciences, and law (58%). Humanities disciplines broke evenly between the two sexes. Males dominated in the traditional "hard" sciences, namely, architec-

TABLE 3.2.
CRC chairs, 2006

	No. of chairs	%	Cumulative %
University of Toronto	227	13.4	13.4
University of British Columbia	138	8.2	34.5
McGill University	123	7.3	20.7
University of Alberta	98	5.8	40.3
Université de Montréal	94	5.6	26.3
Université Laval	78	4.6	44.9
University of Calgary	64	3.8	48.7
McMaster University	61	3.6	52.3
University of Western Ontario	55	3.3	55.6
Queen's University	49	2.9	61.3
University of Ottawa	48	2.8	58.4
University of Waterloo	45	2.7	68.4
University of Manitoba	44	2.6	63.9
University of Guelph	30	1.8	65.7
University of Saskatchewan	27	1.6	70.0
Other universities (58 others)	508	30.1	100.1
Total	1,689	100.1	100.1

Source: CRC Web site, www.chairs.gc.ca, *Program Statistics,* data as of March 13, 2006.
   Note: Percentages may not add to 100 because of rounding.

TABLE 3.3.
Doctoral degrees awarded, by major discipline and sex, 2003

	No.	%	% Male	% Female
Agriculture, natural resources, and conservation	141	3.7	61.7	38.3
Architecture, engineering, and related technologies	537	13.9	83.1	16.9
Business, management, and public administration	105	2.7	64.8	35.2
Education	338	8.8	33.4	66.6
Health, parks, recreation, and fitness	266	6.9	45.5	54.5
Humanities	476	12.3	50.8	49.2
Mathematics and computer and information sciences	208	5.4	82.7	17.3
Personal, protective, and transportation services	4	0.1	100.0	0.00
Physical and life sciences, and technologies	1,039	26.9	65.1	34.9
Social and behavioral sciences, and law	666	17.2	42.3	57.7
Visual and performing arts, and communications technologies	56	1.5	41.1	58.9
Other	25	0.6	40.0	60.0
Total	3,861	100.0		
Average			58.1	41.9

Sources: Statistics Canada, www.statcan.ca; adapted from CAUT 2006, table 3.18, p. 35.

ture, engineering, and related technologies (83%); mathematics and computer and information sciences (83%); physical and life sciences, and technologies (65%); and agriculture, natural resources, and conservation (62%). Males also dominated in the business, management, and public administration disciplines (65%). The four largest discipline groups were physical and life sciences, and technologies (27%); social and behavioral sciences, and law (17%); architecture, engineering, and related technologies (14%); and humanities (12%). The hard science areas[8] accounted for 50% of the total, while the humanities, social sciences, and fine arts accounted for 31%.

The distribution of doctoral degrees corresponds to the distribution of the student population and particularly to the location of the large majority of research-intensive universities (table 3.4). Of the total number of doctorates awarded in 2003, the overwhelming majority were in central Canada (Ontario and Quebec), which accounted for 2,662 (69%), and the West (British Columbia and Alberta), which accounted for a further 907 (24%). Doctoral students are offered support by two principal means: loans and fellowships. Loans come primarily through the Canada Student Loans Program (CSLP), which is jointly funded by the federal and provincial governments. The federal government supports doctoral and postdoctoral students with fellowships across the whole range of disciplines. These fellowships are administered by the national granting councils (CIHR, NSERC, and SSHRC), the National Research Council (NRC), and the CMSF. In 2003 the federal government committed significant new money through the Canada Graduate Scholarships, which are administered by the three councils. The formula for distributing these funds broke with previous practice and used student enrollment by discipline and field rather than the amounts distributed currently by the three granting councils. This has meant that for the first time the humanities and social science disciplines have benefited the most. In addition to the parallel fellowship programs offered in Quebec and Ontario, all universities that offer doctoral programs also run internal fellowship competitions. The level of support ranges from a year-long commitment of between $12,000 and $15,000 to a high of a three-year commitment at $36,000 per year. Aboriginal students are supported separately by the federal government through the programs administered by the Department of Indian and Northern Affairs.

Research funding in the humanities and social sciences remained part of the mandate of the Canada Council[9] until 1977, when Parliament adopted the Social Sciences and Humanities Research Council Act. The act established a separate body (the SSHRC) for those activities, leaving the Canada Council responsible for the arts alone (Canada Council 2004; Doutriaux and Barker 1995). From the beginning, SSHRC adopted a policy of devoting 50% of the grants to individual, investigator-driven or curiosity-driven research, while at the same time taking heed of the federal government's desire to have some grants go to strategic research.

TABLE 3.4.
Doctoral degrees awarded, by province, 2003

Province	No.	%
Newfoundland & Labrador	33	0.9
Prince Edward Island	6	0.2
Nova Scotia	67	1.7
New Brunswick	35	0.9
Quebec	1,134	29.4
Ontario	1,528	39.6
Manitoba	73	1.9
Saskatchewan	78	2.0
Alberta	425	11.0
British Columbia	482	12.5
Canada	3,861	100.1

Sources: Statistics Canada, www.statcan.ca; adapted from CAUT 2006, table 3.15, p. 32.
    Note: Percentages may not add to 100 because of rounding.

Simultaneously, the federal Parliament adopted the Natural Sciences and Engineering Research Council Act, which removed funding of university-based research from the mandate of the NRC, leaving the NRC with the primary role of conducting research in the natural sciences in government laboratories. The purpose of NSERC was "to encourage excellence in research; provide a base of advanced knowledge in the universities; assist in the selective concentration of research activities; aim for a regional balance in scientific capability; maintain a basic capacity for research training; encourage curiosity-oriented research; and encourage research with a potential contribution to national objectives. . . . these objectives . . . [were] intended . . . to ensure long-term coherence in the federal system of university research granting."[10]

On May 1, 1978, SSHRC and NSERC began operations as new, departmental Crown corporations. As noted earlier, this meant that, through these two councils and the MRC, the federal government was now funding all research disciplines represented in Canadian universities. The newly created Ministry of State for Science and Technology urged all three councils to move toward more mission-oriented, strategic research.

During the Progressive Conservative reign, 1984–93, federal policy reflected a neoliberal ideology with an emphasis on shrinking the Keynesian welfare state and freeing the market. Privatization across most OECD countries meant the transfer of costs and responsibility from the public or state sector to the private by selling off Crown corporations to private investors and shifting from taxation to a voluntary, user-pay approach to public services (Stanbury 1989). The need to become more "competitive" became the basis for the Progressive Conservative Party's economic and social policy (Abele 1992).

An enduring initiative in the attempt to soften the boundary separating industry and universities was the Industrial Research Assistance Program (IRAP). Launched in the 1960s, IRAP was one of the first federal attempts to sponsor a collaborative R&D culture. However, most collaborative networks began during the high tide of neoliberalism that marked the policy climate of the 1980s. During this time, according to Niosi (1995), provincial and federal governments launched over one hundred new intersectoral research collaborations.

At the federal level, the Department of Regional Industrial Expansion merged with the Department of Science and Technology in 1984 to form Industry, Science and Technology Canada (subsequently Industry Canada). The new department emphasized industrial partnerships and collaborations. Both NSERC and MRC have actively supported collaborative, targeted research carried out through academy-industry-state partnerships. The opportunities for SSHRC have been less pronounced. NSERC started to fund "big science" networks in the early 1980s—in the earth sciences (Lithoprobe) and integrated circuit design (Canadian Microelectronics Corporation). During 1987–88, the budget year prior to the establishment of the NCE, 15% of NSERC's total budget went to targeted research (Friedman and Friedman 1990).

With the election of the Mulroney Progressive Conservative government and the consequent shift to the "new right," the autonomy of the three national research-funding councils became more tenuous. The new administration wanted to develop Canada's science, technology, and human resources in support of international competitiveness. The government hosted the Forum on Science Policy in 1986. That same year, the government launched a matching-funds policy for the granting councils and an accompanying five-year financial plan. The plan made it clear that the councils were to work in partnership with the private sector in efforts to increase the level of university-industry collaborative activities.[11] The federal and provincial governments announced the broad outlines of Canada's first formal science policy in December 1986; these were formalized in March 1987 as part of the InnovAction strategy (Government of Canada 1987). The central aim was the creation of what has become known as a national innovation system (Archibugi, Howells, and Michie 1999; Nelson 1993).

Even though the federal government has had a long-standing interest in university-industry collaborations, the national research granting councils had no strong mandate to pursue such collaborations until the introduction of the Matching Policy program. The program provided $369.2 million in new funding to the research granting councils over four years in exchange for obtaining matching funding from the private sector (Doutriaux and Barker 1995). The new money was welcomed by the research community, but this good news was tempered by the fact that the federal government had frozen the base funding for the granting councils. Although the government denied this was a freeze—preferring to call it a funding "floor"—it was clearly the government's intention to tie future growth in research funding to private-sector involvement (Savage 1987). Even a subcommittee of the government's handpicked National Advisory Board on Science and Technology (NABST) called the Matching Policy program "a clever way to camouflage a decision to constrain the growth of government funding to the granting councils" (as quoted in CAUT 1989, 2). Established in 1987 by Prime Minister Brian Mulroney's administration to advise the prime minister on national science and technology goals, the NABST consisted of twenty members representing university, industry, labor, and government interests.

Ongoing criticism from the research community and evaluations of the program made it clear that the Matching Policy program was an awkward policy instrument (Doutriaux and Barker 1995). The program was discontinued in 1990–91, and the funds were rolled into the base budgets of the granting councils. This did not mean, though, that the Conservative government had given up on bringing industry and the universities closer together.

The next step in redefining the state's relationship to PSE came in late 1987. Some six hundred delegates, representing a cross-section of Canadian society holding a stake in the future of the country's universities and colleges, met at the National Forum on Post-Secondary Education in Saskatoon, an event cosponsored by the federal and provincial governments (National Forum 1987).[12] This forum was an attempt by the federal government to more clearly define its role in PSE policy and, by extension, science policy. Delegates raised the idea of centers of excellence that would emphasize interdisciplinary research and involve networks of researchers representing several institutions across Canada. In 1988 the Science Council of Canada advised that prosperity depended on integrating the university with the marketplace.[13] Reinforcing this theme, the prime minister's NABST argued that Canadian universities did not adequately exploit intellectual property and recommended that "greater emphasis be given to funding generic pre-competitive research collaboration by university-industry in research consortia" (NABST 1988, 76). This complex of initiatives and recommendations helped to "backfill" the January 1988 decision to launch networks of centers of excellence, partnered with industry. In the same year, the government hosted the National Conference on Innovation and Technology.

Following its reelection in November 1988, the Conservative government renewed its efforts to construct on overall federal science policy. Building on the 1987 InnovAction strategy, the federal government announced in 1989 the creation of the Networks of Centres of Excellence program (Doutriaux and Barker 1995; Office of the Auditor General of Canada 1994).[14] The purpose of this program was to create national research networks involving universities, federal and provincial governments, and the private sector to conduct applied research, to train scientists and engineers, and to speed up the diffusion of new knowledge to industry. Unlike the Matching Policy program, the NCE program was directly funded and did not require private-sector funding (Doutriaux and Barker 1995). The NCE—the forging of a national research capacity under direct federal control—constituted a federal incursion on provincial powers over universities. The NCE offered a major challenge to traditional conceptions of academic autonomy and the public nature of knowledge.

The creation of the NCE in 1989 was arguably the most dramatic change in the nation's science policy since the National Research Council was established in 1916.[15] This policy innovation emerged in the context of the ideological dominance of the market. As the social contract between science and society was rewritten around economic goals, universities and their research programs were discursively repositioned as components of the national system of innovation (Guston and Keniston 1994). The internal normative structure of the "Republic of Science" responded, adapting to government's explicit attempts to close the gap between academy and industry and to make science more commercial (Polanyi 2000).

The model chosen for the NCE program was nongovernmental—the Canadian Institute for Advanced Research (CIAR).[16] The CIAR was designed as a "university without walls," linking researchers across the country through virtual networks. Corporate donations were accepted along with public funding, but neither drove the research program.

The conversion of the CIAR's model of basic-science networks into the NCE model of strategic-science networks was a response to two factors. First, in 1986, came Ontario's Centres of Excellence (COE) program. The concern arose that the province could become a vortex, attracting the best scientists from across the country. Key scientists and policymakers lobbied the federal government to set up a national program. Ottawa was receptive, viewing Ontario's COEs as a possible threat to regional balance. Second was the economic situation, which did not allow for the creation of new fixed COE-type centers.

On January 13, 1988, the prime minister announced his intention "to establish networks of researchers and scientists across the country to conduct world-class research in areas crucial to Canada's long-term competitiveness."[17] The networks were to be part of the wide-ranging science and technology initiatives proposed under the new InnovAction policy. According to one program officer, the prime minister's announcement came "out of the blue and without any consultation."[18] Specifically, the three granting councils responsible for university research were not consulted (Pullen 1990).

In 1990 fifteen different networks (thirteen administered by universities and two by industry) were awarded a total of $240 million over four years (Doutriaux and Barker 1995). The initial group of networks included applied research in fields ranging from aging to telecommunications, to space research, to tele-learning, to pest management, to concrete technology, although it should be noted that the initial announcement did not include a single social science (NCE 2004b).

By 1992 the networks included 35 universities, 66 federal and provincial government bodies, 800 researchers, 1,300 graduate students, 415 postdoctoral fellows, and 173 companies. The networks produced good-quality research and were successful in training the next generation of scientists and engineers. Although the university-industry linkages had improved, the networks still fell short of government's desires for knowledge transfer to the private sector. A 1993 consultant's report found that even though thirteen of the networks had the potential to generate significant economic benefits, only six of the networks had significant potential to result in the private sector's bringing a new product or process to market or to increase industrial efficiency. This report set the next stage of development for the NCE program (Doutriaux and Barker 1995).

This emphasis on private-sector application of university-based research was reinforced in other policy arenas. In the fall of 1991, the Conservative government launched the Prosperity Initiative with two discussion papers: *Learning Well . . . Living Well* and *Prosperity through Competitiveness,* dealing with formal and lifelong learning and the regulatory and policy environment, respectively. These papers were the basis of a consultation process led by the Steering Group on Prosperity, an independent twenty-member panel charged with listening to Canadians and then producing an action plan to secure Canada's economic and social well-being through the 1990s (Prosperity Secretariat 1991a, 1991b; Steering Group on Prosperity 1992).

A significant portion of *Prosperity through Competitiveness* was devoted to matters of science and technology. However, rather than discussing support for university-based research or the mechanisms to induce the private sector to improve its own research capacity, this part of the paper discussed the shortcomings in knowledge transfer to the private sector and the need to make government expenditures on research more effective (Prosperity Secretariat 1991b). In their response, *Inventing our Future: An Action Plan for Canada's Prosperity* (1992), the Steering Group on Prosperity recommended additional government-subsidized support mechanisms to industry for a number of objectives: to introduce management-of-technology courses at the secondary and postsecondary level, to move students and professors off the campuses and into the workplace, to allow additional tax write-offs for capital equipment, to strengthen protection of intellectual property, and to build a high-speed "information highway" (Steering Group on Prosperity 1992). Clearly, the creation of knowledge was to take a backseat to the commercial exploitation of knowledge.

A symbol of the Conservative government's shift away from funding research for its intrinsic worth was evidenced in the November 1992 introduction of legislation that proposed to reorganize certain government agencies for the ostensible purpose of saving administrative costs. Included in this enabling legislation was a proposal to dismantle the SSHRC and to transfer its programs to the Canada Council, undoing the division that had been made in 1978. The move was condemned by all the major organizations in the Canadian academic community. During the February 1993 hearings on the legislation, federal government officials estimated that the merger would result in savings of $1.5 million annually. But by the time the legislation had passed through Parliament in April, this estimate had more than tripled to $5 million, a claim doubted by the academic community (CAUT 1993). As a result of a final lobbying effort aimed at members of the Senate, a group of five Conservative senators voted against their own party and another three Conservative senators abstained from the vote, causing the defeat of the bill on June 10, 1993 (Winsor 1993). The bill was not reintroduced before Canadians went to the polls on October 25, 1993.

Government support for higher education and for academic-initiated research declined during the early 1990s. According to Cameron (1991, 1992), a climate of restraint had already begun to characterize the relationship between government and universities by the late 1980s. The social demand that had once directed the growth of the higher education system gradually gave way to a new, economically driven imperative that placed importance on highly developed human capital, science, and technology to support Canada's needs for economic restructuring and greater international competitiveness. This economic imperative has since been amplified by severe limitations on public expenditures and the emergence of an accountability movement based on a suspicion of public institutions and a belief in the greater efficiency of free-market forces. Further, the federal government decided in 1995 to cut 14% over three years from SSHRC and NSERC. Provincial governments, too, facing their own debt crises, continued to look for ways to reduce unconditional grants to universities.

In the lead-up to the October 1993 general election, the opposition Liberal Party released a comprehensive election platform entitled *Creating Opportunity: The Liberal Plan for Canada*. This document, popularly known as the "Red Book," emphasized employment and economic growth and prescribed closer ties between the education sector and the private sector through targeted job training and increased knowledge transfer—promises consistent with the policy initiatives of the previous nine years (Liberal Party of Canada 1993).

This consistency of approach between the Conservatives and the Liberals was no better illustrated than by the implementation of Phase II of the NCE. In 1994, a budget of $197 million was provided over the following four years for new and renewed networks. Phase II implemented policy changes recommended during the term of the previous government, and in doing so it more narrowly focused the program on the research needs of the private sector and the transfer of knowledge to serve commercial purposes (Doutriaux and Barker 1995). As a consequence, five research networks (aging, space research, chemical physics, pest management, and fish and seafood production) did not have their funding renewed. In their place, four new networks (application of health evidence, tele-learning, sustainable forest management, and intelligent sensing) received funding (NCE 2004a, 2004b). In total, the ten continuing networks received $142 million over four years, and the four new networks received $48 million over the same period (Doutriaux and Barker 1995).

The Liberal government deepened the instrumental emphasis on science policy. Reducing the deficit was paramount. Programs were cut. Those remaining had to return "value for money" or contribute to wealth creation. The Department of Industry, Science and Technology Canada was reorganized almost immediately. By renaming the new department simply Industry Canada, the new government signaled that science was in the service of the economy. Industry Canada assumed an enlarged portfolio and a mandate to foster Canada's competitiveness.

In 1994 the Liberal government embarked on a major review and consultation around science policy. The review process began with the release of a discussion paper, "Building a Federal Science and Technology Strategy," in June 1994 (Industry Canada 2004a). This was followed in the fall by the release of two more discussion papers: "A New Framework for Economic Policy" (Industry Canada 2004c) and "Building a More Innovative Economy" (Industry Canada 2004b). These papers launched the government's "Agenda: Jobs and Growth" process. Both of the latter two papers focused on innovation and science and technology policy (Wolfe 2003). The last step before finalizing the new policy was the release in April 1995 of the NABST report, *Healthy, Wealthy and Wise: A Framework for an Integrated Federal Science and Technology Policy*. The new science and technology policy (Science and Technology for the New Century: A Federal Strategy) was announced in March 1996 (Industry Canada 1996). The government's "national systems of innovation" approach integrated academy, industry, and government research under the rubric of job creation and economic growth (Nelson 1993). In 1996 the NABST was replaced with the Advisory Council on Science and Technology, a body of twelve experts representing industry and the scientific community. Even so, a continuing focus on deficit reduction produced budget cuts for the research councils and reduced general funding for science and technology initiatives.

Just before the election of Prime Minister Jean Chrétien and the Liberals to a second term in June 1997, the government announced what was the beginning of a series of major investments in the infrastructure of R&D and, as a consequence, in universities. Restraint policies aimed at eliminating the deficit were gradually giving way to new investments as the performance of the economy improved. The NCE is a key example of the commercialization agenda. After coming under threat of cancellation in the mid-1990s, it was made permanent in 1997 and is now a central element in the nation's science and technology policy. By the end of the 2000–2001 fiscal year, a total of twenty-nine networks had been funded in areas deemed strategically important to Canada's prosperity and international competitiveness.[19] *Collaboration, partnership,* and *excellence* are key words in the lexicon of this effort to "stimulate leading-edge fundamental and long-term applied research of importance to Canada" (NSERC 1992, 1). What makes this effort unique in Canadian policy history is the explicit attempt to alter the culture of science and to manage research.

In April 1997 the Canadian Foundation for Innovation (CFI) was established under Bill C-93 to fund research infrastructure through partnerships with private and voluntary sectors and provincial governments. The CFI was aimed at universities, colleges, hospitals, and other not-for-profit institutions involved in science and technology development and targeted health, environment science, and engineering needs in Canada. From the outset the CFI was charged with investing all of the allocated funds, not just the interest on the endowment.

The CFI followed in the footsteps of the Canada Millennium Scholarship Foundation (CMSF) as a public foundation. These two ventures mark the beginning of a new type of policy instrument. The CFI is not subject to the scrutiny of Parliament and hence is effectively buffered from changes in government and government policy. The initial endowment of $800 million was boosted in 1998 with a transfer of $200 million, then again in 1999 with an additional $900 million. In February 2001, after the Liberals were returned for a third term in the late fall of 2000, the CFI endowment was again increased by a transfer of $1.25 billion. Between 1997–98 and 2003–4, the federal government invested a total of $3.59 billion in the CFI, a total that includes both grants and operating expenditures (Fisher et al. 2006, appendix 3).[20] The CFI ran its last scheduled competition in September 2005.

In June 1998 the CFI established the College Research Development Fund. The most significant change of policy came in 1999, when the president changed the guidelines to make infrastructure projects in the social sciences and humanities eligible for CFI funding. The original intention was to clearly limit the funding to the natural, applied, and health sciences. While some money did flow to the less "hard" sciences, the overwhelming majority inevitably went to those disciplines that were originally part of the charter. The humanities and social science projects are mostly listed under the science category. They account for only a very small percentage of the total in this category.

The next significant change in the operation of the CFI came in 2000 when the Liberal government created the Canada Research Chairs (CRC) program. This program was funded and administered through the granting councils and the Canadian Institutes of Health Research (CIHR), with infrastructure funds coming from the CFI. The CRC program was part of a series of major initiatives that confirmed the strong commitment of the Liberal administrations to a science policy that placed R&D at the center of their economic strategy. As the only Canadian federal politician in the modern era to have served a third term as prime minister, Chrétien was also conscious of creating a lasting legacy. A series of initiatives began with the CMSF, followed by the CFI, the transformation of the MRC into CIHR in 1999, the CRC in 2000, support for the indirect costs of research in universities in 2001, Canada Graduate Scholarships (CGS) in 2003,[21] and the Canadian Council on Learning in 2004.

The federal Liberals since 1997–98 have embarked on the task of fundamentally reforming the federal-provincial relationship with regard to PSE. In part the initiatives emerged as a response to the 1997 bottom ranking (fifteenth) on the OECD table showing GERD as a percentage of GDP. The Liberal government made a commitment in 1998 to invest in R&D and bring Canada into the top five rankings. The overall strategy was housed in federal science and technology policy and more generally in the emergence of the global knowledge economy. For the most part, the initiatives were undertaken in a unilateral manner at the federal level.

The CFI initiative was a creative way to lever money from provincial jurisdictions. The CFI provided only 40% of the total for each project. Matching funds had to be raised from provincial or private sources. This remarkable venture is slated to last until 2010, although the current endowment was all committed by the end of 2005. The CFI was a response to the underfunding of research infrastructure in Canadian universities. This program, in conjunction with the NCE, more than any other venture, confirmed the preeminence of the federal government in the area of university research. The emphasis on public-private partnerships in the natural, applied, and health sciences confirmed the federal commitment to an applied agenda.

Atlantic Canada, with its relatively weak private sector, faced a special problem when it came to accessing CFI funding. As a means of overcoming this problem, the federal government in 2001 created the Atlantic Innovation Fund (AIF), placed within the Atlantic Canada Opportunities Agency. This policy instrument has been used to provide an additional $300 million per year to fund infrastructure, primarily in the universities. The AIF is a clear example of federal funds being used to compensate for weakness in the private sector.

The most dramatic change in granting council funding arrangements occurred when the Medical Research Council (MRC) was transformed into the CIHR. The transformation was announced in the 1999 budget and then confirmed in legislation the following year. The president of the MRC, Alan Bernstein, had lobbied on behalf of the health sciences. He was able to convince the government that a restructuring of the MRC into a series of thematic research institutes was worthy of a substantial increase in funding. Thirteen distinct institutes were created, each one targeted to particular health problems facing Canadians. The promise was the production of practical and useful knowledge. Between 1999–2000 and 2003–4, the yearly funding for MRC/CIHR in 1988 dollars increased from $238 million to $474 million, an increase of approximately 100% (Fisher et al. 2006, appendix 4).

In 1999 the federal government announced the Canadian Research Chairs program (CRC) in the Speech from the Throne. A one-time investment of $900 million was set aside to fund two thousand chairs over the five-year period from 2000 to 2005. The chairs were to be divided equally between Tier 1 Senior Chairs, worth $200,000 a year, and Tier 2 Junior Chairs, worth $100,000 a year. Matching funding for approved research infrastructure was to be sought from the CFI for each chair. The main criterion for the distribution of these chairs was demonstrated research strength, as measured by the level of funding from the three research councils. As a token with regard to equity, the program also guaranteed that each university would get at least one chair. While the humanists and social scientists through the Canadian Federation for the Humanities and Social Sciences and the Canadian Association of University Teachers (CAUT) argued for more, these disciplines were allocated 20% of the total chairs even though SSHRC only accounted for approximately 12% of the total research council funding. Health sciences were allocated 35% and the natural sciences and engineering 45%. The CRC program is administered by a steering committee composed of the presidents of the three granting councils, the president of the CFI, and the deputy minister of Industry Canada. The committee is advised by a group of international scholars who are appointed to the College of Reviewers.

The CRC marks a clear change in the federal-provincial relationship with regard to PSE. First, the chair holders are expected to do some teaching. This means that for the first time federal funds will be used directly to pay for university teaching positions. Hitherto, such funding had been strictly under provincial and university control. Second, as universities apply for their quota of chairs, they are required to submit detailed research plans, which must be approved by the CRC Steering Committee before the chairs are released to the university. These changes are a clear threat to provincial juridical autonomy and to the academic autonomy of the university.

After extensive lobbying by the Association of Universities and Colleges of Canada (AUCC) and the group of ten major research-intensive universities, the federal government took an unprecedented step by agreeing in February 2001 to provide a one-time grant of $200 million to alleviate pressures from the direct and indirect costs of research. These funds were distributed directly to the universities with no interference from the provinces. Two years later, in February 2003, the Indirect Costs Program was announced with a grant of $225 million per year starting with the 2003–4 fiscal year. A year later this amount was increased to $245 million a year for 2004–5 and 2005–6. The dysfunctional relationship between the universities and the federal granting agencies had long been a major bone of contention, because universities were effectively penalized for being successful in the competition for research funding. The share of these funds going to each university is determined using a sliding proportional scale inversely related to the value of the research grants obtained from the three granting councils. These funds go toward not only the obvious costs of light and heat, but also toward the part of professorial salaries that covers research. The amount was calculated as 40% of the direct costs of research.

This policy decision had been an issue in the debate between the two levels of government and between the federal government and the AUCC. Full funding for both direct and indirect costs of research had been a key recommendation of the Macdonald Commission in 1969. The federal government had made a strategic decision not to fund research costs in the universities when it introduced the new and improved Established Programs Funding in the 1970s transfer system. So the decision to launch the Indirect Costs Program does mark a clear break with past practice and, from the universities' point of view, a positive step. Yet the policy is problematic because it favors institutions with low levels of research costs. In stark contrast to the principles governing other programs, this funding appears to punish success. For example, the University of Toronto, with the largest annual research income, receives reimbursement for overhead at below a 20% level.

In June 2000 the government released *Reaching Out: Canada, International Science and Technology, and the Knowledge-Based Economy* (ACST 2000). This was the report of the Expert Panel on Canada's Role in International Science and Technology, which had been commissioned by the Advisory Council on Science and Technology. Similarly, during a 2000 Conference on Creating Canada's Advantage in an Information Age, Dr. Paul Davenport[22] pointed out that council grants in Canada are less than one-third of their American counterparts and that their low level was restricting innovation and adding to the brain drain. Building on this report and as a response to the advocacy of university presidents, the government of Canada carried out an extensive consultation process over a two-year period, 2001 to 2003, called Canada's Innovation Strategy, which produced two policy papers: *Knowledge Matters: Skills and Learning for Canadians* (HRDC 2002) and *Achieving Excellence: Investing in People, Knowledge, and Opportunity* (Industry Canada

2002). This effort culminated in a national forum in the fall of 2002 and informed federal policy development into the middle of that decade. The new policy aimed at raising Canada's ranking in the R&D league table to the top five and placing the country among the world's leaders in terms of its share of private-sector sales from knowledge innovation and the creation of intellectual property by 2010. Industry set a target of doubling federal investment in R&D and raising the per-capita value of venture capital investments to the prevailing levels in the United States. Human Resources Development Canada focused on training and education and set a series of targets. Three objectives were posited: every high school graduate would have the opportunity to enroll in some form of PSE, 50% of 25-to-64-year-olds would hold a PSE credential, and the number of students admitted to magisterial and doctoral programs would increase by 5% each year.

The AUCC supported the government's vision and in July 2002 responded with *A Strong Foundation for Innovation: An AUCC Action Plan* (AUCC 2002b). The Action Plan committed the universities to the targets set in the Innovation Strategy but made it clear that both levels of government and the private sector would have to substantially increase their investment in university research. In November 2002 the AUCC and the federal government signed an agreement, *Framework of Agreed Principles on Federally Funded University Research* (AUCC 2002a), that confirmed the universities' commitments as outlined in the Action Plan but also included the federal government's acceptance of the responsibility to provide the necessary levels of investment and to make ongoing contributions to the indirect costs of research. Evidence of good faith with regard to the support of graduate studies came in February 2003 with the announcement of the Canada Graduate Scholarships. The federal government committed $25 million in 2003–4, $55 million in 2004–5, and $85 million in 2005–6 in new money. Significantly, the formula for distributing these funds used student enrollment by discipline and field rather than the amounts distributed currently by the three granting councils. This has meant that the humanities and social science disciplines have benefited the most.

From the late 1990s through to the present, the federal government has taken a firm and dominating lead in shaping PSE policy and the academic research enterprise in Canada. As Cameron (2004, 21) notes, the political strategy has aimed at "transforming Canadian universities into more innovative institutions, with closer ties to both government and private industry, more attuned to commercialization of the discoveries, and all the while admitting a growing proportion of Canadians." The overwhelming emphasis has been on research in the health, natural, and applied sciences. Innovation was and continues to be the central ideological plank driving government policy (Doern 2006).

Description and Analysis of the Policy Framework for the Academic Research Enterprise

PUBLIC FINANCING OF RESEARCH IN THE PUBLIC SECTOR

The foregoing sections document clearly the central role played by the federal government and some of its provincial counterparts in funding research in the public sector. The policy commitment has been substantial and has been increasing since the mid-1990s. As federal cash transfers for PSE decreased,[23] the federal government strengthened its commitment to funding R&D in higher education institutions. Yet the picture is complicated by the fact that government science policy, through matching-fund schemes and other incentives, has encouraged the business sector to provide more funding for research. Some differences emerge when we examine the GERD by sector with regard to the proportion of activities funded and the proportions used in carrying out the research. The GERD funded by the government sector hovered in the range of $3.4-$3.8 billion between 1990 and 1996. This amount began to rise slowly in 1997 and took an upward turn in 1999; the trend continues, so that by 2004 the amount was $6.1 billion, a rise of approximately 75% since 1997. At the same time, the amount funded by business between 1997 and 2004 also rose, by approximately 80%, from $7.0 billion to $12.7 billion (Statistics Canada 2007). As table 3.5 illustrates, the proportion of the total GERD funded by the government sector has decreased from a high of 34% in 1990 to a relatively stable level in the range between 22% and 25% since 1996. If anything, the trend over the past decade is slightly down. Over the whole period, the proportion of GERD funded by the business sector has gone up steadily from 39% in 1990 to 49% in 2004, with about half of this increase occurring since 1996. The proportion funded by the higher education sector has remained stable in the mid-teen percentages.

The overall trend is confirmed when we examine the proportion of sponsored research being funded by the federal government. The change is pronounced in the 1980s and 1990s. In constant or real dollars of 2002–3, the federal contribution to the total sponsored research expenditure decreased from 63% ($722 million) in 1981–82 to 51% ($1,006 million) in 1991–92 and to 45% ($1,724 million) in 2001–2 (table 3.6). This is a drop of almost 20% in twenty years. Even though the federal proportion has been decreasing, real funding has been increasing (after discounting inflation); from the 1971–72 level it was up 28% in 1981–82, 39% in 1991–92, and 71% in 2001–2. The apparent proportional federal decrease is due to the failure of federal funding to keep up with the real increase in the overall sponsored research revenue of universities, which has increased since 1971–72 by 43%, 144%, and 380% in 1981–82, 1991–92, and 2001–2, respectively.

TABLE 3.5.
Gross Expenditure on R&D, by source of funds, 1990–2004: Percentage of total expenditure by source

	Business	Government	Higher education	Foreign sources	Private nonprofits
1990	39	34	16	9	2
1991	38	34	16	9	2
1992	39	33	16	9	2
1993	41	31	16	10	2
1994	44	28	14	11	2
1995	46	26	14	12	2
1996	46	25	14	12	3
1997	48	24	13	12	3
1998	46	22	15	16	2
1999	45	23	15	15	2
2000	45	22	14	17	2
2001	50	22	13	13	2
2002	51	23	15	8	3
2003	50	24	15	9	3
2004	49	23	16	9	3

Source: Statistics Canada, www.statcan.ca, CANSIM (using CHASS), table 358-0001.

TABLE 3.6.
Federal contributions to sponsored research in universities, 1971–72 through 2001–2

	Federal source	Sponsored research total	%
In current $			
1971–72	118,333,000	168,028,000	70
1981–82	357,563,000	569,755,000	63
1991–92	833,070,000	1,622,420,000	51
2001–2	1,686,420,000	3,770,529,000	45
In constant 2002–3 $			
1971–72	565,527,189	803,025,382	70
1981–82	722,410,815	1,151,117,912	63
1991–92	1,006,450,051	1,960,081,015	51
2001–2	1,724,089,175	3,854,750,438	45

Source: Canadian Association of University Business Officers, www.caubo.ca, income by fund and by type, all universities, various years.

The GERD carried out by the higher education sector remained relatively stable during the first half of the 1990s, rising from $3 billion in 1990 to $3.7 billion in 1993 and then staying at that level through 1996. Between 1997 and 2004 this amount rose steeply from $3.9 to over $9 billion, by approximately 130% (Statistics Canada 2007). The proportion of the total GERD carried out by the higher education sector also rose over the same period, from 27% to 35% (table 3.7). Throughout the whole period, this proportion was consistently about double the OECD average and above the figures reported for the other seven countries featured in this book (OECD 2007). As the proportion of research carried out in the higher education sector increased during the period 1997 to 2004, the proportion carried out by the business and government sectors decreased from 60% to 56% and from 13% to 9%, respectively (table 3.7).

Overall the GERD data and the sponsored-research evidence indicate that governments did take seriously the policy priority of increasing the resources allocated for research and development, both through their own funding and by increasing the funding coming from the business sector. The commitment to the academic research enterprise is also borne out with the increase in the proportion of research carried out in the higher education sector. The proportional increase in the funding provided by business is most probably a result of the emphasis on commercialization in the science policy pursued by successive Conservative and Liberal governments.

When we place Canada in an international context, the results are quite mixed. As noted above, in relation to other OECD countries, Canada ranks first using the proportion of the total GERD carried out by the higher education sector indicator. In contrast, Canada has been consistently very low in the ranking on the GERD-as-a-percentage-of-GDP indicator. Over the whole period from 1990 to 2004, the OECD mean has been in the range from 2.06% to 2.27% (OECD 2007). In 2001 Canada passed the 2% threshold (table 3.7), but from 2001 to 2004, the OECD figure was in the 2.24% to 2.27% range (table 3.8).

As noted earlier, Quebec and Ontario in recent years have been committed to providing their own funds for R&D. During the period 1990 to 2004, GERD as a percentage of GDP for those provinces has been consistently above the Canadian statistic (fig. 3.1). Quebec overtook the OECD mean in 1998 at 2.2%, and then was substantially above the mean at around 2.75% in the years 2001 through 2004. Ontario overtook the OECD mean in 2001 and for the three years 2002 to 2004 remained at approximately the same level.

When we turn to the distribution of funding between different types of research, the patterns of federal funding of the national granting councils, and more recently of the CIHR, illustrate clearly the commitment by successive governments since the

TABLE 3.7.

Gross Domestic Expenditure on R&D, 1990–2004, in percentages

	Source of funds				
	Business	Higher education	Government	Private nonprofits	Total as % of GDP
1990	50.4	29.6	19.1	1.0	1.51
1991	49.7	30.6	18.7	1.0	1.57
1992	50.6	31.0	17.7	0.6	1.62
1993	52.7	30.0	16.6	0.6	1.68
1994	56.7	27.5	15.1	0.6	1.73
1995	58.1	26.8	14.4	0.7	1.70
1996	57.9	26.8	14.7	0.6	1.65
1997	59.7	26.5	13.2	0.6	1.66
1998	60.2	27.2	12.2	0.5	1.76
1999	59.0	28.8	11.9	0.4	1.80
2000	60.3	28.1	11.3	0.3	1.92
2001	61.6	27.7	10.4	0.3	2.09
2002	57.4	31.7	10.6	0.3	2.04
2003	56.3	33.5	9.9	0.4	2.01
2004	55.5	34.8	9.3	0.4	2.01

Source: OECD 2007, Indicators 2, 17, 18, 19, 20 (updated June 21, 2007).

TABLE 3.8.

Gross Domestic Expenditure on R&D, OECD, 1990–2004, in percentages

	Source of funds				
	Business	Higher education	Government	Private nonprofits	Total as % of GDP
1990	68.6	14.4	14.7	2.3	2.26
1991	68.2	14.9	14.5	2.4	2.19
1992	67.4	15.6	14.5	2.5	2.16
1993	66.3	16.2	14.9	2.6	2.11
1994	66.3	16.5	14.6	2.6	2.06
1995	66.7	16.3	14.5	2.5	2.07
1996	67.6	16.2	13.7	2.6	2.10
1997	68.4	16.1	12.9	2.6	2.12
1998	68.6	16.1	12.7	2.6	2.15
1999	69.0	16.0	12.3	2.6	2.19
2000	69.5	16.0	11.8	2.7	2.23
2001	69.2	16.4	11.9	2.5	2.27
2002	67.8	17.4	12.2	2.6	2.24
2003	67.7	17.5	12.2	2.6	2.25
2004	68.0	17.3	12.1	2.6	2.25

Source: OECD 2007, Indicators 2, 17, 18, 19, 20 (updated June 21, 2007).

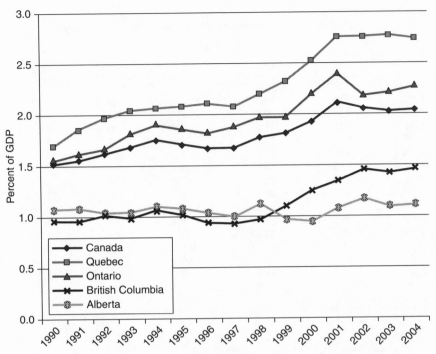

Fig. 3.1. Gross Domestic Expenditures on R&D (GERD) as a proportion of GDP, Canada, 1990–2004. *Sources:* OECD 2007, Indicators 2, 17, 18, 19, 20 (updated June 21, 2007); Statistics Canada CANSIM (using CHASS), table 358-0001 (accessed June 21, 2007).

late 1980s to providing a disproportionate level of support to the natural, applied, and health sciences. The commitment to knowledge production in these disciplines is based not only on the premise that such a commitment is good politics but also on the belief that in a knowledge society, universities should become sites for making a profit. Intellectual property, it is argued, will contribute to the national economy and thereby increase the nation's competitiveness in world markets. Between 1988–89 and 2003–4, federal spending on the four councils (SSHRC, NSERC, Canada Council, and MRC/CIHR) in current or nominal dollars increased from $720.5 million to $1.742 billion (table 3.9). In 1988 constant dollars, the amount increased from $720.5 million to $1.21 billion between 1988–89 and 2002–3 (fig. 3.2). Both table 3.9 and figure 3.2 record a dip in funding during the restraint years of the middle 1990s and then increases beginning in 1997–98. The most dramatic change occurred when the MRC was transformed into the CIHR. Between 1999–2000 and 2003–4, the yearly funding for MRC/CIHR in 1988 dollars increased from $238

TABLE 3.9.
Federal spending on granting councils, 1988–89 to 2003–4 (current $M)

	MRC/CIHR	NSERC	SSHRC	Canada Council	Total
1988–89	188.5	363.7	75.0	93.3	720.5
1989–90	202.1	391.2	80.9	103.5	777.7
1990–91	241.8	465.7	89.9	104.2	901.6
1991–92	247.1	480.3	97.2	106.5	931.1
1992–93	255.6	499.0	101.0	108.0	963.6
1993–94	257.9	492.9	100.0	99.3	950.1
1994–95	264.2	491.0	100.2	98.4	953.8
1995–96	250.6	468.0	98.6	97.9	915.1
1996–97	242.0	451.5	91.2	91.1	875.8
1997–98	236.2	435.1	94.6	114.0	879.9
1998–99	270.3	498.4	101.7	116.2	986.6
1999–2000	309.5	549.6	125.6	116.6	1,101.3
2000–2001	388.4	564.3	140.7	127.4	1,220.8
2001–2	522.4	586.3	148.5	154.3	1,411.5
2002–3	619.6	645.3	164.6	153.8	1,583.3
2003–4	683.3	719.4	185.8	153.4	1,741.9

Source: Fisher et al. 2006, appendix 4.

million to $474 million, an increase of approximately 100%. During the same time period, the comparable increases to NSERC and SSHRC were much less at 18.25% and 34.38%, respectively (see Fisher et al. 2006, appendix 5).

Federal funding has remained focused more on natural sciences and engineering than on social sciences and humanities. The increases to both SSHRC and NSERC between 1978 and 2004 are very similar, standing at approximately 300% (Lee 2005). Two-thirds of these increases occurred between 1998 and 2004, which is in line with the general trends in research funding. Yet the starting point for NSERC in 1978 was five times higher than the funding level for SSHRC, a trend that continued throughout the whole period. Further, while SSHRC was able to maintain its emphasis on basic, researcher-driven funding, it also justified receiving the increases by mounting significant new strategic, applied programs, such as the Initiative on the New Economy, the Major Collaborative Research Initiatives, and the Community University Research Alliances.

Two of the new initiatives added particular weight to the trend in favor of the natural, applied, and health sciences. Between June 1998 and November 2005, the CFI distributed a total of $2.25 billion (table 3.10). Health accounted for the largest proportion at 42.1% ($946.3 million), followed by science at 29.2% ($656.1 million), engineering at 22.5% ($505.4 million), and environmental studies at 6.2% ($140.2 million). The funding in the humanities and social sciences was part of the

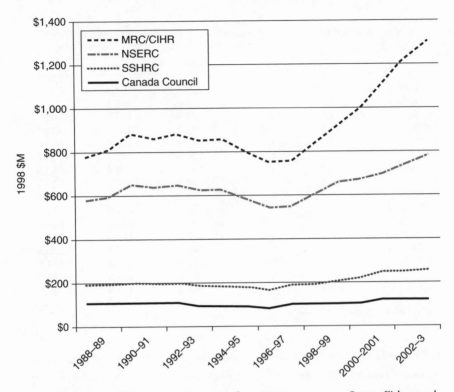

Fig. 3.2. Federal spending on granting councils, 1988–89 to 2002–3. *Source:* Fisher et al. 2006, chart 4, appendix 4.

science entry and, as one might expect, accounted for a very small percentage of this category. By 2006 the CFI had committed a total of $3.75 billion to fund 5,400 projects at 128 research institutions across Canada. The 40% rule means that the CFI has been responsible for levering approximately another $6 billion. Given the criteria for distribution of the chairs, the CRC has added approximately $700 million to the funding of the natural, applied, and health sciences. As noted earlier, the distribution has also furthered the concentration of research funding at the major research-intensive universities.

The overwhelming emphasis in federal research funding as a whole has been upon the natural, applied, and health sciences. Between 1998–99 and 2004–5, the federal government added a total of $9.13 billion in new research funding. Of this total only $1.02 billion, or 11.2%, funded research in the humanities and social sciences (fig. 3.3).

TABLE 3.10.
Canadian Foundation for Innovation (CFI) awards, June 1998–November 2005

Research category	Award	Percent
Engineering	$505,449,579	22.5%
Environmental Studies	$140,167,393	6.2%
Health	$946,282,126	42.1%
Science	$656,052,173	29.2%
Total	$2,247,951,271	100.0%

Sources: Statistics Canada, www.statcan.ca; adapted from CAUT 2006, table 5.2, p. 46.

POLICIES CONNECTING ACADEMIC RESEARCH TO
ECONOMIC DEVELOPMENT

By the 1990s, *marketization* and *commercialization* had become the key watchwords in the political ideology that justified investment in R&D. Marketization is defined as combining market principles of private property, competition, and profit with state interests of authority, public interest, and citizenship. Young (2002) described marketization in Ontario higher education policy as the introduction of market

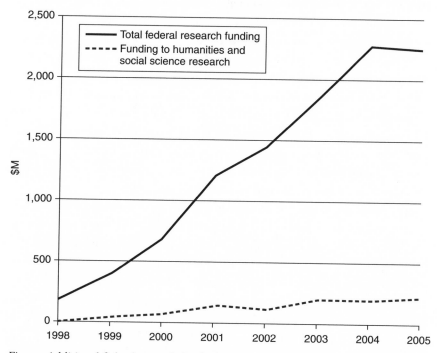

Fig. 3.3. Additional federal research funding, total and humanities and social science, 1998–2005. *Source:* Fisher et al. 2006, chart 17.

mechanisms, not precluding nor increasing government control, but changing government's approach to reform and changing the nature of government control over the system in order to induce universities to adopt priorities identified by the government.

Slaughter and Leslie (1997) and Marginson (1997) believe globalization creates conditions in which nation-state educational policies begin to converge in certain areas, including access, curricula, research, and autonomy for faculty and institutions. Existing institutional structures, values, and beliefs erode as higher education develops to meet a specific economic agenda (Currie 1998; Dudley 1998; Frost and Taylor 2001; Slaughter 1998). In turn, educational markets affect institutional governance models, and the models become increasingly corporatized and managerial (Currie 1998; Fisher and Rubenson 1998; Marginson 2000; Pietrykowski 2001; Schugurensky 2007; Slaughter 1998). Faculty have less input into the decision-making process of the institution as decisions come to be made by an executive core of administrators, who increasingly depend on directors and managers of nonacademic units for advice and guidance on institutional matters.

Financial systems are more complex in quasi-market and entrepreneurial-based environments, with increased accountability requirements, reporting obligations, and budgeting processes. Determining what makes business sense—assessing various projects and their potential to make a financial contribution to the institution—requires business knowledge. Under these conditions it is likely that senior administrators will continue to seek business expertise in dealing with institutional matters (Evans 2005). Governments strategically promote increased efficiency and innovation using education markets (Dill 1997b; Slaughter and Leslie 1997). "Long perceived as a unique characteristic of the US system of higher education, experiments with market competition in academic labor markets, institutional finance, student support, and the allocation of research funds are now evident in the higher education policy of many different nations" (Dill 1997a, 167).

In a neoliberal environment, institutions are often granted more autonomy by central government but are then forced to go into the market, where they must seek sponsorship. In order to gain sponsorship and the resulting funding commitments, they must adapt to new methods and systems. Under resource dependency, the rules are subject to change according to the interdependency that is established. The social controls that accompany government funding differ from the social controls established in a market environment. In a market environment, the behaviors that result include academic capitalism and institutional entrepreneurship (Chan and Fisher 2008; Clark 1998; Marginson 1997; Slaughter and Leslie 1997; Slaughter and Rhoades 2004).

This policy priority is evident across social policy in Canada through retrenchment measures, which began with the federal government when the Progressive Conservative Party was in power in the late 1980s and continued into the 1990s with the federal Liberal government. The shift toward the market was apparent in the opening up of the private sector in previously public-dominated areas, deregulation, downsizing, and rationalization. Increasing industry partnerships are particularly evident in the area of research, where there has also been an increase in investment in research and development, both provincially and federally, with a view to making Canada internationally competitive by building on a strong science and technology infrastructure.

The neoliberal commitment to market ideology and to the reduction in the size of the public sector has led governments in Ontario, Alberta, and British Columbia to introduce legislation that creates the conditions for the establishment of quasi-markets (Marginson 1997; Marginson and Considine 2000). This policy priority in both Ontario and British Columbia is evidenced in the re-regulation of tuition fees, the use of matching funds from the private sector (for student aid, research and development, and the encouragement of specific fields like computing and engineering), and the introduction of legislation that opens up postsecondary systems of education to the market.

The development of research (science and technology) policy occurs at the nexus of three interrelated forces: commodification of knowledge, acceleration of information-based technologies, and the globalization of markets and capital. The production, distribution, and application of knowledge worldwide constitute the logic of new industries such as biotechnology and electronics and saturate traditional industries such as forestry. In Canada, academic science has been positioned as an "engine" of international competitiveness (Industry Canada 1996). Academic science is a key component of policy discourses concerning the "national innovation system" (Fisher, Atkinson-Grosjean, and House 2001; Nelson 1996) and the "knowledge-based economy" (OECD 1996).

In a knowledge-based economy, complex dynamics link the academy and the market. Research universities have adopted an economic mission and have become knowledge entrepreneurs, not only patenting and licensing technologies to the private sector but also spinning off commercial enterprises to exploit their own scientific discoveries. As academic science feeds the market, so the market feeds science, with new questions and funding to maintain the momentum. These interactions and interpenetrations complicate the boundary between academy, industry, and state. They raise important social questions concerning the structure of the scientific enterprise, the privatization of the public knowledge base, and the status of universities as public, knowledge-generating institutions.

These questions foreground the role of technology transfer or industry liaison offices (ILOs), which, as the universities' brokers in the knowledge market, potentially have a large influence on shaping the academic research enterprise. As sites where public and private enterprise converge, these offices occupy an ambiguous space as "boundary objects" (Star and Griesemer 1989) between academic and corporate worlds. The managers are charged with the task of doing "boundary work" (Fisher 1993, 1999; Fuller 1988; Gieryn 1983, 1995, 1999) as they mediate and attempt to increase "boundary permeability" (Borys and Jemison 1989) for the boundaries separating the academy, industry, and the state. Their role is one of "translation" (Callon 1995): translation of research results into concrete commercial applications, translation of academic capital into economic returns, and translation between different social worlds (Kaghan 1998).

As academic science focuses more on application, it becomes a matter to be managed. While the idea of "managing" what was, until recently, a highly individualistic activity may seem counterintuitive (Ziman 1994), research management has become a priority for both the university and the state, as academic science takes on increasing economic significance. The managers are given the task of bridging the gap between two ideal-typical patterns of organizing the university. On the one hand is the "Guild" model (Clark 1983), in which specialized professionals produce alone, teach, undertake basic research, and provide public service. On the other hand is the "Corporate" model (Keller 1983), in which the university takes a business approach and makes profit the central motive. The institution is regarded as an economic enterprise that is part of the knowledge industry. What emerges from this dichotomy is what Clark (1998) refers to as the "entrepreneurial university."

With managerial value systems in the ascendant, corporate structures and extensive, "CEO-style" presidential offices are becoming the norm in Canadian universities. In place of traditional academic leadership, we now see a growing cadre of "professional" managers and administrators (Cassin and Morgan 1992; Newson 1998). Responsibility for research management has been institutionalized in the most senior ranks of the administrative hierarchy, generally under the direction of vice presidents of research. Often, these are scientists who have moved from the laboratory into full-time, senior management, thus emulating the promotion trajectory of their industry counterparts (Ziman 1994).

Some have argued that managerial control and an economic agenda are fundamentally at odds with the autonomy and freedom needed to stimulate research (Hardy 1992). That scientists not only comply with the agenda, but also seemingly endorse it, can be attributed in part to their resource dependency (Slaughter and Leslie 1997). The state is still by far the largest provider of funds for academic research

(Department of Foreign Affairs 1996). When—as over the past two decades—state funding is reduced at the margin or directed into strategic areas, "entrepreneurial science" (Etzkowitz 1997; Etzkowitz and Webster 1994, 1997; Etzkowitz, Webster, and Healey 1998) and "academic capitalism" (Slaughter and Leslie 1997; Slaughter and Rhoades 2004) are the logical results.

To a great extent, therefore, recent entrepreneurial and corporate initiatives in academic science are policy-driven. As in most other OECD nations, science policy in Canada has settled around two key conditions for the funding of academic research: practical and economic utility, with commercialization as the indicator of effectiveness; and partnerships with the private sector to take discoveries from the laboratory to market (Atkinson-Grosjean, House, and Fisher 2001).[24]

Since 1991 the Association of University Technology Managers has conducted an annual survey of the commercial activities of American and Canadian research universities. For the 1997 fiscal year, 16 Canadian and 132 American universities participated, a significant proportion of the major research universities in both countries. Biotechnology and the life sciences were the motors driving commercialization. In 1997 life-science patents accounted for two-thirds of the total number of university patent portfolios. Within this category, biotechnology patents accounted for 82% of the total revenues flowing from life-science patents (AUTM 1998).[25] A major change occurred during the 1990s as the relative proportion of university research funded by industry and the federal government shifted in favor of industry.

In the context of federal policy and the mandate to "manage research," universities began to establish new organizational structures to exploit their discoveries. A few Canadian universities founded "research offices" as early as the 1970s (e.g., McGill, École Polytechnique, and Cape Breton). But ILOs did not appear in their contemporary form until the early-to-mid 1980s, when their frameworks were put in place at eleven key universities (Enros and Farley 1986).[26] Despite the late start, the rate of expansion has been significant. By 1995 some thirty-two universities had established an ILO (Doutriaux and Barker 1995). By 1998 all except the smallest universities and university colleges had active offices (ARA and Brochu 1998; Statistics Canada 1999). Certainly all the major research universities and most of the comprehensive universities now have a central agency responsible for intellectual property management and commercialization.

This cultural shift occurred in part because governments have taken an active role in sponsoring these offices. In the 1980s, federal government assistance for the establishment of the ILOs was available through the Industrial Research Development Agency program. Subsequently, some provinces have provided continuing support directly to ILOs, and the Intellectual Property Management Program of

NSERC has also provided grants. For the sixteen Canadian institutions that were tracked, professional staffing for technology transfer and licensing doubled between 1991 and 1997, while income from commercialization grew fourfold and the number of active patents and technologies approximately tripled (AUTM 1998).

ILOs are Janus-faced, located as they are on the boundary between academic and corporate worlds. Typically, says one ILO officer, they are "criticized by academics for being too aligned with industry, and by industry for being too academic."[27] While all ILOs reach out to the private sector, their degree of embeddedness within the university differs, largely because of historical institutional factors surrounding their establishment. Some are independent, incorporated entities. Others are contained within the vice-presidential domain and budget. Some are located on campus, others are "downtown," and many have satellite offices. All ILOs undertake the protection and licensing of intellectual property. Some also negotiate and administer research contracts, undertake technology start-ups, and launch spin-off companies. This complex of activities is labeled *technology transfer* and defined as "any interaction between the university or its faculty, and industry or the community, that results in the transmittal of intellectual property in any of its forms" (Enros and Farley 1986, 15). Technology transfer includes technology, information, and knowledge (Charles and Howells 1992). Although no single model is "typical," ILOs converge around two broad types and mandates. First is the internal model, in which the ILO is fully integrated into the university's administrative structure. One manager referred to "a kind of a model . . . where technology transfer is basically an integral part of the university's mission and direction."[28] Second is the external model, in which the ILO operates outside the university as a corporation that is either nonprofit or for-profit. In either case, the corporations are owned totally by the university.

The three federal granting councils in Canada do not require disclosure of the intellectual property created with their research grants. It follows that some universities have intellectual property policies, while others do not. The range of operating policies is diverse. According to Statistics Canada (1999), some universities own both the intellectual property and all the royalties, other universities assign both to the researcher, and in many other universities ownership and royalties are shared according to a wide range of formulas.

As early as the mid-1980s, a study initiated by the Science Council of Canada speculated that the establishment of ILOs and research management structures signaled the emergence of a new species of university attuned to the needs of private business, which they labeled the "service university" (Enros and Farley 1986, 16). This concept quickly gained the attention of critics, who saw service to the corporate sector as incompatible with service to the community as a whole. As universities reoriented around commercial rather than academic values, critics predicted that collective knowledge would degenerate into a bundle of private ownership rights in intellectual property and that "relevance" to market demands would determine intellectual and pedagogical direction (Newson 1994). The implications for basic science were particularly troubling, since it has little direct relevance to the marketplace. Business priorities are generally short-term, focused on profit and returns to shareholders. Such priorities have little to do with "high-risk, open-ended, basic science research, in which one discovery leads to another with no clear indication of where it will end and whether it will generate marketable products" (Aguayo and Murphy 1997, 139).

In trying to assess the impact of ILOs, Fisher and Atkinson-Grosjean (2002) concluded that few if any of the ILOs are self-supporting. They estimated that the economics of these offices in 2002 were much as they had been in the mid-1990s when on average 70% of their budgets came from the university, 13% from governments, and only 11% from internally generated income (Doutriaux and Barker 1995). The ILOs are held to a complex and often contradictory interplay of performance, service, and other standards. In order to justify their existence, ILOs use rhetorical strategies to demonstrate their accomplishments to different internal and external audiences. At the same time, Fisher and Atkinson-Grosjean (2002) made it clear that the activities of these offices helped further the commercialization trend and by definition favored the health, natural, and applied sciences.

POLICIES MAINTAINING EFFECTIVE RESEARCH DOCTORAL EDUCATION

"Patchwork federalism" is without doubt an apt description of the federal role with regard to student assistance. Equal access within and between provinces has been the long-standing goal of federal programs. From the mid-1960s through to the mid-1980s, the major source of financial assistance for all students was the joint federal-provincial Canada Student Loans Program (CSLP). Over the past twenty years, the political pressure for balancing the budget, reducing government debt, and paring down the size and scope of government has taken priority over almost all other government activities. Neoliberalism became the basis for federal policy, and as a consequence, in the 1990s federal policy on student financial assistance and related matters moved away from the grand programs of the past toward a more targeted and even niche approach to helping students pay for the rising costs of PSE. In this context, the federal Liberals overhauled the CSLP in 1994; as part of their 1998 Canadian Opportunity Strategy, they created the Canada Millennium Scholarship

Foundation (CMSF) with an endowment of $2.5 billion. The new system of awards was available to all PSE students, including part-time students. The federal government was able to negotiate agreements with all the provinces, even though they saw these fellowships as an invasion into their territory.

As noted earlier, beyond loans the major source of support for doctoral students has been the fellowships awarded by the three national granting councils, parallel fellowship programs at the provincial level, and university fellowships. The federal support became increasingly significant after 1977, as the three councils were made responsible for funding the whole range of disciplines and were given a major increase in available funds. From the beginning, graduate training has been a key function of the granting councils. The new integrated funding structure came on the heels of a massive expansion in the number of graduate programs being offered and the similar rise in enrollment from the mid-1960s. In 2003 the federal government committed significant new money through the Canada Graduate Scholarships.

Total doctoral enrollment remained relatively stable during the 1990s in the mid-to-upper 20,000s and even took a slight dip in the last three years of the decade. Enrollment began to rise in 2000–2001 and then rose dramatically in the next three years from 27,390 to 34,527, by approximately 25% (table 3.11). Yet it should be noted that some of this expansion was due to the rise in the number of full-time international students enrolled in doctoral or equivalent studies, a number that increased significantly from 4,683 in 2001–2 to 6,639 in 2003–4 (CAUT 2006).

The total number of doctoral degrees awarded from the early 1990s through 2003–4 has a different trajectory of expansion. The major expansion happened in the 1990s. Between 1991–92 and 1997–98, the total number of doctoral degrees awarded increased from 3,135 to 3,978, or approximately 27%. This total then declined slightly to a low of 3,708 in 2000–2001 before recovering and rising above the previous high level (of 1997–98) to 4,164 in 2003–4 (table 3.12). The distribution of both enrollment and degrees awarded corresponds to the distribution of the student population and particularly to the location of the large majority of the research-intensive universities. Ontario, Quebec, and British Columbia account for approximately 80% of the total doctoral enrollment and the total number of doctoral degrees awarded throughout the period from 1991–92 to 2003–4 (tables 3.11 and 3.12).

The distribution of doctoral enrollment by major discipline has shifted since the early 1990s (table 3.13). Between 1992–93 and 2003–4, the physical and life sciences and technologies as well as architecture, engineering, and related technologies dropped as a proportion of the total enrollment in the late 1990s, then recovered to their starting levels of approximately 22% and 14%, respectively. Business, manage-

TABLE 3.11.
Doctoral program headcount enrollment, 1991–92 through 2003–4

	Ontario	Quebec	British Columbia	Other provinces	Total
1991–92	10,221	7,788	2,967	4,137	25,113
1992–93	10,452	8,349	3,207	4,467	26,475
1993–94	10,638	8,649	3,333	4,527	27,147
1994–95	10,509	8,850	3,372	4,575	27,306
1995–96	10,356	8,931	3,336	4,575	27,198
1996–97	10,047	8,982	3,354	4,620	27,003
1997–98	9,927	8,829	3,249	4,500	26,505
1998–99	10,182	8,649	3,132	4,530	26,493
1999–2000	10,389	8,631	2,925	4,653	26,598
2000–2001	10,902	8,565	3,048	4,875	27,390
2001–2	11,625	9,141	3,354	5,220	29,340
2002–3	12,537	10,095	3,510	5,862	32,004
2003–4	13,395	11,052	3,867	6,213	34,527

Source: Statistics Canada, www.statcan.ca, CANSIM table 4770013 (using CHASS).

TABLE 3.12.
Doctoral degrees awarded, 1991–92 through 2003–4

	Ontario	Quebec	British Columbia	Other provinces	Total
1991–92	1,350	897	300	588	3,135
1992–93	1,410	885	369	693	3,357
1993–94	1,464	972	387	729	3,552
1994–95	1,506	1,014	474	723	3,717
1995–96	1,605	1,092	474	756	3,927
1996–97	1,578	1,143	477	768	3,966
1997–98	1,545	1,173	498	762	3,978
1998–99	1,569	1,170	501	726	3,966
1999–2000	1,458	1,164	522	717	3,861
2000–2001	1,434	1,092	495	687	3,708
2001–2	1,485	1,038	486	720	3,729
2002–3	1,527	1,134	483	717	3,861
2003–4	1,677	1,215	498	774	4,164

Source: Statistics Canada, www.statcan.ca, CANSIM table 4770014 (using CHASS).

ment, and public administration increased consistently over the period from approximately 3% to 4.2%. The biggest losers were the humanities, which recorded a loss of approximately 4%, from 16.7% to 12.5%. The social and behavioral sciences and law declined from a high of 21% in 2000–2001 to slightly below 20%, which is close to the field's 1992–93 proportion. Visual and performing arts and communication recorded a slight increase, while education dropped slightly from just under 10% to 8.5%.

TABLE 3.13.

Doctoral students in Canadian universities, 1992–93 through 2003–4, in percentages

Major discipline	1992–93	1994–95	1995–96	1996–97	1997–98	1998–99	1999–2000	2000–2001	2001–2	2003–4
Agriculture, natural resources, and conservation	3.5	3.4	3.3	3.4	3.3	3.3	3.5	3.1	3.1	2.9
Architecture, engineering, and related technologies	14.3	14.3	13.6	13.2	12.8	12.3	12.5	12.6	12.9	15.6
Business, management, and public administration	2.6	2.7	2.9	3.1	3.3	3.5	3.6	3.8	4.1	4.2
Education	9.7	9.6	9.8	9.9	10.4	10.3	9.8	9.4	9.4	8.5
Health, parks, recreation, and fitness	4.9	5.2	5.5	5.7	6.1	6.4	6.1	5.9	6.3	6.7
Humanities	16.7	16.2	16.4	16.2	16.0	15.4	14.9	14.7	13.9	12.5
Mathematics and computer and information sciences	5.2	5.2	5.0	4.6	4.6	4.6	4.6	4.7	5.0	5.9
Personal, protective, and transportation services	0.1	0.1	0.1	0.1	0.1	0.1	0.1	0.1	0.1	0.1
Physical and life sciences and technologies	22.0	21.5	21.4	21.4	20.7	20.8	21.5	22.1	22.0	21.7
Social and behavioral sciences and law	19.2	19.7	19.7	19.9	20.1	20.4	20.6	21.3	20.8	19.6
Visual and performing arts and communications technologies	1.1	1.1	1.2	1.3	1.4	1.5	1.6	1.7	1.8	1.7
Other instructional programs	0.6	1.0	1.1	1.2	1.2	1.4	1.2	0.6	0.6	0.6
Total	99.9	100.0	100.0	100.0	100.0	100.0	100.0	100.0	100.0	100.0

Source: Statistics Canada, www.statcan.ca.

Note: Percentages may not add to 100 because of rounding.

TABLE 3.14.
Number of awards and success rate, by council, 1995–2005

	NSERC		SSHRC	
	Awards	Success Rate (%)	Awards	Success Rate (%)
1995	1,325	45.7	563	16.4
1996	1,451	59.7	574	18.0
1997	1,200	58.3	567	19.2
1998	1,461	69.9	612	22.6
1999	1,556	67.1	591	22.7
2000	1,676	67.3	587	21.6
2001	1,689	64.6	575	19.5
2002	1,886	69.6	587	18.2
2003	1,976	69.4	994	25.7
2004	2,075	70.0	975	22.3
2005	2,452	72.9	974	20.2

Sources: NSERC Web site, www.neserc.gc.ca; and SSHRC Web site, www.sshrc.ca; data compiled by Brigitte Gemme.

Note: The years 2003, 2004, and 2005 include Canada Graduate Scholarship Awards.

The number of awards in the natural and applied sciences from NSERC has consistently been higher than in the social sciences and humanities from SSHRC (table 3.14). Between 1995 and 2005, NSERC awarded approximately three times as many graduate awards as SSHRC. In only six of these eleven years did SSHRC's success rate exceed 20%. The three years 2003 to 2005 were all above 20%, the highest, 2003, being 25.7%. In stark contrast, for the eight years from 1998 to 2005, NSERC recorded a rate hovering around 70%. The increases in the number of awards and the success rates for SSHRC since 2002 are a result of the introduction of the Canada Graduate Fellowships.[29]

POLICIES ASSURING THE ACCOUNTABILITY OF PUBLICLY SUPPORTED RESEARCH UNIVERSITIES

The dominant policy paradigm in Canada over our period of study has been accountability. This movement has rejected the idea that the state is primarily responsible for the public-good function of education. Critics called for less taxation, less government interference, more public choice, more deregulation and privatization, and more accountability by government and its subsidiaries to taxpayers (Dale 1997). Within a neoliberal framework, public choice, marketization, and privatization of education are prevalent themes, emphasizing stronger links between industry training needs and the postsecondary sector. These changes manifest themselves in education and other public services in two major ways. First, there is reluctance to use public funds to fund public services; and second, public institutions are encouraged to engage in market behavior in order to fund more of their services. Changes to organizational forms, managerial practices, and institutional cultures result (Deem 2001). Policy changes are accompanied by downloading more financial responsibility onto postsecondary institutions and are characterized by less state funding and an increased emphasis on business practices (Currie 1998). Clark (1998) states that themes of efficiency, effectiveness, excellence, and continuous quality improvement are examples of the thinking that prevails within the organization.

An underlying but consistent theme, both federally and provincially and across party political lines, has been the commitment by governments to make the connections between educational spending and useful outcomes more transparent and understandable to the general public. *Accountability* refers to the possibility of making organizations responsible for their actions. Accountability is defined as the demonstration by an institution that it has met its mandate or is competent to achieve its mission. Along with academic autonomy and a measure of self-governance comes the responsibility to demonstrate accountability to the public and to the government that has determined the institution's mandate. Accountability is actualized through reporting. *Efficiency* is closely linked to accountability and emerged as a priority within the same context. It refers to the achievement of objectives at the best possible cost. In other words, efficiency is an organization's ability to meet objectives while reducing the effort required to do so, or to choose other options that are less costly (not exclusively monetarily) in terms of effort. Similarly, accountability is associated with the quality of both management and outcomes.

Federal governments have made numerous attempts to render provincial governments more accountable. As noted elsewhere in this chapter, the emphasis over the past decade has been to decrease the direct transfer payments in favor of indirect but accountable research funding and grants as well as loans to students. Yet while we can infer that the provinces have responded to federal policy, it remains difficult to isolate the lines of causation. This policy priority has clearly taken on different forms in the three provinces of Ontario, Quebec, and British Columbia. Both Quebec and Ontario have made this a major policy priority. While accountability has been important in British Columbia, the emphasis there has been on the general public-interest aspect rather than on institutions.

Quebec has the longest history of regulating universities. In the 1970s the Ministère de l'Education (MEQ) charged the Conference of Rectors and Principals of the Universities of Quebec with evaluating the "quality" of degree programs. The work of this committee is symptomatic of the way successive governments have eroded the autonomy of the universities. In the 1980s, the Conseil des Universités, which was created in 1968, began publishing sector-based studies on engineering, social science, and education. Across the PSE system, institutions are directly accountable to the MEQ by having to submit "success plans" and annual reports at the college and university levels, respectively. Furthermore, since the early 1990s, and as part of the government's desire to promote both accountability and efficiency, universities are required to do internal evaluations, which are in turn monitored by government. All programs are evaluated with regard to quality and relevance.

In Ontario, institutional accountability has been the major priority. Governments have required institutions to account for public funds and to demonstrate achievements on government-prescribed benchmarks or indicators. Adopting a systemwide accountability perspective has been more problematic in Ontario because of the lack of system-level planning. Moreover, the form of accountability has proven more controversial in the postsecondary system and has reflected the area of most change in this policy priority. Key Performance Indicators (KPIs) were introduced by a Conservative government and were first reported on in 1999. All colleges and universities are required to report on a set of KPIs for each program. This is in addition to existing accountability reporting by academic senates and governing boards of universities. In the 2000 budget, the government started to put money back into the system by way of operating grants.[30] The new funding was tied to performance indicators: enrollment growth [31] and university performance.[32] In addition the government has increased the use of targeted and matching private-sector funds. This combination of funding mechanisms has enabled the government to steer universities.

In British Columbia, the New Democratic Party (NDP) was far more concerned with vocational programs in higher education than with academic research. The only direct interference in doctoral studies came in the late 1990s with the creation of the New Programs Committee to monitor and approve all new degree programs. The Liberals, elected in 2001, adopted a different definition of accountability. Accountability came to mean both quality assurance in the most general sense and a blurring of the boundary between the public and the private sectors. In the latter case, this is direct political accountability to the capital interests that are the main backers of the Liberal Party. As they put in place the Quality Assurance Board in 2003 and dismantled much of the infrastructure created by the NDP to guarantee accountability, they put their faith in the market as the best means of making institutions accountable. In other words, accountability to the marketplace equaled accountability to the public. In the university sector, the government has provided targeted funding for particular occupations and has simply decreed that more students will be educated without an increase in funding.

Conclusion

Alongside the decline in transfers for postsecondary education, we can observe a phenomenal increase in the resources allocated to research and development in Canada. Funds are disbursed primarily using three mechanisms: grants directly to faculty members for research projects, capital funding on a shared-cost basis for infrastructure projects, and grants directly to universities for indirect costs of research. The funding is disbursed by federal agencies on a competitive basis and awarded in accordance with federal criteria, which include merit and national interests. Furthermore, these policy decisions are set within a science and technology policy that emerged from competing definitions of *science, utility,* and the *public good.* At the policy level, the interests of capital are privileged under the guise of serving the national interest. By promoting industry's access to publicly funded research, the science and technology policy recognizes that scientific research is simultaneously fundamental and useful. But the policy also skews the balance in favor of private interests and commercial science.

Until 1977 the federal government's research grants for university scholars were scattered across a range of institutions and programs, starting with the National Research Council and progressing through the Canada Council and the Medical Research Council. With the creation of the SSHRC (Social Sciences and Humanities Research Council) and the NSERC (Natural Sciences and Engineering Research Council), the government of Canada made a clear commitment to fund all research disciplines represented in Canadian universities. This major policy shift was housed in the continuing interest in science and technology symbolized by the creation of the Science Secretariat in 1964, the Science Council in 1966 (Science Council 1968), and the Ministry of State for Science and Technology in 1971. Furthermore, this shift was prompted by the Macdonald Report (1969) and the four volumes of the Lamontagne Report (Lamontagne 1970; Senate Special Committee 1972).[33]

The "matching funding" policy involving government and industry and the focus on the production of intellectual property have resulted in an increase in the proportion of research being funded by nongovernment sources. The proportion of "sponsored research" funded by the federal government has decreased over the past twenty years. At the same time the proportion of research funded by business enterprise has increased. As noted earlier, the commitment of successive federal governments has resulted in a dramatic increase in both public and private funding for R&D.

First the federal Conservatives in 1989 and then the federal Liberals since 1997–98 embarked on the task of fundamentally reforming the federal-provincial relationship with regard to PSE. For the most part, the initiatives were taken in a unilateral manner at the federal level. The Networks of Centres of Excellence program (NCE) was the centerpiece of the federal government's InnovAction science policy and continues to be the flagship initiative in a federal policy framework promoting the commercialization of academic science and academy-industry partnerships. The NCE aimed to enlist academic scientists in a national system of innovation that would translate university research into marketable technologies and enhance Canada's competitiveness in a global knowledge economy. By 2005–6 a total of forty networks (research institutes without walls) had been funded through this program.

The NCE is an ideological policy instrument whose ideological goals have never been hidden: the purpose is to change the research culture itself. Program documents convey a sense that the country can no longer afford researchers who isolate themselves in the academy, pursuing esoteric problems at public expense. Instead, academic researchers must participate in the "national system of innovation" and be encouraged to apply their talents to more immediate ends. "The thrust of the NCE program is to ensure that knowledge is transferred from the generators to the users and applied to benefit the lives of Canadians" (NCE 1997).

From the beginning of the NCE, there was tension between the world of *policy*, represented by government, and the world of *science*, represented by researchers and research council officers. Industry Canada pressed for an emphasis on utility and application, while the granting-council presidents and the appointed scientists held onto basic science and knowledge for its own sake. Industry Canada wanted to privilege capital and private ownership, while the granting councils wanted to privilege academic freedom, believing this would serve the public good. In the first phase the "basic" conception prevailed, as the councils captured the program and instituted peer reviews. The process was transparent, and scientific excellence was the essential criterion. The definition of science was traditional and thus favored the natural and medical sciences. The social science applications were judged to be inferior.

By the second phase, however, the political climate had changed, and with it, the research climate. Programs were being cut to reduce the deficit. Public-sector reform was in the air. No matter how excellent the science, the NCE could no longer be justified as a funding source for fundamental research. Industry Canada insisted on the reinstatement of a demonstrated commitment to "value-added" commercial relevance. Thus, scientific excellence was brought down to the same weight as the other selection criteria. Furthermore, the culture of the NCE was changing through both the second and third phases. Government was successful in pushing the Networks to appoint managers, to place industrialists on their boards, and—importantly—to become corporate entities.

The NCE program was designed to change the traditional ethos of academic science; network science was to be a means whereby a small economy could afford "big science." It was hoped that the program would have the following outcomes: (1) people of multiple disciplinary and institutional backgrounds would come together to resolve research problems; (2) research would be "managed" in the sense that research committees would steer scientific direction and terminate dead ends, while professional managers would pursue commercial opportunities; and (3) networks would be able to leverage their intellectual capital by gaining access to their partners' social, human, and economic capital. Network science thus offered the potential for a scale and scope of investigation on an order of magnitude larger than that currently available from single researchers and small groups.

Policymakers gambled that networks would appeal to scientists, in part because of the elitist discourse that underpins these policies, but also because the idea of the "invisible college" has been part of the lore of science for the past three centuries. In this sense we can see a convergence of interests across government, research administrators, and scientists. The boundaries between academy and industry, science and policy, and basic and applied are likely more permeable because of NCE. The level of interdisciplinary collaboration and the internalization of network and commercial culture have also been furthered through this policy. Government has played a key role in redirecting the culture of Canada's universities toward commercially oriented network science.

Yet the changes provide differential benefit. As more external resources are directed toward elite research programs in the natural and medical sciences, more of the internal resources have to be diverted to support them. Universities and hospitals carry a large and mostly unacknowledged proportion of the costs of the program, subsidizing the bulk of the indirect and infrastructure costs as well as the direct costs of researchers' salaries and benefits.[34]

In effect, the universities and hospitals supplied the essential incubation facilities in which the networks could flourish. If the NCE program conveys an overall impression of fiscal prudence, it is largely because of its ability to distribute the costs of research across space and sector. The program covers only part of the direct costs of research, amounting to $2 million to $4 million per network each year. Industry makes only nominal contributions, despite program claims to the contrary. Other federal and provincial government agencies, tax-sheltered disease foundations, and other nonprofits underwrite a substantial portion. When *all* the public contributions are calculated,[35] it becomes clear that the NCE is built upon state investment.

In the early and mid 1990s, other concepts like *prosperity* and *healthy, wealthy, and wise* were added to the lexicon used to legitimate the science and technology policy. In 1996, as one element of the new federal integrated policy, NABST (the National Advisory Board on Science and Technology) was replaced by the Advisory Council on Science and Technology, a body of twelve experts representing the scientific community and industry. In 1997 the Liberals created the Canadian Foundation for Innovation (CFI). In 1999 the federal government announced the Canadian Research Chairs program (CRC) and once again created a precedent in the federal-provincial PSE relationship. For the first time under the CRC, federal funds were sent directly to individual universities to pay for positions. Further, in order to qualify for their quota of chairs, universities were required to submit research plans to a federal committee for approval. These changes are a clear threat to provincial juridical autonomy and to the academic autonomy of the university. These two programs have had a dramatic impact on the academic research enterprise (Polster 2002).

Against the background of policy papers referring to the knowledge economy, the federal government in 2001 took another unprecedented step when it decided to begin offering grants to the universities to alleviate the direct and indirect costs of research. This policy initiative has since become a permanent program distributing over $200 million a year directly to the universities without any interference from the provinces. The current rate of distribution is $300 million per year. When all the funding changes and interventions are taken into account and we examine the total federal program spending on PSE and research, we can see the balancing impact of the R&D agenda. As noted earlier, between 1997–98 and 2004–5, the federal government added a total of approximately $11 billion in university research spending. Approximately 70% of the funding was received in the past four years (see AUCC 2005, 22). We estimate that of this total, only $1.4 billion, or 12%, funded research in the humanities and social sciences.

The Canadian political establishment and the university community on the whole accept the position taken by the OECD (2003) that increased investment in R&D will enhance economic development. In support of this position, Canada has dramatically increased R&D spending in the higher education sector since 1996. Canada recorded the fastest growth in R&D spending in the G7 between 1997 and 2003. The federal government's support of university R&D in 2005–6 at $2.5 billion for the first time exceeded its own internal R&D expenditures. In 2004 the federal government's intramural spending on R&D was almost $2.3 billion, a figure that had remained fairly constant over the previous six years. Even so, Canada's GERD as a percentage of GDP in 2004 was below the OECD mean of 2.25% and was only half the score of the leader, Sweden (Council of Canadian Academies 2006).

Canada does record some success when publications and citations are used to measure the impact and outcomes of research investment. King (2004) used biblio-metric data on thirty-one countries between 1997 and 2003 to obtain a measure of productivity and quality of science. Canada stood sixth both by rank of the top 1% of cited publications and by share of total publications. Yet once King focused on the G7, Canada ranked as follows: environmental sciences, tied for second; preclini-cal medicine and health, third; biology, fifth; clinical medicine, fifth; mathematics, tied for fourth; physical sciences, tied for sixth; and engineering, seventh (Council of Canadian Academies 2006). But when bibliometric data is used to measure re-search intensity (SI) and research output quality (ARIF) in world science, Canada shows broad strength. Of the 125 areas analyzed over the period 1997 to 2004, 38% were above the world average for both SI and ARIF. Only 10% were below the world average on both indicators. The greatest strength was shown in psychology and psychiatry, earth and space sciences, biology, biomedical research, and clinical medi-cine (Council of Canadian Academies 2006). The general results correlate positively with Archambault and Gingras's (2004) bibliometric studies. They concluded that between 1997 and 2001, Canada ranked sixth in the world with regard to science and engineering publications. Within the G7, Canada ranked second, behind the UK, in total publications per capita and fourth in publications per university re-searcher.

Canada's medium ranking on the GERD as a percentage of GDP is explained in part by the nation's relatively low business expenditure on research (BERD) as compared to that of other OECD countries. This is in spite of the very generous tax credits offered as an incentive to make these investments. Furthermore, while Canada's proportion (44%) of the population aged 25 to 64 with postsecondary education was the highest in the OECD for 2003, Canada's ranking for PhD gradu-ates in science and engineering in 2002 was ninth (Council of Canadian Academies 2006). So even with the massive investment since 1997 in the natural, applied, and health sciences by governments, Canada is not moving ahead when output measures are taken into account.

Overturning a fragmented science-policy history, both the Conservative and Liberal administrations crafted a climate of commercialization by applying a mul-titude of mutually reinforcing policy instruments. Available data indicate that their efforts to drive science to the market have been successful. Business support of university research appears to be advancing more rapidly in Canada than elsewhere. However, the Canadian business sector remains a low performer, suggesting that Canada's industries continue to rely on publicly supported research rather than developing their own infrastructure. The federal government has promoted this tendency by providing the most favorable tax regime for investment in public R&D of any country among the G7.

The key development in the academic research enterprise over the past fifteen years has been the emergence of a clear stratum of research-intensive universities. These universities first began identifying themselves as the "Group of 10." More recently the group has expanded to become the "Group of 13." While historically it has always been possible to group Canadian universities into a status hierarchy, a degree from any university was regarded until recently as being a relatively standard credential. As research funding from both federal and provincial sources has increased dramatically since 1996, this relatively small group of universities, between 10 and 15 out of a total of 83, have maintained their hold on a very large proportion of the total university research income.

Given the combined impact of the major federal research policy instruments (NCE, CFI, CRC, research councils, and Indirect Costs) as well as the provincial R&D policies in Ontario, Quebec, British Columbia, and Alberta, we can be certain that the trend has favored the natural, applied, and health sciences. Furthermore, given the links between R&D policy and commercialization and economic development, we can conclude that these research-intensive universities have been pushed closer to the boundary separating the academy from industry and the market.

Federal R&D policy has been successful on two fronts. The increase in public funding has guaranteed that a high proportion of the research continues to be conducted in the higher education sector. At the same time, the emphasis on commercialization and academy-industry partnerships has resulted in an increase in the proportion of GERD that business enterprise funds. Alongside these intended consequences are the unintended consequences of increased stratification of regions and institutions. The emphasis on competition and knowledge production has created quasi-markets within Canada as provinces and universities compete for resources and status.

NOTES

1. This section draws heavily on Fisher et al. 2006.
2. The federal spending power draws on the historical prerogative of the Crown to make gifts to its citizens (Cameron 2004, 7).
3. Note that all monetary figures used in this chapter refer to Canadian dollars.
4. It should be noted that various federal ministries, such as Industry Canada, Human Resources and Skills Development Canada, Social Development Canada, and Heritage Canada, do directly fund a small amount of academic research.

5. Some of the established networks began in the mid-1990s.

6. See also Canadian Foundation for Innovation (CFI) Annual Reports, 1997–98 to 2005–6, www.innovation.ca.

7. The fifteen universities in order of the amount of research funds received are Toronto, McGill, Montréal, British Columbia, Alberta, Laval, Calgary, McMaster, Western Ontario, Ottawa, Queen's, Manitoba, Guelph, Waterloo, and Saskatchewan.

8. The "hard sciences" are the three groups of disciplines identified in the previous paragraph plus agriculture, natural resources, and conservation.

9. The Canada Council for the Encouragement of the Arts, Letters, Humanities and the Social Sciences was established by the federal government as a Crown corporation in 1957. The council was provided with $50 million to be spent in ten years on university capital grants and another $50 million of which it could spend only the interest. The latter funds were meant to stimulate and help voluntary organizations in the above fields, to provide scholarships, and to foster Canada's cultural relations abroad (Fisher 1991).

10. Comments from the Honorable Hugh Faulkner, then minister of state for science and technology, December 13, 1976, as quoted in NSERC 2004, para. 6.

11. These goals were part of the ideologically driven reform ("new public management") of public-sector institutions in OECD countries generally at this time.

12. Invitations were sent to representatives from universities and colleges (30%), employers and users of research (25%), labor and employee organizations (10%), federal and provincial governments (15%), nongovernment organizations and special interest groups (aboriginal groups, visible minorities, student groups, etc.) (15%), and the general public (5%).

13. The Science Council of Canada, established in 1966 to provide independent advice on the formulation and implementation of federal science policy, was abolished in 1992 as part of a massive restructuring of the country's R&D system.

14. This whole section relies heavily on two articles: Fisher, Atkinson-Grosjean, and House 2001; and Atkinson-Grosjean, House, and Fisher 2001.

15. For an account of the role of the NRC in the development of the national innovation system, see Doern and Levesque 2002.

16. The CIAR was created in 1982 by Fraser Mustard, a distinguished medical scientist and policy actor. Prior to the NCE, this model was utilized by the Ontario Centres of Excellence program.

17. Brian Mulroney, cited in an unpublished paper of the ARA Consulting Group Inc., Toronto, ON, January 1997, p. 9.

18. Program officer, interview by Donald Fisher, October 2, 1999, 2; transcript at University of British Columbia. See Fisher and Atkinson-Grosjean 2002.

19. In 2001 eighteen networks were pursuing research on arthritis, aquaculture, stroke, bacterial diseases, genetic diseases, protein engineering, vaccines, telecommunications, geomatics, health evidence systems, robotics, structural sensing systems, mathematics of complex systems, wood pulps, microelectronics, sustainable forest management, distance learning, and photonics. In March 2001, after a year-long targeted competition, four new networks were announced in stem cell genomics, childhood literacy, "clean water," and advanced automotive engineering. Earlier networks that lost funding at some stage of the renewal cycle studied neuroscience, respiratory disease, concrete, aging, molecular facial dynamics, fisheries, space

science, and pest control in agriculture. The network on aging chose not to apply for renewal for reasons related to the status of the social sciences in the early years of the NCE program.

20. See also Canadian Foundation for Innovation (CFI) annual reports, 1997–98 to 2005–6, www.innovation.ca.

21. With the establishment of the CGS, the federal government committed $25 million in 2003–4, $55 million in 2004–5, and $85 million in 2005–6 in new money.

22. Paul Davenport is president and vice chancellor of the University of Western Ontario. His research in economics has centered on the theory of economic growth, analysis of the productivity slowdown in Canada over the past two decades, and federal-provincial fiscal arrangements (see Davenport 2000). He is also an advocate of the values of higher education, with a particular commitment to maintaining excellence in university teaching and research.

23. Between 1983–84 and 1992–93, federal cash transfers for PSE as a percentage of GDP declined from 0.56% to 0.41%. By 1998–99 this figure had dropped to 0.25%, and then to 0.19% in 2004–5 (CAUT 2006, fig. 1.1, p. 14).

24. In recognition of the importance of the changes occurring in universities, the prime minister's Advisory Council on Science and Technology (ACST) in late 1998 set up an expert panel to study the commercialization of university research. At the same time, Statistics Canada initiated the first survey of university commercialization, asking all members of the AUCC to report their activities for 1997 (Statistics Canada 1999). The expert panel published a series of working papers early in the year and a final report and recommendations in May 1999 (ACST 1999).

25. While not without weaknesses, the AUTM survey is the only source of centralized longitudinal data in either country.

26. Alberta (1981), British Columbia (1980), Calgary (1982), Guelph (1982), Laval (1984), Manitoba (1980), Queens (1985), Ryerson (1982), Simon Fraser (1984), Toronto (1984), and Waterloo (1981).

27. Program Officer, interview by Donald Fisher, March 17, 1999, 10; transcript at University of British Columbia.

28. Interview Program Officer, May 20, 1999, 15.

29. While the NSERC numbers include all graduate awards and the SSHRC numbers only doctoral awards, the distinctions drawn are still valid, since SSHRC only began to offer magisterial awards in 2003 with the introduction of the CGS program.

30. It has also created the SuperBuild Fund, which allows for capital expansion (upgrades, renovations of existing buildings, and new residences) but does not contribute to operating costs.

31. Enrollment growth is determined by growth of first-entry undergraduate programs and second-entry professional and graduate programs.

32. Performance is determined by institutions' graduation and employment rates and student loan default rates.

33. In 1968, Senator Maurice Lamontagne was appointed chair of the parliamentary Committee on Science Policy. The committee's brief was to consider and report on the science policy of the federal government: its priorities, its budget, and its efficiency in comparison with the governments of other industrialized countries, in light of the requirements of the new scientific age (Fisher 1991).

34. Nominally 10% of a researcher's time is spent on network activities, but the percentage is often considerably more, due to board, committee, and leadership commitments.

35. For an examination of the relative proportions of public and private funding in the NCE program, see Janet Atkinson-Grosjean, Excellence, Networks, and the Pursuit of Profit: Academic Science and Public Policy in Canada, paper presented at the Society for the Social Studies of Science, San Diego, CA., November 1, 1999.

REFERENCES

Abele, F. (1992) The Politics of Competitiveness. In F. Abele (ed.), *How Ottawa Spends, 1992–93: The Politics of Competitiveness,* 1–22. Ottawa: Carleton University Press.

Advisory Council on Science and Technology (ACST), Expert Panel on Canada's Role in International Science and Technology. (2000) *Reaching Out: Canada, International Science and Technology, and the Knowledge-Based Economy.* Ottawa: ACST, http://acst-ccst.gc.ca/intel/report-web2/rw2_title_e.html.

Advisory Council on Science and Technology (ACST), Expert Panel on the Commercialization of University Research. (1999) *Public Investments in University Research: Reaping the Benefits.* Ottawa: ACST, http://acst-ccst.gc.ca/comm/rpaper_html/report_title_e.html.

Aguayo, A. J. and R. A. Murphy. (1997) Canada's Crisis: Can Business Rescue Science? *Science* 275 (5297): 139–140.

ARA and Brochu Consulting Group Inc. (1998) Approaches of Canadian Universities to the Management and Commercialization of Intellectual Property: Diversity and Challenges. Discussion paper prepared for the Symposium on Intellectual Property, November 26, AUCC, Ottawa.

Archambault, É. and Y. Gingras. (2004) Opinion Leader: The Decline of Canadian Science. *Re$earch Money* 18 (14).

Archibugi, D., J. Howells, and J. Michie. (1999) *Innovation Policy in a Global Economy.* Cambridge: Cambridge University Press.

Association of Universities and Colleges of Canada (AUCC). (2002a) *Framework of Agreed Principles on Federally Funded University Research.* Ottawa: AUCC, www.aucc.ca/_pdf/english/reports/2002/frame_cadre_e.pdf.

———. (2002b) *A Strong Foundation for Innovation: An AUCC Action Plan.* Ottawa: AUCC.

———. (2005) *Momentum.* Ottawa: AUCC.

Association of University Technology Managers (AUTM). (1998) *Licensing Survey, Fiscal Year 1997, Annual Survey of Member Institutions.* Deerfield, IL: AUTM.

Atkinson-Grosjean, J., D. House, and D. Fisher. (2001) Canadian Science Policy and Public Research Organizations in the 20th Century. *Science Studies* 14:3–25.

Bakvis, H. (2002) A Checkerboard Federalism? Labor Market Development Policy in Canada. In H. Bakvis and G. Skogstad (eds.), *Canadian Federalism: Performance, Effectiveness, and Legitimacy,* 197–219. Don Mills, ON: Oxford University Press.

Borys, B. and D. B. Jemison. (1989) Hybrid Arrangements as Strategic Alliances: Theoretical Issues in Organizational Combinations. *Academy of Management Review* 14:234–249.

Callon, M. (1995) Four Models of the Dynamics of Science. In S. Jasanoff, G. E. Markle,

J. C. Peterson, and T. Pinch (eds.), *Handbook of Science and Technology Studies,* 29–63. Thousand Oaks, CA: Sage.

Cameron, D. M. (1991) *More Than an Academic Question: Universities, Government, and Public Policy in Canada.* Halifax: Institute for Research on Public Policy.

———. (1992) Higher Education in Federal Systems: Canada. In D. Brown, P. Cazalis, and G. Jasmin (eds.), *Higher Education in Federal Systems,* 45–67. Kingston, ON: Queen's Institute of Intergovernmental Relations.

———. (2004) Collaborative Federalism and PSE: Be Careful What You Wish For. Paper presented at the Higher Education in Canada conference, Kingston, ON, February 13–14, http://jdi.econ.queensu.ca/Files/Conferences/PSEconferencepapers/Cameronconference paper.pdf.

Canada Council for the Arts. (2004) *A Brief History of the Canada Council for the Arts.* Ottawa: Canada Council for the Arts, www.canadacouncil.ca/aboutus.

Canada Millennium Scholarship Foundation (CMSF). (2001) *Does Money Matter?* Montreal: CMSF.

———. (2003) Trickle-Down Fiscal Federalism: The Use of the Federal Spending Power for Income Support, unpublished paper, CMSF, Montreal.

Canada Research Chair Program (CRC). (2006) *Program Statistics,* March, www.chairs.gc.ca . (accessed May 5, 2006).

Canadian Association of University Teachers (CAUT). (1989) Planning and Direction: PC Research Policy and Funding. *CAUT Bulletin,* November.

———. (1993) *Brief to the Standing Committee on National Finance on Bill C-93.* Ottawa: CAUT.

———. (2006) *Almanac of Post-Secondary Education in Canada.* Ottawa: CAUT.

———. (2007) *Almanac of Post-Secondary Education in Canada.* Ottawa: CAUT.

Cassin, A. M. and J. G. Morgan. (1992) The Professoriate and the Market-Driven University: Transforming the Control of Work in the Academy. In W. K. Carroll, L. Christiansen-Ruffman, R. F. Currie, and D. Harrison (eds.), *Fragile Truths: Twenty-Five Years of Sociology and Anthropology in Canada,* 247–260. Ottawa: Carleton University Press.

Chan, A. S. and D. Fisher (eds.). (2008) *The Exchange University: Corporatization of Academic Culture.* Vancouver: University of British Columbia Press.

Charles, D. and J. Howells. (1992) *Technology Transfer in Europe.* London: Belhaven Press.

Clark, B. R. (1983) *The Higher Education System: Academic Organization in Cross-National Perspective.* Berkeley: University of California Press.

———. (1998) *Creating Entrepreneurial Universities: Organizational Pathways of Transformation.* Oxford: IAU Press.

Council of Canadian Academies, Committee on the State of Science and Technology in Canada. (2006) *The State of Science and Technology in Canada.* Ottawa: Council of Canadian Academies.

Currie, J. (1998) Globalization Practices and the Professoriate in Anglo-Pacific North American Universities. *Comparative Education Review* 42:15–29.

Dale, R. (1997) The State and Governance of Education: An Analysis of the Restructuring of the State-Education Relationship. In A. Halsey, H. Lauder, P. Brown, and A. S. Wells (eds.), *Education, Culture, Economy, Society,* 272–282. Oxford: Oxford University Press.

Davenport, P. (2000) Universities, Innovation, and the Knowledge-Based Economy. Paper presented at the CERF/IRRP Conference on Creating Canada's Advantage in an Information Age, Ottawa.

Deem, R. (2001) Globalization, New Managerialism, Academic Capitalism, and Entrepreneurialism in Universities: Is the Local Dimension Still Important? *Comparative Education* 37:7–20.

Department of Foreign Affairs and International Trade. (1996) *Canada's Research and Development Infrastructure.* Ottawa: Department of Foreign Affairs and International Trade, www.dfait-maeci.gc.ca/english/html/60/60150.htm.

Dill, D. (1997a) Higher Education Markets and Public Policy. *Higher Education Policy* 10:167–185.

———. (1997b) Markets and Higher Education: An Introduction. *Higher Education Policy* 10:163–166.

Doern, G. B. (ed.). (2006) *Innovation, Science, and Environment: Canadian Policies and Performance, 2006–2007.* Montreal: McGill-Queen's University Press.

Doern, G. B. and R. Levesque. (2002) *The National Research Council in the Innovation Policy Era: Changing Hierarchies, Networks, and Markets.* Toronto: University of Toronto Press.

Doern, G. B. and M. Sharaput. (2000) *Canadian Intellectual Property: The Politics of Innovation, Institutions, and Interests.* Toronto: University of Toronto Press.

Doutriaux, J. and M. Barker. (1995) *The University-Industry Relationship in Science and Technology.* Ottawa: Industry Canada, http://strategis.ic.gc.ca/epic/internet/ineas-aes.nsf/vwapj/op11e.pdf/$FILE/op11e.pdf.

Drucker, P. F. (2002) *Managing the Next Society.* New York: St. Martin's Press.

Dudley, J. (1998) Globalization and Education Policy in Australia. In J. Currie and J. Newson (eds.), *Universities and Globalization: Critical Perspectives,* 21–44. Thousand Oaks, CA: Sage.

Enros, P. and M. Farley. (1986) *University Offices for Technology Transfer: Toward the Service University.* Ottawa: Science Council of Canada.

Etzkowitz, H. (1997) The Entrepreneurial University and the Emergence of Democratic Corporatism. In H. Etzkowitz and L. Leydesdorff (eds.), *Universities and the Global Knowledge Economy,* 141–152. London: Cassell Press.

Etzkowitz, H. and A. Webster. (1994) Science as Intellectual Property. In S. Jasanoff, G. E. Markle, J. C. Peterson, and T. Pinch (eds.), *Handbook of Science and Technology Studies,* 480–505. Thousand Oaks, CA: Sage.

———. (1997) *Capitalising Knowledge.* New York: SUNY Press.

Etzkowitz, H., A. Webster, and P. Healey. (1998) *Capitalizing Knowledge: New Interactions of Industry and Academe.* Albany: SUNY Press.

Evans, K. (2005) Going Global with the Locals: Examining the Impact of Internationalization on University Colleges in British Columbia. PhD diss., University of British Columbia.

Fisher, D. (1991) *The Social Sciences in Canada: Fifty Years of National Activity by the Social Science Federation of Canada.* Waterloo, ON: Wilfrid Laurier University Press.

———. (1993) *Fundamental Development of the Social Sciences: Rockefeller Philanthropy and the United States Social Science Research Council.* Ann Arbor: University of Michigan Press.

———. (1999) Harold Innis and the Canadian Social Science Research Council: An Experiment in Boundary Work. In C. R. Acland and W. J. Buxton (eds.), *Harold Innis in the New Century: Reflections and Refractions*, 135–158. Montreal: McGill-Queen's University Press.

Fisher, D. and J. Atkinson-Grosjean. (2002) Brokers on the Boundary: Academy-Industry Liaison in Canadian Universities. *Higher Education* 44:449–467.

Fisher, D., J. Atkinson-Grosjean, and D. House. (2001) Changes in Academy/Industry/State Relations in Canada: The Creation and Development of the Networks of Centres of Excellence. *Minerva* 39:299–325.

Fisher, D. and K. Rubenson. (1998) The Changing Political Economy: The Private and Public Lives of Canadian Universities. In J. Currie and J. Newson (eds.), *Universities and Globalization: Critical Perspectives*, 77–98. Thousand Oaks, CA: Sage.

Fisher, D., K. Rubenson, T. Shanahan, C. Trottier, J. Bernatchez, R. Clift, G. Jones, J. Lee, M. MacIvor, and J. Meredith. (2006) *Canadian Federal Policy and Post-Secondary Education*. Vancouver: Centre for Policy Studies in Higher Education and Training.

Friedman, R. S. and R. C. Friedman. (1990) The Canadian Universities and the Promotion of Economic Development. *Minerva* 28:272–293.

Frost, N. and R. Taylor. (2001) Patterns of Change in the University: The Impact of "Lifelong Learning" and the "World of Work." *Studies in the Education of Adults* 33:49–59.

Fuller, S. (1988) *Social Epistemology*. Bloomington: Indiana University Press.

Gieryn, T. F. (1983) Boundary Work and the Demarcation of Science from Non-Science: Strains and Interests in Professional Ideologies of Scientists. *American Sociological Review* 48:781–95.

———. (1995) Boundaries of Science. In S. Jasanoff, G. E. Markle, J. C. Peterson, and T. Pinch (eds.), *Handbook of Science and Technology Studies*, 393–443. Thousand Oaks, CA: Sage.

———. (1999) *Cultural Boundaries of Science: Credibility on the Line*. Chicago: University of Chicago Press.

Government of Canada. (1987) *InnovAction: The Canadian Strategy for Science and Technology*. Ottawa: Queen's Printer.

Guston, D. H. and K. Keniston. (1994) Introduction: The Social Contract for Science. In D. H. Guston and K. Keniston (eds.), *The Fragile Contract: University Science and the Federal Government*, 1–41. Cambridge, Mass.: MIT Press.

Hardy, C. (1992) Managing the Relationship: University Relations with Business and Government. In J. Cutt and R. Dobell (eds.), *Public Purse, Public Purpose: Autonomy and Accountability in the Groves of Academe*, 193–218. Ottawa: Institute for Research on Public Policy and Canadian Comprehensive Auditing Foundation.

Human Resources Development Canada (HRDC). (2002) *Knowledge Matters: Skills and Learning for Canadians*. Ottawa: Government of Canada.

Industry Canada. (1996) *Science and Technology for the New Century: A Federal Strategy*. Ottawa: Government of Canada.

———. (2002) *Achieving Excellence: Investing in People, Knowledge, and Opportunity*. Ottawa: Industry Canada, www.innovationstrategy.gc.ca/gol/innovation/interface.nsf/vSS GBasic/in04142e.htm.

———. (2004a) Building a Federal Science and Technology Strategy. Discussion paper, Government of Canada, Ottawa.

———. (2004b) Building a More Innovative Economy. Discussion paper, Government of Canada, Ottawa.

———. (2004c) A New Framework for Economic Policy. Discussion paper, Government of Canada, Ottawa.

Jones, G. A. (1996) Governments, Governance, and Canadian Universities. In J. C. Smart (ed.), *Higher Education: Handbook of Theory and Research,* 11:337–371. New York: Agathon Press.

Kaghan, W. N. (1998) Court and Spark: Studies in Professional University Technology Transfer Management. PhD diss., University of Washington.

Keller, G. (1983) *Academic Strategy: The Management Revolution in American Higher Education.* Baltimore: Johns Hopkins University Press.

King, D. (2004) The Scientific Impact of Nations. *Nature* 430:311–316.

Lamontagne, M. (1970) *A Critical Review: Past and Present.* Vol. 1 of *A Science Policy for Canada: Report of the Senate Special Committee on Science Policy.* Ottawa: Queen's Printer for Canada.

Lee, J. (2005) Access to Postsecondary Education: A Comparative Study of British Columbia and Ontario. PhD diss., University of British Columbia.

Liberal Party of Canada. (1993) *Creating Opportunity: The Liberal Plan for Canada.* Ottawa: Liberal Party of Canada.

Macdonald Report. (1969) *The Role of the Federal Government in Support of Research in Canadian Universities.* Special Study No.7. Ottawa: Science Council.

Marginson, S. (1997) *Markets in Education.* St. Leonards, New South Wales, Australia: Allen and Unwin.

———. (2000) Rethinking Academic Work in the Global Era. *Journal of Higher Education Policy and Management* 22:23–37.

Marginson, S. and M. Considine. (2000) *The Enterprise University.* Cambridge: Cambridge University Press.

National Advisory Board on Science and Technology (NABST). (1988) *Report by the University Committee.* Ottawa: Government of Canada.

———. (1995) *Healthy, Wealthy, and Wise: A Framework for an Integrated Federal Science and Technology Policy.* Ottawa: Government of Canada.

National Forum on Post-Secondary Education, Saskatoon. (1987) *National Forum Proceedings,* October 25–28. Halifax, Nova Scotia: Institute for Research on Public Policy.

Natural Sciences and Engineering Research Council (NSERC). (1992) *Networks of Centres of Excellence: Powerful Partnerships.* Ottawa: Minister of Supply and Services, Canada.

———. (2004) *NSERC's History.* Ottawa: NSERC, www.nserc.gc.ca/about/history.htm.

Nelson, R. R. (1993) *National Innovation Systems: A Comparative Analysis.* New York: Oxford University Press.

———. (1996) *The Sources of Economic Growth.* Cambridge, MA: Harvard University Press.

Networks of Centres of Excellence Program (NCE). (1997) *Annual Report.* Ottawa: Minister of Supply and Services, Canada.

———. (2004a) *Currently Funded Networks.* Ottawa: Government of Canada, www.nce
.gc.ca/nets89_e.htm (accessed September 22, 2004).

———. (2004b) *Previously Funded Networks.* Ottawa: Government of Canada, www.nce
.gc.ca/nets89_e.htm (accessed September 20, 2004).

———. (2007) *Currently Funded Networks.* Ottawa: Government of Canada, www.nce
.gc.ca/nets_e.htm (accessed November 13, 2007).

Newson, J. A. (1994) Subordinating Democracy: The Effects of Fiscal Retrenchment and
University-Business Partnerships on Knowledge Creation and Knowledge Dissemination
in Universities. *Higher Education* 27:141–161.

———. (1998) The Corporate Linked University: From Social Project to Market Force.
Canadian Journal of Communications 23:107–124.

Noël, A. (2002) Without Québec: Collaborative Federalism with a Footnote? In T. McIntosh
(ed.), *Building the Social Union: Perspectives, Directions, and Challenges,* 17–18. Regina:
Saskatchewan Institute of Public Policy.

Niosi, J. (1995) *Flexible Innovation: Technological Alliances in Canadian Industry.* Montreal:
McGill-Queen's University Press.

Office of the Auditor General of Canada. (1994) *Science and Technology: Overall Management
of Federal Science and Technology Activities.* Report of the Auditor General of Canada.
Ottawa: Office of the Auditor General of Canada, www.oag-bvg.gc.ca/domino/reports
.nsf/html/9409ce.html.

Organisation for Economic Co-operation and Development (OECD). (1996) *The Knowledge-Based Economy: Science, Technology, and Industry Outlook.* Paris: OECD.

———. (2003) *The Sources of Economic Growth in OECD Countries.* Paris: OECD.

———. (2006) *Main Science and Technology Indicators, 2006–1.* Paris: OECD.

———. (2007) *Main Science and Technology Indicators, 2007–1.* Paris: OECD.

Pietrykowski, B. (2001) Information Technology and the Commercialization of Knowledge:
Corporate Universities and Class Dynamics in an Era of Technological Restructuring.
Journal of Economic Issues 35:299–307.

Polanyi, M. (2000) The Republic of Science: Its Political and Economic Theory. *Minerva*
38:1–21 (reprinted from *Minerva* 1:54–73).

Polster, C. (2002) Break from the Past: Impacts and Implications of the Canada Foundation
for Innovation and the Canada Research Chairs Initiatives. *Canadian Review of Sociology
and Anthropology* 39:275–299.

Prosperity Secretariat. (1991a) *Learning Well . . . Living Well.* Ottawa: Prosperity Secretariat.

———. (1991b) *Prosperity through Competitiveness.* Ottawa: Prosperity Secretariat.

Pullen, J. W. (1990) *Centres of Excellence.* Report prepared by the Canadian Centre for Management Development. Government Catalogue No. SC93-2/2-1990E. Ottawa: Minister
of Supply and Services, Canada.

Savage, D. (1987) Frank Oberle: Toward a National Science Policy. *CAUT Bulletin,* May.

Schugurensky, D. (2007) Higher Education Restructuring in the Era of Globalization: To-
ward a Heteronomous Model? In R. Arnove and C. Torres (eds.), *Comparative Education:
The Dialectic of the Global and the Local,* 257–276. 3rd ed. Lanham, MD: Rowman and
Littlefield.

Science Council of Canada. (1968) *Towards a National Science Policy for Canada.* Ottawa: Queen's Printer and Controller of Stationery.

———. (1988) *Winning in a World Economy: University-Industry Interaction and Economic Renewal in Canada.* Ottawa: Minister of Supply and Services.

Senate Special Committee on Science Policy. (1972) *Targets and Strategies for the Seventies.* Vol. 2 of *A Science Policy for Canada: Report of the Senate Special Committee on Science Policy.* Ottawa: Information Canada.

Simeon, R. and I. Robinson. (2004) The Dynamics of Canadian Federalism. In J. Bickerton and A. G. Gagnon (eds.), *Canadian Politics,* 101–126. 4th ed. Peterborough, ON: Broadview Press.

Slaughter, S. (1998) National Higher Education Policies in a Global Economy. In J. Currie and J. Newson (eds.), *Universities and Globalization: Critical Perspectives,* 45–70. Thousand Oaks, CA: Sage.

Slaughter, S. and L. Leslie. (1997) *Academic Capitalism: Politics, Policies, and the Entrepreneurial University.* Baltimore: Johns Hopkins University Press.

Slaughter, S. and G. Rhoades. (2004) *Academic Capitalism and the New Economy: Markets, State, and Higher Education.* Baltimore: Johns Hopkins University Press.

Stanbury, W. T. (1989) Privatization in Canada: Ideology, Symbolism, and Substance. In P. W. MacAvoy, W. T. Stanbury, G. Yarrow, and R. J. Zeckhauser (eds.), *Privatization and State-Owned Enterprises: Lessons from the United States, Great Britain, and Canada,* 273–329. Boston: Kluwer.

Star, S. L. and J. R. Griesemer. (1989) Institutional Ecology, "Translations," and Boundary Objects: Amateurs and Professionals in Berkeley's Museum of Vertebrate Zoology, 1907–39. *Social Studies of Science* 19:387–420.

Statistics Canada. (1999) *Survey of Intellectual Property Commercialization in the Higher Education Sector.* By M. Bordt and C. Read, 88F0006XPB No. 01; ST-00-01. Science and Technology Redesign Project. Ottawa: Government of Canada.

———. (2007) CANSIM using CHASS, Table 382-0002, *GDP-Expenditure Based-Provincial Accounts,* www.statcan.ca (accessed June 21 2007).

Steering Group on Prosperity. (1992) *Inventing Our Future: An Action Plan for Canada's Prosperity.* Ottawa: Steering Group on Prosperity.

Winsor, H. (1993) Tory Senators Snub Mulroney Government. *Globe and Mail* (Toronto), June 11.

Wolfe, D. A. (2003) Innovation Policy for the Knowledge-Based Economy: From the Red Book to the White Paper. In G. B. Doern (ed.), *How Ottawa Spends, 2002–2003,* 137–156. Toronto: Oxford University Press.

Young, J. and B. Levin. (2002) *Understanding Canadian Schools: An Introduction to Educational Administration.* 3rd ed. Toronto: Thompson Nelson.

Young, S. (2002) The Use of Market Mechanisms in Higher Education Finance and State Control: Ontario Considered. *Canadian Journal of Higher Education* 32 (2): 79–101.

Ziman, J. (1994) *Prometheus Bound: Science in a Dynamic Steady State.* Cambridge: Cambridge University Press.

CHAPTER FOUR

Japan

AKIRA ARIMOTO

This chapter analyzes the Japanese academic research enterprise (ARE) based on three sociological concepts: social condition, social structure, and social function (Arimoto 2003). After providing an overview of the current state of the ARE, the chapter presents the policy framework, including discussion of the Japanese ARE in key stages of its history, important research-related policies, the status and context of doctoral education, and current issues in knowledge transfer. It concludes with an assessment of the ARE, including unintended consequences of recent policies and some directions for reform.

The Academic Research Enterprise

For more than a century, Japan's policy for science and technology has influenced its national policy for the higher education system and the system's reform. Current university reforms reflect the following trends:

- Science and technology have had a great impact on university reforms.
- An increasingly knowledge-based society has driven demand for university reforms.
- The greater importance of knowledge has heightened expectations for higher education and for the development of human resources.
- The demands of globalization and a knowledge-based economy have heightened expectations through international competition.

The first of the sociological concepts on which this analysis of the Japanese ARE is based, social condition, refers to the effects of social change on the ARE from three sources: the knowledge function, social change, and national government policy. The first source of influence on the social condition is the knowledge function, because knowledge, as the basis of academic work, is the major determinant of university structure and function. A series of studies in the sociology of science have approached academic work in relation to the function and production of knowledge in the scientific and academic community (Arimoto 1987, 1996; Becher 1981, 1989; Becher and Trowler 2001; Bleiklie and Henkel 2005; Gibbons et al. 1994; Light 1974; Merton and Storer 1973; Parry 2007; Shinbori 1985; Spiegel-Rösing and de Solla Price 1977). From this perspective academic work, with regard to learning, research, teaching, and service, is based on knowledge, including the application of knowledge, scientific knowledge, academic discipline, and related factors. The function of knowledge is therefore divided into five main parts: understanding, discovery and invention, dissemination, application, and control of knowledge. Accordingly, attention should be paid to each of these functions in the study of the ARE, even though the enterprise is mostly involved in the knowledge functions of research.

The second influence on the social condition is social change, such as globalization, marketization, and an orientation toward an increasingly knowledge-based society. Social change exerts pressure on academia to respond with reform and innovation. As discussed later in this chapter, a series of national government policies for the university and the ARE reflect these social changes.

Social changes affect the social condition directly and, through policy, indirectly. Thus, the third source of influence on the social condition is national government policy, which shapes both universities and the ARE and gives Japan's ARE some country-specific characteristics.

In addition to its relationships to the social condition, the ARE should also be understood in terms of its social structure and function. The social structure of the ARE refers to the society within the ARE, including the structure of social power, prestige, and hierarchy. The social function of the ARE is based on the function of knowledge previously mentioned, because the ARE is fundamentally concerned with the research function of knowledge.

Globalization has many aspects. "Globalization encompasses global financial markets, growing global interconnectedness, global and regional trade agreements, media, information systems, labor markets, telecommunication, etc." (Maassen and Cloete 2006, 16). It is likely to emphasize global interconnectedness; to promote integration, centralization, and standardization of education and culture; and to challenge higher education systems by bringing about some level of monoculturization of all systems worldwide. Specifically, the trend that enables the World Trade Organization and Global Agreements on Tariffs and Trade to view education as a commodity has become a source of pressure leading to global standardization of higher education. With its connection to globalization, marketization became dominant worldwide in the fields of politics and economy in the 1980s, moving gradually over time into the domain of culture and education (Arimoto 2002, 2005a).

In addition to these social changes, an orientation to lifelong learning, a declining population (National Institute of Population and Social Security Research 2004), and economic retrenchment are also important external pressures on Japanese universities and colleges. Academic reforms in response to these pressures are of increasing public interest because the reforms are expected to affect the development of society and learning directly and indirectly.

The university was originally conceptualized as an institution that prioritized the development of knowledge. As an enterprise uniquely based on knowledge as its raw material, academia is expected to develop the function of knowledge and to contribute to societal development by adding to scientific development (Clark 1983). Accordingly, an inevitable problem for academia is to ensure the quality of the academic work at the core of its knowledge function. Knowledge can be expressed through discovery, dissemination, application, and control, or in other words, through research, teaching, service, and management. Research and teaching are generally considered the most important. Because of its contribution to the development of society and to learning, the academic research function is accorded the highest esteem.

This section presents the ARE's current structure, which results from some aspects of the policy framework. Three important elements of the structure are the number and types of higher education institutions, the human resources engaged in research at universities, and university research funding.

The higher education institutions in Japan can be grouped into three sectors: national, public, and private. Table 4.1 documents the size of each sector and the increasing share of the private sector over the past half century.

The private sector quantitatively occupies a majority share of the market with 580 (76.7%) of the total of 756 institutions as of 2007. The national sector with 87 (11.5%) and the public sector with 89 (11.8%) provide the smaller share (MEXT 2008). Table 4.2 provides a recent (2005) breakdown of university offerings by sector and course of study. The traits of the present structure of Japan's ARE seem to originate from the process of catching up with the advanced countries in academic productivity. One reason is related to social stratification, in which the national sector prevails.

The hegemony of the national university is also evident in the fact that a great share of the research university group is occupied by the national universities. The

TABLE 4.1.
Number of universities in Japan, by sector, 1955–2007

	Total	National	Public	Private	% Private
1955	228	72	34	122	53.5
1960	245	72	33	140	57.1
1965	317	73	35	209	65.9
1970	382	75	33	274	71.7
1975	420	81	34	305	72.6
1980	446	93	34	319	71.5
1985	460	95	34	331	72.0
1990	507	96	39	372	73.4
1995	565	98	52	415	73.5
2000	649	99	72	478	73.7
2001	669	99	74	496	74.1
2002	686	99	75	512	74.6
2003	702	100	76	526	74.9
2004	709	87	80	542	76.4
2005	726	87	86	553	76.2
2006	744	87	89	568	76.3
2007	756	87	89	580	76.7

Source: MEXT 2008.

term *research university* (*kenkyu daigaku*) was originally used by the Carnegie Classification in the United States (Carnegie Council on Policy Studies in Higher Education 1976) and was introduced into Japan in the early 1980s when Amano (1984) made the first Japanese version of a classification for the ARE in higher education institutions. At that time, research universities numbered 24 (15 national, 4 public, and 5 private institutions), or 5.4%, among 443 total universities (Amano 1984). If the percentage of universities classified as research universities had remained constant, one would expect about 40 of the current 756 (in 2007) four-year institutions to be classified as research universities (MEXT 2008). As we will see later, research universities actually make up a somewhat smaller share, due to funding patterns and increased stratification.

Table 4.3 provides data on the human resources dedicated to R&D (research and development) in Japan's universities by sector. Despite the private sector's overall larger size, the national sector, because it dedicates a greater proportion of resources to research, has more total resources engaged in research (Ministry of Internal Affairs and Communications 2004).

Table 4.4 presents the number of persons engaged in R&D by field. Natural sciences and engineering fields have the most total resources as well as the greatest proportion of resources engaged in research. Of the 222,458 persons working in natural sciences research, medical sciences (with 116,368 engaged) and engineer-

TABLE 4.2.
Number of universities, by sector and course of study, 2005

Course of study	Total	National	Public	Private	% Private
Evening	115	35	8	72	62.6
Master's	540	87	71	382	70.7
Doctor's	409	75	52	282	68.9
Professional degree	92	27	3	62	67.4
Correspondence and mass media (universities)	35 (32)	0	0 (32)	35	100.0
Correspondence and mass media (graduate schools)	19 (17)	0	0 (17)	19	100.0

Source: MEXT 2006b.

Note: Figures in parentheses refer to those providing regular courses as well as correspondence courses.

TABLE 4.3.
Persons engaged in R&D at universities, by sector and type of profession, 2003

Profession	Total	National	Public	Private	% Private
Total engaged at universities	684,275	256,388	48,473	379,414	55.4
Total engaged in research at universities	335,983	161,510	24,533	149,940	44.6
% engaged in research	49.1	63.0	50.6	39.5	—
Total university researchers	284,330	131,081	21,963	131,286	46.2
Doctoral students	68,476	48,637	3,811	16,028	23.4

Source: Ministry of Internal Affairs and Communications 2004.

Note: Total engaged in research includes assistant research workers, technicians, and clerical and other R&D support workers in addition to researchers. Researchers include teachers, medical staff, and doctoral students.

ing (66,373) represent the largest share. Within the humanities and social sciences, literature (22,471) and economics (18,113) have the most people engaged in research. Compared to the overall population of researchers, doctoral students are more concentrated in the natural sciences and engineering (Ministry of Internal Affairs and Communications 2004).

National policy for higher education in the postwar period emphasized reliance on private expenditure by restricting public expenditure. Therefore, government expenditure on higher education as a proportion of GDP remained as small as

TABLE 4.4.
Persons engaged in R&D at universities, by field of science, 2003

Profession	Total	Natural science and engineering	Humanities and social science	Other	% Natural science and engineering
Total engaged at universities	684,275	409,279	181,379	93,617	59.8
Total engaged in research at universities	335,983	222,458	77,100	36,425	66.2
% engaged in research	49.1	54.4	42.5	38.9	—
Total university researchers	284,330	184,978	68,118	31,234	65.1
Doctoral students	68,476	48,090	15,877	4,509	70.2

Source: Ministry of Internal Affairs and Communications 2004.

Note: Total engaged in research includes assistant research workers, technicians, and clerical and other R&D support workers in addition to researchers. Researchers include teachers, medical staff, and doctoral students.

0.5%, which is lower than that of other countries in the Organisation for Economic Co-operation and Development (OECD). On the other hand, private expenditure on higher education amounts to 2.0% of GDP, which is equivalent to that of the United States and higher than that in the advanced countries of Europe (Arimoto 2006). These statistics represent Japan's long-term pattern of higher education spending.

The University Council (UC) and the Central Education Council (CEC) have regularly pointed out that the share of national expenditure on higher education was among the smallest of advanced countries (UC 1998; CEC 2005a). In fact, if we look at the trend in the government-financed share of R&D expenditure, Japan ranked lowest in FY2003 in a group of selected countries: France (42.1%), EU-25 (34.7%), EU-15 (34.0%), Germany (31.5%), the United Kingdom (31.3%), the United States (31.0%), and Japan (20.2%) (MEXT 2006a).

The distribution of Japan's R&D expenditures by performance sector in 2003 was as follows: government research institutions, 8.7%; universities and colleges, 19.4%; industry, 70.0%; and private research institutions, 1.9%. Restricted to only the natural sciences (engineering, technology, and medical science), the distribution was 9.1%, 13.2%, 75.8%, and 1.9%, respectively. The proportion of industry expenditures was consistently high. In universities and colleges in 2003, R&D expenditures per researcher varied according to the sector: national sector, ¥21.32 million; other public sector, ¥14.13 million; and private sector, ¥17.97 million (MEXT 2006a).

The R&D expenditures by character of the work show greater expenditures on development than on basic and applied research: in Japan in 2003, the proportions were basic research, 15.0%; applied research, 23.0%; and development, 62.0%. Similar recent statistical data for the United States (in 2003, basic research, 19.1%; applied research, 23.9%; and development, 57.1%), France (in 2001, basic research, 23.3%; applied research, 33.5%; and development, 43.2%), and Germany (in 1993, basic research, 21.2%; and applied research/development 78.8%) show that these countries spend more on basic research (MEXT 2006a).

Those research universities with high research achievement are now allocated much higher funding. Among the 744 four-year universities and colleges throughout Japan in 2006, the category *research university* included approximately 4%, or 30 institutions. Among these 30 institutions, the number of research universities in each sector can be estimated as follows: 7 former imperial universities, 10 other national universities, 3 public universities, and 10 private universities (MEXT 2006a).

By source of funding for R&D expenditures at universities and colleges in FY2002, the government accounted for about 50% of the total (which includes the full costs of academic and support staff), with the remainder of funding coming from donations and fees (Polt et al. 2001). The total R&D expenditures at universities and colleges that year had increased by 1.5% over the previous fiscal year to ¥3,282.1 billion, accounting for about 19.4% of Japan's total R&D expenditures. One trend in R&D expenditures by university sector shows national and private universities registering year-by-year increases. In 2003 the sectors had the following total expenditures: private, ¥1.671 trillion; national, ¥1.41 trillion; and other public sector, ¥181.4 billion. Similarly, all fields of study within the natural sciences registered year-by-year increases to give the following totals: health sciences, ¥861.3 billion; engineering, ¥748.2 billion; physical sciences, ¥312.0 billion; and agricultural sciences, ¥131.3 billion (MEXT 2006a). Still, the share of ARE expenditures directed to the natural sciences has been classified as "rather low" (Polt et al. 2001). Compared with other OECD countries, Japan devotes a relatively high share (34%) of academic research to humanities and social sciences. However, the lack of public-sector research enterprises engaged in this type of research means that virtually all of Japan's research in the humanities and social sciences takes place in the ARE (Polt et al. 2001). Investment of funds and resources in key areas in natural sciences is now perceived to be more advantageous than in other areas. In the 21st Century Centers of Excellence (COE) program, which was started in 2002, much more weight was put on provision of resources to the natural sciences than to the humanities and social sciences fields.

Policy Framework

The policy framework of Japan's ARE can be characterized by key historical periods, several recent innovations in research policy, modernizing trends in doctoral education, and increased impact in the area of knowledge transfer.

HISTORY

This section discusses national research policy on ARE reform by focusing on the current situation in Japan. It is noteworthy that three distinctive stages—the prewar, the postwar, and the contemporary eras—are recognizable in ARE reform policy. A brief consideration of the characteristics of the prewar era is based on a previous work by the author (Arimoto 2007).

Following the institutionalization of the university in society, university education has been assumed to constitute the major purpose of higher education. Yet, the commitment of teachers and students to research has become increasingly strong since the modern university was established in the nineteenth century. Today, the university is also expected to contribute to learning through research. The graduate school, invented in the United States in the nineteenth century, has subsequently become a central component of centers of learning throughout the world (Clark 1993, 1995; Oleson and Voss 1979). The concept of the graduate school was imported into Japan during the postwar period with the introduction of a two-tier system of undergraduate and graduate levels. Graduate schools, considered to be a core of the ARE, were gradually institutionalized in higher education institutions throughout the country. Research conducted within the university forms a core part of graduate schools.

Science and technology, including advanced models of higher education, were introduced to Japan from the advanced Western countries as a means of rapidly modernizing the country and catching up with developments already established overseas. Implementation of advanced Western models in its system presented Japan with significant conflicts. Accordingly, Japan was led to create its own model of higher education. This became especially evident in the postwar years. The German Humboldtian model and the American model guided a persistent and intense search, over the more than sixty-year span of the postwar period, for a suitable Japanese model for the higher education system. Some of the current higher education reforms are largely based on the American model of higher education. However,

Americanization has not been successful, in spite of endeavors to institutionalize it spanning at least half a century (Amano 2006).

In general, Japan has successfully caught up with advanced models to the extent that it has reached the level of other centers of learning, or centers of excellence, by penetrating the barrier from the periphery of learning in the sense described by Ben-David in the 1970s (Ben-David and Carnegie Commission on Higher Education 1977). This is evidently true in the fields of natural sciences and engineering, though many problems may remain in the humanities and social sciences. The gap between these two kinds of fields may be due to conflicts between the imported and the native cultures.

National government policy for the academic research organization basically maintained the prewar structure from 1945 to 1990, when a market mechanism became clearly evident. Van Vught (1989) distinguished two types of national policy: national control and self-regulation. According to these categories, national control operated exclusively throughout the history of higher education in Japan until 1990, when, accompanied by much conflict, it was superseded by self-regulation. Market failures observed over recent years are likely to result in reconsideration of the current problematic situation.

Asonuma (2003) analyzed the postwar structure of funding in the Japanese national university in five time periods: the end of World War II to 1955, 1956 to 1973, 1974 to 1980, 1981 to 1991, and 1992 to the present. Based on this analysis, what follows is a brief description of trends in national government policy for academic organization from 1945 to the present.

The End of World War II to 1955: Formation of the Postwar Structure. National government policy for the ARE in the postwar period paid attention mainly to recovery from the devastation left by the war. Strengthening the prewar structure provided a basis for the national universities. The concept of academic autonomy was increasingly strengthened as a result of government policy emphasizing the initiative of academic staff and their organizations. Consequently, the structure established during this time was closed to society at large. Asonuma (2003, 155) named it "the postwar structure in the research budget of the national university."

1956 to 1973: Development of the Postwar Structure. In this period, government policy continued to increase the emphases on recovery, autonomy, and initiative. Meanwhile, during the period of campus turmoil throughout the country, criticism and doubts arose about academic autonomy. Social expectations for academic research had been diminished by the violence and deficiencies evident in the campus conflicts.

1974 to 1980: Transformation I of the Postwar Structure. In this period, the postwar structure could not cope with rapid social change, and the university research organization began to decline. Allocations of research funds to individual academic staff began to decrease, though the academic research subsidies from the Ministry of Education increased. In this period, provision for "big science" began to be established outside the university. In this regard, the university was reduced to a marginal component in the total social research system.

1981 to 1991: Transformation II of the Postwar Structure. Over this decade, national government policy for the organization of academic research, which had focused entirely on the national university, began a new stage that differed significantly from the previous policy. In particular, a market mechanism was newly evident in the form of changes to the academic research budget and in private sources of funding.

A number of indicators showed this trend clearly and consistently. Allocation of funds to academic staff decreased, causing a change that devastated academic faculty in the national universities so much that the universities were called the "coffin of knowledge." Simultaneously, funding through the special education and research budget was increased, though the noncompetitive allocation to individual academic staff decreased rapidly. Academic research funding increased, but it was accessible only on the basis of competition among researchers.

It is also notable that a substantial expansion of business research in the 1980s resulted in the recruitment of many researchers from universities to industry. These transitions brought about many personal linkages between industry and the ARE (Polt et al. 2001).

1992 to the Present: Transformation III of the Postwar Structure. In the most recent years, operation of a market mechanism has become even more evident, replacing that of the postwar structure and encouraging competition among universities. In general, this period is known as the age of education and teaching reforms, or even of revolution, in Japanese higher education. In particular, the national government shifted its policy focus from reform to revolution after 1995, when the Science and Technology Basic Law (discussed in depth later in this chapter) was introduced to place priority on scientific and academic productivity.

In contrast with transformations I and II, research funding for academic staff has increased during this period. The large increase of funds for academic research has been remarkable and has resulted from a continuous increase in the resources of the special education and research budget. The commercial economic depression has restricted the possibility of any rise in private funding, and there has been little increase of funding for research institutes. A social demand for strengthening the ability to compete internationally has resulted in a policy of promoting the sciences and engineering and brought about the Science and Technology Basic Plan of 1996.

This period has also seen changes to the legal framework of careers for university researchers. For example, during this time it has become possible for universities to make term rather than indefinite appointments and to base pay on performance rather than seniority.

The history of higher education over the past century explains the existence of such a well-developed national research enterprise. The national government has concentrated many resources in the universities, particularly the research universities, and in the fields of science, engineering, agriculture, and especially health in the national sector.

RESEARCH POLICY

In the past ten years, the Ministry of Education, Culture, Sports, Science, and Technology (MEXT) has introduced a corresponding series of policies and plans: the Jutenka policy, a policy of selecting key graduate institutions in various fields of study; the Science and Technology Basic Plan, which started in 1996; the Toyama Plan, which was issued by Atsuko Toyama, the minister of MEXT, in 2001; the 21st Century COE program in the area of research and the Good Practices program in the area of teaching and education, both of which were initiated in 2002; the Global COE program succeeding the 21st Century COE program; and the establishment of the National University Corporations in 2004. These national policies have had a considerable impact on the research universities.

Jutenka Policy. A national policy of Jutenka, which was introduced in 1991, shifted the institutional emphasis from members of the faculty at the undergraduate level (*gakubu*) to faculty in the graduate school (*kenkyuka*) with the provision of an additional 25% of annual funding. The policy was adopted initially by the faculties of law in only two institutions, the University of Tokyo and the University of Kyoto. These two faculties shifted their emphases from the undergraduate to the graduate level in 1991. Subsequently, all faculties in these two institutions gradually underwent similar shifts over the next ten years. The other former imperial universities, the universities of Tohoku, Osaka, Kyushu, Hokkaido, and Nagoya, have now all followed suit in almost all faculties with permission from the MEXT. Some of the other major research universities, including Hitotsubashi University, the Tokyo Institute of Technology, the University of Tsukuba, and Hiroshima University, were also allowed to implement such shifts. In implementing a process of Jutenka, three groups of institutions were differentiated: Jutenka with 25% additional funds; *bukyokuka,* or a false Jutenka, without 25% additional funds; and no Jutenka.

In this way, the graduate school has been promoted to an independent internal organization instead of a subordinate organization to the undergraduate faculty.

Some sixty years after the postwar educational reform of 1945, the separation of the two tiers, which was nominally made then, has now been substantially realized. The shift of the graduate school to a research orientation is now established both nominally and substantially, although in the Japanese university a strong research orientation is retained at the undergraduate as well as at the graduate level.

During the same period, there have been other policy changes related to graduate education. As the CEC pointed out in 2005, the trend of political change for graduate education, as a core part of the ARE, has affected both the institution and the organization. A number of clear trends can be identified: the growth of new types of graduate schools, such as universities serving only graduate students (12 universities as of 2004); correspondence graduate schools (17 universities and 23 graduate courses); part-time and evening graduate schools (22 universities and 28 graduate courses); flexibility in conditions for enrollment and the length of graduate study; reinforcement of teaching functions; establishment of professional schools (93 universities and 122 courses as of 2005); graduate schools linked to business and industry (105 universities and 206 graduate courses as of 2004); and an increase in the number of graduate students (from 87,476 in 1988 to 244,024 in 2004) (CEC 2005a). The national graduate universities, including the Japan Advanced Institute of Science and Technology and the Nara Institute of Advanced Science and Technology, have as their mission the promotion of economic development through advanced interdisciplinary research with close ties to industry (Polt et al. 2001).

The Science and Technology Basic Plan. A policy for a creative country with intensive promotion of scientific and technological productivity (*Kagaku Gijutsu Souzou Rikkoku*) was introduced accompanying the Science and Technology Basic Law (*Kagakugijutsu Kihon-hou*) in 1995. The law was the basis for the establishment of the 1996 Science and Technology Basic Plan, aimed at shaping a nation based on the creativity of science and technology.

To encourage comprehensive and systematic policies for the promotion of science and technology, such as the promotion of scientific research activities at universities, the plan was formulated to create the science and technology five-year policy (from FY1996 to 2000) with the following ten years in view (MEXT 2006a). The first term lasted five years, from 1996 to 2000, and the second term extended another five years, from 2001 to 2005. A third term for a further period of five years began in 2006. Governmental expenditure was prepared to finance each term by ¥17 trillion, ¥22 trillion, and ¥24 trillion, respectively. This financing meant that money could be spent freely, both directly on the plan and indirectly on the research universities and research doctoral courses.

During these three terms, the amount of competitive funding has been extended dramatically. In FY2000, the final year of the First Basic Plan period, the amount available was ¥296.8 billion, or 2.4 times the amount in FY1995. In FY2005, the final year of the Second Basic Plan period, the amount was ¥467.1 billion, or 1.6 times the amount in FY2000. These funds have fostered a competitive environment at R&D sites (MEXT 2004a).

The third plan contains several recommendations. Some of the key recommendations relate to the employment of researchers: greater transparency in selection and appraisal processes; support for a diverse community of researchers, including women, non-Japanese, elderly, and early-career scholars; and improvements in mobility, especially for international researchers. Other recommendations in the plan relate to funding, suggesting that greater use should be made of competitive awards; and to interactions with industry, such as conducting more private-sector research within universities and expanding university-industry internship programs (Government of Japan 2006).

Toyama Plan. The Toyama Plan is a policy focusing on three targets. First, the plan pursues mergers and integration of national universities to decrease their number, with the aim of revitalization through a policy of "scrap and build." Second, private-sector management approaches are being rapidly introduced in the national universities, leading to new national university corporations. Third, a competitive principle is being introduced in universities and colleges to create a group of thirty top universities—in national, public, and private sectors—ranked among the top universities worldwide based on a third-party evaluation system.

Putting greater priority on support for the research universities is perhaps the first stage in a series of national plans related to higher education reconstruction. As a result, the national research universities have considerably increased their academic productivity. In the process, some basic principles, such as rationalization, efficiency, and accountability, have been emphasized to ensure that funding is used as effectively as possible. Some provisions, including mergers, integrations, and restructuring of institutions, were necessarily introduced into higher education institutions throughout the country. For example, some single-faculty medical schools were asked to merge with other institutions, such as adjacent comprehensive universities.

The 21st Century Centers of Excellence Program. The 21st Century COE program, introduced in 2002, is a policy of developing research bases that can attain worldwide competitiveness in academic productivity. According to the Japan Society for the Promotion of Science (JSPS), 113 programs were selected in 2002, 133 in 2003, and 28 in 2004. The total funds invested in them were ¥17.6 billion in 2002, ¥15.8 billion in 2003, and ¥31.0 billon in 2004 (JSPS 2006).

In the first year, for example, 113 among the 464 applicants were selected in five fields (life sciences, chemistry/material science, information sciences/electrical and electronic engineering, interdisciplinary/combined fields/new disciplines, and humanities) with support for their programs indicated to continue for five years from 2002 to 2007. The total amount of support for the first year, ¥17.6 billion, was distributed across fields and sectors (table 4.5). These bases were intended to be identified as important research centers both inside and outside the academic world. Among the 113 programs, the national sector claimed the largest share with 84 (74.3%) followed by the private sector with 25 (22.1%) and the public sector with 4 (3.5%) (JSPS 2006). The program sought to construct COEs in each disciplinary field to increase Japan's international competitiveness.

The Global Centers of Excellence Program. Having started in 2007 as a modification of the 21st Century COE program, which lasted five years, from 2002 to 2007, the new program covers the same five fields and puts more weight on graduate education to train researchers competitive in the international research environment. A similar philosophy is also observable in the third Science and Technology Basic Plan, as expressed in the MEXT annual report for 2006: "It is essential to develop capable researchers who enthusiastically pursue the creation of scientific knowledge and to promote their activities. If Japan produces world-class, capable researchers, they will be a good target for young human resources, and enthusiasm for new creation will be elevated. Therefore, the second basic plan set a goal of producing as many as 30 Nobel laureates in 50 years, aiming to increase the number of Japanese who win international prizes in science to the level of major European countries" (MEXT 2006a, chap. 1–2 [1]).

In the first year, 63 of the 281 project applicants were selected in five fields (life sciences, chemistry/material science, information sciences/electrical and electric engineering, interdisciplinary/combined fields/new disciplines, and humanities), with support for their programs scheduled to continue for the five years from 2007 to 2012. In other words, 28 of 111 institutions that had project applicants were selected. The total amount of support for the first year, ¥15.0 billion, was made available across a range of academic disciplines.

Table 4.6 shows the composition of applicant and adoption populations for the Global COE program by university sector. The national sector has strong representation. Among 281 applicants, 200 (71%) belong to the national sector (the former 7 Teidai and others), and among 63 adoptees, 50 (79%) belong to the national sector. In particular, the former imperial universities (Teidai) are well represented, as indicated by the presence of 75 applicants (27%) and 32 adoptees (51%). The strong representation of the former imperial universities is more pronounced in the adoptees than among the applicants.

TABLE 4.5.
*Recipients of first-year funding from the 21st Century
Centers of Excellence (COE) program, 2002*

Field	Total	National	Public	Private
Life science	28	21	1	6
Chemistry/material science	21	18	0	3
Information sciences/ electrical and electronic engineering	20	15	0	5
Interdisciplinary	24	17	2	5
Humanities	20	13	1	6
Total	113	87	4	25

Source: JSPS 2006.

TABLE 4.6.
*Application and adoption of Global Centers of Excellence
program, by university sector, 2007*

	Application		Adoption	
National				
Former 7 Teidai	75	27%	32	51%
Others	125	44%	18	29%
Public	22	8%	3	5%
Private	59	21%	10	16%
Total	281	—	63	—

Source: JSPS 2007b.

The National University Corporation. A massive reform of Japan's public higher education system, sometimes described as a "Big Bang," took place on April 1, 2004. The objectives of the reforms were primarily to reduce costs, increase transparency, and increase agility (Goodman 2005). As part of this reform effort, ninety-nine national universities were designated as eighty-seven national university corporations. Their management was drastically changed from a bottom-up to a top-down mode by empowering the presidents and adding external representatives as members of new trustee committees responsible for governance and management. This reform transformed the national universities into independent agencies, removed civil-servant status from university staff, and transformed the role of university presidents from appointed academic executives to something more like chief executive officers (CEOs). Under the new system, national universities can now hire and fire staff, set budgets, adjust pay, and pursue external funding sources (Hatakenaka 2005).

The radical change from the previous pre-evaluation by a chartering committee of the national government to a new post-evaluation system has started, leading to evaluation of the outcome of achievement related to each university's individual six-year plan. Additionally, the national government has started to allocate block funding to each university on the basis of its post-evaluation. Through this process, higher education institutions will become increasingly differentiated.

These reforms represent a significant increase in autonomy compared to past practice in Japan, but the level of autonomy is still low by OECD standards. Further, while the reforms are described as a "Big Bang," implementation has been gradual and incomplete (Hatakenaka 2005).

DOCTORAL EDUCATION POLICY

As far as the ARE is concerned, the share of the private sector is still high, with 282 (68.9%) of the total of 409 doctoral courses in 2005 (see table 4.2). However, universities in the national sector enroll and graduate a large majority (70%) of doctoral students.

The number of PhDs graduating from the national sector is much greater than from the private sector, especially in the field of natural sciences. In 2005 new entrants to doctoral courses numbered 17,553, of which 11,937 (68.0%) were enrolled in the national sector, 4,525 (25.8%) in the private sector, and 1,091 (6.2%) in the public sector. The courses these students enrolled in are categorized as humanities, 1,621; social science, 1,571; science, 1,621; engineering, 3,359; agriculture, 1,057; health, 5,696; home economics, 94; education, 410; fine arts and music, 183; and others, 1,941 (MEXT 2006b). From a comparative perspective, Japan has relatively low rates of participation in graduate education, except for master's students in engineering. In 2002 Japan's graduate student population was 8.9% of the undergraduate population, compared to 13.3% in Korea, 13.7% in the United States, 21% in the United Kingdom, and 21% in France (MEXT 2004b). In a report by the Ministry of Education, Culture, Sports, Science, and Technology (MEXT 2004b), the authors note low individual returns on education in Japan, with the returns on a PhD not usually justifying its opportunity cost. However, PhD student enrollment has increased from 43,774 in 1995 to 74,907 in 2005. In 2004 postdoctoral students numbered 12,583. As a result of its larger share of doctoral education, the national sector represents the center of learning in the Japanese ARE.

PhD programs in Japan are gradually moving away from an apprenticeship model, criticized for excessive specialization and a deficit in practical training in the management aspects of research, toward a model based on international best practices (CEC 2005b). Doctoral programs in Japan are heavily research-focused, with PhD students spending 70.9% of their time on research (MEXT 2006c). Hayashi and Tomizawa (2006) suggest that the growing number of doctoral students may be driving enhancements in Japan's research productivity, particularly with respect to increases in the number of published papers. (See tables 4.3 and 4.4 for additional data about doctoral students engaged in research.)

The present situation of national research grants and fellowships for research doctoral students is briefly described below. The Japan Student Service Organization (JASSO) offered scholarships to 1,091,607 students at a cost of ¥788.8 billion in 2006; in 1998 the comparable numbers were 499,121 students and ¥266.5 billion (JASSO 2006). Both numbers of students and costs continuously increased over these eight years. The number of fellowships for research doctoral students in graduate schools has also increased to some extent. The program of the Japan Society for the Promotion of Science (JSPS) intensively supports graduate students in their research and so contributes greatly to the development of academic research. The JSPS has described its fellowship programs in this way:

> The growing need to foster young researchers who will play an important role in future scientific research activities is recognized by JSPS through two special programs under which fellowships are granted to 1) young Japanese postdoctoral researchers to conduct research activities at Japanese universities or research institutions on a non-employment basis; and to 2) graduate students who conduct research in Japanese university doctoral programs. The fellowship recipients may apply for a research grant of up to ¥1.5 million per year. If it becomes necessary to advance their work, they may, for a stipulated period of time, conduct research at other research institutions including those overseas. (JSPS 2007a)

National fellowships are available for studies both in the natural sciences and in the humanities and social sciences. However, provision for doctoral students does not cover all necessary costs and entails many difficulties during their period of training as researchers. In 2000 doctoral students in the national, public, and private sectors on average received ¥2.73 million, consisting of ¥0.53 million (19%) from family, ¥0.89 million (33%) from scholarships, ¥0.68 million (25%) from part-time jobs, and ¥0.63 million (23%) from permanent jobs and other sources. Their average expenditure was ¥2.73 million, consisting of ¥0.74 million for tuition charges and fees, ¥1.51 million for living expenses, and ¥0.48 million for other expenses (Science and Technology Council Personnel Committee 2002).

JASSO (the Japan Student Service Organization) surveyed the support available to 45,222 students (excluding international students) in 2000 with the following results: of all students (45,222), 2,924 (6%) had support as JSPS special researchers; 4,450 (10%) from research assistantships; 8,140 (18%) from teaching assistantships; 20,943 (46%) from JASSO scholarships; and approximately 9,000 to 21,000 (20% to 46%) from other sources (Science and Technology Council Personnel Committee 2002).

KNOWLEDGE TRANSFER POLICY

Until twenty years ago, linkages between the university and industry in Japan were almost nonexistent: there was a kind of taboo forbidding such linkages among people inside academia. Further, personnel mobility was low. Among elderly scientists and engineers working in universities, only 45% had changed employers during their career. More common was temporary mobility based on industry researchers visiting universities (Polt et al. 2001). This climate was created during the time of campus turmoil that prevailed from the 1960s to the early 1970s and lasted more or less until the 1980s. During this period, many companies invested research funds in foreign research universities, especially those in the United States, rather than in their counterparts in Japan. Recently, however, all universities in Japan have begun attempts to strengthen the linkage between university and industry.

There are several programs intended to encourage direct interaction between the ARE and researchers in business and industry. Perhaps the earliest such programs were the University-Industry Research Cooperation Committees, begun in 1933. Other such programs include the Program for Joint Research with the Private Sector, the Innovative Industrial Technology R&D Promotion Program, Core Research for Evolutionary Science and Technology, the Precursory Research for Embryonic Science and Technology program, the Cooperative Research Centers Program, the Venture Business Laboratories Program, the Research for the Future Program, and the Network-Structured Center of Excellence (Polt et al. 2001).

One such program, the Exploratory Research for Advanced Technology Program (ERATO), has been identified as a good practice (Polt et al. 2001). ERATO promotes "the creation of advanced science and technology while stimulating future interdisciplinary scientific activities and searching for better systems by which to carry out basic research" (Japan Science and Technology Agency 2007). ERATO's finite research projects are unique in Japan for their personnel mobility. Projects focus on unexplored and precompetitive technologies and science questions and address emerging and challenging themes. The projects are now international and borderless, led by a director and made up of a diverse team of fifteen to twenty researchers.

The results of ERATO projects are impressive. By August 1996, they had produced 1,107 patents (925 in Japan and 182 overseas) and 5,672 papers and presentations (3,335 in Japan and 2,337 overseas) (Kusunoki 1998).

While such formal programs for collaboration are gradually gaining momentum, most industry-university collaboration in Japan has historically been informal (Hane 1999). Other types of interactions include informal collaboration, under which industry provides personnel and resources to university labs to establish recruiting channels and university researchers act as advisers to industry through industry and professional organizations (Hicks 1993).

The following laws represent some of the more recent developments in the Japanese innovation system. In 1998 the Law for Promoting University-Industry Technology Transfer was passed in Japan. In 1999 the Industrial Revitalization Law, partly modeled after the Bayh-Dole Act in the United States, was passed. In 2000 the Industrial Technological Ability Strengthening Law was passed, allowing national university professors and other staff to hold a job in a private company while continuing to be employed by the university (Rissanen and Viitanen 2001).

Various kinds of linkages are now under development to reform the structure for disseminating information and to create research exchanges aimed at strengthening coordination among industry, academia, and government. By 2004 systems to manage and make use of intellectual property had been established for 76% of the national university corporations (MEXT 2006c). In recent years, universities have begun to extend their activities beyond generating the seeds that are the source of knowledge to actively establishing university start-ups to develop new goods and services utilizing their own research results. The number of university-based venture start-ups grew from 128 in 2000 to 1,141 in 2005 (MEXT 2006a). The number of invention disclosures by the national university corporations increased rapidly, from 3,832 in 2002 to 6,787 in 2003. This increase was followed in 2004 by corresponding increases in patent applications, patent registrations, licensing agreements, and licensing income (MEXT 2006c).

Although the intermediary structure may be less extensive than elsewhere, several intermediary organizations facilitate interaction between industry and the ARE (Polt et al. 2001). Intermediaries include the Japan Society for the Promotion of Science, the Japan Science Council, the Council for Science and Technology, the Science and Technology Agency, and the Japan Science and Technology Corporation. Other intermediaries include science and technology parks, technology licensing associations, industrial and professional associations, and databases on science and technology.

Assessment

Japanese universities, which started as modern universities after the Meiji Restoration of 1868, have developed gradually, reaching the level of other advanced countries within a century. That they could enter the core group of advanced countries in terms of academic productivity is indicated, among other criteria, by the number of publications, Science Citation Index scores, Nobel laureates, and patents. Of course, it is undeniable that Japan's status is generally not considered comparable to Europe and the United States: "Japan was perhaps the nation where performance in science and technology came closest to Europe and the United States, but even their competitiveness in basic science remained modest even under its most successful period as an economic superpower in the 1970s and 1980s, although Nobel Prizes were high on the research policy wish list, and some were also received" (Sörlin and Vessuri 2007, 5). Further, the university sector plays a relatively smaller role in Japanese research (compared to other OECD nations) due to the significant research focus of Japan's industrial sector (OECD 2005). However, it is also undeniable that Japan's ARE has gained considerable quality and reputation starting from a far weaker position than those of Europe and North America.

Study of academic productivity, a concept derived from the scientific productivity concept used by Merton and Storer (1973), is inevitable because the main role of the ARE is to raise academic productivity (Arimoto 2006). In due course, as the logic of academic productivity suggests, study of the centers of learning has been developed since Ben-David first introduced this concept (Ben-David and Carnegie Commission on Higher Education 1977). In his study, Japan was identified as located on the periphery of the centers of learning in the 1960s, and subsequently it has progressed gradually from the periphery toward the center.

THE CURRENT SITUATION IN COMPARATIVE PERSPECTIVE

The 2006 Global Innovation Scorecard Report (Hollanders and Arundel 2006) placed Japan fourth in a group of seven countries designated as global innovation leaders, based on Global Summary Innovation Index (GSII) scores. In the categories of indicators that make up the GSII, Japan ranked as follows among the thirty-four countries considered: fifth in innovation drivers, sixth in knowledge creation, seventh in applications, fourteenth in diffusion, and eighteenth in intellectual property. Cluster analysis of absolute and relative performance on the indicators that make up the GSII showed Japan to be similar to other global innovation leaders: Germany, Switzerland, Finland, Sweden, and Israel. This cluster was characterized by strengths in knowledge creation and intellectual property as well as high labor productivity and per capita measures of GDP, number of researchers, scientific articles, and patents. However, the report did describe Japan as losing momentum in innovation from the late 1990s to the early years of the 2000s.

PUBLICATIONS AND PRODUCTIVITY

Publication trends are another way to assess the status of Japan's ARE. As is shown in table 4.7, which presents changes over time in the number of published papers among scientists worldwide, Japan was placed in a group with three other countries (the United Kingdom, Germany, and France) that lagged far behind the United States on these indicators. Japan's share of published papers, for example, amounted to 10.2% of 735,000 in 2002, while the United States contributed 32.0%.

Some analyses of science linkage—for example, the number of academic studies cited in patent applications—have been interpreted to indicate that Japanese academic research is not used as extensively in new technology patents as is research in the United States and the European Union (MEXT 2001). Tamada et al. (2004) analyzed filings with the Japan Patent Office and found patterns in the level of science linkage similar to the pattern found by Michel and Bettels (2001) for the United States and the European Union: biotechnology fields had the most linkage, followed by information technology. While the overall average number of citations per patent was lower in the Japanese data—0.5 nonpatent citations compared to Michel and Bettels's 0.85 for the European Union and 2.98 for the United States—the inclusion of citations in patent applications is influenced by differing legislative requirements. In Japan, patents resulting from joint university-industry research are usually filed for by the business enterprise, due to the burden of the application process and a lack of incentives for university researchers to pursue patents. Professors are typically named in the application and may be compensated with a donation or a nonfinancial award (Polt et al. 2001).

It is not unexpected that the recent series of new scientific policies should yield

TABLE 4.7.
Selected countries' share of published papers, 1992–2002, in percentages

	Total publications (in thousands)	Japan %	US %	Germany %	France %	UK %	Others %
1992	603	9.3	36.9	7.9	6.1	8.0	31.9
1997	678	10.0	34.2	9.0	6.7	9.0	31.2
2002	735	10.2	32.0	9.0	6.5	8.7	33.7

Source: Thomson Scientific, Institute for Scientific Information, National Science Indicators, 1981–2002.

significant effects on academic productivity and Japan's status in the world scientific community. One of the most effective policies in recent years has been the Science and Technology Basic Plan, which was introduced in 1996. Many believe it has been fundamental to the development of the ARE.

With respect to personnel mobility, policy changes are having only a modest and gradual impact; only 6.0% of appointments in the national universities were term appointments in 2005 (NISTEP 2005). Researchers report finding it difficult to secure new employment at the end of a term appointment, and the structure of the pension system also discourages mobility. Practices such as failure to publicly announce vacancies and same-school hiring are also still common. It is also rare for researchers to be recruited from outside Japan, as they can only receive fixed-term appointments, and language issues and high costs make family relocation difficult (NISTEP 2003, 2005).

A relatively small number of national universities have shown great initiative in academic productivity. This outcome has been accelerated rapidly by recent academic policies, which have been directed toward sustaining the past achievements of the research universities and have gradually increased differentiation between the research and nonresearch universities.

INTEGRATION OF TEACHING AND RESEARCH

The hierarchy of the higher education system was originally created politically and deliberately from the start of the modern higher education system in Japan (Amano 1986). It has been constantly maintained through the postwar era, notwithstanding the process of massification of higher education. There are status gaps between the national and private sectors, the research and nonresearch university sectors, and key and nonkey institutions, and among individual research universities. Reflecting this kind of policy, culture, and climate, many institutions have chosen to pay more attention to the research function of the university while giving less attention to its teaching function. There are also many gaps internally between an institution's research orientation and a lack of teaching orientation (Altbach 1996; Arimoto and Ehara 1996).

In the twenty-first century, it is necessary for us to seek an integration of research, teaching, and learning by reconsidering the role of scholarship. According to the Carnegie International Survey on the Academic Profession in 1992, Japanese academic staff showed a much stronger orientation to research than to teaching, and correspondingly a rather weak orientation to learning (Altbach 1996; Arimoto and Ehara 1996). In connection to this matter, there are two problems to be resolved as soon as possible. The first problem is related to students and prefaculty development (FD). Teaching outcomes in doctoral education have not been realized because there has been little development of the pre-FD that should be undertaken for graduate education. This failure is not necessarily restricted to the Japanese context. For example, Harvard University's Task Force on General Education noted that "[t]he more varied their teaching experience is during graduate school, the more resourceful and effective our PhD students will be when they develop courses of their own as professors" (Harvard University Faculty of Arts and Sciences 2006, 38). In this context, the accountability of doctoral programs should be questioned as seriously in teaching as in research. Research orientation is prevailing to the extent that graduate students are not prepared for teaching in universities and colleges after their graduation. This historical shortcoming should be addressed.

The second problem is related to faculty members. Based on Boyer's (1990) concept, an integration of research and teaching should be realized as soon as possible. However, for the next stage of FD, at the level of theory and practice in the twenty-first century, a component of learning should be included in the composition of scholarship, so that an integration of research, teaching, and learning can supplement the existing integration of research and teaching (Clark 1995; Nicholls 2005).

UNINTENDED CONSEQUENCES OF POLICIES

As described above, a series of intensive research policies was implemented to establish graduate schools, especially in the research universities. However, such policies have also brought about some unintended, negative results in addition to the intended benefits.

For example, the policy of strengthening graduate schools, Jutenka, was expected to improve the quality of the national universities, especially the former imperial universities. As a result, differentiation between the research universities and the nonresearch universities, and even among the research universities themselves, has increased. The distinctions that had emerged in the prewar period became somewhat ambiguous because of a series of national government policies in the postwar period that were nominally designed to realize equality among institutions. Nevertheless, the current policies, drastically emphasizing the distinction between institutions, evoke the differentiated academic society deliberately established by the national government in the prewar period.

As has already been discussed, the Science and Technology Basic Plan invested extensively in important fields by supporting designated research universities with a focus on specific advanced fields of study, particularly those related to the sciences, engineering, agriculture, and health. Support was not increased so much in the fields of the humanities and social sciences or to other research universities.

As a result, basic research has not developed as well as applied and developmental research.

The Toyama Plan brought about an institutionalization of the national university corporation and has decreased the research funds available for internal allocation to faculty members in each institution. In the former national universities, faculty members received an allocation automatically from the national government, without any control or filter at the governance level of the president and trustees. In contrast, in the new system, allocation of the block grant is undertaken at the level of governance so that faculty members receive competitively distributed allocations.

There are several categories of chairs identified in universities and colleges for the purposes of determining the amount of research funding allocated: a diagnosis chair (*rinsho-koza*), an experimental chair (*jikken koza*), and a nonexperimental chair (*hijikken koza*). Under the new provisions, reclassification of faculty members as researchers has forced some to move from experimental chairs to nonexperimental chairs, with substantial changes to their funding. There are great differences in the total amounts of support allocated to an experimental chair and a nonexperimental chair; the holder of an experimental chair might receive three times the support that a nonexperimental faculty member receives. Additionally, it is considered easier to get competitive funds in the field of natural sciences than in the fields of the humanities and social sciences. Accordingly, the differentiation between these two sectors has increased considerably.

The series of national policies that had been implicitly practiced for many years during the postwar period has become explicit and evident in recent years. These policies are now related to an international environment of competition among institutions in which every country seeks to achieve a peak of academic institutional stratification at a worldwide COE level. COE programs can be found in several Asian countries, including China and South Korea as well as Japan (Altbach and Umakoshi 2004).

In this context, the 21st Century COE program has brought about great differences in access to research funds between research and nonresearch universities. Among research universities, the successful ones have achieved greater visibility and prestige in scientific and academic society and also in society at large. For example, the University of Tokyo, which was selected as a COE institution in a number of areas, has become uniquely powerful and visible.

DIRECTIONS FOR REFORM

There are many problems relevant to reform of the ARE. For more than a century, Japanese research universities have devoted a great deal of effort to catching up with the research system in the advanced countries. As a result, Japan's research universities became comparable to the dominant centers of learning in the world in some disciplinary fields. However, the gap between the COEs of the world, the United States, and Japan is still thought to be great. One reason for this gap may be the insufficient preparation of policies, plans, and reforms in connection with the institutionalization and enhancement of graduate education (Clark 1995), though it has rapidly accelerated in the past ten years.

As already mentioned, the national government proposed new academic research policies successively in 1998 and 2004, increasing the priority accorded to the research university and its graduate education (Arimoto 2005b; CEC 2005a; UC 1998). These policies have generated a number of problems. In considering the international dimension, Japan needs an academic system capable of responding to worldwide competition in academic productivity. As has been described above, the reformed centers of academic learning were built on the base of the former imperial universities. More support was provided to these established institutions to raise their level of competitiveness. Formation of the array of COE institutions and other universities was determined by a series of policies. On the basis of the traditional research institutions that had existed since the prewar period, the COE institutions were developed by a combination of the 21st Century COE program, the Science and Technology Basic Plan, and the Good Practices program. These policies stress the advanced sciences in order to establish a top tier of key institutions, but it is also necessary to expand the range of basic research.

Recent policies have paid much more attention to the fields of natural sciences, engineering, agriculture, and medicine than to the humanities and social sciences. It is true to say that in Japanese graduate schools the humanities and social sciences have been neglected in comparison with the areas of sciences, engineering, and medicine in terms of the number of doctoral degrees, the quantity and quality of academic productivity, and the exchange and mobility of researchers among institutions (Clark 1993, 1995). Of course, both kinds of fields are important for the development of academic studies and learning, and more consideration should be paid to the humanities and social sciences if the differences between the two fields are not to be expanded.

Compared with that of other countries, the hierarchy of Japanese higher education institutions is seen as a steep, stratified structure rising to a pinnacle (Clark 1983). Transformation from this pinnacle structure to a mountain range with several peaks is expected to result in a better environment for effective competition among institutions.

The increase of differentiation between the "hard" sciences and the humanities and social sciences derives from the gap in resource allocation among and within institutions. It is understandable that the costly hard-science sector needs much greater funding than the humanities and social sciences, but the widening differences in funding between the two sectors causes a crisis of decay for researchers in the humanities and social sciences and also for the research graduate schools. The consequent decay, in the long run, will impose fundamental constraints not only on the development of learning, but also on the development of society itself.

Third-party evaluation processes have usually worked on the basis of peer review, and such a mechanism is expected to be employed in the newly introduced institutional accreditation system. JABEE (the Japan Accreditation Board for Engineering Education), established in 1999, is a nongovernmental organization that examines and accredits programs in engineering education in close cooperation with professional engineering associations and societies. Procedures similar to those used by JABEE should be considered in other areas to raise academic productivity and the quality of teaching to an international standard.

Conclusion

This chapter concludes with five considerations for further reform and understanding of the Japanese ARE. First, the research university enterprise in Japan was delayed in its development compared to its counterparts in the West, especially the United States, due to the delayed establishment of a graduate school system. As a result, it is now rapidly pursuing reforms of its graduate schools. It is clear that the reforms have made great progress in regard to the research universities, academic staff, nonacademic staff, students, and funding.

Second, the research university enterprise imported foreign models and pursued academic reforms to catch up with centers of learning elsewhere in the world. In the prewar period, the concept of the research graduate school was not well-developed, because the German system used as a model had no graduate school. Accordingly, research developed under a traditional chair system at the undergraduate level. A few of the then imperial universities, with "organizing chair systems," assumed the roles of research universities and made up an elite system in the development of higher education.

Third, in the postwar period, higher education itself shifted from an elite stage to a massified stage modeled on the American university system. Based on the former imperial universities of the prewar period, several research universities were deliberately developed into a graduate school system. The graduate school developed smoothly in the sciences, though in the humanities and social sciences it achieved little development and an undergraduate-centered structure persisted, just as in the prewar period.

Beyond the 1990s, higher education policy sought to strengthen graduate education and invested much funding in its development. As a result, differentiation between the research university and the nonresearch university extended considerably. At the same time, an institutional hierarchy in the stratification of research universities rapidly increased. A further stratification between the sciences and the humanities and social sciences also emerged at the institutional level.

Fourth, the ARE must respond to social changes as well as to a logic of academic study to realize its proper development. In other words, it has to accommodate social changes such as an increasingly knowledge-based society, globalization, and market mechanisms as well as structural change in relation to reconstruction of knowledge. In this context, it faces key problems of realizing social imperatives at the same time as accommodating reconstructed knowledge.

These problems are clearly related to the issues of higher education policy; governance, administration, and management of the university institution and organization; quality assurance of academic work in the research university enterprise; and reconstruction of the academic profession. Choosing an approach to academic reform is an issue whenever we attempt to coordinate external pressures from outside academia with internal pressures from inside academia.

Fifth, this chapter has sought to analyze the ARE in Japan. Based on these tentative observations, more study is needed to yield the greater detail that would fully develop the given theme from an international comparative perspective.

NOTE

I wish to thank Ms. Jennifer M. Miller for her kind and substantial contribution in editing the original paper for publication.

REFERENCES

Altbach, P. G. (1996) *The International Academic Profession: Portraits of Fourteen Countries.* Princeton, NJ: Carnegie Foundation for the Advancement of Teaching.

Altbach, P. G. and T. Umakoshi (eds.). (2004) *Asian Universities: Historical Perspectives and Contemporary Challenges.* Baltimore: Johns Hopkins University Press.

Amano, I. (1984) Method of University Classification [in Japanese]. In T. Keii (ed.), *Study of University Evaluation.* Tokyo: University of Tokyo Press.

———. (1986) *Koto Kyoiku No Nihonteki Kozo* [Japanese structure of higher education]. Tokyo: Tamagawa University Press.

———. (2006) *Sociology of University Reform.* Tokyo: Tamagawa University Press.

Arimoto, A. (1987) *Merton Kagaku-Shakaigaku No Kenkyu* [A study of Merton's sociology of science: Formation and development of its paradigm]. Tokyo: Fukumura.

——— (ed.). (1996) *Gakumon No Chushinchi Ni Kansuru Kenkyu: Sekai To Nihon Niokeru Gakumon No Seissansei To Sono Jouken* [A study of "Centers of Learning": Academic productivity and its conditions in the world and Japan]. Tokyo: Toshindo.

———. (2002) Globalization and Higher Education Reforms: The Japanese Case. In J. Enders and O. Fulton (eds.), *Higher Education in a Globalising World: International Trends and Mutual Observations: A Festschrift in Honour of Ulrich Teichler,* 127–140. Dordrecht: Kluwer Academic Publishers.

———. (2003) Recent Higher Education Reforms in Japan: Consideration of Social Conditions, Functions, and Structure. *Higher Education Forum* 1:71–87.

———. (2005a) Globalization, Academic Productivity, and Higher Education Reforms. In A. Arimoto, F. Huang, and K. Yokoyama (eds.), *Globalization and Higher Education,* 1–21. Hiroshima: Research Institute for Higher Education, Hiroshima University.

———. (2005b) National Research Policy and Higher Education. *Journal of Educational Planning and Administration* 19 (2): 175–198.

———. (2006) Structure and Function of Financing Asian Higher Education. In Global University Network for Innovation, *Higher Education in the World 2006: The Financing of Universities,* 176–187. New York: Palgrave Macmillan.

———. (2007) National Research Policy and Higher Education Reforms with Focus on Japanese Case. In S. Sörlin and H. M. C. Vessuri (eds.), *Knowledge Society vs. Knowledge Economy: Knowledge, Power, and Politics,* 175–197. New York: Palgrave Macmillan.

Arimoto, A. and T. Ehara (eds.). (1996) *Daigaku Kyojushoku No Kokusai Hikaku* [International comparison of the academic profession]. Tokyo: Tamagawa University Press.

Asonuma, A. (2003) *Sengo Kokuritsudaigaku Niokeru Kenkyuhihojo* [Research budget subsidy at the national university in the postwar period]. Tokyo: Taga.

Becher, T. (1981) Towards a Definition of Disciplinary Cultures. *Studies in Higher Education* 6:109–122.

———. (1989) *Academic Tribes and Territories: Intellectual Enquiry and the Cultures of Disciplines.* Bristol, PA: Society for Research into Higher Education and Open University Press.

Becher, T. and P. Trowler. (2001) *Academic Tribes and Territories: Intellectual Enquiry and the Culture of Disciplines,* 2nd ed. Buckingham, UK: Society for Research into Higher Education and Open University Press.

Ben-David, J. and Carnegie Commission on Higher Education. (1977) *Centers of Learning: Britain, France, Germany, United States: An Essay.* New York: McGraw-Hill.

Bleiklie, I. and M. Henkel. (2005) *Governing Knowledge: A Study of Continuity and Change in Higher Education—A Festschrift in Honour of Maurice Kogan.* Dordrecht: Springer.

Boyer, E. L. (1990) *Scholarship Reconsidered: Priorities of the Professoriate.* Princeton, NJ: Carnegie Foundation for the Advancement of Teaching.

Carnegie Council on Policy Studies in Higher Education. (1976) *A Classification of Institutions of Higher Education.* Rev. ed. San Francisco: Jossey-Bass.

Central Education Council (CEC). (2005a) *Future Vision of Higher Education in Japan* [in Japanese]. Tokyo: Central Education Council.

———. (2005b) *Graduate Education in the New Era: Towards Constructing Internationally Attractive Graduate Education Systems.* Tokyo: Central Education Council.

Clark, B. R. (1983) *The Higher Education System: Academic Organization in Cross-National Perspective.* Berkeley: University of California Press.

———. (1993) *The Research Foundations of Graduate Education: Germany, Britain, France, United States, Japan.* Berkeley: University of California Press.

———. (1995) *Places of Inquiry: Research and Advanced Education in Modern Universities.* Berkeley: University of California Press.

Gibbons, M., C. Limoges, H. Nowotny, S. Schwartzman, P. Scott, and M. Trow. (1994) *The New Production of Knowledge: The Dynamics of Science and Research in Contemporary Societies.* London: Sage.

Goodman, R. (2005) W(h)ither the Japanese University? An Introduction to the 2004 Higher Education Reforms in Japan. In J. S. Eades, R. Goodman, and Y. Hada (eds.), *The 'Big Bang' in Japanese Higher Education: The 2004 Reforms and the Dynamics of Change,* 1–31. Melbourne: Trans Pacific Press.

Government of Japan. (2006) The 3rd Science and Technology Basic Plan [Provisional translation]. Tokyo: Bureau of Science and Technology Policy, www8.cao.go.jp/cstp/english/basic/3rd-Basic-Plan-rev.pdf.

Hane, G. (1999) Comparing University-Industry Linkages in the United States and Japan. In M. Branscomb, F. Kodama, and R. L. Florida (eds.), *Industrializing Knowledge: University-Industry Linkages in Japan and the United States,* 20–64. Cambridge, MA: MIT Press.

Harvard University Faculty of Arts and Sciences. (2006) *Preliminary Report.* Cambridge, MA: Task Force on General Education, Harvard University.

Hatakenaka, S. (2005) The Incorporation of National Universities: The Role of Missing Hybrids. In J. S. Eades, R. Goodman, and Y. Hada (eds.), *The 'Big Bang' in Japanese Higher Education: The 2004 Reforms and the Dynamics of Change,* 52–75. Melbourne: Trans Pacific Press.

Hayashi, T. and H. Tomizawa. (2006) Restructuring the Japanese National Research System and Its Effect on Performance. *Scientometrics* 68:241–264.

Hicks, D. (1993) University-Industry Research Links in Japan. *Policy Sciences* 26:361–395.

Hollanders, H. and A. Arundel. (2006) 2006 Global Innovation Scoreboard (GIS) Report. In European Commission (ed.), *TrendChart: Innovation Policy in Europe.* Maastricht: Maastricht Economic and Social Research and Training Centre on Innovation and Technology (MERIT).

Japan Science and Technology Agency. (2007) *Outline of ERATO.* Tokyo: Japan Science and Technology Agency, www.jst.go.jp/erato/basics.html.

Japan Society for the Promotion of Science (JSPS). (2006) *Overview of Centers of Excellence (COE) Program.* Tokyo: JSPS.

———. (2007a) *JSPS Research Fellowships for Young Scientists.* Tokyo: JSPS, www.jsps.go.jp/english/e-pd/pddc.htm.

———. (2007b) *Results of Judgments of the Global COE Program in 2007.* Tokyo: Committee of Global COE Program, JSPS.

Japan Student Services Organization (JASSO). (2006) *Enhancement of Scholarship Project.* Kanagawa: JASSO.

Kusunoki, T. (1998) Dynamic Network of Basic Research: Organizational Innovation of ERATO [in Japanese]. In H. Itami, T. Kagono, M. Miyamoto, and S. Yonekura (eds.), *Case Book, Nihon Kigyo No Koudou,* vol. 3, *Innovation to Gijutsu Chikuseki,* 253–284. Tokyo: Yuhikaku.

Light, D., Jr. (1974) Introduction: The Structure of the Academic Professions. *Sociology of Education* 47:2–28.

Maassen, P. and N. Cloete. (2006) Global Reform Trends in Higher Education. In N. Cloete, P. Maassen, R. Fehnel, T. Moja, T. Gibbon, and H. Perold (eds.), *Transformation in Higher Education: Global Pressures and Local Realities,* 7–33. Dordrecht: Springer.

Merton, R. K. and N. W. Storer (eds.). (1973) *The Sociology of Science: Theoretical and Empirical Investigations.* Chicago: University of Chicago Press.

MEXT. *See* Ministry of Education, Culture, Sports, Science, and Technology.

Michel, J. and B. Bettels. (2001) Patent Citation Analysis: A Closer Look at the Basic Input Data from Patent Search Reports. *Scientometrics* 51:185–201.

Ministry of Education, Culture, Sports, Science, and Technology (MEXT). (2001) *Annual Report on the Promotion of Science and Technology, FY 2001.* Tokyo: MEXT.

———. (2004a) *Annual Report on the Promotion of Science and Technology, FY 2004.* Tokyo: MEXT.

———. (2004b) *Japan's Education at a Glance.* Tokyo: MEXT.

———. (2006a) *Annual Report on the Promotion of Science and Technology, FY 2006.* Tokyo: MEXT.

———. (2006b) *Gakko Kihon Chosa* [The basic statistics of school education]. Tokyo: MEXT.

———. (2006c) *OECD Thematic Review of Tertiary Education: Country Background Report of Japan.* Tokyo: MEXT.

———. (2008) *Gakko Kihon Chosa* [The basic statistics of school education]. Tokyo: MEXT.

Ministry of Internal Affairs and Communications, Statistics Bureau. (2004) *Report on the Survey of Research and Development 2004.* Tokyo: Ministry of Internal Affairs and Communications.

National Institute for Science and Technology Policy (NISTEP). (2003) *Research and Educational Environments and Effective Measures to Nurture World Class Scientists.* Research Material No. 102. Tokyo: NISTEP.

———. (2005) *Comparative Analysis on Abilities and Careers of HRST (Human Resources in Science & Technology) between Japan and the U.S.: Career Paths for Doctoral Recipients.* Report No. 92. Tokyo: NISTEP.

National Institute of Population and Social Security Research. (2004) *Population Projections for Japan: 2001–2050.* Tokyo: Department of Population Dynamics Research, National Institute of Population and Social Security Research, www.ipss.go.jp/index-e.html.

Nicholls, G. (2005) *The Challenge to Scholarship: Rethinking Learning, Teaching, and Research, Key Issues in Higher Education.* London: Routledge.

Oleson, A. and J. Voss (eds.). (1979) *The Organization of Knowledge in Modern America, 1860–1920.* Baltimore: Johns Hopkins University Press.

Organisation for Economic Co-operation and Development (OECD). (2005) *Main Science and Technology Indicators.* Paris: OECD.

Parry, S. (2007) *Disciplines and Doctorates, Higher Education Dynamics.* Dordrecht: Springer.

Polt, W., C. Rammer, H. Gassler, A. Schibany, N. Valentinelli, and D. Schartinger. (2001) *Benchmarking Industry-Science Relations: The Role of Framework Conditions.* Vienna/Mannheim: European Commission, Enterprise DG.

Rissanen, J. and J. Viitanen. (2001) *Report on Japanese Technology Licensing Offices and R&D Intellectual Property Right Issues.* Tokyo: Finnish Institute in Japan.

Science and Technology Council Personnel Committee. (2002) *Towards Training World-Class Top-Level Researchers.* Tokyo: Science and Technology Council.

Shinbori, M. (ed.). (1985) *Evaluation of Academic Achievement: Eponymy Phenomena in Science* [in Japanese]. Tokyo: Tamagawa University Press.

Sörlin, S. and H. Vessuri (eds.). (2007) *Knowledge Society vs. Knowledge Economy: Knowledge, Power, and Politics.* New York: Palgrave Macmillan.

Spiegel-Rösing, I. and D. de Solla Price (eds.). (1977) *Science, Technology, and Society: A Cross-Disciplinary Perspective.* London: Sage.

Tamada, S., Y. Naito, K. Gemba, F. Kodama, J. Suzuki, and A. Goto. (2004) *Science Linkages in Technologies Patented in Japan.* RIETI Discussion Paper Series, No. 04-E-034. Tokyo: Research Institute of Economy, Trade, and Industry (RIETI).

University Council (UC). (1998) *A Vision of Universities in the 21st Century and Reform Measures: To Be Distinctive Universities in a Competitive Environment* [in Japanese]. Tokyo: University Council.

Van Vught, F. A. (1989) *Governmental Strategies and Innovation in Higher Education.* London: Jessica Kingsley.

CHAPTER FIVE

The European Union

FRANS A. VAN VUGHT

For centuries, the European universities have contributed significantly to the social, economic, and cultural development of Europe. Since the creation of the European Union, the universities have gained a special position as objects of European-level policymaking. This chapter discusses and analyzes that special position. It presents both a historical overview and an analysis of EU policies with respect to the European academic research enterprise (ARE).

The European policy domains of research, higher education, and innovation are embedded in the broader European integration process. Analyzing these policy domains requires an understanding of the main phases and dimensions of this integration process. As a short introduction to our further analyses, let us therefore first look at the broader European political context.

In the aftermath of World War II and during the onset of the cold war, the desire to create peace and stability in Europe became a common target, and the idea of pooling European countries' interests seemed highly attractive. The 1950s in Europe were a time of reconstruction, reorientation, and reconciliation. In this context, European visionaries like Jean Monnet and Robert Schuman conceived of and took the first steps toward an integrated Europe.

However, these first steps were not taken easily. Winston Churchill's dream of launching a "United States of Europe," which he had outlined directly after World War II, was discussed at a European Congress chaired by him in May 1948 in The Hague. But a unified and integrated Europe appeared to be a goal too difficult to realize. The individual interests and sovereignties of the national European states allowed only for intergovernmental cooperation.

The objective to create an integrated Europe, able to act in a supranational way, apparently had to be addressed in a pragmatic, step-by-step fashion. The first steps resulted in the first three community treaties, creating the foundation of what is now called the European Union.

The treaties are effectively the basic constitutional texts of the European Union. They set out the objectives of the Union and establish the various institutions intended to achieve the objectives. These three founding treaties of the European Union are the Treaty of Paris, establishing the European Coal and Steel Community, which was signed in April 1951, took effect July 23, 1952, and expired July 23, 2002; the treaty establishing the European Atomic Energy Community (Euratom), which was signed in March 1957 and took effect January 1, 1958; and the treaty establishing the European Economic Community (EEC), which was signed—along with the Euratom Treaty—in Rome on March 25, 1957, and took effect January 1, 1958. This third treaty is usually referred to as the Treaty of Rome.

With the establishment of these communities, six continental governments agreed to work together and to create a common market.[1] In particular, the EEC Treaty has been an important source for European research policy. Before the 1980s, European activities in this policy domain were undertaken on the basis of Article 235 of this treaty, which allowed European policy initiatives when they were necessary to realize a community objective.

The first treaties were essentially economic in scope and were basically pragmatic. However, they created the first supranational policy context in Europe. In contrast with the pure intergovernmental approach, the new community method began to focus on a true European integration process in key policy domains, with the European Commission (EC), established in the EEC Treaty in 1957, as the major supranational institution.

History shows that further important milestones were the Single European Act (signed in 1986 and becoming effective July 1, 1987), which led to the single market strategy, and the Maastricht Treaty, which paved the way for the Economic and Monetary Union. The Maastricht Treaty (signed on February 7, 1992 and becoming effective November 1, 1993) changed the name of the European Economic Community to simply "the European Community." With the establishment of the two other so-called pillars (the Common Foreign and Security Policy and the Police and Judicial Cooperation Policy), the European Community has become known as the European Union, an entity with political and economic scope. It was also the Maastricht treaty (also called the Treaty on European Union) in which, after some debates on the interpretation of the competences of the community in the EEC Treaty's articles in the 1980s, the subsidiarity principle was formulated. This principle ensures that decisions are made as closely as possible to the citizens of Europe and that actions at the community level are evaluated to determine whether they are justified. It is assumed that the European Union will not take action except in the areas that fall within its exclusive competence, unless the member states cannot themselves sufficiently achieve the intended results.

A further milestone of European integration was the Treaty of Amsterdam, signed on October 2, 1997, and entering into force on May 1, 1999. This treaty includes a whole chapter on research and technological development, which are considered crucial for the competitiveness of European business and industry and the employment of Europe's citizens as well as for consumer protection and environmental policy.

In general political terms, the European integration process has moved slowly since the Amsterdam Treaty. When the European Council (the heads of state or government and the president of the EC) met in Nice in December 2000, it agreed on a review of the existing treaties. In an annex to the Treaty of Nice (signed on February 26, 2001 and taking effect February 1, 2003), the Council declared that it intended to hold an open debate on the future of the Union. When the Council met again in Laeken (Belgium) in December 2001, it established the European Convention, a body of 105 members chaired by former French president Valéry Giscard d'Estaing, which was given the task of producing a draft European constitution.

The draft constitution was signed by the heads of state or government at a ceremony in Rome on October 29, 2004. However, before it could enter into force, it had to be unanimously ratified by each member state, a process that was expected to take around two years. Following the rejection of the constitution in the referenda in France and the Netherlands in 2005, the European Council extended the deadline for ratification. Finally, after a period of uncertainty and confusion, in June 2007 the European political leaders reached an agreement that there would be no constitution; a review of the existing treaties would be established in the Treaty of Lisbon in 2008. It was also agreed that the decision-making processes of the Union and the composition of the Commission would be simplified; there would be more authority for the European Parliament and the national parliaments; and member states could withdraw from the Union if they wished to do so. However, after the negative outcome of the referendum in Ireland in June 2008, the ratification of the Lisbon Treaty has again become problematic, although the EC still expects the treaty to enter into force (EC 2008). Once again, Churchill's dream and the intentions of the early European visionaries appear to be far removed from political reality.

The most crucial recent phase in the European integration process that has had a major impact on developments in the policy domains of research, higher education, and innovation is the "Lisbon process." When the EU leaders met in Lisbon in 2000, they decided to boost the Union's competitiveness and growth. They wanted to create "a Europe of knowledge" and formulated the goal that by 2010 the European Union should be "the most competitive and dynamic knowledge-based economy in the world, capable of sustainable economic growth, with more and better jobs, and greater social cohesion" (European Council 2000, para. 5). The ambition formulated by the European political leaders created an additional context for European policymaking, not so much with classic policy instruments such as directives and regulation, but rather under an "Open Method of Coordination" (OMC), by which the governments of the member states themselves agreed to peer review and benchmarking on a number of relevant policy indicators (EC 2000a).

The OMC radically changed European policymaking. It provided a new platform to discuss national policies and their outcomes at the European level without further impinging on national competences. The objective was to benchmark the performances of the member states on common concerns and priorities and to discuss and compare their progress on reaching the goals of the Lisbon agenda.

Unfortunately, as the evaluation report of a special high-level group showed (European Communities 2004), at midterm (in 2005) the ambitious political goals of the Lisbon summit appeared to be very difficult to reach. Of course, the weak economic growth in the larger member states has been a major factor. But the fact that the design and implementation of the policy actions to reach the European goals rely strongly on the efforts of the member states and industry has also been identified as a major reason for the failure of the Lisbon process (Weber 2006).

The European Commission deliberately restarted the process. During the 2005 Spring European Council, the Commission launched the New Lisbon Partnership for Growth and Jobs (EC 2005e), which resulted in the singling out of "knowledge and innovation for growth" as one of the three main areas of action. In addition, it developed integrated guidelines for the preparation of the three-year so-called National Reform Programmes (NRPs) of the member states as well as the Community Lisbon Programme, which consisted of a set of Actions for Growth and Employment (EC 2005a), thus building a new, overarching partnership between the community and member states for the Lisbon agenda.

With this new partnership, the EC took a major step forward. The policy domains of research, higher education, and innovation had never been higher on the European policy agenda.

The European Academic Research Enterprise

The EU research system is a rather complex one, with research being carried out in industry, the universities, and other public and private research institutions. These three main types of institutional actors complement each other, and all contribute to the processes of knowledge production, dissemination, and application. The research sponsors in the European Union are predominantly industry and governments. Research sponsors appear to play increasingly important roles and to become more diversified as time passes; they include public authorities, industry, financial institutions, foundations, and so forth. This increasing diversity of stakeholders clearly adds to the complexity of the EU research system.

The total gross domestic expenditure on R&D (research and development) by the European Union in 2005 was \$226,827.5 million (purchasing power parity), larger than that of almost all other OECD countries. Only the United States has a higher expenditure level (more than one-third larger), but the EU investment in R&D is nearly two times that of Japan (OECD 2006). Of these investments, 53.4% comes from industry and 35.7% from government. The EU academic research enterprise (ARE) represents only a limited portion of the overall EU research system. Industry performs 63% of the total research activities. Of industry research, nearly a quarter is performed by small and medium-sized enterprises (SMEs). The university sector performs 22.4% and government 13.6%. The proportion of EU R&D performed by the European ARE is well above the OECD average of 17.3%.

The EU ARE is the main producer of scientific knowledge in Europe and the most important training ground for researchers. There are no clear and formal data about the size of the European ARE. The European University Association (EUA) assumes that the number of PhD-granting universities in the European Union is about one thousand, a large number compared to the number in the United States. In 2004 the EU universities employed 36.6% of all European researchers, which is also a large proportion compared to the percentages in the United States (14.7% in 2000) and Japan (25.5% in 2003) (EC 2007b, 49). The addition of twelve new EU member states since 2000 has increased both the demand for R&D outcomes in the European Union and the overall R&D capacity. The enlargement has also increased diversity in terms of scientific culture, specialization patterns, and investment levels.

As has been indicated, research in the European Union is performed by business and industry and by public and private research organizations, in addition to the university sector. Business and industry carry out nearly two-thirds of all EU R&D. The various public and private research organizations provide R&D, technology, and innovation services to business, governments, and other clients. In many cases they were created by government and may have started their activities with a publicly ordained mission paid for with public funds. However, many of these organizations have evolved into contract institutes and consultancy firms. A limited number are involved in academic research and sometimes even research training.

The European Union is the world's leading market in terms of purchasing power and demand for knowledge-intensive products and is likely to remain so in the near future. Demand for science- and technology-intensive products is a major driver for decisions regarding investments in and location of R&D activities worldwide. The problem that the European Union faces is that a single EU market for knowledge-intensive products does not yet exist. The European Union's market is confronted with barriers like different national legislation and different technical standards and specifications in local markets (European Communities 2006a).

From the early 1950s to the beginning of the 1970s, the European Union witnessed sharp labor productivity growth. The GDP per capita increased and appeared to be catching up with that of the United States. However, from the 1970s on, things changed. After the mid-1970s, the European Union continued to catch up with the United States in terms of labor productivity. Yet, because of the European Union's increasing employment and decrease in average working hours per capita, the gap in GDP per capita levels between the European Union and the United States did not narrow. Even more crucial, the European Union stopped catching up in terms of labor productivity in the mid-1990s, and the European Union's labor productivity growth rate fell below that of the United States. The policy conclusion drawn from these trends was that this development implies a serious threat to the international competitiveness of business activities in Europe. To address the threat, Europe will have to strengthen its ability to create and apply knowledge and increase its potential for innovation. Europe will have to become a knowledge society. Increased investments in R&D and education are crucial for Europe's future (European Communities 2005).

However, although the EU investments in R&D are substantial, they are also problematic. The R&D intensity of the European Union (Gross Domestic Expenditure [GERD] as a percentage of GDP) showed slow but continued growth between 1996 and 2002 and a slight decrease between 2002 (1.89%) and 2005 (1.85%) (fig. 5.1). Figure 5.1 also shows that the R&D investment gap between the European Union and the United States has been increasing since 2002. In addition, the data predict that if the current trend persists, China will have caught up with the European Union by 2009.

As is discussed later, the European Union has set itself a policy target of a 3%-of-GDP R&D investment level, of which two-thirds should come from the private sector. As figure 5.1 shows, this 3% objective is still far from being reached.

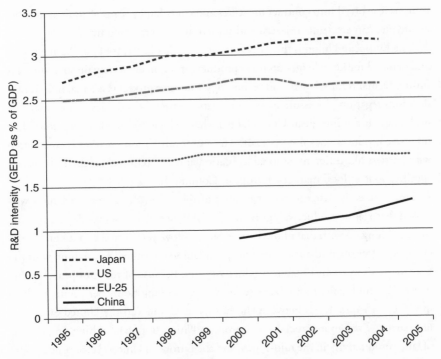

Fig. 5.1. R&D Intensity in the EU-25, US, Japan, and China, 1995–2005. *Source:* DG Research Data: Eurostat Database, OECD. *Notes:* For the US there is a break in the series between 1998 and previous years, and for Japan a break in the series between 1996 and previous years. Japan's GERD was adjusted by OECD for the years 1991 to 1995, inclusive. The China data do not include Hong Kong.

Government funding of R&D has been stable since the 1990s at around 0.64% of GDP, which is too low to reach the 3% target. There also is little progress toward the two-thirds objective for investment by the private sector. Until 2000, business funding of R&D grew at a high rate, but after 2001 the economic slowdown translated into a sharp reduction in the growth of business-funded R&D. See table 5.1, in which growth of R&D expenditure by business is compared with GDP growth.

The enlargement of the European Union has increased the diversity in R&D investment levels in the Union. The discrepancies among the EU member states in terms of R&D intensities ranged from 0.4% in Cyprus to 3.86% in Sweden in 2005. The discrepancies among subnational regions are even more dramatic, with R&D intensity ranging from 0.01% in Severozapadan in Bulgaria to 7.11% in Braunschweig, Germany, in 2002 (EC 2007b).

The number of researchers in full-time equivalents (FTE) per thousand labor

TABLE 5.1.
GERD financed by business and GDP in the EU-25: Real growth per annum,
1998–2005, in percentages

	1998	1999	2000	2001	2002	2003	2004	2005
GERD	5.0	8.5	5.8	2.9	−0.5	−0.3	2.9	1.8
GDP	3.0	3.1	3.9	2.0	1.2	1.4	2.5	1.8

Source: Eurostat Database, OECD.

force participants amounted to 5.4 in the European Union in 2003, compared to 10.1 in Japan and 9.0 in the United States. Nonetheless, this number grew at an annual average rate of 2.8% between 1997 and 2003. This deficit in the share of researchers in the labor force compared to Japan and the United States is mainly found in the business sector. Of the 1.18 million researchers (FTE) in the European Union in 2003, 49% were employed by the business sector, whereas the comparable percentages were 67.9% in Japan and 80.5% in the United States (European Communities 2005). In contrast, the EU ARE is relatively larger than those in the United States and Japan, with nearly 37% of the researchers employed in the EU university sector, compared to about 15% in the United States and 25.5% in Japan (EC 2007b, 49).

Especially in the earliest member states, the aging of the research labor force is becoming a concern. In 2003, 34.7% of highly qualified science and technology employees were in the 45-64-year-old age group, compared to 30.8% in the 25–34-year-old age group. In addition to the need to expand the research workforce, there also is a clear need to ensure a sufficient replacement rate for it. Fortunately, the European Union produces a substantial number of science and technology graduates: in 2003, 24.2% of all degrees awarded in the European Union were in the science and engineering fields of study, compared to 18.5% in the United States and 23.1% in Japan. Women are still underrepresented among both science and engineering researchers and graduates. Their share in the total number of researchers in 2002 was below 50% in nearly all member states (EC 2005d).

The intra-European transnational mobility of researchers and doctoral candidates is poor. Only around 5% of doctoral candidates and at most 10% of researchers at the postdoctoral level are involved in mobility processes. In addition, intersectoral mobility (between the private and public sectors) is still underdeveloped, largely because of cultural differences, but also because of practical issues like pension buildup (EC 2007b).

There is a considerable drain of EU graduates and researchers, particularly to the United States. The number of EU researchers working in the United States amounts to some 5% to 8% of the total EU researcher population. Most of these researchers are reluctant to return to Europe, primarily because of a lack of attractive research conditions and career prospects.

The overall situation in terms of quality of the ARE in the European Union can be characterized as "generally good on average, but with a very limited basis of universities at world-level" (EC 2007b, 50). In terms of both total number and world share of scientific publications, the European Union is the world leader. In 2004 the European Union's world share was 38%, compared to 33% for the United States and 9% for Japan. China ranked fourth with 6%.[2] However, the picture changes when publications are compared to population. Then the United States leads with 809 publications per million population, followed by the European Union with 639 and Japan with 569 (European Communities 2005).

Evidence clearly shows that the European Union's scientific impact lags behind that of the United States in almost all disciplines. Using data on the field-normalized Citation Impact Score per scientific discipline, figure 5.2 shows that the European Union's scientific impact is around or below world average in almost all disciplines. The European Union demonstrates a score above world average in only 6 out of the 37 fields. Also, the European Union shows lower scores than the United States in 35 of the 37.

An institutional citation impact analysis per discipline shows that only 26% of EU universities are world leaders in at least one discipline, compared to 81% of US universities. In addition, the number of disciplines in which an EU university is the world leader is on average substantially lower than that for US universities. A number of EU universities are considered among the top universities in the world, but their top is generally less broad than that of US universities (EC 2007b).

The European Union's performance in the exploitation of scientific knowledge remains problematic. The European Union's share of triadic patents (30%) is below that of the United States (36%) (fig. 5.3).

A broader view of innovation performance is provided by the European Innovation Scoreboard. The 2006 and 2007 results of this instrument show that the European Union is still not performing at the level of the United States and Japan but that the gaps in innovation performance, particularly between the European Union and the United States, are decreasing (Innovation Scoreboard 2006, 2007).

The EU ARE is seen as a key system for the European innovation agenda. Given the fact that the European Union's innovation performance is still considered to be internationally too limited, the European ARE is assumed to be an important object of EU policy. The prominent role of the EU universities for the further social and economic development of Europe has become an important reason why the

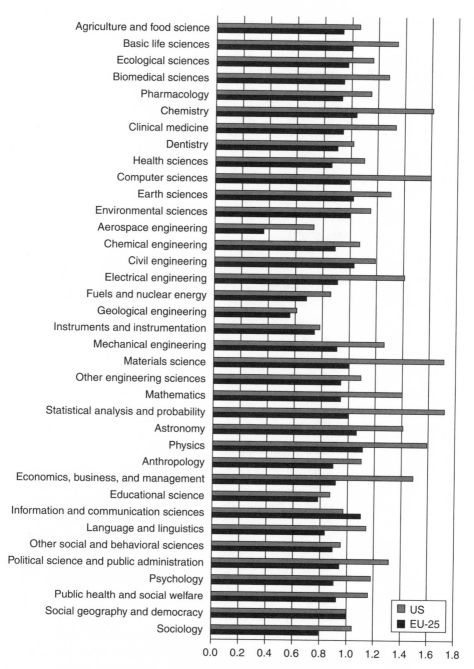

Fig. 5.2. Field-normalized Citation Impact Score (CPP/FCSm) per scientific discipline: The EU-25 versus the US, 2002–4. *Source:* DG Research Data: Thomson Scientific, processed by CWTS/Leiden University. *Note:* This figure is based on scientific articles published in 2002 and citations that occurred in 2002, 2003, and 2004.

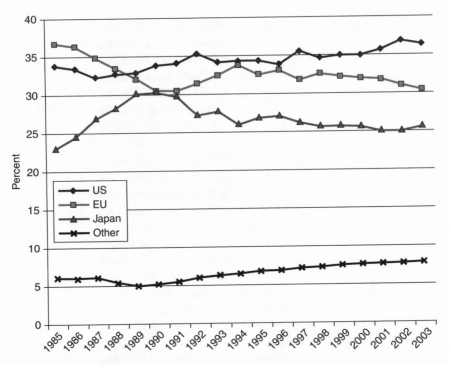

Fig. 5.3. World share of triadic patents. *Source:* DG Research, with data provided by the OECD Database.

European Union has designed and implemented a number of crucial policies in the domains of research and higher education.

The Policy Framework

HISTORY

The European research policy domain has fully developed since the 1980s. The Treaty of Rome (1957), the Single European Act (1987), and the Maastricht Treaty (1993) have created important foundations for European policy on research and technological development.

The first three European integration treaties provided specifically for coal and steel research, nuclear energy research, and agricultural research. In Article 235, the Treaty of Rome offered the community a formal base for action in the research policy domain. On this basis, an incremental diversification of research policy initiatives in other fields, such as environmental, materials, and biomedical research could develop (Caracostas and Muldur 2001).

In the early 1980s the European Commission initiated major collaborative technology projects in information technologies (ESPRIT) and communication technologies (RACE). The Single European Act introduced a special title with respect to research and technological development. The Maastricht Treaty provided an explicit basis for research activities in all fields in which the European Union is competent.

A crucial step was the creation of the first multiannual research and technological development framework program (FP). The first FP (1984–87) was designed to strengthen strategic areas of European competitiveness. It funded the research efforts of business and industry as well as of higher education and research institutions, and it stimulated the creation of research networks spanning organizational and national boundaries. After this first FP, a series of other FPs followed. FP2 (1987–90) was deliberately designed as the basis and instrument of European research and technology policy. FP3 (1990–94) regrouped the research activities around three strategic areas and emphasized quality as a major criterion for selection. FP4 (1994–98) emphasized the importance of consistency between national and community policies. FP5 (1998–2002) focused on a problem-solving approach and user needs and involvement (Geuna 1999). FP6 (2002–6) and FP7 (2007–13) have added major new policy elements to the general instrument of the FP. These elements will be discussed in more detail.

RESEARCH POLICY

The FPs have developed into the central EU instrument in the policy domain of research and technology. The FPs have become *the* strategic documents describing the broad strategic EU research priorities, each to be implemented through specific programs. In addition, they address the overall EU budget to be spent for the duration of the program, the breakdown of this budget into priority areas, and the ways and modalities according to which funding is made available to projects (Caracostas and Muldur 2001). The FPs are a medium-term planning instrument, indicating the priority areas and the financial scope of European activities in research and technological development. As such, they have become a dominant factor in the European research policy.

It should be pointed out, however, that other European programs have an impact in this domain as well. Particularly the regional and social policy activities funded by the EU Structural Funds (the Regional, Social, and Agricultural Guidance Funds) should be mentioned here. The Structural Funds have as their political objective to strengthen economic and social cohesion in the European Union, in particular in less-favored and declining regions. In the overall European integration policy, economic and social cohesion is seen to be as crucial as the development of the monetary union and the single market, expressing the political commitment by the European Union to bridge the gaps between the more advanced and the less-favored regions of Europe. Throughout the 1980s and 1990s, the Structural Funds were increasingly used for interventions by means of research and innovation-related activities, hence creating an extra funding base for research and technology policy. In a communication in 2004, the EC argued for a reform of this cohesion policy and for making the Lisbon agenda "one of the main bases for Structural Fund intervention" (EC 2004b, 10). Complementary funding from the Structural Funds and the FP is assumed to strengthen the process of reaching the strategic Lisbon objectives.

In addition, it should not be forgotten that the research and technology policy domain in Europe is a comprehensive, multiactor environment in which a multiplicity of intergovernmental associations and organizations exist. Examples are the EUREKA initiative, launched in 1985, which finances precompetitive projects according to a bottom-up industrial cooperation process; ESA (the European Space Agency); CERN (the European research center for particle physics); EMBO (the European molecular biology laboratory); and the European Science Foundation, which brings together a substantial number of networks in many European countries around a large number of research programs.

Furthermore, although the financial and political strengths of the FPs are considerable, the proportion of their financial research investments on a Europe-wide scale is limited. In the sixth framework program, this proportion was only 5%. The other 95% invested in European research comes from the member states. Of course, because these national resources often cover infrastructure, salaries, and running costs of European projects, the impact of the FP funding reaches much further than the 5% invested. Nevertheless, the overall European research landscape suffers from fragmentation and unnecessary duplication of efforts and resources (Andersson 2006). The major challenge in the European research and policy domain is to create critical mass and joint investment schemes. This is the challenge that is being addressed in the proposals for the European Research Area (ERA).

The European Research Area. The heads of state or government of the European Union decided at their Lisbon summit in 2000 that their common and national investments in research and technological development had to be increased. Aware that the European investments were considerably more limited than those of the United States and Japan, and keeping in mind that since their Amsterdam Treaty they were on the way to the "European Knowledge Society," they proposed to create the European Research Area. They proposed that the ERA should be able to better integrate national research policies, to encourage researchers to work together at the

European level, to stimulate cooperation between universities and industry, and to lower the political and administrative barriers to that cooperation.

The creation of the ERA has its own history. Since the 1980s, the FPs had developed substantially. By the end of the 1990s, not only had the overall funding levels more than quadrupled, but the priorities in terms of support for specific research fields had clearly evolved in the direction of a knowledge economy. In addition, the relationships with industrial and societal needs were substantially intensified.

Nevertheless, an overall assessment of the FPs in the mid-1990s (chaired by former European commissioner Viscount E. Davignon), while praising the quality of research undertaken under the programs and the strength of the networks developed as a result, also identified a major weakness of the FP instrument. The detailed decision-making procedures (in particular, the codecision mechanism by the European Parliament and the Council of Ministers as well as the member states unanimity rule) slowed down the further development of a common European research policy, often leading to unsatisfactory compromises. European research policy needed a streamlined and better-managed research strategy, set in the broader context of innovation and the increase of Europe's global competitiveness (Davignon 1997).

During the 1990s, several initiatives were undertaken to speed up the consolidation of the European innovation agenda, eventually leading to the decision by the heads of state or government during the 2000 Lisbon summit mentioned previously. The Amsterdam Treaty (1999) clearly spoke of the importance of research and technological development for the economy and social future of Europe. Regarding the European research policy, the fifth FP reflected the results of the Davignon evaluation. FP5 was to focus on a limited number of strategic priorities in four thematic programs that would redirect the European research efforts toward addressing the major economic and societal issues of Europe. In addition, FP5 was to strengthen the structural consolidation of a truly European system of research and innovation through three so-called horizontal programs aimed at confirming the international role of community research, promoting innovation of and encouraging the participation of SMEs, and improving human potential and the socioeconomic knowledge base. The horizontal programs were the new coordinating mechanisms in the program. They were designed to allow an integrated pursuit of better structural relationships between the European research policy and the overall innovation agenda.

The ERA was formally launched in 2000, when it was put on the agenda by means of a communication of the EC (EC 2000b). The subsequent conclusions of the Lisbon summit of that year endorsed the idea of the ERA, making it a key component of the Lisbon strategy.

However, the ERA did not take further shape until 2002. In the communication *The European Research Area,* the EC noted that a specific European research policy still did not exist in the full sense of the term, and the EC called for a more ambitious approach and greater cooperation. The ERA should be a major vehicle to implement the European Union's declared ambition to achieve a genuine common research policy. The objective was "to move into a new stage by introducing a coherent and centered approach at Union level from which joint strategies could be developed." The ERA was intended to reflect the political will expressed by the political leaders at the 2000 Lisbon summit. "Without this political will, Europe is condemned to increasing marginalisation in a global world economy. With the ERA, on the other hand, Europe gives itself the resources with which to fully exploit its exceptional potential and to become—in the words of the Lisbon European Summit of March 2000—'the world's most competitive and dynamic economy'" (EC 2002a, 9).

The EC noted that European research represented a jigsaw of (then) fifteen often very different national scientific and technological policies. The FPs appeared to be no more than a "sort of 16th research policy, coming on top of national effects, but not dynamic enough to have a truly integrating effect." The result was compartmentalization, dispersion, and duplication as well as the failure to assemble the critical mass of human, technological, and financial resources that major scientific advances demand today. Europe also still largely lacks a proper market for knowledge capital and technological development. Creating such a market calls for a genuine European research policy (EC 2002a, 8).

The Commission pressed for a concerted effort and suggested that FP6 be designed to do exactly that. FP6 would promote integrated, cross-border projects that would benefit from shared resources and critical mass. In this sense, FP6 would create "European added value," a principle thought to be more or less synonymous with that of subsidiarity, implying that action undertaken by the Union is intended to complement that undertaken by the member states. In the research policy domain, this principle of European added value could apply in several areas: where the "critical mass" of a research project, in terms of financial and human resources, exceeds the means of a single country; where cooperation is economically meaningful (economies of scale) and offers positive effects in terms of stimulating private research; where complementary national skills can be combined, in particular in interdisciplinary situations; where joint research is of interest given the cross-border nature of the problem; and where the research links in with the Union's priorities and implementation of its policies. FP6, in this sense, was to have three main areas of action: integrating European research, structuring the ERA, and strengthening the foundations of the ERA.

FP6 had a substantial budget: over €19 billion, compared to nearly €15 billion for FP5. At a more operational level, FP6 contained various action lines, either existing ones that were strengthened in terms of support for researcher training and mobility, or new lines for development aid and for creating the scientific and technological infrastructures of the ERA. Substantial parts of the program were also devoted to identifying future European science and technology priorities and to the coordination and reciprocal opening up of national research programs. In addition, three major new instruments were introduced: support for cross-national networking of thematic centers of excellence in universities, research organizations, and business enterprises; support for integrated projects involving a critical mass of scientific and industrial partners and directed toward significant products, processes, or service applications; and participation in specific science and technology cooperation programs set up jointly by certain governments or national research organizations.

In its communication *More Research for Europe* (2002b), the EC stated that the only way to reach the ambitious targets was to increase the general investment in research to 3% of GDP and that a substantial part of this effort should come from business and industry. The Commission challenged the member states, showing that this target would imply an increase of the national expenditure levels for research of 6% to 10% on average per year. Nevertheless, during a political summit in Barcelona in March 2002, the 3%-of-GDP target (of which two-thirds was expected to come from private funding) was accepted and was to be reached by 2010. The European Union took its research area seriously.

Unfortunately, the 3%-of-GDP target for 2010 appeared to be very hard to reach. The knowledge gap between Europe and the United States remained large. In particular, the European R&D expenditure by business and industry lagged behind (Van Vught 2004).

Therefore, it is not surprising that, at midterm between 2000 and 2010, the European Union was far from its target. The special high-level group chaired by former Dutch prime minister Wim Kok that evaluated the progress toward the Lisbon and Barcelona objectives noted that the targets were still far away. It concluded that "halfway to 2010 the overall picture is very mixed and much needs to be done in order to prevent Lisbon from becoming a synonym for missed objectives and failed promises" (European Communities 2004, 10). Similarly, another high-level expert group, chaired by former Finnish prime minister Esko Aho, concluded in 2006 that there was a large gap between the political rhetoric about the knowledge society and the realities of budgetary and other priorities, and that action was urgently needed "before it is too late." The group suggested forming a pact for research and innovation made up of political, industrial, and social leaders to really build the Europe of innovation (European Communities 2006a).

A number of other pacts had meanwhile been established in the so-called technology platforms. These platforms have been set up largely at the initiative of industry; they bring together companies, research institutions, the financial world, and the regulatory authorities at the European level. The objective is to develop a common strategic research agenda per platform that could mobilize a critical mass of public and private funding at the national and European levels (EC 2004b). The technology platforms are bottom-up processes uniting stakeholders around a joint vision and approach for a strategic technology. They define an overall strategy in a specific technological field to be translated into an operational program. In 2008 there were thirty-four European technology platforms spinning a wide range of technologies. The platforms' objective is to influence industrial and research policy at the EU, national, and regional level, while encouraging public and private investments in key technological fields. The EC sponsored some of the platforms through specific support actions in FP6. It encouraged the platforms to apply for funding in FP7, but it indicated that they should also look elsewhere for financial support. In FP7, the Commission introduced the Joint Technology Initiatives (JTIs), which are eligible for FP funding.

When the sixth FP came to its end, the ERA was still under construction. Of course, FP6 could not realize the full ERA by itself. The European Union needed a policy framework that would create incentives for its member states to contribute beyond what the Union is able to offer. "Without active involvement of Member States, the Commission cannot succeed in enacting an effective global strategy for science and technology" (European Research Advisory Board 2006, 7).

In developing the seventh FP, the EC took this suggestion to heart. It noted that in FP6 successful efforts were made to improve the coordination of national research programs but that these efforts must be strengthened. The networking of national programs (through the so-called ERA-NET activities) should receive more resources, and more attention should be given to the mutual opening up of national research programs (EC 2004b).

During the summer of 2006, the leaders of the European Union reached an agreement on FP7, with a budget of €53.2 billion. FP7, which started in January 2007, is a major program for realizing the "re-launched" Lisbon agenda. It is the chief instrument for funding European research and innovation and creates a new policy context for the ERA.

Subtitled "Building the European Research Area of Knowledge Growth," FP7 is designed to respond to the competitiveness and employment needs of the European

Union. It is based on the assumption that knowledge is Europe's greatest resource and that growth and competitiveness can be given a new impetus by this resource. The program places greater emphasis than before on research that is relevant to the needs of European industry through the technology platforms and the new Joint Technology Initiatives (JTIs), which will develop research projects in fields identified through dialogue with industry. The program also, for the first time, provides support for the best European "frontier research," with the creation of a European Research Council (ERC). The ERC funds the best European research activities, as assessed by international academic peer review.

The Commission acknowledges that the technology platforms, led by industry, have been able to create more focus in the European research activities. In FP7, it introduces the JTIs as a new funding scheme, offering a framework for particularly ambitious research and technology agendas that require public and private investment at the European level. JTIs are dedicated legal structures that coordinate the mobilization of large-scale public and private investments and substantial research resources. JTIs are assumed to accelerate the generation of new knowledge, to enhance the uptake of research results into strategic technologies, and to foster the necessary specialization in high-tech sectors that may determine the European Union's future industrial competitiveness. Initially, six JTIs have been identified.

FP7 is organized into four specific programs. In the Cooperation Program, the objective is to gain European leadership in key areas through cooperation of industry and research institutions (such as the JTIs). Support is given to research activities carried out in transnational cooperation, from collaborative projects and networks to the coordination of national research programs. In the Ideas Program, the objective is to strengthen the science base of Europe by funding European-wide competition. The autonomous ERC (with a budget of €7 billion) supports frontier research by individuals or partnerships in all scientific and technological fields. The People Program has as its objective to reinforce the career prospects and mobility of European researchers. Support is available for training, mobility, and the development of European research careers. In the Capacities Program, the objective is to develop research and innovation capacities throughout Europe, so that European researchers will have excellent facilities at their service. Support is available for research infrastructures, regional clusters, research for and by SMEs, "science in society" issues, and horizontal international cooperation.

FP7 is a continuation of FP6 and continues to develop the ERA. But FP7 is intended to be less bureaucratic and simpler in its operation. In addition, FP7 has international cooperation as an integrated dimension in each of its four programs; it has a focus on the development of the "regions of knowledge" (strengthening the research potential of regions); and it comprises a Risk-Sharing Finance Facility aimed at fostering private investment in research by improving access to the European Investment Bank.

In the spring of 2007, the EC took an important next step in developing the ERA. It published a green paper, entitled *The European Research Area: New Perspectives* (2007c) and opened a discussion on how the current research system of Europe can be improved. The Commission intends to further develop the idea of the ERA. It argues that, since the ERA's introduction in 2000, the context of European research has evolved. Globalization has accelerated, various new socioeconomic challenges have grown (climate change, aging, the risks of infectious diseases), and the European research landscape has changed (notably with the launching of new measures such as the ERC and the European Institute of Innovation and Technology). Within this changing context, the ERA concept itself has been subject to gradual changes. Its scope has broadened from a focus on how to improve the effectiveness and efficiency of the fragmented European research landscape, to the awareness that more public and private investment in research is needed, and to the view that research policy should be related to other EU policies in order to achieve coherence and synergies in the context of the overall Lisbon strategy. According to the Commission, the expanded ERA must comprise six features: (1) an adequate flow of competent researchers with high levels of mobility among institutions, disciplines, sectors, and countries; (2) world-class research infrastructures, accessible to all; (3) excellent research institutions engaged in public-private cooperation, involved in clusters and virtual communities, and attracting human and financial resources; (4) effective knowledge-sharing between the public and private sectors and with the public at large; (5) well-coordinated research programs and priorities; and (6) the opening of the ERA to the world, with special emphasis on neighboring countries.

Researcher mobility is clearly a priority throughout the green paper. It suggests that the movement of knowledge is crucial for the future of the European Union. The movement of knowledge should become a "fifth freedom" within the European Union, complementing the four freedoms of the Treaty on European Union, which protects the free movement of goods, services, capital, and labor (EC 2007c).

The 2007 green paper once more reflects the idea that knowledge is Europe's best resource. According to the EC, investing in knowledge is the most important way to foster economic growth and to create more and better jobs while at the same time ensuring social progress and environmental sustainability. The EU research policy plays an important role in delivering these goals. Research has come to be at the core of the European Union's renewed ambition to stimulate growth and employment. It is the key component of a broader European innovation policy.

HIGHER EDUCATION POLICY

Generally speaking, higher education has come only slowly onto the supranational European agenda. Although some educational activities were developed at the European level during the 1970s (in particular in the field of vocational training and the education of migrant workers' children), the education sector was for a long time "taboo" for European policy initiatives (Neave 1984, 6). The European Community had not been given competence in the field of education by the national governments.

Not until the period 1985–93 did a first group of policy initiatives come about. Well-known European programs in the late 1980s and early 1990s are the Community Programme for Education and Training in Technology (Comett), the European Community Action Scheme for the Mobility of Students (Erasmus), and the Trans-Mobility Programme for University Students (the interuniversity cooperation program with nations of central and eastern Europe—Tempus) (European Communities 2006b). The Maastricht Treaty is an important further milestone in this context. This treaty is a landmark in the history of the European policy domain of education and training, creating for the first time a clear and legal basis for European initiatives in this field.

In the Treaty of Nice (2001), it was decided that the European Union would be able to contribute to the development of quality education by encouraging cooperation between member states through a wide range of actions, such as promoting the mobility of citizens, designing joint study programs, establishing networks, exchanging information, and teaching languages for all citizens of the European Union. The basic idea was that, although the competence for education in general and higher education in particular remains at the level of the member states, the Union has a complementary role to play by adding a European dimension to education, by helping to develop quality education, and by encouraging lifelong learning.

The main tool for putting this ambition into practice became the Socrates program. The first phase of this program ran for the period 1995–99, and the second phase during the years 2000–2006. The Socrates II program supports European cooperation in eight areas, from school to higher education and from new technologies to adult learners. The higher education section of the program continues the older Action Scheme for the Mobility of the Students, called the Erasmus program, established in 1987. As the higher education action of Socrates II, the Erasmus program intends to enhance the quality of and reinforce the European dimension of higher education by encouraging transnational cooperation between universities, by boosting mobility, and by improving the transparency and recognition of studies and qualifications.

The European Higher Education Area. However, the roots of the current European higher education policy lie, in a special way, in the history of the European universities. In May 1998, the French minister of education used the eight-hundredth anniversary of the Sorbonne as an occasion to design a joint declaration of the ministers of France, Germany, Italy, and the United Kingdom on the "harmonization of the architecture of the European higher education system." The objective was to create a European higher education area by means of a common two-cycle degree structure, the mutual recognition of degrees, and increased student mobility. The ministers stressed that "the Europe we are building is not only that of the Euro, the banks, and the economy, it must be a Europe of knowledge as well" (Sorbonne Declaration 1998, 1).

The Sorbonne Declaration proved to be a "quantum leap" in the development of European higher education policy (Witte 2006, 124). The four ministers agreed to organize a follow-up meeting in Italy and called on other European countries to join their initiative. The result was astonishing. The eagerness of the other European education ministers to participate in this initiative became visible in the twenty-nine signatures of the European ministers under the Bologna Declaration, designed in a city with an even older European university. The Bologna Declaration formulates the wish to construct the European higher education area (EHEA), to promote mobility and employability, and to increase the compatibility and comparability of the European higher education systems. It also emphasizes the need to increase the "international competitiveness" of Europe's higher education and its "worldwide degree of attraction" (Bologna Declaration 1999).

After the Bologna conference, the process accelerated. Follow-up conferences were held in Prague (2001), Berlin (2003), Bergen (2005), and London (2007). The "Bologna ministers" (an expanding group made up of forty-six nations in 2007) added new actions on lifelong learning, on a common framework of qualifications, on a coherent quality assurance and accreditation mechanism, and on an additional focus on the doctorate level (the third cycle) in the Bologna process. In addition, they increasingly focused on the alignment of the EHEA with the ERA.

The Bologna process rapidly became a central dimension in the emerging higher education policy of the European Union. In particular, since the formulation of the Lisbon agenda in 2000, higher education rapidly moved up the ladder of the European Union's policy concerns (Corbett 2005; Shaw 1999).

In 2003 the EC opened a debate on the "place and role of European universities in society and the knowledge economy" (EC 2003b, 4). The EC had not reflected on the European universities since the early 1990s, when it had published a memorandum on higher education in Europe (EC 1991). However, since the European universities at the beginning of the twenty-first century are at the heart of the Europe of Knowledge, being responsible for 80% of Europe's fundamental research, the EC intended to explore the conditions under which Europe's universities would be better able to effectively play their role in the knowledge society and economy (EC 2003b).

The analysis by the Commission is stern: "The European university world is not trouble-free, and the European universities are not at present globally competitive." They should realize that the traditional model of Wilhelm van Humboldt no longer fits the current international context and that the high degree of fragmentation of the European university landscape prevents Europe from responding to new global challenges. These challenges go beyond national frontiers and have to be addressed at a European level. "More specifically, they require a joint and coordinated endeavour by the Member States . . . , backed up and supported by the European Union" (EC 2003b, 10).

In its contribution to the Bergen conference of the Bologna process (2005), the Commission stated clearly that "from the EU perspective, the Bologna process fits into the broader Lisbon Strategy" (EC 2005b, 2). But it again emphasized that Europe and its universities face stronger competition than ever before. The figures tell us "that the situation is alarming."

Only 21% of the EU working-age population in 2002 had achieved tertiary education, significantly lower than the percentage in the United States (38%), Canada (43%), Japan (36%), and South Korea (26%). In the European Union, 52% of the relevant age group was enrolled in higher education; this was slightly more than in Japan (49%) but less than Canada (59%) and certainly the United States (81%) and South Korea (82%). The number of researchers per 1,000 employees in the European Union was 5.4, marginally less than Canada or South Korea, but far below the United States (9.0) or Japan (9.7) (EC 2005d, 3). Only a handful of European universities were found in the top fifty of the world. Indeed, the situation is alarming, and profound reforms are needed.

According to the EC, the European universities have so far failed to unleash their full potential to stimulate economic growth, social cohesion, and improvement in the quality and quantity of jobs. In a policy paper in 2005, the EC identifies the following bottlenecks: a tendency to uniformity and egalitarianism in many national higher education systems, too much emphasis on monodisciplinarity and traditional learning and learners, and too little world-class excellence (EC 2005d). European higher education remains fragmented into medium or small clusters with different regulations and languages; it is largely insulated from industry; graduates lack entrepreneurship; and there is a strong dependency on the state. European higher education also is overregulated and therefore inefficient and inflexible. In addition, European universities are underfunded; EU countries spend only 1.9% of GDP on research while the United States, Japan, and South Korea are close to 3% in such spending; EU countries spend only 1.1% of GDP on higher education, far less than Canada (2.5%), the United States (2.7%), and South Korea (2.7%); underfunding leads to low enrollment rates, failure to prepare students for the labor market, and difficulties in attracting and retaining top talent.

In the view of the Commission, the quality and attractiveness of the European universities need to increase, human resources need to be strengthened, and the diversity of the European higher education system needs to be combined with increased compatibility (EC 2004a). The Commission launched in 2004 the Integrated Lifelong Learning Programme (2007–13), with the general objective of contributing to the European knowledge society. The Lifelong Learning Programme consists of four subprograms, one of which is the Erasmus program. A crucial aim of this program is to reinforce the contribution of higher education institutions to the process of innovation. For this aim, the autonomy of and investments in universities should be increased. The Commission urges the member states to establish a new partnership with their universities, moving from state control to accountability, and to acknowledge that addressing the severe funding deficit in higher education is a core condition for achieving the Lisbon ambitions (EC 2005d).

The political leaders of Europe lent a ready ear to the Commission's analysis and suggestions. During the Spring European Council of 2006, they decided that all member states should try to reach the overall EU target for 2010 for R&D spending of 3% of GDP and that the investments in higher education should rise to at least 2% of GDP by 2010. The Commission produced another communication in which it identified four priority actions for more growth and jobs, "requiring a strong impetus from the highest political level and which should be implemented no later than the end of 2007." One of the four actions is "investing more in knowledge and innovation," including the urgent promotion of excellence in both research and education, "particularly world-class universities with adequate funding streams and closer links with business" (EC 2006e, 14–17).

It appears that by 2006 higher education had evolved as a crucial policy concern at the supranational, European level.

INNOVATION POLICY

Innovation is a concept that has rapidly risen on the political agenda of Europe. Already FP1, which was intended to strengthen strategic areas of European competitiveness, has stimulated the creation of research networks in which both universities and industry participate.

However, it was not until 1995 that the EC published a green paper on innovation, launching a vast debate on innovation in all member states (EC 1995). Based on this, the Commission drew up an action plan for innovation in which three objectives were singled out: to develop a true culture of innovation; to adapt the administrative, legal, financial, and fiscal environment; and to strengthen the links between research and innovation.

The first objective was intended to improve the generic societal perception of innovation and the mobility of researchers, teachers, and students. The second was to design the various administrative service systems that could stimulate research and innovation (intellectual property issues, venture capital, etc.). The third was to create better links between research and innovation and was reflected in FP5 as well as in other Community instruments (in particular the Structural Funds).

The fifth and sixth framework programs made explicit reference to the importance of innovation. FP5 was intended to consolidate a truly European system of innovation through the strengthening of the horizontal programs for innovation, international distinction, and human capital formation. FP6 was adopted to help create a proper European market for knowledge capital and technological development in order to allow the Union to further develop its innovation agenda.

The seventh framework program is clearly designed to address the competitiveness and innovation needs of the European Union. FP7 puts greater emphasis on the linkages between knowledge institutions and industry than ever before. It acknowledges that knowledge is Europe's main asset and that the production and application of knowledge are crucial processes in achieving the goals of increased economic growth and more and better jobs.

As indicated previously, in 2005 the EC launched the New Lisbon Partnership for Growth and Jobs (EC 2005e). The Spring European Council of that year singled out "knowledge and innovation for growth" as one of the three main areas of action for the Union. Innovation and research policies were assumed to be the central elements of the new strategy, as together they cover the full spectrum of issues regarding the production and exploitation of knowledge.

The new partnership specifies the actions for the member states (based on the partnership's integrated guidelines) regarding the preparation of the three-year National Reform Programmes as well as the actions for the Community (the Community Lisbon Programme). The overall objectives are to achieve sustainable global competitiveness and to create a vibrant European knowledge economy. A coherent and integrated set of policies is assumed to be crucial to reach these ambitious goals.

The European policy domain of research and technology is a priority area in this broader policy set. It has become part of the integrated research and innovation agenda that focuses on both knowledge creation and knowledge utilization. The Commission's action plan *More Research and Innovation* (European Innovation 2005) makes this very clear. It addresses the full research and innovation spectrum, including nontechnological innovation. It makes explicit the commitments made by the Community Lisbon Programme, by detailing the measures in support of research and innovation that will be undertaken. It outlines ambitious actions, reaching beyond the 3% target for 2010. It strengthens the links between research and innovation, with research policy focusing more on developing new knowledge and innovation policy focusing on transforming knowledge into economic value and commercial success (European Innovation 2005).

Research and innovation clearly have moved to the heart of EU policymaking. In this central focus of policy, the Commission promises better and more effective regulation (the "better regulation" initiative) with respect to research and innovation, but it also urges the member states to support research and innovation wherever possible. The Commission also encourages cross-border research cooperation, public-private partnerships, research dissemination strategies, and joint European research projects. It intends to further develop the European protection of intellectual property and to continue exploring a "European patent." It wants to create an open and competitive European labor market for researchers and to stimulate research career paths at the transnational level. It points at the benefits of public procurement for innovation (public authorities as launching customers for innovative products and services) and at the positive effects of R&D tax incentives. In its "industrial policy" the EC provides special innovation policy tools, like the Innovation Scoreboard for benchmarking national and regional innovations performance; the Network of Innovating Regions for the exchange of best practices; Europe IN-NOVA, a communication platform for professionals in various clusters and sectors: the Innovation Relay Centers for stimulating knowledge transfer in a wide range of markets; and the IPR Helpdesk, an online information tool on intellectual property rights for SMEs.

In short, the EC intends to create a truly European innovation agenda. It asks the member states to identify research and innovation as key challenges in their National Reform Programmes (NRPs) and to report annually on their targets and policy progress in these fields. In the context of its research policy, the Commission focuses on the new instruments in the seventh framework program, in particular the

JTIs, the Risk-Sharing Finance Facility, the extra funding for the research and innovation capacity of SMEs, and the Regions of Knowledge initiative. In addition, the Commission promotes the use of the Structural Funds and the Rural Fund to improve the European knowledge and innovation base and the use of financial risk facilities for SMEs. The focus on SMEs is particularly found in the European Union's Competitiveness and Innovation Programme (CIP), 2007–13, which is designed to support actions to help business and industry to innovate, particularly SMEs.

A special initiative by the EC concerns the European Institute of Innovation and Technology (EIT). The proposal to establish an EIT was put forward as part of the midterm review of the Lisbon strategy in 2005 and has since been further developed. The suggestion is that the EIT could be an "education, research, and innovation operator" and that it be structured to integrate these three areas. The EIT should be a "knowledge flagship." It should set out "to attract and keep the best talents in students, researchers and staff in Europe, to work side by side with leading edge business in the development and exploitation of knowledge and research, and to enhance research and innovation skills generally" (EC 2006b, 2).

Realizing that the European Union's funding to promote innovation and research activities represents only a small fraction of the overall public European effort (only about 5%), the Commission also presses for the mobilization of national funding in support of European research and innovation activities. It proposes to extend the Community's instruments to further stimulate transnational cooperation and coordination by providing direct Community support for joint research and innovation programs between member states. The new European innovation agenda indeed is designed as a new partnership between the European Union and the member states.

The coordination of national and regional research programs in the EU context already had been a focus of attention since early 2002, when the European research ministers acknowledged the importance of the progressive opening-up of national research programs as an important step toward the construction of the ERA. For the first time, in FP6, a specific action line was designed to address this issue: the ERA-NET scheme. ERA-NET is the network of national research councils, working together on a voluntary basis for more cooperation. ERA-NET is essentially a bottom-up process in which national research programs are coordinated and mutually opened up. The objective of the ERA-NET is to increase the cooperation and coordination of national and regional research programs in the member states through the networking of these programs. In FP7, the ERA-NET is reinforced by a new module (called ERA-NET Plus) that provides a financial incentive by supplementing joint transnational funding with Community funding.

As a result of the EU innovation policy, the positions of the European universities in research and innovation are clearly reinforced. The European research policy domain has become a cornerstone of a full European innovation policy, and the European universities and other public research institutions are being challenged to contribute to implementing that policy.

DOCTORAL EDUCATION POLICY

A crucial dimension of the overall policy framework of the European academic research enterprise is the increasing attention to and importance of doctoral education. The topic of doctoral education was addressed at the European level during the Bologna summit in Berlin (2003) and later in Bergen (2005) and London (2007). In Berlin, the education ministers decided that it was "necessary to go beyond the present focus on two main cycles of higher education and to include the doctoral level as the third cycle in the Bologna process." They emphasized the importance of research and research training in enhancing the competitiveness of European higher education and called for increased mobility at the doctoral level and stronger interinstitutional cooperation (Berlin Communiqué 2003, 7). During the Bergen meeting, the ministers urged the European universities "to ensure that their doctoral programs promote interdisciplinary training and the development of transferable skills, thus meeting the needs of the wider employment market." Also, the number of doctoral candidates should be increased to contribute to the needs of the knowledge society (Bergen Communiqué 2005, 4). At their London meeting, the ministers invited the universities to reinforce their efforts to embed doctoral programs in their institutional strategies and to develop career paths for doctoral candidates and early-stage researchers (London Communiqué 2007).

In 2003 the EC also paid attention to doctoral education. In its communication *Researchers in the European Research Area: One Profession, Multiple Careers*, the Commission discussed the recruitment, training, and career opportunities of researchers (EC 2003a). In particular, it argued that the competencies and skills of doctoral candidates should focus on a wider labor market perspective than only academic careers.

Doctoral education is beginning to take a higher position on the European research and education agendas. The days of the Humboldtian doctorate as the entrance to an academic career appear to have passed. Doctoral education is assumed to be able to play a major role in creating the highly trained labor force for the knowledge society, which is understood to need knowledge professionals who have the competencies to work in highly complex, knowledge-intensive environments.

Europe indeed seems to have discovered the full potential of the third cycle in higher education (Bartelse and Huisman 2008).

Doctoral education is considered to be the major link between the Bologna and the Lisbon agendas (Aghion et al. 2008), and more specifically between the European higher education and research areas. Not only has it become an official part of the European political agenda in the Bologna process, but it also is a crucial point of attention in the EU innovation strategy. The EC presses for an open, single, and competitive labor market for researchers with attractive career prospects and incentives for mobility. In the near future, it may be assumed that doctoral graduates will find their careers not only in academia, government, and private-sector R&D laboratories but also in general management positions.

As a consequence, doctoral education appears to be entering a phase of further innovation and diversification. The European University Association (EUA) has set up a special Council on Doctoral Education with the objective of contributing to the development, advancement, and improvement of doctoral education and research training. European universities also have recognized the need to offer doctoral candidates a broader experience than core research disciplinary skills based on individual training by doing research. They increasingly introduce courses and modules offering transferable skills-training and preparing candidates for career opportunities in sectors beyond academic institutions (EUA 2005a). European universities appear to be accepting the challenge to diversify their doctoral training programs. Relatively new concepts like professional doctorates, industrial doctorates, taught doctorates, and practice-based doctorates have made their entry into the European discussions. In the years to come the traditional Humboldtian doctorate may very well be supplemented with a variety of new European doctorates (Scott 2006).

KNOWLEDGE-TRANSFER POLICY

In the 2006 Spring European Council, the European member states indicated that they expected to increase their R&D spending. The Commission urged them to implement their NRPs (National Reform Programmes) and to set ambitious expenditure targets for R&D and higher education, but it also indicated that Europe needs to continue to improve its knowledge infrastructure and to reinforce its capacity for knowledge transfer. More investments in knowledge and innovation are needed, and "the quality of the European innovation systems requires particular attention" (EC 2006e, 16). Excellence in both research and higher education needs to be further promoted. But stronger links with business and industry are also needed, and the knowledge-transfer processes need to be strengthened.

The 2006 Council called on the EC to present a "broad based innovation strategy for Europe that translates investments in knowledge into products and services" (European Council 2006, 6). In September 2006 the Commission published its communication *Putting Knowledge into Practice: A Broad-Based Innovation Strategy for the EU* (EC 2006d).

The strategy is a comprehensive European innovation policy, with a clear emphasis on knowledge transfer. It is intended to frame policy discussions on innovation at national and European levels. It outlines the most important planned and ongoing initiatives, identifies new areas of action, and in particular introduces a more focused strategy to facilitate the creation and marketing of new innovative products and services in promising areas—"the lead markets" (EC 2006d, 3).

In April 2007 the Commission published its communication on knowledge transfer and a set of voluntary guidelines for research organizations to help them improve their links with industry. The document uses a broad definition of the knowledge-transfer concept: "Knowledge transfer involves the process of capturing, collecting and sharing explicit and tacit knowledge, including skills and competence. It includes both commercial and non-commercial activities such as research collaborations, licensing, spin-off creation, researcher mobility, publication, etc." (EC 2007d, 2). The general objective of the communication is to serve as a starting point for increased cooperation among the member states and the European Union in this field, leading to a common European approach to knowledge transfer.

According to the Commission, major barriers to greater knowledge transfer exist in the European Union. They include cultural differences between the academic and the business communities, legal barriers, fragmented markets, and lack of incentives. Some member states have set up initiatives to promote knowledge transfer, but these largely ignore its international dimensions (EC 2007d).

The document highlights the importance of a number of measures, such as creating a workforce of skilled knowledge-transfer staff in universities (and a professional qualification and accreditation scheme), developing a more entrepreneurial mind-set in universities, and allowing for exchanges of staff between research organizations and industry. It also emphasizes the importance of financial support for knowledge transfer. In addition to FP7, European funds mentioned include the Regional Development Fund and the Social Fund. The suggested voluntary guidelines to help improve knowledge transfer cover issues like intellectual property management, incentives for researchers to participate in knowledge-transfer activities, and the development of knowledge-transfer resources (EC 2007f).

During the 2007 Spring European Council of the heads of state or government, the current level of European competitiveness was discussed. The member states invited the Commission to push forward the implementation of the general innovation policy. The meeting concluded that the member states are determined to improve the conditions for innovation, such as competitive markets, and to mobilize additional resources for research, development, and innovation. They also underlined the importance of increased attention to knowledge transfer. In this context, the European ARE is seen as a crucial actor for the revitalization of the knowledge and innovation capacity of Europe. It is urged to use its qualities not only to create, but also to "activate" knowledge (Soete 2005). It is invited to play a pivotal role in the further development of the European knowledge society.

Assessment
INTEGRATED SUBSYSTEMS

In the preceding paragraphs, it has become clear that during the past few decades a supranational European innovation policy has been developed that now includes a number of interrelated policy fields. Generally speaking, these policy fields can be divided into two large policy domains: research policy (including knowledge-transfer policy) and higher education policy (including doctoral education policy). Although these two policy domains have their own origins and histories, they increasingly appear to have come together under the broader umbrella of the all-embracing Lisbon agenda.

In particular, since the relaunch of this agenda in the New Lisbon Partnership in 2005, the EC has tried to develop a general strategy that could form a solid base for the further development of the European Union. The Union faces fierce economic competition on a global scale and sees it as a major task to develop a comprehensive innovation agenda. The higher education and research policy domains have become crucial elements of this broader agenda.

The Commission's communication to the Spring European Council in 2005 makes these points clearly: "Just think what Europe could be. Think of the innate strengths of our enlarged Union. Think of its untapped potential to create prosperity and offer opportunity and justice for all its citizens. Europe can be a beacon of economic, social and environmental progress to the rest of the world" (EC 2005e, 3). Referring to the midterm evaluation by the Kok high-level group, which stressed the widening growth gap with North America and Asia, the Commission rose to the challenge to present its proposals for the Union's strategic objectives:

> Europe's performance has diverged from that of our competitors in other parts of the world. Their productivity has grown faster and they have invested more

in research and development. We have yet to put in place the structures needed to anticipate and manage better the changes in our economy and society. We need a dynamic economy to fuel our wider social and environmental ambitions. This is why the renewed Lisbon Strategy focuses on growth and jobs. In order to do this we must ensure that: Europe is a more attractive place to invest and work; knowledge and innovation are at the beating heart of European growth; we shape the policies allowing our business to create more and better jobs. (EC 2005e, 4)

The higher education and research policy domains are interpreted as subsystems of a larger overall European innovation policy. To allow Europe to focus on its two principal tasks—delivering stronger, lasting growth and creating more and better jobs—the Union needs a knowledge and innovation policy, consisting of a number of elements:

[P]ublic authorities at all levels in the Member States must work to support innovation, making a reality of our vision of a knowledge society; more investments by both the public and private sector spending on research and development are needed; a major reform of State Aid policy should be realized to allow Member States to support research and innovation; our universities should be able to compete with the best in the World through the completion of the EHEA; a "European Institute of Innovation and Technology" should be created; Innovation Poles designed to help regional actors bring together the best scientific and business minds should be supported; the promotion of eco-innovation should be intensified; partnering with industry should be fostered by European Technology Initiatives. (EC 2005e, 9)

The renewed Lisbon agenda is an ambitious innovation agenda to which it is assumed the higher education and research policies will contribute. The Commission's 2005 action plan "More Research and Innovation" argues that the relationship between research and innovation is the key factor for economic growth (European Innovation 2005). Similarly, the Commission's communication to the 2006 Spring European Council stresses the importance of promoting excellence in both higher education and research as well as further strengthening the links between academia and industry (EC 2006a).

During that conference, the European heads of state or government asked the Commission to design a broad-based European innovation strategy. The Commission's answer was its 2006 communication *Putting Knowledge into Practice,* in which it argues that "the EU can only become comprehensively innovative if all actors become involved and in particular if there is a market demand for innovative products." For this, "education is a core policy" and a "two-way communication between researchers and business" is required. (EC 2006d, 4, 8–9). In its innovation strategy, the Commission emphasizes the importance of the integration of the various policy domains. The innovation strategy "presents a framework to take innovation forward bringing together different policy areas that have a bearing on innovation" (EC 2006d, 3).

In the broader context of the Union's innovation strategy, the European research and higher education policies have not only been subsumed under a more comprehensive policy; they also have become more integrated than ever before. FP7, the Modernization Agenda for the European Universities (EC 2006a), the communications on doctoral education and knowledge transfer, and the green paper on the ERA (EC 2007c) breathe the same spirit. As major documents in their policy domains, they convey the same message: the European universities have crucial roles to play in further developing the European knowledge society.

A MULTI-ECHELON SYSTEM

The integrated European innovation policy system is a rather complex phenomenon. It encompasses a broad array of issues, addresses a variety of actors, and makes use of a diversified set of instruments.

It appears that the European innovation policy system can be described as a multi-echelon system. Such a system consists of a family of interacting subsystems, of which many have decision-making competences that allow them to influence one another. The concept of echelons refers to the mutual relationships between the decision-making levels comprising the overall system. The echelons have their own authorities and competences but are mutually interdependent. The very nature of multi-echelon systems is this interdependency of the various levels of a system. In a multi-echelon system the higher-level units condition, but do not completely control, the goal-seeking activities of the lower-level units. The lower-level units have the freedom to (at least partly) select their own courses of action. Their decisions might be, but are not necessarily, the ones that the higher level would select (Mesarović, Macko, and Takahara 1970).

In the European innovation policy system, at least four crucial echelons can be distinguished: the European Union (with its various actors), the member states (with their political actors), the universities, and the groups of individual academics working in these universities. All of these four echelons have their own competences and decision-making authorities. But in the European innovation policy system, these echelons also are interrelated and interdependent.

In the increasingly integrated policy domains of research and higher education, the European Union addresses both the member states and the public research organizations (including the individual academics in these organizations). In addition, in the broader context of its innovation policy, the Union reaches out to business and industry and to other societal actors (regional and local authorities, civil society organizations, trade unions, and consumers). The European Union calls for "a wide partnership for innovation where supply of new ideas and demand for new solutions both push and pull innovations" (EC 2006d, 3–4).

As we have seen, the European Union has strong competences in the research policy domain. Since the mid-1980s, the European Union has used its FPs as a crucial, strategic, financial policy instrument to influence the overall research agenda of Europe. Arguing that the Union's policies in this field create "European added value," the European Union has emphasized building focus and critical mass and fostering cooperation and excellence.

The FPs address research organizations, research groups, and individual researchers directly. They are strong and attractive funding schemes that have a considerable influence on the dynamics of European research. Many European universities and researchers have acquired EU research funding from the successive programs.

Regarding the member states, the European research policy initiatives promote the coordination and reciprocal opening up of national research programs. Through the ERA-NET scheme (implemented in FP6 and expanded in FP7), the European Union seeks to increase the cooperation and coordination of national research programs, using its financial capabilities to provide incentives for the organization of joint initiatives by member states. In addition, also in this domain, the Open Method of Coordination (OMC) is being used to create a Union-wide policy forum for comparing and discussing national efforts and results.

Another initiative by the European Union that addresses the other echelons of the innovation policy system is the introduction of an integrated European human resources strategy. To enhance the public recognition of the role of researchers in the European knowledge society, the European Union has published a recommendation on the European Charter for Researchers and on the Code of Conduct for the Recruitment of Researchers (EC 2005c). The intention for the Charter and the Code is to give individual researchers the same rights and obligations wherever they work in the European Union. In the 2007 green paper on the ERA (EC 2007c), the Commission outlines a broad strategy for the free movement of knowledge, of which researcher mobility is a crucial element.

In the innovation policy system, several policy echelons coexist. The European Union tries to influence and coordinate the decisions and activities of the other echelons by using the competences it has at its disposal. It does so by means of a diversified set of policy instruments.

EUROPEAN POLICY INSTRUMENTS

The European Union has several categories of policy instruments at its disposal. In accordance with the literature on policy analysis, I distinguish three basic categories of European policy instruments (Hood 1983; Mitnick 1980; Van Vught 1994).

The first category consists of the legal policy instruments of the European Union. Here it is important to keep in mind that the European Union is formally neither a federal government nor an intergovernmental organization. The European Union is based on the agreements among its member states.

Generally speaking, there are three types of European legal instruments. The first is the so-called primary legislation type. Primary European legislation basically consists of the various European treaties and their annexes and protocols. Treaties are legally binding and have to be ratified by the parliaments of the member states. The treaties form the "constitutional structure" of the European Union and provide the formal contexts in which research and higher education policies can be developed.

In addition to the treaties (primary legislation), secondary European legislation involves the various legal instruments that the European Union uses to develop and implement its policies. These instruments are

- Regulations: legislative acts of the European Union that have general scope, are obligatory in all of their elements, and are directly applicable in all member states.
- Directives: legislative acts of the European Union that require the member states to achieve a particular result without dictating the means of achieving that result. Directives are binding only on the member states to which they are addressed. In practice, they are addressed to all member states.
- Decisions: binding legislative acts of the European Union that are not of general application and apply only to their particular addressees.
- Recommendations: non-binding EU acts aiming at the preparation of legislation by the member states.

The three binding policy instruments of secondary legislation (regulations, directives, and decisions) are strong and powerful forms of EU law. They are applied in the research policy domain but only minimally in the context of higher education. In the research policy domain in particular, the FPs are implemented through decisions and their more detailed elaboration (in the so-called Rules of Participation) through regulations.

The third type of European legal instrument includes the decisions by the European Court of Justice and the Court of First Instance. The European Court of Justice is the European Union's Supreme Court, which adjudicates on matters of interpretation of EU law, most commonly claims by member states that the EC has exceeded its authority or by the Commission that a member state has not implemented a binding legislative act. The Court of First Instance (which is an independent Court attached to the European Court of Justice) hears and determines nearly all direct actions brought by individuals and member states, including actions against acts as well as inactions by the Community institutions.

In the research and higher education policy domains, several examples of this type of legal instrument can be found. Well-known is the landmark decision by the European Court of Justice in 1985 (the Gravier case) that access to a university is covered under European primary legislation and that any discrimination based on nationality would be against European law. As a result, EU students can be charged only the same (if any) study fees as national students. Another well-known case is the one in 2004 of a French student claiming financial support from the British government on the basis of the nondiscriminatory principle (the Bidar case). The Court judged in favor of the claim. In the research policy domain, many cases are found in which either one or more member states or the Commission asks for the Court's judgment. It appears that in the majority of cases the Court rules in favor of the Commission's position.

Nearly all policy issues in the European higher education policy domain are implemented by means of intergovernmental conventions and resolutions, without any legal authority at the level of the Union. In the context of the Bologna process, for example, the action lines (like the three-cycle structure, the Bologna process qualifications framework, and the coordination of the quality assurance processes) all are the result of interministerial agreements. The European Union does not have the authority to make use of its legal instruments in this policy domain. Given the subsidiarity principle, it necessarily limits itself largely to its instruments of information and communication.

The second category of European policy instruments are the financial instruments. Most prominent in this category is the subsidy instrument, which basically refers to the power of signing checks. Subsidies are payments made to individuals or organizations under the condition that the recipient supplies a particular product or service. In this sense, subsidies are contracts, under which payments are made when the recipient accepts the conditions set by the provider of the funds.

In the EU research policy domain, the FPs employ this policy instrument. Similarly, the CIP (Competitiveness and Innovation Programme) uses this instrument to stimulate actions and outcomes that support the objectives of the renewed Lisbon strategy. In the higher education policy domain, the Life Long Learning Programme and the Socrates/Erasmus mobility program provide examples of the application of the subsidy instrument.

In addition to the subsidy instrument, other financial instruments, particularly loans and warranties, appear to be increasingly considered in the European context. In the seventh framework program, the Risk-Sharing Finance Facility is a clear example. In the CIP, a financial risk facility instrument for SMEs has been suggested.

The third and final category of European policy instruments are those of information and communication. Two crucial instruments in this category are the communication and the OMC.

The EC can publish communications in areas wherein it is not assigned authority to use legal instruments. In practice, the Commission uses this policy instrument for agenda-setting and as a means to share its views on certain issues. Communications are usually preceded by consultations with the relevant stakeholders or expert groups, or both.

Several examples of the use of the communication instrument have already been presented. Let me repeat a few. The relaunch of the Lisbon strategy started with a communication to the Spring European Council in 2005 (EC 2005e). The 2006 broad EU innovation strategy, with its ten-point program, was presented as a communication from the Commission to the European Council, the European Parliament, the European Economic and Social Committee, and the Committee of the Regions (EC 2006d). In the research and technology policy domain, the Commission used the communication instrument, among others, to develop the ERA and to set the 3%-of-GDP target for R&D investments (EC 2002a). In the area of knowledge-transfer policy, the Commission used the communication instrument to suggest guidelines for effective knowledge transfer (EC 2006d, 2007f). In the higher education policy domain, the Commission used the communication instrument to suggest a modernization agenda for the European universities (EC 2006a), to launch the EIT, and to start a discussion on the labor market skills of doctoral graduates.

The OMC is a relatively new European policy instrument, which was created at the 2000 Lisbon Council as an instrument for intergovernmental policy analysis. The OMC works with information and communication mechanisms, such as indicators, benchmarking, and the sharing of best practices. Generally speaking, the OMC works in stages. First, a Council of Ministers agrees on a set of (often broad) policy goals. These are then translated by the member states into national and regional policies. Next, specific benchmarks and indicators to measure best practice are agreed upon. Finally, the results are monitored, compared, and evaluated.

The OMC is a "soft" policy instrument that has a decentralized approach. The agreed policies are implemented by the member states and supervised by the European Council. The Commission has primarily a monitoring role, but in practice it appears to have considerable scope for agenda-setting and persuading member states to increase their efforts to reach agreed policy objectives. The OMC indeed allows the Commission to use peer pressure and "naming and shaming" processes to create stronger member state involvement in European policy processes.

In the policy domains of research and higher education, several applications of the OMC are found. The European Innovation Scoreboard monitors the member states' innovation performance. In the education policy domain, the progress reports analyze the progress toward the educational Lisbon objectives and provide a platform to discuss education policies at the European level.

All in all, the European policy instruments show a rather large variety in their contexts and conditions of application. The EU policy level is increasingly equipped with European policy instruments.

EUROPEAN POLICY TRENDS

The overall European innovation policy has a number of major influences on the European academic research system. Because of the integration of the European research and higher education policies (including the doctoral education and knowledge-transfer policies), the European ARE has clearly opened up and become more integrated. The level of fragmentation in the European ARE has decreased, although it certainly still exists.

In addition, the innovation policy appears to have created a growing awareness of the European policy echelon. Clearly different from the days before the Lisbon agenda, the European Union has become a major research and higher education policy actor, and many universities and academics have experienced its conditions and effects.

Two crucial trends appear to be related to European-level policymaking. These trends relate particularly to the changing conditions for the European ARE. They are presented here as fairly general developments in European policymaking regarding research and higher education.

A first general trend is the growing importance of the supranational European policy echelon. The Bologna Ministers Conferences regarding the EHEA and the EU political summits regarding the (renewed) Lisbon agenda (and the ERA as part of that agenda) leave their traces in the various research systems of Europe. Not only national politicians but also academics and research managers keep an eye on the European policy processes. They are aware of the available budgets and of the contexts in which these can be obtained; they look for partnerships in order to be eligible for funding; and they design consortia and networks to implement their cooperation strategies.

Generally speaking, as the 2005 Glasgow Declaration of the EUA, for instance, shows, the European universities accept their role as contributors to the "Europe of Knowledge." They emphasize that interinstitutional cooperation is increasingly important in a globalized and competitive environment. They are willing to reinforce the European dimension in various ways: benchmarking curricula, developing joint degrees, and enhancing intercultural and multilingual skills. They also urge European politicians to view the Bologna and Lisbon policy agendas together for each to be successful in the long term. "Universities acknowledge that European integration must be accompanied by strengthened international cooperation based on a community of interests" (EUA 2005b, 2).

The successive studies of the implementation processes of the Bologna agenda also show that there is general acceptance among European universities of the need for these reforms and that many universities have made great efforts to internalize the European reform process. Of course, moving toward a comprehensive three-cycle system throughout Europe is a complex cultural and social transformation. But it appears that considerable progress is being made and that a major innovation process of European higher education is currently taking place (EUA 2007; Reichert and Tauch 2005).

In the context of the ERA, the impact of the European (in particular EU) policy instruments is considerable. With a budget of €19.2 billion, FP6 has been one of the largest R&D programs in the world. In 2003 more than 16,000 proposals were submitted for funding, involving nearly 160,000 participants from more than fifty countries. Some 2,600 of these proposals—with 27,000 participants—were selected for funding (EC 2005e). The substantial budget increase of FP7 to €53.2 billion, which is 40% higher than the FP6 budget, will create an even larger impact in the European research field.

The European FPs provide a vital opportunity for universities in countries with limited research funding (Geuna 1999). But even in countries where substantial research funds are available, the FPs appear to be increasingly attractive. As was indicated before, the amounts of research funding that are directly awarded by the various national funding entities are still far larger than those allocated through the EU FPs. Yet, for many universities and other public research institutions, the EU funding for collaborative research projects—comprising teams that can be institutionally situated anywhere within the European Union—is a key element in their pursuit of international academic reputation. The European FPs have created a competitive research funding context at the European level, triggering and prompting European researchers to try to obtain these prestigious research funds.

The second trend regards the increasing emphasis on the alignment of the European policies and the various national policies in the domains of higher education and research. Although the EU competence in the higher education policy domain is still limited, the reemphasis on the European innovation strategy appears to offer a way to try to strengthen the alignment in this field. The integrated guidelines of the relaunched Lisbon strategy (2005), the benchmarking of the NRPs, and the ten-point program of the broad-based innovation strategy (2006) appear to create extra pressure on the member states to relate their national policy efforts to the European agenda. In its 2006 communication on the innovation strategy, the Commission argued that the member states "should ensure that there is sufficient availability of key skills to support innovation. Education must move with the times. As already agreed within the Integrated Guidelines for Growth and Jobs, Member States are invited to set, as a matter of priority, ambitious targets in their National Reform Programmes that address weakness in these areas" (EC 2006d, 5). The Commission also made it clear that it would continue to use the OMC to facilitate the modernization and restructuring of the national education systems. It invited the member states to significantly increase the share of public expenditure for education, to tackle the obstacles in their education systems for innovation, and particularly to implement the Commission's recommendation regarding the Modernisation Agenda for Universities.

In the research policy domain, the alignment is addressed in, on the one hand, the benchmarking of the national targets set for research and, on the other hand, the reinforcement of the ERA-NET instrument in FP7. The national targets, first of all, consider the level of R&D investments that, if met, would raise the level of EU R&D expenditure from 1.9% GDP to 2.6% in 2010. In addition, the member states are invited to explicitly take the transfer of knowledge into account in their national innovation policies; to earmark a large proportion of the cohesion policy funds for investing in knowledge and innovation; to create an open, single, and attractive European labor market for researchers; and to target their state aid on market failures preventing research and innovation activities. In the renewed Lisbon strategy, the Commission presses for a kind of "multilateral surveillance." "The Commission will, when assessing the [national] progress reports on the implementation of the National Reform Programmes, assess carefully Member States' reforms and policies addressing the innovation system and report on this in its Annual Progress Report" (EC 2006d, 15).

Using the new ERA-NET Plus Module in FP7, the Commission hopes to provide an incentive for joint, transnational research funding. Combining funding from member states and the Community appears necessary in order to be able to really create the ERA. Alignment of national and European funding policies therefore is a major target. Without such an increased alignment, the high European ambitions cannot be reached.

However, it should be noted that the intergovernmental European cooperation in R&D has so far hardly resulted in more coordination and integration of national research policies. In this sense, a common European research space is still far away. As Marimon and Graça Carvalho (2008) argue, at national and regional levels in the European Union, there seems to be both a lack of willingness and a lack of coordination capacity to really develop a joint European Research Area. The former because of the absence of incentives to collaborate, the latter because national research agencies are not designed to organize funding on a larger scale (Marimon and Graça Carvalho 2008).

Nevertheless, growing importance of the supranational European policy echelon and the increasing pressure for alignment of the European Union's and the member states' policies show that European research and higher education policies have indeed developed at the supranational European level. Although the European Union is not a federal government, it does have its own policymaking capacity in these fields.

EUROPEAN POLICY RESULTS AND EFFECTS

What are the results of the EU innovation policy and its various subpolicies? Which effects are observable? Which were intended? Which unintended? A number of results and effects will be discussed. In addition, conceptual frameworks will be offered to interpret some of these effects.

Let us first look at the *results* of the EU policies. These are monitored by the various assessment and benchmarking instruments that the Commission has developed since the introduction of the Lisbon strategy; by now they offer a rather comprehensive statistical overview of the performances of both the member states and the European Union.

The European Innovation Scoreboard is the main statistical tool for monitoring innovation performance. The Scoreboard was developed after the Lisbon European Council in 2000 and has been published since 2001. It shows the progress of the individual EU member states regarding the Lisbon ambitions and compares and ranks them on a number of indicators.

In the sixth and seventh editions of the Scoreboard (2006, 2007), innovation performance is measured by combining a set of twenty-five indicators organized in five broad categories: innovation drivers (input), including population with tertiary education per hundred population aged 25–64 and broad-band penetration rate; knowledge creation (input), including public and private R&D expenditure; innovation and entrepreneurship (input), including innovation expenditures and early-stage venture capital; application (output), including employment in high-tech services and export of high-tech products as share of total products; and intellectual property (output), including European Patent Office (EPO) patents per million population and triadic patent families per million population.

A summary innovation index and an average growth rate of this index allow an assessment of the relative strengths and weaknesses of the innovation performance of the member states and a comparison of the EU performance with that of the United States and Japan. The comparisons show four groups of countries: innovation leaders, innovation followers, moderate innovators, and catching-up countries.

The 2006 and 2007 Scoreboard editions show that some Scandinavian countries (Finland and Denmark), Germany, the United Kingdom, and non-EU-member Switzerland are the European innovation leaders. Most of the other "old" member states appear to be followers. The "new" members and the southern European countries are either moderate innovators or catching up. At the same time, there appears to be a process of convergence in innovation performance in Europe. The countries showing below-average EU innovation performance are closing the gap with the innovation leaders and followers.

The United States and Japan are still performing better in innovation than the European Union, but the gaps, particularly with the United States, are decreasing. The United States leads in 11 of the 15 indicators for which comparable data are available. Japan leads the European Union in 12 out of 14 such indicators. Compared to the United States, the European Union has improved its performance in, among other indicators, number of science and engineering graduates (13% of population aged 20–29; United States has 10.6%) and employment in medium-high- and high-tech goods industries (7% of total workforce, compared to 4% in the United States). The European Union is lagging behind the United States and Japan in other areas, including business expenditure for R&D (1.2% GDP in European Union, 1.9% in United States, and 2.4% in Japan), information and communication technology expenditure (6.4% in European Union, 6.7% in the United States, and 7.6% in Japan), and tertiary education attainment level (23% of the population in European Union, 39% in the United States, and 40% in Japan) (Innovation Scoreboard 2007).

The results regarding the EU innovation policy are getting better but are still too limited. The performances of the member states vary widely. On average, the European innovation results are still disappointing. There appears to be good reason to quickly implement the new broad-based EU innovation strategy.

With respect to education and training, the Commission publishes its so-called Progress Reports. The first report was adopted by the Commission in 2004 and analyzed the progress of thirty European countries (including the then fifteen EU countries) toward the Lisbon objectives in the field of education and training. The second report (EC 2006c), adopted by the Commission in 2005, delivered a number of strong political messages to the European Spring Council of 2006 (when this Council reviewed the revised Lisbon strategy for the first time). The report stated that reforms were moving forward but that more substantial efforts were required and that education and training must be viewed as a priority for investments.

The 2006 Progress Report (EC 2006c) offered (among many other things) a mixed picture of the results in the policy domain of higher education. On the one hand, it appears that the European Union is on and even ahead of schedule regarding the objective to have 1 million students graduating in mathematics, science, and technology every year by 2010 (compared to 755,000 in 2003). On the other hand, the European Union still suffers from underinvestment in higher education. In 2002 public spending on tertiary education in the European Union amounted to 1.14% of GDP, compared to 1.40% in the United States. A considerably larger gap existed in private spending on higher education: 0.2% in the European Union and 1.42% in the United States. To match the US level of public-plus-private expenditure, the European Union would have to spend an additional €140 billion per year (EC 2006c).

Furthermore, in the EU higher education context, access to higher education is not improving sufficiently. Although there have been increases, further progress regarding the participation in tertiary education is still needed. Also, most EU students are still not taught two languages (an objective formulated during the 2002 European Council), and the mobility of students within the Erasmus program would have to more than double to reach the target of affecting 10% of the student population (EC 2006c).

The results thus far in the (geographically broader) Bologna process show that the European universities are going through a remarkable process of change. In less than a decade, the universities have become engaged in processes to adapt their curricula and degree systems, to implement quality assurance systems, to develop their governance models, and to professionalize their management. The European universities have taken responsibility for the emerging EHEA and support the underlying ideas of student-centered and problem-based learning (EUA 2007). Nevertheless, whether these changes will create the EHEA still remains to be seen. Much will depend on the ways in which the highly cherished diversity of the European higher education systems can be combined with increased cooperation, harmonization, and transparency on the European scale (Floud 2006).

The results in the policy domain of research and technological development are included in the independent five-year assessments of the research FPs. The third assessment reviews the implementation and achievements of the FPs over the period 1999–2003 (FP5 and partly FP6).

The review shows that, on the one hand, the FPs have played an important role in developing the European knowledge base. "The strength of emphasis on information and communication technologies and on life sciences has, for example, been instrumental in strengthening European capabilities. There has been strong interest from industry, universities, and other research institutes. The FPs have played an important part in the generation and diffusion of new knowledge and the formation and reinforcement of inter-organisational networks." But on the other hand, the achievements of the programs, in terms of direct contribution to innovations with the potential to dominate global markets, are still limited (Assessment FPs 2004, II).

Regarding FP6, the assessment refers to the review by the Marimon Panel (a panel of high-level experts, chaired by R. Marimon), which praised the new instruments of FP6 (the Networks of Excellence and the Integrated Projects) for their ambition and their emphasis on transnational collaborative research, but which also pointed out the relatively high costs and risks of participation in these instruments for industry partners, notably SMEs, and the need for more flexibility and simplification (Marimon 2004).

The working document (EC 2007b) accompanying the 2007 green paper on the ERA provides an evaluation of the progress regarding the ERA so far. It mentions the following issues:

- The ERA-NET instrument has made a start at addressing the inefficiency and the fragmentation of the European research system. However, the volume involved is still marginal and the national and regional program owners are still reluctant to further develop genuine joint research programs.
- Regarding research infrastructure, good progress has been made. A first milestone was the adoption of the European Strategy Forum for Research Infra-

structures (ESFRI) Roadmap. However, new legal, institutional, and financial tools need to be developed for implementation of the Roadmap.

- In the area of international research cooperation, the European Union has demonstrated that it is able to show leadership to address global challenges. The International Thermonuclear Experimental Reactor (ITER) is a showcase. However, these initiatives are far from systematic and often poorly coordinated with those of the member states.
- Although there is some success in better exploiting human resources, Europe still lacks an open, competitive, and attractive market for researchers. Some researchers are still leaving the European Union. Others cannot enter research careers in Europe.
- Private investment in research is still far too limited. Europe's business-funded research intensity has not increased since 2000, and the gap between the European Union and the United States has not been reduced.
- The research policies of the member states have certainly evolved, but the question is whether the pace of national policy reform is sufficient.
- Some convergence in national policymaking is materializing, largely through the communicative instruments of the Commission.
- Although transnational cooperation is an element of member state research policy, there is little evidence that national policymakers have taken ownership of the ERA concept. (EC 2007b, 8–9)

The conclusion regarding the results of the EU policies with respect to research and technological development is that while some results are visible, the innovation policy objectives certainly have not been reached. The policy instruments (in particular the FPs) can be further developed (as has been done in FP7), and the relationships between research policy and innovation policy can be further intensified. The progress on the ERA is limited, and there is still much to do. This is particularly true with respect to addressing the "governance void" in the ERA, which implies a lack of sufficient incentives for the implementation of strong and joint intergovernmental research funding (Marimon and Graça Carvalho 2008, 14)

But there also is reason for optimism. In its annual policy strategy for 2008, the Commission argued that the Lisbon strategy "is beginning to yield results and has contributed to the economic performance of the European Union. The challenge now is to capitalise on the current upturn in order to press ahead with further reforms" (EC 2007a, 5). A 2007 review of progress in the Lisbon Strategy also showed that finally, in the seventh year of the Lisbon agenda, some of the objectives seem within reach. In an analysis of the original E15 countries, Europe is reported to be doing better economically than in any year since 2000. Economic growth in the E15 reached a level of 2.8% in 2006, and this growth appears to be more stable than before (Lisbon Council 2007). Realizing the Lisbon ambitions for the expanded European Union of twenty-seven member states remains a major challenge.

The 2007 European Competitiveness Report also reveals that in 2006 the European Union's GDP increase was the highest since 2000. The revised Lisbon strategy is judged to be beginning to bear fruit. The report identifies the key drivers contributing to this growth and argues that "increased investment in R&D can significantly increase productivity growth, especially if the elements of the knowledge triangle, R&D, innovation and education and training, are well integrated, including as concerns the provision of scientific personnel" (EC 2007e, 4).

Another recent economic analysis criticizes the weak European progress in R&D, particularly the cost of patenting (still five times as high in Europe as in the United States) and the failure to reverse the "brain drain" from the European Union to the United States (London School of Economics 2006). In this context, the recent EU initiative to introduce a European "blue card for talented knowledge workers," comparable to the US "green card" (which establishes legal permanent residency), can be mentioned. It may be expected that this initiative will have a positive influence on the reported brain drain.

A third study (Begg 2007), based on an analysis of the implementation reports of the NRPs, concludes that the relaunched Lisbon strategy is more successful than Lisbon I. Although the economic improvement of the European Union cannot be completely assigned to the Lisbon achievements up to now, real advances appear to have been made and a coherent and effective strategy is developing.

However, the global financial turmoil also has its impacts on the European Union, requiring the Union to increase its innovation efforts. In its 2009 policy strategy paper, the EC indicates that it will continue to pursue its broad-based innovation strategy and to "deepen the European Research Area" (EC 2008, 4).

In addition to the results of the EU policies, a set of basically *intended effects* of European policymaking in the domains of research and higher education can be distinguished. Four intended effects can be indicated.

A first and clearly intended effect concerns the reinforcement of the relationships between the European ARE and European society. The very heart of the European innovation strategy is to create stronger linkages between academia and European society to further develop Europe as an innovation-friendly society.

According to the EC, "Europe has to become a truly knowledge-based and innovation-friendly society where innovation is not feared by the public but welcomed, is not hindered but encouraged, and where it is part of the core societal values and understood to work for the benefit of all its citizens" (EC 2006c, 3–9). For education, this implies that talent and creativity should be promoted from an early stage and that key skills that support innovation need to be nurtured. The European universities can contribute to these processes, and they can do so more efficiently than they have done so far. For research, the innovation strategy emphasizes the importance of knowledge transfer:

> The knowledge economy relies on the transfer of knowledge from those who generate it to those who use it and can build on it. The transfer of knowledge between public research organisations and third parties (including industry and civil society organisations) needs to improve. . . . Doing so will help to build new market opportunities on research. Public research organisations. . . have a particularly important role to play in this. All of the many forms of knowledge transfer— contract research, collaborative and co-operative research, licensing, publications and exchanges of skilled researchers between the public and private sectors—need to be further developed and better managed. (EC 2006d, 3–9)

The EU innovation strategy implies a shift in the orientation of the activities of the European universities. In their educational programs, they are urged to focus more intensely on entrepreneurial skills and to develop joint training activities with business and industry. In their research programs, they are prompted to address not only knowledge creation but also knowledge diffusion processes.

As a result, the basic functions of the European universities appear to be changing. The general goal of the ERA is to bring researchers together to facilitate both knowledge creation and knowledge distribution. The centers of excellence and networks create critical mass and synergies that improve the productivity of knowledge creation. Networking also intends to facilitate both the geographic diffusion of new knowledge and its industrial application.

It may be argued that EU research funding entails a few related strategic purposes. One certainly is to improve European industry's international competitiveness by inventing and developing new products and processes and by forging links between academic and industrial research groups. Another is to foster the "cohesion of Europe" by reducing disparities in its regions' R&D capabilities (David and Keely 2003, 261). Yet another purpose is to use the investment in R&D to further develop the extensible knowledge infrastructure, thus enabling it to continue to produce new knowledge (Geuna, Salter, and Steinmueller 2003).

These strategic European funding purposes have their impact on the ways the European universities address their research functions. Because of the financial and in particular the prestige-related aspects of EU funding, they appear to adapt to the European emphasis on the application of knowledge as a criterion for funding. The policy goal to increase the application of scientific knowledge in industry and other parts of society has pushed universities to develop, at least partly, a closer and more instrumental role in advancing Europe's economic competitiveness. The European universities are increasingly held responsible for generating not only intellectual but also economic and social capital. As Martin (2003, 25) argues, the social contract between (European) society and the universities has changed: "there are now much more explicit and direct expectations that, in return for public funding, universities and researchers should endeavor to deliver greater and more direct benefits to society."

A second observable and clearly intended effect is the reinforcement of the knowledge-transfer capacity of the European Union. In the research and technology policy domain, the networking efforts in the FPs have increased Europe's knowledge-diffusion capacity. The formation and strengthening of collaborative networks have had a major influence on the European R&D landscape. The Networks of Excellence and the Integrated Projects of FP6 have brought together groups of researchers and specialists from academia and industry. In particular, the technology platforms (although not an instrument of the FPs) have developed important strategic technology agendas with a strong knowledge-sharing and -diffusion capacity. These platforms unite all relevant stakeholders, with industry taking the lead role. They have built bridges between industry, the academic community, the financial world, regulators, and consumers. FP7 is intended to capitalize on these platforms, helping them translate their strategic research agendas into concrete actions. The Cooperation program of FP7 provides the means, via the JTI scheme, to implement the agendas that have a European dimension as well as a strong industrial relevance. The Structural Funds, the European Investment Bank, and the new Risk-Sharing Finance Facility in FP7 provide further financing options. It may be expected that, with the reinforcement of the research networking instruments and the increased emphasis on knowledge transfer, the European knowledge-diffusion capacity will continue to grow.

For the process of knowledge transfer, a recent (rough) comparison between the European Union and the United States shows that the European universities lag behind their US counterparts regarding invention disclosures as well as patent applications and grants but do better in regard to licenses granted and start-ups established (EC 2007b). Apparently, despite less effort, the European Union is relatively successful in the actual use of public R&D results by the business sector, which again shows that the European knowledge-diffusion capacity has been reinforced.

A special aspect of the reinforcement of the knowledge-diffusion capacity of the European Union is the increasing mobility of students. In the higher education policy domain, the 2004 decision of the European Parliament and of the Council on a single framework for the transparency of qualifications and competences (Europass) can be mentioned here as well as the development of the credit transfer systems for academic and vocational education and training. And although the Erasmus mobility program has not yet reached its target of affecting 10% of the European student population, between 1987–88 and 2004–5 more than 1.3 million students studied abroad under the aegis of this program, and 87% of all European universities participated (EC 2006c). Generally speaking, the harmonization activities in the Bologna process are creating a comprehensive system of easily readable and comparable degrees, with the potential to further integrate the various national higher education systems. As an effect, the knowledge-sharing and -diffusion capacity of the European Union is increasing.

A third effect could be described as "the rise of European social networking" in the European ARE. In the European research system, collaborative networks have become the major "instruments for collective action" (Foray 2003, 372). In the FPs, networks (of various kinds) have proven to be the main mechanisms of exchange among researchers and between research and industry.

However, it should be observed that in EU research and innovation policies there appears to be a strong emphasis on *establishing* academic-industry linkages. Geuna, Salter, and Steinmueller (2003, 399) argue that this European approach to social networking for research and innovation differs from the US approach. "In the case of the US, it was the combination of high industrial demand for research and the relative high quality of the US science system's output that helped to generate the new networks bridging science and innovation. It was demand that created the new networks, rather than the networks that created the demand. In the case of Europe, policy has often created networks that are in search of demand." The EU research policy intends to create and formalize social networks that promise to develop and implement research and innovation agendas but that have, at the time of their application for funding, often hardly begun to do so. The social networking processes in the European research policy domain are triggered and stimulated by the instruments that are being applied in the FPs, in particular. The basic assumption is that the relationships between academia and industry are limited and thus have to be increased. For this reason, the European social networking strategy focuses on the generation of new research networks rather than on the recognition of existing ones. In this sense, the EU social networking strategy also is a top-down rather than a bottom-up process. The exceptions are the technology platforms. These are largely initiated and led by industry and are clearly demand-driven. The recent recognition of these platforms as crucial instruments for strategic knowledge creation and application agenda setting indicates that the EU social networking approach is slowly changing.

It may be concluded that the social networking approach has become an increasingly dominant mechanism in European research. Over the years, the FPs have grown substantially. The financial support for the various research networks has become a major and increasingly prestigious factor in the European research efforts (Geuna 1999).

The fourth effect has to do with the changing governance model in the European ARE. Generally speaking, for decades now the trend in European higher education and research systems has been to move from state control to self-regulation and accountability (Van Vught 1989). The autonomy of universities has increased in Europe over the past twenty-five years, and detailed governmental planning strategies have lost ground to supervision and accountability strategies. The move to accountability has brought along the recognition of stakeholders' needs and interests, and hence the acceptance by universities of their social embeddedness and their relationships with and dependencies on various political and economic organizations.

On the European scene, this trend toward a new diversified governance model is recognizable in the increasingly instrumental role that universities are assumed to play in the European knowledge society (Geuna, Salter, and Steinmueller 2003). They are increasingly being challenged to focus on their new roles of advancing the competitiveness of Europe and contributing to a European culture of innovation, rather than on their traditional academic role.

The EU universities' modernization agenda provides a clear example of this changing role. It suggests that the member states should move from a control to an accountability strategy and that public research organizations should be stimulated to contribute to the overall innovation strategy of Europe.

In the research domain, universities are being incited to intensify their relationships with industry. Based on the belief that Europe is afflicted by an interaction deficit in its innovation system, they are encouraged to develop and participate in collaborative networks with industry in order to jointly create better conditions for the overall European innovation strategy.

The result is the emergence of a new, multistakeholder governance model in the European ARE with multiple funding sources, a stronger focus on autonomy combined with accountability, and a pressure to deliver innovation-relevant outcomes (Enders et al. 2006; Van der Ploeg and Veugelers 2008). The governance systems of the European higher education and research systems appear to have changed during the past decades, mainly as a result of the social and economic expectations regarding their processes and outputs.

Finally, in addition to the results and intended effects mentioned, there appears to be some empirical evidence for a few general *unintended effects.*

The first unintended effect occurs as the combined result of two processes and can be described as a stratification effect in the overall European ARE. The two processes that appear to create this stratification effect are the changing participation processes of the European universities in the FP programs and the occurrence of a counterproductive consequence of the reinforcement policy regarding the interaction between universities and industry.

Regarding the first process, it has been pointed out that larger and older universities have a higher participation rate in the networks of the FPs than other universities. In addition, there appears to be (in the FPs) an increased homogenization of research institutions. The variety of participating institutions is decreasing over time. Furthermore, past success appears to be an important indicator for future participation (David and Keeley 2003; Geuna 1999). What appears to be occurring is the well-known Matthew Effect: "research groups that are successful in finding external funding for their research have a higher priority of producing publishable research, which improves their probability of getting funds in the future" (Geuna 1999, 117). Universities with successful research groups, in terms of receiving EU funding, appear to have a higher probability of also receiving future funding, because their earlier successes are seen as an indication of their high quality.

The effect is a slowly increasing stratification between universities that are more successful and those that are less successful in receiving FP funding. Given the fact that the variety of applying institutions appears to be decreasing and that larger and older universities are on average more successful in getting EU funding, the EU research funding policy appears to be contributing to the creation of a European category of research universities that are distinguishable from other higher education institutions.

The other process contributing to this stratification effect is the occurrence of a counterproductive force as a result of the push in EU innovation policy toward closer links between universities and industry. Geuna (1999) argues that this counterproductive effect occurs particularly in universities with a relatively weak financial situation. "Constrained to accept industrial funds for developing routine contract research, and faced with the impossibility of charging the real cost of the research, their collaboration with industry results not in a contribution to the wealth of society, but in an exploitation for private profit of a public investment" (Geuna 1999, 173).

Again, the effect of the further weakening of already financially vulnerable universities appears to be increased stratification in the overall European higher education system. While universities that are successful in getting EU funds appear to increase their potential for success in acquiring future funds (and thus in principle reinforce their financial situation), financially weaker universities that increase their links with industry run the risk of further weakening their financial situation. The outcome of this increasing differentiation between financially stronger and weaker universities is a growing diversity in terms of reputation. As in any other higher education system around the world, the European universities are engaged in a "reputation race" (Van Vught 2006). Universities are first and foremost driven by their wish to maximize their academic prestige and to uphold their reputations. In this race, universities are constantly trying to create the highest possible reputation for themselves, and for this they need all the financial resources they can find. A weakening of a university's financial position has a negative effect on its reputation-building capacity, and there will be stratification consequences at the overall systems level.

A second unintended effect is a growing regional diversification in the European ARE. This effect appears to result from policy tensions between the Lisbon strategy, on the one hand, and the European cohesion policy, on the other. The general Lisbon innovation agenda focuses on an internationally competitive Europe and on the strengthening of its collective capacities for innovation and research. In contrast, the cohesion programs aim to reduce income disparities between the European Union's poorest regions and the rest of Europe. The main instrument of the cohesion policy is the Structural Funds, specifically devoted to regions with per capita incomes that are less than 75% of the EU average.

The objectives of the two policies appear to be incompatible. The Lisbon strategy appears to generate disproportionate benefits for the richer regions relative to the poorer regions, simply because they have larger concentrations of researchers and are better able to attract academic talent and quality. The result is an increasing "innovation gap" between Europe's most and least advanced regions, a process that the cohesion policy exactly tries to minimize. In other words, there appear to be trade-off effects between the European Union's innovation and cohesion policies (Clarysse and Muldur 2001; Sharp 1998).

The EC sees this differently. The Commission argues that the Structural Funds enable weaker regions to strengthen their knowledge bases, allowing them to participate more frequently and successfully in the programs of the ERA, and hence to decrease their innovation gaps. According to the Commission, the cohesion policy is compatible with the Lisbon innovation strategy.

However, three interrelated processes have been identified that appear to create the unintended effect of a growing regional diversification as a result of the European Union's general innovation policy (Frenken, Hoekman, and Van Oort 2008). The first one is the preference of researchers in "excellent regions" to collaborate with each other, rather than with colleagues in lagging regions. Particularly, the EU research policy appears to stimulate concentration of talent in the richer and academically better equipped regions. Lagging regions find it difficult to participate in successful European research networks and appear to have to pass a threshold of quality and size before they can do so.

Secondly, the ERA policy objective of the free movement of people appears to lead not only to an increased mobility of researchers but also to the concentration of talent in a selected number of excellent regions. The most talented researchers compete for the positions at the most prestigious universities, thus rendering it difficult for the lagging regions to retain talent within their borders.

Thirdly, the sectoral structure of the poorer European regions is usually characterized by a dominance of low-tech and medium-tech activities that hardly fit the thematic priorities of the ERA. The FPs almost exclusively concern high-tech sectors, thus creating a situation in which the research subsidies are becoming concentrated in the richer regions.

The result is the unintended but nevertheless real effect of regional diversification. The geography of the European ARE is changing from one based on the priority of national borders into one based on the clustering of talent. Wealthier regions are increasingly able to profit from the general European innovation policy, while poorer regions are left with the resources of the cohesion policy.

Conclusion

During the first years of the twenty-first century, the research and higher education policy domains have moved to the heart of EU policymaking. Since the March 2000 Lisbon European Council meeting, and in particular since the relaunch of Lisbon strategy in 2005, the two policy domains have become integrated into an overall and comprehensive European innovation strategy. In this innovation strategy, knowledge is singled out as Europe's most important resource. Knowledge is to be the driving force of the European lifelong learning society; it is the basis for Europe's future economic, social, and environmental development; it is the fundamental resource for the innovation of products and services. Universities are Europe's most crucial knowledge institutions. The European ARE has become the cornerstone of the European innovation policy.

The history of European research and higher education policies shows that the

European universities have become one of the most important categories of policy addressees for the EU institutions. They are increasingly being addressed as crucial actors for the provision and exploitation of knowledge and hence for the realization of the European innovation strategy.

In the policy domain of research and technology, the EU policy intentions are to encourage researchers to work together at the European level and to stimulate cooperation between universities and business and industry. The European Union seeks to increase the investments in R&D (to 3% of GDP) and to create stronger coordination of national research policies. Research is seen as the core factor in the European Union's ambition to create growth and employment. Particularly the seventh FP is offered as a major instrument to fully develop the ERA and, by doing so, to respond to the competitiveness and employment challenges of the European Union. In this policy domain, the European universities are seen as major knowledge institutions, the positions and roles of which need to be reinforced to better enable them to contribute to the renewed Lisbon agenda.

In the policy domain of higher education, the recent EU Lifelong Learning Programme (2007–13) is intended to contribute to the European knowledge society by fostering cooperation between the various national education systems. In its higher education subprogram (Erasmus), it is designed to reinforce the contribution of higher education to the overall European innovation strategy by supporting the realization of the EHEA. In the context of the Bologna process, the European Union intends to create a "consistent, compatible and competitive" European higher education system fitting into the broader Lisbon strategy (EC 2003b, 11). The program emphasizes the importance of the three-cycle structure, a European qualification framework, and doctoral education. In addition, it emphasizes that the universities have a crucial place and role in the European knowledge society and should modernize themselves to become globally competitive institutions.

Analysis of the historical development of the two policy domains of research and higher education leads to the conclusion that, since the turn of the century, the European Union indeed has developed an innovation policy. In this policy, the European higher education and research areas have become integrated, and a coherent set of policy objectives and instruments involving various policy actors has been designed and implemented. The EU innovation policy is being developed and implemented in a multi-echelon policy system in which several decision-making levels, each with its own authorities and competences, exist and are interrelated. The European universities are addressed both directly and indirectly by European policy initiatives. They are addressed indirectly by the coordination of the policy actions of their own nation-states that are increasingly aligned with the overall European ambitions and strategies. They are addressed directly by EU initiatives in both the research and the higher education policy fields (for example FP7, the European Research Council, the Erasmus program, and the European qualifications framework).

In this multi-echelon policy system, the EU policy actors make use of a whole array of policy instruments: legal instruments, financial instruments, and information and communication instruments. In particular this last category of policy instruments has increased policy capacity at the EU level. The EC is increasingly using the soft OMC instrument (benchmarking and naming and shaming) to create stronger member state involvement in EU policy processes.

The European policymaking processes appear to have contributed to two crucial trends that may be expected to continue to be relevant for the future dynamics of the European higher education and research systems. Generally speaking, these trends can be described as the growing importance of the supranational European policy echelon and the increasing alignment of European and national policies. These general trends once more underscore the emergence of an EU policy context in the domains of research and higher education. And, although the European Union is not a federal government, they show that a European-level EU innovation policy has developed and is creating its effects. In this sense, a supranational policy echelon has indeed come about "gradually and smoothly" (de Wit 2003; Huisman and Van der Wende 2004, 355). In the increasingly integrated policy domains of research and higher education, the EU member states have accepted this supranational policy echelon and have even started to develop, coordinate, and implement national policies that fit the European innovation policy agenda.

Because this policy is still rather young, the results of the EU innovation policy are currently hard to assess. On average they still appear to be disappointing. The policy results in both the research and the higher education policy domains are so far still limited. But there also is some reason for optimism, since some of the Lisbon objectives seem within reach.

At a general level, a set of basically intended effects of European policymaking in the research and higher education domains can be distinguished: the reinforcement of the relationships between the European ARE and European society, the reinforcement of the knowledge-diffusion capacity of the European Union, the rise of European social networking in research and higher education, and the changing higher education and research governance model. These effects mark the considerable changes that are taking place in the European ARE. The European universities are clearly influenced by the EU policies, and they are adapting to the new conditions that are being created by these policies.

Two unintended effects can also be identified. The first is an increasing stratification process in European higher education. As a combined result of the European universities' participation in the FPs and of the reinforcement policy regarding the interaction between universities and industry, a reputation differentiation appears to be emerging, which will have its impact on the dynamics of the overall European higher education and research system and which may in the future very well lead to a growing diversification of this system.

The second unintended effect is a growing regional diversification in the European ARE, with the richer regions profiting more from the general EU innovation policy than the poorer regions. The result is a concentration of talent and quality in the wealthier regions of Europe and a growing "innovation gap" between Europe's most and least advanced regions.

Finally the general conclusion must be that the overall European innovation policy and its various subpolicies increasingly have their effects on the dynamics of the European ARE. The European policy echelon is having its impacts on the public research organization of Europe. In the future, the European ARE will certainly also be challenged to continue to contribute to the further development of the "Europe of Knowledge."

NOTES

1. The six countries that created the first three European treaties were Belgium, France, Germany, Italy, Luxemburg, and the Netherlands. In 2008 the European Union consists of twenty-seven member states and, with around 490 million people, has the world's third-largest population after China and India.

2. It is sometimes argued that a so-called European Paradox appears to exist, that is, the conjecture that the European Union is the global leader in terms of scientific output but lags behind in the ability to convert this strength into relevant innovations. However, as Dosi, Ilerena, and Labini (2005) clearly show, this paradox has no empirical foundation. The European Union lags behind the United States both in the relative number of scientific publications and in their impacts.

REFERENCES

Aghion, P., M. Dewatripont, C. Hoxby, A. Mas-Colell, and A. Sapir. (2008) *Higher Aspirations: An Agenda for Reforming European Universities*. Vol. 5. Brussels: Breugel Blueprint Series.

Andersson, B. (2006) European Research Policy: Towards Knowledge and Innovation or Trivial Pursuit. In L. E. Weber and J. J. Duderstadt (eds.), *Universities and Business: Partnering for the Knowledge Society*, 79–86. London: Economica.

Assessment FPs. (2004) *Five-Year Assessment of the European Union Research Framework Programmes, 1999–2003*. Brussels: European Communities.

Bartelse, J. A. and J. Huisman. (2008) The Bologna Process. In M. Nerad and M. Heggelund (eds.), *Towards a Global PhD? Forces and Forms in Doctoral Education Worldwide*, 101–113. Seattle: CIRGE/University of Washington Press.

Begg, I. (2007) *Lisbon II, Two Years On: An Assessment of the Partnership for Growth and Jobs.* Brussels: Centre for European Policy Studies.

Bergen Communiqué. (2005) The European Higher Education Area—Achieving the Goals. Communiqué of the Conference of Ministers Responsible for Higher Education, May 19–20, Bergen, Norway.

Berlin Communiqué. (2003) Realising the European Higher Education Area. Communiqué of the Conference of European Ministers Responsible for Higher Education, September 19, Berlin.

Bologna Declaration. (1999) The European Higher Education Area. Joint Declaration of the European Ministers of Education, June 19, Bologna, Italy.

Caracostas, P. and U. Muldur. (2001) The Emergence of a New European Union Research and Innovation Policy. In P. Larédo and P. Mustar (eds.), *Research and Innovation Policies in the New Global Economy: An International Comparative Analysis*, 157–204. Cheltenham, UK: Edward Elgar.

Clarysse, B. and U. Muldur. (2001) Regional Cohesion in Europe? An Analysis of How EU Public RTD Support Influences the Techno-Economic Regional Landscape. *Research Policy* 30 (2): 275–296.

Corbett, A. (2005) *Universities and the Europe of Knowledge: Ideas, Institutions, and Policy Entrepreneurship in European Union Higher Education, 1955–2005.* Houndmills, UK: Palgrave Macmillan.

David, P. A. and L. C. Keely. (2003) The Economics of Scientific Research Coalitions: Collaborative Network Formation in the Presence of Multiple Funding Agencies. In Geuna, Salter, and Steinmueller 2003, 251–308.

Davignon, E. (1997) *Five-Year Assessment of the European Community RTD Framework Programmes.* Luxembourg: Office for Official Publications of the European Communities.

de Wit, K. (2003) The Consequences of European Integration for Higher Education. *Higher Education Policy* 16:161–178.

Dosi, G., P. Llerena, and M. S. Labini. (2005) Science-Technology-Industry Links and the "European Paradox": Some Notes on the Dynamics of Scientific and Technological Research in Europe. LEM Working Paper Series, Laboratory of Economics and Management, Sant' Anna School of Advanced Studies, Pisa.

Enders, J., J. File, B. Stensaker, H. de Boer, I. Marheim Larsen, P. Maassen, S. Nickel, H. Vossensteyn, and F. Ziegele. (2006) *The Extent and Impact of Higher Education Governance Reform across Europe.* Enschede, Netherlands: Center for Higher Education Policy Studies.

European Commission (EC). (1991) *Memorandum on Higher Education in the European Community.* COM (1991). Luxembourg: EC.

———. (1995) *The Green Paper on Innovation.* COM (95), 688. Luxembourg: Office for Official Publications of the European Communities.

———. (2000a) *Development of an Open Method of Co-ordination for Benchmarking National Research Policies: Objectives, Methodology, and Indicators.* SEC (2000), 1842. Brussels: EC.

———. (2000b) *Toward a European Research Area.* COM (2000), 6. Brussels: EC.

———. (2002a) *The European Research Area: An Internal Knowledge Market.* Luxembourg: Office for Official Publications of the European Communities.

———. (2002b) *More Research for Europe.* COM (2002), 499. Brussels: EC.

———. (2003a) *Researchers in the European Research Area: One Profession, Multiple Careers.* COM (2003), 436. Brussels: EC.

———. (2003b) *The Role of the Universities in the Europe of Knowledge.* COM (2003), 58. Brussels: EC.

———. (2004a) *Proposal for a Decision of the European Parliament and the Council Establishing an Integrated Action Program in the Field of Lifelong Learning.* SEC (2004), 971. Brussels: EC.

———. (2004b) *Science and Technology, the Key to Europe's Future: Guidelines for Future European Union Policy to Support Research.* COM (2004), 353. Brussels: EC.

———. (2005a) *Common Actions for Growth and Employment: The Community Lisbon Program.* COM (2005), 330. Brussels: EC.

———. (2005b) *Contribution to the Conference of European Higher Education Ministers.* Bergen, May 19–20.

———. (2005c) *The European Charter for Researchers: The Code of Conduct for the Recruitment of Researchers.* DGR, EUR 21620. March 11. Brussels: EC.

———. (2005d) *Mobilising the Brainpower of Europe: Enabling Universities to Make Their Full Contribution to the Lisbon Strategy.* COM (2005), 152. Brussels: EC.

———. (2005e) *Working Together for Growth and Jobs: A New Start of the Lisbon Strategy.* COM (2005), 24. Brussels: EC.

———. (2006a) *Delivering on the Modernisation Agenda for Universities: Education, Research, and Innovation.* COM (2006), 208. Brussels: EC.

———. (2006b) *Developing a Knowledge Flagship: The European Institute of Technology.* COM (2006), 77. Brussels: EC.

———. (2006c) *Progress Towards the Lisbon Objectives in Education and Training, Report 2006.* SEC (2006), 639. Brussels: EC.

———. (2006d) *Putting Knowledge into Practice: A Broad-Based Innovation Strategy for the EU.* COM (2006), 502. Brussels: EC.

———. (2006e) *Time to Move Up a Gear: The New Partnership for Growth and Jobs.* COM (2006), 30. Brussels: EC.

———. (2007a) *Annual Policy Strategy for 2008.* COM (2007), 65. Brussels: EC.

———. (2007b) *Commission Staff Working Document* (accompanying the Green Paper *The European Research Area: New Perspectives*). SEC (2007), 412/2. Brussels: EC.

———. (2007c) *Green Paper—The European Research Area: New Perspectives.* COM (2007), 161. Brussels: EC.

———. (2007d) *Improving Knowledge Transfer between Research Institutions and Industry, across Europe: Embracing Open Innovation.* COM (2007), 182. Brussels: EC.

———. (2007e) *Raising Productivity Growth: Key Messages from the European Competitiveness Report 2007.* COM (2007), 666. Brussels: EC.

———. (2007f) *Voluntary Guidelines for Universities and Other Research Institutions to Improve Their Links with Industry across Europe.* SEC (2007), 449. Brussels: EC.

————. (2008) *Annual Policy Strategy for 2009.* COM (2008), 72. Brussels: EC.

European Communities. (2004) *Facing the Challenge: The Lisbon Strategy for Growth and Employment.* Report from the high level group, chaired by Wim Kok. Luxembourg: Office for Official Publications of the European Communities.

————. (2005) *Key Figures 2005: Towards a European Research Area: Science, Technology, and Innovation.* Luxembourg: Office for Official Publications of the European Communities.

————. (2006a) *Creating an Innovative Europe.* Report of the Independent Expert Group on R&D and Innovation, chaired by Esko Aho. Luxembourg: Office for Official Publications of the European Communities.

————. (2006b) *The History of European Cooperation in Education and Training: Europe in the Making: An Example.* Luxembourg: Office for Official Publications of the European Communities.

European Council. (2000) *European Council Presidency Conclusions.* No. 100/1/00. Lisbon: European Council.

————. (2006) *Presidency Conclusions.* May 18. Brussels: European Council.

European Innovation. (2005) *More Research and Innovation: Investing for Growth and Employment: A Common Approach.* December. Brussels: Enterprise and Industry DG, EC.

European Research Advisory Board. (2006) *International Research Cooperation: Final Report.* EURAB 05.032. June, http://ec.europa.eu/research/eurab/pdf/eurab_05_032_wg9_final report_june06_en.pdf.

European University Association (EUA). (2005a) *Doctoral Programs for the European Knowledge Society.* Brussels: EUA.

————. (2005b) *Glasgow Declaration: Strong Universities for a Strong Europe.* Brussels: EUA.

————. (2007) *Trends V: Universities Shaping the European Higher Education Area.* Brussels: EUA.

Floud, R. (2006) The Bologna Process: Transforming European Higher Education. *Change* 38 (4): 9–15.

Foray, D. (2003) On the Provision of Industry-Specific Public Goods: Revisiting a Policy Process. In Geuna, Salter, and Steinmueller 2003, 360–374.

Frenken, K., J. Hoekman, and F. Van Oort. (2008) *Towards a European Research Area.* Rotterdam: Netherlands Institute for Spatial Research, NAI.

Geuna, A. (1999) *The Economics of Knowledge Production: Funding and the Structure of University Research.* Cheltenham, UK: Edward Elgar.

Geuna, A., A. J. Salter, and W. E. Steinmueller (eds.). (2003) *Science and Innovation: Rethinking the Rationales for Funding and Governance.* Cheltenham, UK: Edward Elgar.

Hood, C. C. (1983) *The Tools of Government.* London: Palgrave Macmillan.

Huisman, J. and M. van der Wende. (2004) The EU and Bologna: Are Supra- and International Initiatives Threatening Domestic Agendas? *European Journal of Education* 39 (3): 349–357.

Innovation Scoreboard. (2006) *European Innovation Scoreboard 2006: Comparative Analysis of Innovation Performance: Trendchart.* Brussels: EC, www.proinno-europe.eu.

————. (2007) *European Innovation Scoreboard 2007: Comparative Analysis of Innovation Performance: Trendchart.* Brussels: EC, www.proinno-europe.eu.

Lisbon Council. (2007) *European Growth and Jobs Monitor: Indicators for Success in the Knowledge Economy.* Frankfurt/Main: Allianz Dresdner Economic Research.

London Communiqué. (2007) Towards the European Higher Education Area: Responding to Challenges in a Globalised World. Communiqué of the Conference of European Ministers Responsible for Higher Education, London.

London School of Economics and Political Science. (2006) *Boosting Innovation and Productivity Growth in Europe: The Hope and the Realities of the EU's "Lisbon Agenda."* Paper No. CEPPA007. London: London School of Economics and Politics, Centre for Economic Performance.

Marimon, R. (2004) *Evaluation of the Effectiveness of the New Instruments of Framework Program VI.* Report of a high-level expert panel chaired by Professor Ramon Marimon, June 21. Brussels: European Commission, DG Research.

Marimon, R. and M. da Graça Carvalho. (2008) Governance and Co-ordination of S&T Policies in the European Research Area. Discussion paper presented during the 10th meeting of the Expert Group on Knowledge for Growth, Llubljana, Slovenia, March 27–28.

Martin, B. R. (2003) The Changing Social Contract for Science and the Evolution of the University. In Geuna, Salter, and Steinmueller 2003, 7–29.

Mesarović, M. D., D. Macko, and T. Takahara. (1970) *Theory of Hierarchical, Multilevel Systems.* New York: Academic Press.

Mitnick, B. M. (1980) *The Political Economy of Regulation: Creating, Designing, and Removing Regulatory Reforms.* New York: Columbia University Press.

OECD, (2006) *Main Science and Technology Indicators.* Vol. 2006/2. Paris: Organization for Economic Cooperation and Development.

Neave, G. (1984) *Education and the EEC.* London: Trentham Books.

Prague Communiqué. (2001) Towards the European Higher Education Area. Communiqué of the Meeting of the European Ministers in Charge of Higher Education, Prague, Czech Republic.

Reichert, S. and C. Tauch. (2005) *Trends IV: European Universities Implementing Bologna.* Brussels: European University Association.

Scott, P. (2006) Global Context of Doctoral Education. Paper presented at the joint seminar of the European University Association and the US Council of Graduate Schools, Salzburg, Austria, September 3–5.

Sharp, M. (1998) Competitiveness and Cohesion—Are the Two Compatible? *Research Policy* 27 (6): 569–588.

Shaw, J. B. (1999) From the Margin to the Centre: Education and Training Law and Policy from Casagrande to the "Knowledge Society." In P. Craig and G. de Burca (eds.), *The Evolution of EU Law,* 555–595. Oxford: Oxford University Press.

Soete, L. (2005) Activity Knowledge. Discussion paper prepared for the UK Presidency of the European Union, United Nations University, Maastricht.

Sorbonne Declaration. (1998) Joint Declaration of the Architecture of the European Higher Education System. May 25, Paris.

Van der Ploeg, F. and R. Veugelers. (2008) Towards Evidence-Based Reform of European Universities. *CESifo Economic Studies* 54 (2): 99–120.

Van Vught, F. A. (1989) *Governmental Strategies and Innovation in Higher Education.* London: Jessica Kingsley.

———. (1994) Policy Models and Policy Instruments in Higher Education. In J. C. Smart (ed.), *Higher Education: Handbook of Theory and Research,* 10:88–126. New York: Agathon Press.

———. (2004) Closing the European Knowledge Gap? Challenges for the European Universities of the 21st Century. In L. E. Weber and J. J. Duderstadt (eds.), *Reinventing the Research University,* 89–106. London: Economica.

———. (2006) Higher Education Systems Dynamics and Useful Knowledge Creation. In L. E. Weber and J. J. Duderstadt (eds.), *Universities and Business: Partnering for the Knowledge Society,* 63–76. London: Economica.

Weber, L. E. (2006) European Strategy to Promote the Knowledge Society as a Source of Renewed Economic Dynamism and of Social Cohesion. In L. E. Weber and J. J. Duderstadt (eds.), *Universities and Business: Partnering for the Knowledge Society,* 3–18. London: Economica.

Witte, J. (2006) *Change of Degrees and Degrees of Change: Comparing Adaptations of European Higher Education Systems in the Context of the Bologna Process.* Enschede, Netherlands: CHEPS.

CHAPTER SIX

Finland

SEPPO HÖLTTÄ

During the past fifteen years, Finland's higher education and research policies have been closely linked to economic development and industrial policy. Research and the development of strong research-oriented universities have received increased attention recently. These higher education and research policies are explicitly seen in the framework of the Lisbon Strategy of the European Union, with its focus on five policy areas: the knowledge society, the internal market, the business climate, the labor market, and environmental sustainability.

A special feature of the Finnish higher education and research system is that government has a comprehensive role in policy design and implementation. Although a major trend since the 1960s has been the transition from a very centralized and controlled university system toward deregulation and decentralization, government has introduced new methods of steering and coordination to make the university system an even more essential part of the national infrastructure. Research and higher education are seen as the most essential part of the national innovation system and as key elements for improving the competitiveness of Finland in the global knowledge-based economy. The Finnish academic research enterprise (ARE), its policy framework, and future directions are discussed in this chapter.

The Academic Research Enterprise
OVERVIEW OF THE NATIONAL INNOVATION SYSTEM

Finland has consistently been identified as an innovation leader. In 2007, Finland was ranked third in the European Union on the European Innovation Scoreboard's Summary Innovation Index. Finland ranked near the top on all dimensions of the index: innovation drivers, knowledge creation, innovation and entrepreneurship, applications, and intellectual property. Relative to population, Finland has one of the highest rates of scientific publication among OECD nations (Ahlbäck 2005). Finland is also one of the leading EU countries in early-stage venture capital (UNU-MERIT 2008). Global competitiveness is another area in which Finland is highly ranked. The nation was ranked number one by the World Economic Forum (WEF) in 2003, 2004, and 2005 and ranked number six in the 2007–8 report (WEF 2004, 2005, 2008). Finland's strengths in global competitiveness are identified as efficient institutions and excellent higher education and training systems (WEF 2008). "The national innovation system approach used in Finland stresses that the flows of information and technology between people, enterprises, and research institutions are the main sources of innovations" (Ahlbäck 2005, 1).

Finland was first among the European nations to declare "knowledge intensity and technological superiority" as strategic policy objectives. This declaration was accompanied by the establishment of the Finnish Funding Agency for Technology and Innovation (TEKES) in 1983 and the Science and Technology Policy Council in 1987. For many years, Finland's strategy has been to give government a major role in the process of building up a comprehensive research and educational infrastructure for the nation and its industries. Finland is one of the strongest examples in the world of central and regional governments' involvement in building up the innovation infrastructure. In the 1990s, Finland's private sector also began to be heavily involved in R&D (research and development), and with the incorporation of the concept of a national innovation system (NIS) into policy, government's focus shifted to networking and innovation policy. With respect to support for higher education, the focus shifted from a welfare system rationale to one based on increasing output (Ahlbäck 2005; Ormala 2001).

From a structural perspective, Finland's NIS is directed by three policymaking bodies: Parliament, the Cabinet, and the Science and Technology Policy Council, chaired by the prime minister. Ministries also play important roles in coordination, funding, supervision of R&D, and oversight of public funding organizations. The Ministry of Education, which oversees the Academy of Finland and the higher education system, and the Ministry of Employment and the Economy (formerly Trade and Industry), which oversees TEKES and the Technical Research Center, are the ministries that have the greatest involvement with the NIS. Nearly 80% of government R&D funding (table 6.1) is channeled through these two ministries. Sitra, an independent, public funding agency that operates in conjunction with the Bank of Finland, is under the oversight of Parliament. Of these public funding agencies, the Academy of Finland focuses on basic research, TEKES on generic technology,

TABLE 6.1.

State-funded research and development funding in Finland, by recipient, 1999–2006, at current prices (€M)

	1999	2000	2001	2002	2003	2004	2005	2006	2006 (%)
Universities	323.3	346.4	349.8	377.7	393.3	407.9	416.7	427.5	25.4
University hospitals	60.5	59.4	56.7	56.7	48.7	48.7	40.7	48.7	2.9
Academy of Finland	155.5	153.8	187.1	184.9	188.6	214.6	223.5	257.4	15.3
Technology Development Center	410.8	390.8	399.4	398.5	412.7	430.0	448.4	478.2	28.5
State-owned research institutes	208.0	215.8	219.8	234.4	239.5	250.0	259.4	272.6	16.2
Other public funding	117.0	129.7	139.6	136.5	169.9	183.9	208.0	195.6	11.6
Total	1,275.1	1,295.9	1,352.4	1,388.7	1,452.7	1,535.1	1,596.7	1,680.0	99.9

Source: Statistics Finland Database, www.stat.fi/til/kou_en.html.

Note: Percentages may not add to 100 because of rounding.

and Sitra on exploratory research. Finland's regulatory institutions are sensitive to their potential impacts on innovation and pay particular attention to the issues of patents, branding, competition law, and environmental regulation (Ahlbäck 2005; Oksanaen, Lehvo, and Nuutinen 2003; Ormala 2001).

Finland's NIS is also characterized by a large number of intermediaries, such as science and technology parks, technology transfer companies, industry liaison offices, innovation centers, and business incubators. Many of these intermediaries are located near universities to facilitate collaboration (Ahlbäck 2005).

HIGHER EDUCATION

The universities make up the core of the national research system. The quality and volume of research are the key considerations for the higher education system in the broader context of improving the competitiveness of the Finnish economy and, in an even broader European context, of realizing the Lisbon Strategy. The main actors and their policies and programs related to the development of higher education and research in the context of the national innovation system are discussed in this chapter. The Finnish approach to building up excellence in research and strong research universities is based on the context of a national innovation system and illustrates the importance of a holistic approach, the commitment of top-level national leadership, and the development of R&D networks (Ahlbäck 2005; Oksanaen, Lehvo, and Nuutinen 2003).

The Finnish higher education system is made up of two parallel sectors: universities and polytechnics. Historically, scientific research has been concentrated in the universities. In 2005, there were 20 universities in Finland: 10 multifaculty institutions, 3 universities of technology, 3 schools of economics and business administration, and 4 art academies (table 6.2). All universities are research universities and have the right to award doctorates. The university network covers the whole country, including the sparsely populated areas in the northern and eastern parts of the country (Ministry of Education 2005). A series of mergers is under way in the system, led by the establishment of a foundation-based innovation university, Aalto University, from the Helsinki University of Technology, the Helsinki School of Economics, and the University of Art and Design. Aalto University will begin operation in January 2010 with approximately 21,000 students. Other mergers of regional universities and polytechnics are also planned (Dobson 2008; Ministry of Education 2007a, 2008b).

The provision of higher education in Finland is extensive. Each year, approximately 65% of the 19-to-21 age group is enrolled in the higher education system. In 2007, there were 176,300 students enrolled in degree programs at universities

TABLE 6.2.
Personnel at universities, 2005

	Teaching staff		Research staff		Other staff		Students at graduate schools	
	Number	%	Number	%	Number	%	Number	%
Åbo Akademi University	370	4.4	242	3.7	597	3.8	84	5.3
University of Helsinki	1,790	21.2	1,499	23.1	3,726	26.2	388	24.3
University of Joensuu	425	5.0	194	3.0	567	4.0	51	3.2
University of Jyväskylä	785	9.3	422	6.5	988	7.0	173	10.8
University of Kuopio	367	4.3	323	5.0	752	5.3	81	5.1
University of Lapland	213	2.5	63	1.0	339	2.4	12	0.8
University of Oulu	853	10.1	532	8.2	1,426	10.0	145	9.1
University of Tampere	659	7.8	358	5.5	1,003	7.1	100	6.3
University of Turku	847	10.0	498	7.7	1,195	8.4	157	9.8
University of Vaasa	191	2.3	43	0.7	221	1.6	7	0.4
Helsinki University of Technology	487	5.8	1,134	17.4	1,425	10.0	214	13.4
Lappeenranta University of Technology	247	2.9	300	4.6	297	2.1	25	1.6
Tampere University of Technology	365	4.3	674	10.4	745	5.2	122	7.6
Helsinki School of Economics	160	1.9	73	1.1	248	1.7	18	1.1
Swedish School of Economics and Business Administration	104	1.2	17	0.3	92	0.6	8	0.5
Turku School of Economics and Business Administration	118	1.4	74	1.1	131	0.9	6	0.4
Academy of Fine Arts	31	0.4	0	0.0	29	0.2	0	0.0
Sibelius Academy	221	2.6	8	0.1	145	1.0	4	0.3
Theatre Academy	58	0.7	1	0.0	88	0.6	1	0.1
University of Art and Design Helsinki	159	1.9	45	0.7	253	1.8	3	0.2
Total	8,450	100	6,500	100.1	14,267	99.9	1,599	100.3

Sources: Ministry of Education, KOTA Database, www.kotaplus.csc.fi; Statistics Finland Database, www.stat.fi/til/kou_en.html.

Note: Percentages may not add to 100 because of rounding.

(Statistics Finland 2008c). See table 6.3 for data on the number of degrees awarded. The higher education system expanded very rapidly in the 1990s. In 1990, the total number of entrants to universities was 16,000. The corresponding figure in 1998 was 19,400. The growth did not continue at that pace in the next decade, because the national target of the provision of a study place at universities and polytechnics for 60% of the relevant age group had been reached. The number of entrants in 2004 was 20,500. The number of entrants in the field of engineering grew from 3,000 to 4,200 in the 1990s, and in the natural sciences from 1,100 to 2,300, reflecting the changing focus in higher education and research policy and the policy of improving the knowledge base for Finnish industrial development. In 2004, the number of entrants in engineering was 3,900 and in natural sciences 3,600. The variation in figures is partially the consequence of a special funding and development program in these fields during the 1990s and partially due to the growth of degree goals, agreed upon between the ministry and the universities. With respect to graduate education, the university system produced 13,884 master's and 1,524 PhD degrees in 2007. The corresponding figures were 11,515 and 1,156 in 2000 and 8,423 and 490 in 1990.

Polytechnics are higher education institutions with a professional focus and the responsibility to conduct applied R&D for teaching and vocational purposes. The first five Finnish polytechnics gained permanent status in 1996 and the most recent five became permanent in 2000. There were twenty-nine polytechnics in 2005, dropping through reorganization to twenty-six in 2008. Polytechnic degrees are bachelor's degrees in professional fields such as engineering, business, and health care that take 3.5 to 4.5 years to complete (Dobson 2008; Ministry of Education 2008a).

Finnish universities also have numerous specialized multidisciplinary research institutes. These institutes provide chargeable services in research, training, consulting, and so forth. Institutes' budgets are independent of the university and often include industry and other customer financing. The research carried out at such institutes may be either basic or applied and may be aligned with national or regional objectives. One successful example is the Digital Media Institute at Tampere University of Technology (Polt et al. 2001).

Research expenditure in the tertiary education sector (universities, polytechnics, and university central hospitals) has grown accordingly from some €400 million in the mid 1990s to about €1,079.2 million in 2006 (Finnish Science and Technology Information Service 2007a). The majority of research personnel in higher education are in the fields of natural sciences, engineering, and medicine—fields with a high potential value for industrial innovation (Polt et al. 2001).

TABLE 6.3.
Academic degrees awarded in universities, 1991–2005

	1991	1995	2000	2001
Master's level degrees	8,410	9,819	11,515	11,581
Postgraduate students, total	11,839	15,927	20,537	21,008
Licentiates	604	793	748	695
Doctorates	524	765	1,156	1,206
	2002	2003	2004	2005
Master's level degrees	12,075	12,411	12,588	12,920
Postgraduate students, total	21,937	22,960	22,105	22,196
Licentiates	654	606	558	533
Doctorates	1,224	1,257	1,399	1,422

Source: Ministry of Education, KOTA Database, www.kotaplus.csc.fi.

All Finnish universities are state-run institutions and are primarily financed from the national higher education budget. The polytechnics are either municipally or privately run and are cofinanced by the government and local authorities. By law, the institutions of higher education are not allowed to charge tuition fees. Parliament passes educational legislation and decides on the overall lines of education and research policy. The universities are governed by the Universities Act and the polytechnics by the Polytechnics Act. The main policy guidelines and development targets are determined at a general level in the Development Plan for Education and Research, which is adopted by the government for a six-year period and revised every four years (Ministry of Education 2004b, 2005).

The most important goals for universities are the annual average numbers of master's and PhD degrees to be awarded during the contract period. The goals are set for each broad study field, such as the humanities, social sciences, natural sciences, and engineering. The master's degree goals—and performance—are aimed at reflecting the educational productivity of universities. The PhD goals have been thought to reflect research productivity at universities. More qualitative national goals are also agreed upon. They are related to main political goals, based on the Development Plan for Higher Education and Research approved by Parliament, and are set for the whole higher education system. Development programs and projects of national importance, nowadays based on the implementation of national programs and their funding, are also agreed upon in these contracts. These programs represent program steering, and they are, in some cases, linked to more general national policy goals and cooperation with other ministries (Höltta 1998).

Funding of the national programs, as agreed in the performance contract, constitutes another central steering element related to funding. The national programs for the contract period of 2004–6 included the development of studies (the two-cycle degree structure, student selection, and quality assurance), the virtual university, the development of teacher training, a national health project, language technology, business know-how, biotechnology, the Russia action program, the information industry, and regional development. The priority in national programs has gradually transferred to research-related programs (Ministry of Education 2005).

Also, from the very beginning, the budget has included a performance component, which has allocated funding to universities on the basis of performance indicators. The indicators used have changed from one contracting period to another and have reflected the changing priorities of the national higher education policy. In addition to the use of PhD degrees in the funding model to measure research productivity, quite a few of the performance indicators have been linked to research output. Such indicators have included the publication rate and the ability to attract external funding, in particular from the Academy of Finland. The following criteria were used for the period of 2004–6 in the allocation of performance funding: quality of research and artistic activity, including designation of Centers of Excellence in research by the Academy of Finland; Academy of Finland financing; other external research funding; designation of Centers of Excellence in artistic activity by the Arts Council of Finland; quality, efficiency, and effectiveness of education; designation of Centers of Excellence in education by the Higher Education Evaluation Council; designation of Universities of Excellence in adult education by the Higher Education Evaluation Council; progress in studies (credit accumulation, the proportion of bachelor-level graduates, and duration of studies before graduation); graduate placement (the unemployment rate and changes in it); and internationalization (student exchanges, balance in exchanges, and degrees awarded to foreign students).

The first concrete steps in the development of the new relationship between government and universities were taken toward increased academic and financial autonomy: transferring power in academic affairs such as curriculum development to universities, and enhancing financial autonomy in the form of lump-sum-budgeting development. However, the funding model, based directly on the performance agreements with the universities, has constituted a powerful centralized tool for government to implement the national goals within the higher education system. In fact, although the universities have the power to allocate the budget internally as they wish, the increasing share of the national program in the performance contracts has gradually decreased the flexibility created by the block grant. Institutional autonomy has become very much conditional. In principle, the universities have extensive formal autonomy to use their educational and research resources for the desired purpose, but they have had to be committed to the national goals agreed upon with the ministry in the performance contract. The establishment of new fields of study at institutions is an essential variable that is still in the direct control of government, and one that affects the research profiles of universities quite directly.

Another element of the current steering system, representing self-regulation, is the system of evaluation of higher education, research, and institutions of higher education. It represents qualitative regulation based mainly on the authority of the academic community itself, balancing the quantitative steering of the ministry in the form of management-by-results, where the contracting and funding models have the most visible roles.

The first practical element in the introduction of market coordination to the new steering system of higher education and research has been realized in the form of incentives encouraging the universities to enter competitive markets in research and adult education. The first element in the transition toward a more flexible environment, in terms of using resources in the Finnish university system, was the implementation of the lump-sum-budgeting system in the early 1990s and, soon after that, the system of flexible workload. The system of flexible workload made it possible to include externally funded assignments in the annual work plan with the purpose of creating new revenues for the departments and providing a more flexible environment for research (Höltta 1998).

Another element aiming at improving the possibilities for universities and their departments to enter the marketplace was the net budgeting system. Instead of the long-established system, in which the incomes from external services were transferred to the centralized government account and the government budgeted an approximately corresponding amount to the university's account, the universities were given the right to keep the incomes, including profit, to create incentives for contract research and adult education. Government is also encouraging universities to increase income-generating activities by setting institutional goals for net incomes. Currently, the goal for the annual return for each university is part of the performance agreement. It is stated in law that universities are not allowed to subsidize any services provided to external customers. Since fall 2005, the universities have been allowed to own shares of firms, making it possible for them to operate in the markets for educational and research services, but this opportunity has not yet been used by the institutions.

Already the increased financial autonomy—making operation in the marketplace possible and more attractive—has resulted in increased external funding. The total amount of external funding of universities has increased since 1994 from €283 million to €564 million in 2000 and €690 in 2004, including national research funding from the Academy of Finland and TEKES, resulting in growth from 30.6% to 35.8% of total funding. The amount of funding from outside the higher education and research budget grew from €377 million to €545 million between 2000 and 2004. In 2004, the total amount of funding from the private sector was €201 million, representing 10.4% of the revenues to universities. There are remarkable differences between institutions in these figures, reflecting the funding structure (table 6.4). Some technical universities receive almost half of their incomes from outside the higher education budget (Ministry of Education 2008b; Statistics Finland 2007).

Today, Finland is one of the countries that has kept the welfare society principles as a cornerstone of economic and social development. Within the educational sector this means, for example, that education is free up to and including the level of universities, and government subsidizes the living costs of the students quite heavily by a package composed of a grant, a housing subsidy, and a loan (Höltta 2000). A special characteristic of the Finnish higher education development policy, reflecting the importance given to higher education and research, has been that the growth of resources has been guaranteed by legislation since the mid-1960s.

As indicated above, gradual changes have taken place in the direction of introducing market discipline into the Finnish higher education system, but the political climate is not yet favorable to tuition fees. A ministerial working group has made a proposal about tuition fees for non-European students, but implementation requires a change to the University Act. However, it is quite possible that financial autonomy might be increased radically within a couple of years, which may improve the opportunities for Finnish universities to operate in the marketplace in continuing education and applied research without the present restrictions, such as restrictions on salaries paid. All these changes would also increase the opportunities for universities to operate as part of the innovation system, which has been under extensive development during the past ten years or so.

The Policy Framework

This section describes the policy framework of the Finnish ARE, starting with the history of the higher education and innovation systems. It then discusses research policy, including support for higher education and research; doctoral education; and knowledge transfer policy, including key institutions and intermediaries.

TABLE 6.4.
Expenditures and Academy of Finland funding decisions at universities, 2005

	Expenditure (budget funding)		Expenditure (external funding)		Academy of Finland funding	
	€M	%	€M	%	€M	%
Åbo Akademi University	47.6	3.8	28.5	4.1	5.4	3.2
University of Helsinki	304.0	24.1	190.9	27.5	61.1	36.8
University of Joensuu	56.8	4.5	20.7	3.0	5.9	3.5
University of Jyväskylä	95.2	7.5	50.5	7.3	16.6	10.0
University of Kuopio	51.3	4.1	47.5	6.8	8.1	4.9
University of Lapland	30.9	2.4	10.6	1.5	1.2	0.7
University of Oulu	125.7	10.0	56.7	8.2	10.5	6.3
University of Tampere	84.4	6.7	43.8	6.3	9.1	5.5
University of Turku	107.1	8.5	52.2	7.5	14.2	8.5
University of Vaasa	22.0	1.7	6.1	0.9	0.4	0.2
Helsinki University of Technology	116.0	9.2	91.1	13.1	25.2	15.2
Lappeenranta University of Technology	35.1	2.8	21.5	3.1	0.9	0.5
Tampere University of Technology	68.2	5.4	44.3	6.4	4.8	2.9
Helsinki School of Economics	24.8	2.0	11.1	1.6	1.1	0.7
Swedish School of Economics and Business Administration	12.8	1.0	2.8	0.4	0.6	0.4
Turku School of Economics and Business Administration	15.0	1.2	6.2	0.9	0.8	0.5
Academy of Fine Arts	3.8	0.3	0.2	0.0	n/a	n/a
Sibelius Academy	23.4	1.9	2.5	0.4	0.02	0.2
Theatre Academy	10.7	0.8	0.6	0.1	n/a	n/a
University of Art and Design Helsinki	27.1	2.1	6.3	0.9	0.3	0.2
Total	1,261.9	100	694.1	100	166.2	100.2

Sources: Ministry of Education, KOTA Database, www.kotaplus.csc.fi; Statistics Finland Database, www.stat.fi/til/kou_en.html; Academy of Finland Database, www.aka.fi.

Notes: Academy of Finland funding includes estimates of the costs associated with research posts. Detail may not add to total and percentages may not add to 100 because of rounding.

HISTORY

History of the Higher Education System. The Finnish higher education system has similar roots to those of other Nordic systems. The current University of Helsinki was established originally in Turku in 1640 by the Royal Government of Sweden, when Finland belonged to the Kingdom of Sweden. The concept of higher education in Finland is based on the German university model. The influence of the Humboldtian assumption about the unity of teaching and research has strongly characterized the development of the Finnish higher education system and its structures (Ministry of Education 2005).

The status given to higher education, research, and the academic profession by the state is illustrated by the fact that a right of autonomy was granted to the University of Helsinki in the constitution of independent Finland in 1919. While some autonomy was constitutionally guaranteed early on, achieving the optimal level of university autonomy is still an active policy issue (Ministry of Education 2005, 2007b). For example, "The new Universities Act, promulgated in 1998, increased university autonomy by delegating many matters previously regulated by separate Acts and Decrees to the university decision-making bodies. The distribution of educational responsibilities between the universities is still determined by decree. In return for the larger latitude in resource use, the universities agree on target outcomes with the Ministry, monitor the achievement of the targets and report on them to the Ministry" (Ministry of Education 2005, 9).

The history of Finland's higher education system can be understood in terms of three phases of expansion. In the first phase, motivated by the desire to implement a post–World War II rebuilding program in the context of a Nordic welfare society, Finland established a comprehensive planning system, and higher education was given a central role in the development policy.

The second phase of expansion, with a strong emphasis on regional policy, began in the 1950s. Expansion was fuelled by change in the country's economic structure, high demand for an academically trained labor force, an increase in the number of matriculated students, and a general rise in the standard of living. The University of Oulu was established in northern Finland in 1956. Today, there is a major concentration of high-technology industries cooperating closely with the university research laboratories in Oulu (Ministry of Education 2005).

Several major decisions on expansion and placement were subsequently made in the 1960s and 1970s. The network of research-based universities was established to cover the whole country. Higher education policy was tightly linked to regional development policy, with the aim of keeping the whole country economically viable and preventing the population from excessive concentration in the south (Ministry of Education 2005).

This second phase of expansion was accompanied by further development of the government planning system and the system of centralized higher education administration. The importance of higher education and research was emphasized in national development. In the 1960s, the Ministry of Education became one of the most significant government actors, leading to national educational, science, technology, and cultural policies; planning machinery; and a policy implementation

bureaucracy. The role of the educational sector was seen as important in supporting the developments of the new industrial society while guaranteeing the balanced
development of the rural regions in the form of regional development policy. The
process of establishing a network of universities covering even the most sparsely
populated areas was the most visible feature of this policy (Ministry of Education
2005).

In this second expansion phase, Finland implemented the transition toward a
mass higher education system, that is, the provision of wide access and equal opportunities for young people in all parts of the country, by using the very heavy,
centralized implementation bureaucracy concentrated in the strong Ministry of
Education. The consequent national higher education system, although made up
of autonomous universities, was characterized by the features of a single multicampus national university, as expressed by an OECD (Organisation for Economic
Co-operation and Development) review group in the early 1980s (OECD 1982).
The system was marked by extreme homogeneity in terms of infrastructure and
resources, all defined at the level of legislation, the Development Act on Higher
Education of 1975–86. However, the advantage of strong government control and
the supervisor-subordinate relationship between the ministry and the universities
had been that, during the period when the current institutional and educational
structures were created, national needs and political priorities could be effectively
implemented in the system through the administrative machinery. Although the
roots of the higher education system were in the Humboldtian model, research was
not a priority and was a function very much based within individual universities.

The new higher education policy, formally initiated as the Higher Education
Development Act of 1987–96, and the associated Decision of Government restored
trust in universities by allocating additional resources to them. Simultaneously, it
established the basis for the development of new steering instruments and accountability measures to replace the direct-control machinery. Government required that
the new resources mainly serve to improve the conditions of research and postgraduate education and to increase efficiency of performance. Furthermore, the conditions for goal-oriented management in higher education were improved, greater opportunities were provided for the institutions of higher education to independently
decide how they would use the funds they received, and all institutions of higher
education were required to implement a system of performance evaluation that
produced sufficient and comparable data on the output and costs of research and
teaching. In allocating new resources, the results achieved in research and teaching
were to be taken into account, and existing resources were to be redeployed according to changing needs (Hölttä 1988, 1998).

This set of goals for the development of government steering instruments for the higher education system was in line with the ideas of New Public Management (Lane 2000), and corresponding measures were developed for all sectors of public administration in Finland. (This framework also characterized the developments in the 1990s. In fact, the execution of many of the policy goals decided in the 1980s had to wait until the 1990s for implementation.) The policy had to be implemented as part of the national survival strategy after the recession of the early 1990s. The policy instruments, which were developed on the basis of the above-mentioned decision of government, were a mix of deregulation and centralized steering instruments in line with the practices of other European countries and characterized by the conditions of self-regulation (Neave and Van Vught 1991).

The main policy instruments used in the implementation of the new higher education policy goals were contracting with the Ministry of Education system and a funding model that linked funding to a set of goals. Also, a quality assurance system was established, and a high priority was given to the development of an information provision infrastructure for a new kind of interaction and dialogue between government and the universities (Hölttä 1998).

Contracting between the Ministry of Education and each university is the ministry's main policy instrument. All other instruments of the present steering system are closely connected to it. The following quotation illustrates the targets set for the new steering system and budgetary process and the practice applied in the late 1980s and the beginning of the 1990s: "The Ministry of Education and the higher education establishment collaborate to introduce a certain form of management by objectives. The ministry negotiates the objectives to be set and the funds needed to implement them with institutions. Later, the institutions will report on how well the objectives were reached and what the costs were. In dealing with the institutions of higher education, the Ministry is gradually abandoning its former policy of administrative supervision and increasingly replacing it by dialogue with institutions" (Jäppinen 1989, 336).

The core of the new steering and planning practice is based on the model of steering-by-results. Based on national policy goals from the Development Plan for Higher Education and Research submitted by Parliament for a period of six years, the three-year plans of the universities, and proposals from institutions describing the institutional goals, the representatives of the ministry and the rector of each university negotiate and finally sign a performance agreement confirming funding and goals for the institution (Hölttä and Rekilä 2003).

The third stage of the expansion of the national higher education system took place in the mid-1990s in the form of establishing the polytechnic sector. By the early 1990s, Finland was one of the few European countries with a uniform higher education system consisting of universities only. The main reasons for the late establishment of the nonuniversity sector of higher education can be found in the academic tradition of Finnish higher education and from the policy decisions made in the 1960s and 1970s concerning the regionalization of universities. The idea of higher education based on the Humboldtian unity of research and teaching has been deeply rooted in the Finnish academe. This view has had concrete implications for the institutional structure of the system throughout the history of Finnish higher education. The Finnish solution of establishing a mass higher education system in the 1960s and 1970s was different from that of its Nordic neighbors with similar regional challenges, that is, large sparsely populated and less-developed areas outside the southern industrialized areas. In the 1960s and 1970s, using the same regional policy arguments as Finland in their university reforms, Sweden and Norway set up nonuniversity sectors composed of regional colleges without a research function to extend higher education to the less-developed areas in the north (Hölttä 1999).

At the outset, polytechnics were considered mainly as teaching organizations, but their role in research has increased very rapidly over the past few years. Their statutory tasks now include R&D. On the whole, the polytechnics' role in the national research system remains quite small. They still have two main problems in research. The first is a lack of basic funding for research activities, and the second is a lack of tradition. Genuine research communities in polytechnics are only slowly emerging. However, the regional development programs, in the context of creating the regional innovation systems in which the universities have a major role, are providing the polytechnics a basis for applied research advancing local industries (Ministry of Education 2005).

History of the Innovation System. The Finnish economy underwent major changes in the 1990s. The reason for the deep recession in 1991–92 can be found partly in international economic fluctuations, but what made the international recession particularly difficult in Finland was the simultaneous collapse of the Soviet Union, with which Finland had extensive trade relations, importing energy and exporting consumables and technology. As a result, many traditional industries broke down in Finland. For example, the government had to save the banking sector from a total disaster by extremely heavy subsidies. The main implications affecting higher education were high unemployment (20% at the highest, after a growth period with almost full employment) and the need to restructure Finnish industries (Lavery 2006; Singleton and Upton 1998). Budget cuts for the educational institutions and the ministerial result-oriented steering strategy pushed universities to earn a higher share of funding from external sources (Ahlbäck 2005).

The post–World War II history of Finland shows, however, that political parties have consistently preserved the main structures and functions of the Nordic welfare society whenever they have been threatened. The establishment of the national recovery strategy, based on the restructured welfare state instead of another discussed alternative—extreme market orientation or "the New Zealand model"—was a major achievement of government. A central element of the subsequent government policy was to take high-technology and knowledge-intensive production as the national priority in the economic recovery program. This was an obvious choice after the policy directions already established in the 1980s.

In 1994, preceded by an initiative by the prime minister and government strategy documents concerning the general national development strategy and the role of information technology in the process of realizing the information society, the Finnish Ministry of Education set up an expert committee to prepare a national strategy for education, training, and research in the information society. This strategy process was closely integrated with corresponding strategy development processes of other ministries, all within the framework of the objective set by the government, with the grand goal that Finland should become the leading nation in applications of information technology (Ministry of Education 1995).

The mid-1990s were important to Finland in many ways. Finland joined the European Union in 1995. This changed the operating environment for Finnish institutions of higher education, although Finland had already been participating in European higher education programs. The European student-exchange programs and research funds became fully available to Finnish universities and polytechnic institutions. Since the early 1990s, they had established and developed programs in English to be able to satisfy the demands of international students. The national coordinating body, the Centre for International Mobility, was established to support the change process at the institutional level. The structural development funding of the European Union became available for regional development, and the universities and polytechnics had a major role in the implementation of the programs. The government supported the process of more effectively integrating the knowledge created at higher education institutions with the development of the business sector by restructuring regional administration. The main regional functions of different ministries were merged at the regional level. Also, a network of centers of expertise was established by the government to integrate expertise at higher education institutions into the regionally specified industrial and business clusters to promote economic growth. The network meant that universities and polytechnics were given a new key role in the implementation of regional policy in this new framework, in particular, in the implementation of the EU-funded development programs. They were seen as the main sources of academic and professional expertise. The centers are linked with the science and technology parks established in university towns. The establishment of technology centers has been part of the policy of government and regional authorities since the mid-1980s.

For Finland, the adaptation of higher education policy to the framework of the Lisbon Strategy of the European Union has been quite natural. The Lisbon agenda involved many reforms, such as the establishment of effective internal markets, an improvement of the educational system, and a more productive innovation and research base. This European strategic framework has provided Finnish higher education and research policy a natural continuation, linking it even more closely to economic and industrial policies.

RESEARCH POLICY

Finland's ARE can be understood in terms of the nation's research council, the Academy of Finland; its Science and Technology Policy Council; and the country's financial and human resources devoted to research.

The Academy of Finland. The Academy of Finland, the Finnish research council organization, was established in its present form in 1970. The Academy takes care of the central research administration and finances a major part of university research. Its mission is to advance top-quality research, and it uses competitive funding, reliable evaluation, and expertise in science policy to achieve its goals. Other priorities of the Academy include enhancing the multidisciplinary and international nature of research in Finland. Its 2007–15 strategy emphasizes the dependence of welfare and social development on the advancement and application of scientific research (Academy of Finland 2007a, 2007b).

The Academy primarily funds individual researchers working at universities. The total budget of the Academy was €264 million in 2007, a 10% increase over the year before. Of this total, 39.5% went to research projects, 12.5% to postdoctoral researchers, 11.2% to fund research posts, 6.7% to membership in international organizations, and 6.4% to research programs. Approximately 80% of the Academy's funding was allocated to universities in 2007. The Academy's share of Finland's total public research funding was 16% in 2007 (Academy of Finland 2007b).

The Academy has four research councils, each financing research in its disciplines: biosciences and environment, culture and society, natural sciences and engineering, and health. Council members are appointed by the government for three-year terms. Research programs are the key instruments of the Academy of Finland for allocating research funding to support the national higher education policy goals. The policy change since the 1980s has been characterized by the transition from a project-by-project funding model to thematic programs. New programs are launched annually. A research program consists of a number of interrelated research projects focused on a specific subject area. The promotion of multidisciplinarity, interdisciplinarity, and international cooperation are emphasized in the design of the programs, which aim to create and strengthen the knowledge base, promote professional careers in research, facilitate the networking of researchers, intensify researcher training, and support the establishment of creative research environments (Academy of Finland 2007b).

Research programs are usually scheduled to run for three or four years. Other funding bodies, both Finnish and foreign, are also often involved in the programs. In 2007, the Academy had fourteen research programs. Two programs newly introduced that year were "Power and Society in Finland" and "Substance Use and Addictions." The Academy also accepted applications for 2008 funding in "Sustainable Energy" and "The Future of Work and Well-Being." The "Sustainable Energy" program will be the first to directly incorporate funding from business enterprises (Academy of Finland 2007b).

The Academy has a strong focus on internationalizing Finnish research. This includes maintaining Finland's connection to research initiatives at the Nordic and EU levels. For example, the Academy has responsibility for several programs and subprograms under the EU Framework Program FP7, participates in the European Research Council and the European Science Foundation, and provides funding to five Nordic Centers of Excellence. In addition to involvement in these regional initiatives, the Academy interacts with nations around the globe, connecting the Finnish research enterprise to Japan, China, India, the United States, and other countries. The Academy of Finland is also responsible for the national system of graduate schools, that is, national PhD programs, which is discussed in the doctoral education section.

Jointly with TEKES, the Academy funds international researchers (twenty-eight in 2007) at Finnish universities and research institutes through the Finland Distinguished Professor (FiDiPro) program (Academy of Finland 2007b). Making the international academic labor market work better in Finland has been a national concern. The FiDiPro funding program is jointly managed by the Academy of Finland and TEKES with the aim of recruiting professor-level top researchers, either foreign or Finnish, who have long worked abroad. As expressed by the Embassy of Finland (2006, para. 6): "The goal of the funding program is to strengthen scientific and technological knowledge and know-how in Finland as well as to add a more international element to the Finnish research system, generate added value into the national innovation system and support research-driven profiling of universities and

research institutes. Another aim is to create a new kind of international cooperation between basic and applied research as well as between corporate research and
development activities."

Another important task for the Academy is to evaluate research. The Academy of
Finland undertakes a review of Finnish research every three years in accordance with
the Government Resolution on the Development Plan of Education and Research
and the performance agreements between the Academy and the Ministry of Education. In addition, the Academy organizes disciplinary reviews that look at the inputs
made in the field, research processes, and results. The Academy of Finland has the
responsibility to organize the national program for the evaluation of research. The
Higher Education Evaluation Council has responsibility for developing and coordinating the evaluation of the institutions of higher education and study programs.
Additionally, as the universities have been developing their capacities in quality
assurance, the Evaluation Council is now running the program of auditing the institutional quality assurance systems. Research evaluation has been strongly focused
on fields that have been seen as most important to the social and economic developments of Finland, like biotechnology and certain fields of engineering (Academy of
Finland 2007b).

Research Funding in Figures. The development of R&D spending in Finland
is portrayed in table 6.5. Finland's input into R&D has grown substantially from
€1,711 million in 1991 (Statistics Finland 2007). Comparing 2005 data on R&D
expenditure as a percentage of GDP, Finland's 3.48% was the second-highest among
the OECD countries after Sweden's 3.89%. The average figure for the EU27 was
1.74%, and the percentage for the United States was 2.62% (OECD 2008). In 2006,
the total R&D expenditure in Finland was €5,761 million, representing 3.45% of
GDP. The growth of investments in R&D can mainly be explained by the rapid
increase of private-sector investments, but the public sector has also increased its
funding significantly. Business enterprises account for 70% of R&D expenditure
(Statistics Finland 2007). The substantial increase in R&D in the business enterprise sector has been mainly due to the information and communication technologies (ICT) sector and more specifically to certain major actors in the field (e.g.,
Nokia).

Government R&D expenditure, as a proportion of overall government spending exclusive of debt servicing, stands at 4.4% (Statistics Finland 2008a). In the EU
countries, the public R&D funding as a percentage of GDP was highest in Finland,
1.03 % in 2005 (Wilén 2008). Out of the R&D budget, 25.2% is allocated directly to
the universities, 16.5% to the Academy of Finland, and 29.3% to TEKES. The main
recipients of the remaining 29% are state research institutes and university central

TABLE 6.5.
R&D spending, by sector, 1991–2006

	Total (€M)	R&D as % of GDP	Enterprise		Public		Higher education	
			€M	%	€M	%	€M	%
1991	1,711	2.0	975	57	358	21	378	22
1992	1,747	2.1	992	57	372	21	384	22
1993	1,796	2.2	1,049	58	380	21	368	21
1994	2,008	2.3	1,250	62	380	19	379	19
1995	2,172	2.3	1,373	63	374	17	425	20
1996	2,504	2.5	1,657	66	395	16	452	18
1997	2,905	2.7	1,917	66	409	14	580	20
1998	3,354	2.9	2,253	67	444	13	658	20
1999	3,879	3.2	2,644	68	470	12	765	20
2000	4,423	3.4	3,136	71	497	11	789	18
2001	4,619	3.4	3,284	71	501	11	834	18
2002	4,830	3.5	3,375	70	530	11	926	19
2003	5,005	3.5	3,528	70	515	10	962	19
2004	5,253	3.5	3,684	70	530	10	1,040	20
2005	5,388	3.5	3,770	70	538	10	1,080	20
2006	5,761	3.5	4,108	71	574	10	1,079	19

Source: Statistics Finland Database, www.stat.fi/til/kou_en.html, Science and Technology Statistics.
Note: Detail may not add to total because of rounding.

hospitals (Statistics Finland 2008a). In 2005, the Ministry of Education's allocation to research was €640 million, representing 40% of total government funding for R&D. The Ministry of Education research funding went primarily to universities and the Academy of Finland (Ministry of Education 2005).

Total university funding, including funding to education, was €2,101 million in 2007, including funding from the state budget and external financing. In 2007, budgetary funding accounted for 64.7% of total funding. Broken down by performance areas, total university spending in 2007 was as follows: teaching, 36.2%; research and artistic activity, 54.1%; and societal services, 9.7%. The university core funding was teaching, 48.0%; research and art, 42.3%; and societal services, 9.7%.

Compared with the core funding of the universities and government research establishments, competitive research funding from the Academy of Finland and TEKES has increased much faster. Funding from the Academy to universities and university hospitals continued to grow rapidly, from €150.9 million in 2001 to €211 million in 2007 (Academy of Finland 2001, 2007b). Funding from TEKES to universities, research institutes, and polytechnics was €185 million in 2007 (Helminen et al. 2007). Consequently, the Academy of Finland and TEKES are the most important sources of external research funding in universities.

Science and Technology Policy Council. The coordination of the national innovation system in Finland is in the hands of the highest political leadership. The national science, technology, and innovation policies are formulated by the Science and Technology Policy Council, chaired by the prime minister. The Council advises government and its ministries on questions relating to science and technology, the general development of scientific research and researcher training, and Finnish participation in international scientific and technological cooperation.

In its reviews, the Science and Technology Policy Council has stressed the importance of higher education institutions' cooperation with business. According to the Council, effective contacts and flexible cooperation forms have a significant effect on the spread of research findings and, with it, on the emergence of knowledge-intensive business. Such connections further strengthen the capacity of the research system to react to changes in business life and to respond to research challenges arising from business and industry. Therefore, universities, research institutes, and polytechnics must be developed as active and dynamic cooperation partners of business and industry with balanced development of research organizations' resources. The rules of the game in cooperation must be clarified and developed to provide more incentives. These principles have also established the goals for the development of academic research at universities in recent years (Science and Technology Policy Council 1987, 1990, 2000).

Human Resources in the Finnish Higher Education and Research System. In 2006, there were nearly 80,000 persons employed in R&D in Finland, with over 40,000 of those employed by business enterprises (Finnish Science and Technology Information Service 2007b). In universities, the teaching staff was 8,450 FTEs in 2005. Other personnel numbered 21,000 in all, of whom 27% were professional researchers and 7% were full-time PhD students. Figures on research personnel by sector in 1995–2004 are displayed in table 6.6.

The number of foreign researchers working within R&D is considerably lower in Finland than in the other EU countries (Davies et al. 2006). With a view to attracting foreign researchers, which is an important objective for Finland, measures are being taken to promote a high level of R&D and high-standard research conditions and to eliminate obstacles to the recruitment of foreign researchers (Academy of Finland 2008; Ministry of Education 2005).

The proportion of R&D staff with a PhD increased during the early 1990s, jumping from 8.8% to 10.5% between 1991 and 1993. However, this proportion has remained stable since 1993. In 2001, with 7,400 research jobs held by researchers with PhDs, the share remained at 10.5% (Academy of Finland 2003).

In 2001, over 60% of R&D staff who had a PhD worked in universities. The number of R&D staff with a PhD outside the university sector was 2,854. Both

TABLE 6.6.
R&D headcount, by sector, 1995–2004

	1995	1997	1998	1999	2000	2001	2002	2003	2004
Business enterprise	24,243	29,139	32,430	36,406	38,169	37,971	39,239	40,089	40,674
Public	8,902	9,666	10,295	10,523	10,096	10,300	10,756	10,635	10,715
Higher education	14,721	16,685	18,165	20,036	20,548	21,517	23,126	24,049	25,297
Total	47,866	55,490	60,890	66,965	68,813	69,788	73,121	74,773	76,686

Source: Statistics Finland Database, www.stat.fi/til/kou_en.html.

the absolute number of people with a PhD and the proportion of R&D staff in private-sector companies with a PhD have increased. In 1991, 452 (11%) PhDs in R&D worked in business enterprises. By 2001, such employment had increased in absolute and relative terms to 1,030 (14%) (Academy of Finland 2003).

Finland had the highest number of researchers per thousand labor force participants in the OECD, 16.5 in 2005, followed by Sweden, Japan, and New Zealand. The average figure for EU27 countries was 6 (OECD 2008). Long-term follow-up data indicate that PhD placement tends to concentrate in the public rather than the private sector (Husso 2002). Part of the national policy is to enhance conditions for professional researcher careers. In 2004, the Ministry of Education appointed a committee to propose a strategy for developing professional researcher careers, to make research an attractive career choice, and to ensure varied expertise for the public and private sectors.

The committee identified the following concerns as the most significant barriers: "short terms of employment, obstacles to intersectoral mobility, difficulties in combining external research funding and career development, career advancement of women researchers, a low degree of international mobility, a small number of foreign researchers in Finland, attractiveness of research career, economic position of researchers, and the volume of researcher training" (Vuorio, Halinen, and Ylikarjula 2006, 6). Limited intersectoral mobility primarily refers to barriers in movement from industry to the public or university sector due to limited salary flexibility. Finland actually has a relatively high level of mobility from science to industry, attributed to "long-term oriented and stable relations between enterprises or industrial sectors and universities in graduates' mobility; . . . close co-operation in graduate education between universities and industry (including placements); and . . . the existence of coordinating structures for considering industry needs and changes in industry demand, in university education programmes" (Polt et al. 2001, 111). Such mobility is often facilitated by personal contacts resulting from joint research.

Sponsored and invited professorships and adjunct instructors funded by industry are emerging approaches to intersectoral mobility (Polt et al. 2001).

The committee framed its recommendation in terms of a four-stage research career model encompassing young researchers writing doctoral dissertations in the first stage, researchers with recently completed doctorates in the second stage, independent researchers and educators providing academic leadership in the third stage, and professorship as the fourth stage (Lehikoinen and Kaunismaa 2008).

Academic labor markets are quite poorly developed in Finland because the universities have not had the authority to make salary decisions. The formal status of the faculty and staff of universities is that of civil servants (Polt et al. 2001). For example, salaries are based on centralized salary scales that have been negotiated for the whole public sector by government (as an employer) and the unions. Because Finnish universities have not been able to use financial rewards in recruitment, the institutions have quite restricted means to compete for talent.

A new performance-based salary system that evaluates individual performance in research, teaching, and service is now in the implementation phase. The final salary will be composed of two components: one that measures the requirements of the job and another that measures the performance of the person in comparison to colleagues. Research output is a significant factor in this assessment. It is the duty of the department head to negotiate with each faculty and staff member in his or her department. Finally, any differences in scales used in different departments will be harmonized at the institutional level by the representatives of the employer and of unions.

DOCTORAL EDUCATION POLICY

Training of researchers is also one of the focus areas for strengthening research in Finland. The graduate school (national doctoral program) system has been evaluated very recently by an international peer group. The group believes that the graduate school system strengthens the Finnish doctoral education system and should be continued. From a US perspective, the use of the term *graduate schools* as applied in Finland is, however, a misnomer. The graduate schools in Finland are not graduate schools in the US meaning of the term, but rather collaborative doctoral programs. The tradition of the "mentor-apprentice" system of doctoral education, as well as the academic authority vested in the individual professor and the department or faculty, is still a visible feature of doctoral education in Finland (Dill et al. 2006).

In 2007, Finnish universities awarded 1,500 PhDs (Statistics Finland 2008b). The Development Plan for Education and Research for 2003–8 set a target of 1,600 doctorates in 2008 (Ministry of Education 2004a). "In 2000, the number of new doctorates completed per thousand of the population aged 25–34 was 1.1 in Finland, the second highest in the European Union. Sweden's ratio per 1,000 population aged 25–34 was 1.2, and the EU average was 0.56" (Academy of Finland 2003, 10).

Problems were identified in Finnish postgraduate education in the beginning of the 1990s. Consequently, in 1995, a new graduate school (doctoral program) system was introduced to improve researcher training and the national research infrastructure. Smaller institutions, both old and new, had reported difficulties in managing doctoral programs with only their own resources. In public debate, the representatives of industries criticized the long training period needed for a doctorate and the separation of studies from the production sector. The main goals in establishing the graduate school system were to improve the relationship between research and work and to raise the quality and efficiency of postgraduate education, making it possible for students to obtain their doctorates more rapidly than before. For example, the median age at dissertation was about thirty-seven years in the early 1990s (Ministry of Education 1995).

The graduate schools are mainly based on the network model. For each school, there is one coordinating university and its department that has the main responsibility for the program. It also takes care of the school's administration. The students of other cooperating universities in the respective field participate in the program and perform the degree at the home university. The graduate school programs are linked to centers of excellence in research at universities and to high-standard research projects forming comprehensive national networks. The centers of excellence are nominated by the Ministry of Education based on the evaluations and proposals by the Academy of Finland (Oksanaen, Lehvo, and Nuutinen 2003).

The system has gradually been expanded, and the number of schools is now double what it originally was. At the beginning of 2006, there were 124 graduate schools funded by the Ministry of Education, with 1,458 researcher-students supported by the ministry. All in all, over 4,000 postgraduates with funding from various sources are pursuing full-time studies in these programs. In 2000, 320 postgraduate student places were aimed at information industry branches and roughly the same number at biotechnology branches (Academy of Finland 2003).

In 2000, 60% of new PhD graduates took less than four years of full-time work to complete their degree. Around 30% of PhDs graduating from graduate schools received their doctorates before age 30. The average age of graduating PhDs had dropped to 32.4 years (Academy of Finland 2003).

TEKES, Sitra, and the Centres of Expertise are key institutions of Finland's knowledge-transfer policy. These and other institutions contribute to strong local innovation systems.

TEKES, the Finnish Funding Agency for Technology and Innovation. While the Academy of Finland mainly concentrates in its funding policy on support for academic research, TEKES is the main public funding organization for research and development that more closely supports industrial development. TEKES funds both industrial projects and projects at universities and other research organizations. At universities, funding from TEKES is specifically targeted to engineering and natural sciences projects. In particular, it promotes innovative, risk-intensive projects. Innovativeness of high-growth companies, regional development based on skills and knowledge, international cooperation, and industrial renewal are the key words characterizing its policy to increase the productivity of the Finnish economy and to promote the well-being of its citizens. The funding model of TEKES is pushing universities to be more closely linked to the national research system and industrial companies (TEKES 2007).

The activities and funding of TEKES have been mainly in the form of technology programs. Technology programs are used to promote development in specific sectors of technology or industry and to pass on results of the research work to business in an efficient way. The programs promote cooperation and networking between companies, universities, and research institutes; strengthen technology transfer; and support international expansion (TEKES 2007).

In 2007, there were twenty-nine ongoing TEKES programs. Formerly known as technology programs, these were renamed to reflect the agency's broadened focus and expertise. TEKES focused its program funding on projects with small and medium-sized enterprises (SMEs) and those that advanced the internationalization of research. New programs starting in 2007 and 2008, reflecting the current priorities and characterizing the policy focus, were "New Biomass Products," "Concepts of Operations," "Digital Product Process," "Fuel Cell," "Functional Materials," "Pharma—Building Competitive Edge," "Safety and Security," "Sustainable Community," and "Embedded ICT." In addition to these programs, TEKES is participating in other research programs run by the Academy of Finland and ministries. In 2007, the TEKES programs had 2,721 instances of business participation and 1,136 instances of research unit participation (TEKES 2007).

TEKES program evaluations also make a valuable contribution to knowledge transfer. In 2007, the most significant of these evaluations was a review of sixty programs that were conducted during the period 1995–2005. This evaluation reviewed the programs' management and strategy work and identified program steering group commitment to implementation as the factor most predictive of program success. Other major evaluations performed that year looked at the FinnWell program for the future of health care, commercialization of biomaterials, and interactive computing (TEKES 2007).

Consistent with TEKES's focus on internationalization, 47% of the projects under its programs included international cooperation, defined as "organised international exchange or acquisition of information, researcher mobility to or from Finland, purchase of intellectual property rights or joint projects with foreign partners." Some of the most important recent developments in TEKES internationalization efforts have been participation in the networking of national research programs in the European Union through the ERA-NET Scheme, in European joint research programs, and in initiating cooperation with partners in China (TEKES 2007, para. 10).

Sitra, the Finnish National Fund for Research and Development. Sitra is an independent public foundation under the supervision of the Finnish Parliament. Financially, Sitra is outside the government budget. Its operations are funded with endowment capital and returns from capital investments. Through its activities, Sitra aims "to promote stable and balanced development in Finland, the qualitative and quantitative growth of its economy and its international competitiveness and co-operation. [Its] operations are governed by a vision of a successful and skilled Finland."[1]

Sitra's operations have been focused on programs in the areas of health care, food and nutrition, and energy as well as a growth program for the mechanical industry. The programs utilize a variety of methods, including research and education, innovative projects, business development, venture-capital investments, and other corporate funding.

Centre of Expertise Programme. The Centre of Expertise Programme (OSKE) is a fixed-term government program that encourages the regions to utilize their top expertise to strengthen the competitiveness of the region. For the 2007–13 term, the program was organized into a cluster-based model with thirteen national Clusters of Expertise and twenty-one regional Centres of Expertise. The thirteen clusters are Cleantech, Digibusiness, energy technology, food development, forest industry future, HealthBIO, health and well-being, intelligent machines, living business, maritime, nanotechnology, tourism and experience management, and ubiquitous computing.[2]

One of the special characteristics of Finland is that the innovation network has also been built to cover the whole country, taking the regional differences, needs, and strengths into account to make regions more attractive places for domestic and foreign R&D investment. The Centres of Expertise seek success through exploiting local excellence and growth potential.

The program seeks to enhance the innovation environments in regions by improving cooperation between the research sector and local business and industry, in particular building around the knowledge base of the universities and polytechnics in the region. The Centres of Expertise are mostly run by the local science or technology parks. The centers also implement one of the entrepreneurial programs of TEKES, which seeks to identify research-based business ideas, especially in universities, polytechnics, and research institutes. The program offers expert services to research teams and researchers in commercializing ideas.

Local Innovation Systems. The relationships among universities, industry, the programs discussed above, and other institutions in Finland have been examined in the MIT Industrial Performance Center's Local Innovation Systems (LIS) Project (Lester 2005). In comparison with the United States, Finland's universities are more closely linked to industry (Chakrabarti and Lester 2002). For example, TEKES requires that all of its projects include industry-university collaboration and that they grant industry substantial control of resulting intellectual property. At Finland's technical universities, it is common for students to complete theses and dissertations focused on corporate problems. With its university-based business incubators and industry clusters, the Finnish ARE tends to be strong on context-specific problem-solving. "The institutional infrastructure in Finland has developed the appropriate social capital conducive for development of business ventures" (Chakrabarti and Rice 2003, 10).

The LIS Project has produced several case studies of university-industry interaction in Finland. The case of Tampere has been examined with respect to its transition to a knowledge economy, the upgrading of its mechanical engineering industry, and its new digital media cluster (Kolehmainen 2003; Kostiainen and Sotarauta 2002; Martínez-Vela and Viljamaa 2004). Turku is considered a successful example of universities' growing "third role" as agents of economic development, especially with respect to the region's biotechnology cluster (Srinivas and Viljamaa 2003, 2008). Oulu, in northern Finland, has been profiled with respect to its medical-device development industry and its environment conducive to informal industry-university collaboration, which includes Europe's first science park. The science park was formed with seed funding from the city, the university, and local industry (Nummi 2006).

Assessment

The Science and Technology Policy Council (2006) identified challenges, drivers of change, and opportunities for Finland and also established a development program for 2007–11. The Council found that Finnish science, technology, and innovation (STI) policy is characterized by a number of strengths, including active international and public-private partnerships; long-term STI policy implementation; a high priority placed on R&D investment, much of it competitively funded; effective systems of education, research, and innovation; gender diversity and talent retention among the highly skilled; a large proportion of researchers in the workforce, with correspondingly high research volume and quality; strong international patent activity; and retention of knowledge businesses.

However, the Council also noted some limitations. Finland is relatively dependent on global trends, yet its remote location puts it far from many global markets. Its small domestic market and language area and extreme climate are potential barriers to growth. By European standards, Finland's level of internationalization is low. The number of highly educated foreign researchers in Finland is low, and the country fails to attract foreign direct investment. Its intellectual and economic resources, while proportionately high, are small in absolute terms and are often highly concentrated. From a business perspective, venture capital availability; spin-off businesses from universities and research institutes; and competence in marketing, business, knowledge management, and innovation are limited. Research activities are somewhat fragmented, and growth is concentrated in a small number of enterprises (Science and Technology Policy Council 2006).

Opportunities and future directions identified include building on the NIS's effectiveness to promote the competitiveness and internationalization of its activities; continuing to enhance the knowledge base and R&D environment by attracting investment and talent; expanding international cooperation, including cooperation beyond the European Union; using cooperation strategically to compensate for Finland's small size and remote location; targeting limited resources; and "enhancing foresight activities and their linkage with decision-making and strategic steering" (Science and Technology Policy Council 2006, 43). More specifically, the development plan calls for Finland to increase R&D spending to 4.1% of GDP by 2011, an increase of approximately €100 million per year, and to administer the funds in a way that reduces fragmentation. Improving the structures that administer such funding was also identified as a priority, to ensure that "resources shall be allocated for creating larger entities, promoting networking, and making management and evaluation of activities more effective" (39). With respect to the ARE, the Council also notes: "The most critical part from the perspective of the entire public research system is structural reform of the higher education system so that it can function fully as a high-level element of the national innovation system and the international education and science community. Renewal must be accelerated with performance-based steering and other means of funding and structural development, including increasing universities' and polytechnics' opportunities and encouraging self-initiated development activities" (Science and Technology Policy Council 2006, 40).

Directions for reform of the higher education system are discussed further in the following sections.

It can be clearly seen that the importance of higher education and research has been growing even within the past fifteen years or so. Higher education and research have been given a special role in the national development strategies formulated since the 1990s, based on the key role of knowledge and technology development as national success factors in global markets. In fact, it was a widely accepted political conclusion that, after a deep recession in the early 1990s, the future of Finland would be built upon the foundations of research and knowledge. That decision led to an important program aimed at strengthening the research base in information technology, electronics, and telecommunication and encouraging research cooperation with rapidly growing telecommunication and electronics industries (Höltta and Malkki 2000).

Today, higher education and research policy in Finland is seen very much in the context of the Lisbon Strategy. The involvement of universities and research in developing a strong knowledge-based economy is not new to Finland. Finland had already totally reassessed its national development strategy and the role of higher education and research in the mid-1990s.

As discussed earlier, higher education and research have continuously been in the spotlight as the national development strategies have been established. Within the past ten or more years, the Finnish policy has been to construct an extensive system of research and innovation, in which the universities have a central role as the producers of basic knowledge. But the developments of the past fifteen years or so can be characterized by the development of an extensive mediating system of actors and programs aimed at transforming and disseminating this basic knowledge into a form usable by national business and industry. In terms of the modes of knowledge production, the emphasis has clearly been moving toward Mode 2 knowledge production (Gibbons et al. 1994), in which the questions for research increasingly come from society, and research itself is characterized by trans-disciplinarity and cooperation with universities, research institutions, and research units of private companies.

During recent years, the research role of universities has been subject to strong public debate. Partly this is due to international ranking lists, like the Shanghai List, which do not place Finnish universities at the top. The University of Helsinki was seventy-third on the list in 2007 (Institute of Higher Education, Shanghai Jiao Tong University 2007). However, there is also a strong belief that intensifying research is the key to the country's economic success. In this context, the questions concerning the development of world-class research universities have become the focus of the national higher education policy and institutional development in Finland. Industry representatives have also expressed their worries about the competitiveness of Finnish universities in research.

INITIATIVES AND GUIDELINES FOR THE FUTURE

The higher education and research policies adopted during the second half of the 1990s and the deep commitment of government to the policy goal of improving the competitiveness of the European Union provide the framework for the future of Finnish research and higher education policy. So far, all the development activities have taken place in a fixed institutional framework. Now, a serious question has been raised about the disadvantages of the system structures, that is, whether it is possible to create excellence in research in a system of twenty universities located around a large, sparsely populated country. The ministerial policy of pushing universities to create effective and interactive networks of their disciplines and research groups has worked only partially, and the next steps in the structural development are currently being discussed. Also, the polytechnics and the regional innovation structures are under political scrutiny.

IMPROVING DOCTORAL EDUCATION

A number of challenges have been identified for doctoral education. One thing that seems to be missing is the collective mechanism of the university, as in the US system, for assuring the quality of research and training in all doctoral degrees. A recent review (Dill et al. 2006) also recommends that the primary criterion for the award of a graduate school should be the existence of a strong program of research. Along these lines, the authors recommend that new and existing graduate schools be associated with centers of excellence identified by the Academy of Finland. A second criterion should be an effective process for assuring quality in doctoral education within the graduate schools. A third criterion should be evidence of effective organization and leadership. A fourth criterion might be involvement in regional or international networks (Dill et al. 2006). Other suggestions include improving graduate schools' research and innovation environments, avoiding overlap while encouraging cooperation, and supporting specialization and differentiation. The importance of ensuring adequate resources for student support, facilities, and supervision has also been noted. Finally, there is concern about improving the situation for recent PhD graduates, especially providing support structures for women graduates and encouraging the most talented graduates to remain in Finland (Oksanaen, Lehvo, and Nuutinen 2003).

This chapter concludes with a discussion of two recent landmarks that guide the development of the national research and research training systems: the reports of a one-person committee on the structural conditions for research, along with the subsequent ministerial structural development initiative; and an ongoing project in which the financial autonomy of Finnish universities is being assessed to improve their potential to operate in the markets for research and education.

Strengthening of the Research System. A one-person committee appointed by the Ministry of Education to review the structure of university and polytechnic research submitted its recommendations in 2005. The ministry had set two questions: (1) What kind of operational challenges will universities and polytechnics, as part of the public research system, have to tackle as a result of growing international competition and national and regional development needs? (2) What structural development measures are needed to enable universities and polytechnics, together and individually, to respond to the challenges?

The starting point in the report (Rantanen 2004) and its recommendations are that research excellence, active and productive international cooperation, and concentration of research resources in strong and socially relevant areas constitute the basis of the competitiveness of Finland in the future.

The larger multidisciplinary universities could concentrate on international top-level research and on the enhancement of their international competitiveness. This entails strengthening basic research through the development of more top-level units, international research alliances, increased researcher exchanges, and participation in international research projects.

The policy of increasing the integration of research done at the universities into the national and regional innovation systems is supported in the report. The report also supports the policy that universities should sharpen their academic profiles and concentrate more on the development of their strengths. Furthermore, it includes the thought that the efforts to develop excellence of a high international level should be mainly concentrated in larger, multidisciplinary universities, and that the policy focus in this effort should take into account improving the competitiveness of the Finnish economy. Basic research should be concentrated in the strategic areas of greatest relevance to Finland.

This principle should not, however be seen as contradicting the efforts to develop excellence in smaller universities. The smaller institutions should pay more attention to regional development activities and foster network-like cooperation with each other. Some universities could focus on supporting national and regional development. These could also aspire to top quality. Universities are also expected to directly support corporate R&D and especially the creation and development of SMEs. These activities could be stepped up within the societal mission of the universities in collaboration with polytechnics and other regional players.

It is also recommended that universities form networks with a view to preventing fragmentation. These could be based on regional, operational, thematic, or international activities and objectives. Over the longer term, it would be useful to pilot new forms of university organization based on larger groupings. University administration should be developed toward corporate governance by means of strategic planning and administrative and management training. The personnel structure should be developed toward greater flexibility. Studies should be undertaken to investigate the applicability of the public corporation model for university administration.

After extensive policy discussion, the Ministry of Education reserved funding to be used in 2007–9 for structural development of universities and cooperation between universities and polytechnics. The creation of strong research units and universities was the main policy goal. The 2008–11 plan for structural development in higher education aims to create a system with fewer but stronger institutions of higher education. A Vision 2020 plan calls for no more than fifteen universities and seventeen polytechnics, with four to five strategic university-polytechnic alliances (Ministry of Education 2008c). It looks as if the universities have chosen primarily defensive strategies, reflecting the slow development of academic values. Universities are emphasizing their uniqueness and trying to convince the ministry of the superiority of the models based on loose, voluntary programs and research-project-level cooperation.

Other research reform efforts relate to intellectual property and internationalization. A new national strategy for intellectual property rights, begun in 2007, was expected by the end of 2008. Also in 2008, a new strategy was developed for the internationalization of Finnish universities. While the strongest emphasis of the strategy is on improving the attractiveness of Finnish universities to foreign students and researchers, the strategy is also concerned with international mobility (European Commission 2008).

Extension of Financial Autonomy of Finnish Universities. Another ongoing effort to improve the conditions for research at Finnish universities is the assessment of their financial autonomy. A concrete step toward more extensive financial autonomy was taken in 2005 in the form of legislation making it possible for universities to establish "university companies," whose initial capital was to come from the government higher education budget. The institutions have, however, been slow to materialize. In preparation for reform, universities are being given more autonomy in financial matters, including the ability to have the status of a foundation under private law or that of a legal person under public law. These reforms will move universities outside the scope of government budgeting. Other recommendations relevant to the financial autonomy of universities include separating the functions of the board of directors and the university senate from those of the executive management, having the board of directors appoint rectors for a fixed term, having the university rector or a designee hire university personnel, and managing posts and tenures under labor contract legislation rather than civil service regulations (Ministry of Education 2007a, 2007b). As described earlier in this chapter, the autonomy of universities to decide on the use of their budget funding and human resources has progressed rapidly from a model of extreme control to one in which all internal allocation decisions fall within the authority of the universities.

Conclusion

Finland has shown leadership in the development of its academic research and innovation systems. Recent reforms have been proposed to build on existing strengths and address some identified weaknesses as well as to respond to challenges in the operational environment, such as globalization and shifting demographics (Ministry of Education 2007a). The period of reform implementation is planned to last through 2012, and international assessment of the results can be expected in 2020 (Ministry of Education 2008c). The success of these reform efforts will shape the development and future competitiveness of the Finnish ARE.

NOTES

I want to express my sincere appreciation to Ms. Jennifer M. Miller for her significant assistance in revising this chapter for publication.

1. The Sitra Web site, www.sitra.fi/fi, is the source of both this quotation, which comes from paragraph 2 there, and the remainder of the discussion of Sitra.

2. This discussion of the Centre of Expertise Programme is based on information from the OSKE Centre of Expertise Programme Web site, www.oske.net/en/, accessed in 2008.

REFERENCES

Academy of Finland. (2001) Academy of Finland Annual Report 2001. Helsinki, www.aka.fi/
Tiedostot/Tiedostot/Julkaisut/Annual%20Report%202001%20(pdf).pdf.

———. (2003) *PhDs in Finland: Employment, Placement, and Demand.* Helsinki: Academy
of Finland.

———. (2007a) *Academy of Finland in Brief.* Helsinki, www.aka.fi.

———. (2007b) Annual Report, 2007. Helsinki, www.aka.fi/Tiedostot/Tiedostot/Julkaisut/
Akatemia%20vsk_LR_ENG.pdf.

———. (2008) *The Researcher's Mobility Portal: Finland.* Helsinki, www.aka.fi/eracareers/.

Ahlbäck, J. (2005) *The Finnish National Innovation System.* Helsinki: Helsinki University
Press.

Chakrabarti, A. K. and R. K. Lester. (2002) Regional Economic Development: Comparative
Case Studies in the US and Finland. MIT IPC Local Innovation Systems Working Paper
02-004, MIT, Cambridge, MA, http://web.mit.edu/lis/papers/LIS02-004.pdf.

Chakrabarti, A. K. and M. Rice. (2003) Changing Roles of Universities in Developing Entrepreneurial Regions: The Case of Finland and the US. MIT IPC Local Innovation
Systems Working Paper 03-003, MIT, Cambridge, MA, http://web.mit.edu/ipc/publica
tions/pdf/03-003.pdf.

Davies, J., T. Weko, L. Kim, and E. Thulstrup. (2006) Thematic Review of Tertiary Education: Finland Country Note. OECD Directorate for Education and Training Policy Division, www.oecd.org/dataoecd/51/29/37474463.pdf.

Dill, D. D., S. K. Mitra, H. S. Jensen, E. Lehtinen, T. Mäkelä, A. Parpala, A. Pohjola, M.
A. Ritter, and S. Saari. (2006) *PhD Training and the Knowledge-Based Society: An Evalu-
ation of Doctoral Education in Finland.* Tampere: Finnish Higher Education Evaluation
Council Publications.

Dobson, I. (2008) Finland: Merger Fever Hits Universities. *University World News,* April 20.

Embassy of Finland. (2006) Finland Hot Spot for Top Research. News release, February 28.

European Commission. (2008) *ERAWATCH Research Inventory Report For: Finland.* Brussels:
EC.

Finnish Funding Agency for Technology and Innovation (TEKES). (2007) Annual Review
2007. TEKES, Helsinki, www.tekes.fi/eng/tekes/annuals/2007.html.

Finnish Science and Technology Information Service. (2007a) *R&D Expenditure in the Higher
Education Sector, 2002–2006, Million Euros.* www.research.fi.

———. (2007b) *R&D Personnel by Sector.* www.research.fi.

Gibbons, M., C. Limoges, H. Nowotny, S. Schwartzman, P. Scott, and M. Trow. (1994)
*The New Production of Knowledge: The Dynamics of Science and Research in Contemporary
Societies.* London: Sage.

Helminen, S., S. Karvonen, P. Mörk, A. Palkamo, and L. Peltonen. (2007) *TEKES Online
Annual Review 2007.* www.tekes.fi/eng/tekes/annuals/english/index.html.

Hölttä, S. (1988) Recent Changes in the Finnish Higher Education System. *European Journal
of Education* 23:91–103.

———. (1998) Funding of Universities in Finland: Towards Goal-Oriented Government

Steering. In *Financing Higher Education: Innovation and Changes,* 33:55–63. Special issue, *European Journal of Education.*

———. (1999) Regional Cooperation in Postsecondary Education in Europe: The Case of the Nordic Countries. *Journal of Institutional Research in Australasia* 8 (2).

———. (2000) From Ivory Towers to Regional Networks in Higher Education Governance. *European Journal of Education* 35:465–474.

Hölttä, S. and P. Malkki. (2000) Response of Finnish Higher Education Institutions to the National Information Society Programme. *Higher Education Policy* 13 (3): 231–243.

Hölttä, S. and E. Rekilä. (2003) Ministerial Steering and Institutional Responses: Recent Developments of the Finnish Higher Education System. *Higher Education Management and Policy* 15:73–92.

Husso, K. (2002) Tohtoreiden Työllistyminen, Sijoittuminen ja Liikkuvuus Työmarkkinoilla. Esitelmäpaperi Suomen Akatemian ja Opetusministeriön Järjestämään "Tutkijanuran Haasteita ja Mahdollisuuksia"—Seminaariin. [The hiring, placement, and mobility of PhDs on the employment market, a lecture paper for the seminar "The challenges and opportunities of a research career," arranged by the Academy of Finland and the Ministry of Education], Suomen Akatemia, Helsinki.

Institute of Higher Education, Shanghai Jiao Tong University. (2007) Academic Ranking of World Universities—2007, http://ed.sjtu.edu.cn/ranking.htm.

Jäppinen, A. (1989) University and Government in Finland: Complementary Roles. *Higher Education Management* 5:333–336.

Kolehmainen, J. (2003) Territorial Agglomeration as a Local Innovation Environment: The Case of a Digital Media Agglomeration in Tampere, Finland. MIT IPC Local Innovation Systems Working Paper 03-002, MIT, Cambridge, MA, http://web.mit.edu/lis/papers/LIS03-002.pdf.

Kostiainen, J. and M. Sotarauta. (2002) Finnish City Reinvented: Tampere's Path from Industrial to Knowledge Economy. MIT IPC Local Innovation Systems Working Paper 02-002, MIT, Cambridge, MA, http://web.mit.edu/lis/papers/LIS02-002.pdf.

Lane, J-E. (2000) *The New Public Management.* London: Routledge.

Lavery, J. E. (2006) *The History of Finland.* Westport, CT: Greenwood Press.

Lehikoinen, A. and E. Kaunismaa. (2008) *Neliportainen Tutkijanura* [The four-stage research career model]. Helsinki: Ministry of Education.

Lester, R. K. (2005) Universities, Innovation, and the Competitiveness of Local Economies. MIT Industrial Performance Center Working Paper 05-010, MIT, Cambridge, MA, http://web.mit.edu/lis/papers/LIS05-010.pdf.

Maastricht Economic and Social Research and Training Centre on Innovation and Technology (UNU-MERIT). (2008) *2007 Global Innovation Scoreboard (GIS): Comparative Analysis of Innovation Performance.* Maastricht, Netherlands: UNU-MERIT.

Martínez-Vela, C. and K. Viljamaa. (2004) Becoming High-Tech: The Reinvention of the Mechanical Engineering Industry in Tampere, Finland. MIT IPC Local Innovation Systems Working Paper 04-001, MIT, Cambridge, MA, http://web.mit.edu/lis/papers/LIS04-001.pdf.

Ministry of Education. (1995) *Education, Training, and Research in the Information Society.* Helsinki: Ministry of Education.

————. (2004a) *Education and Research, 2003–2008: Development Plan.* Helsinki: Helsinki University Press.

————. (2004b) *Management and Steering of Higher Education in Finland.* Helsinki: Ministry of Education.

————. (2005) *OECD Thematic Review of Tertiary Education: Country Background Report for Finland.* Helsinki: Ministry of Education.

————. (2007a) *Education and Culture: Annual Report.* Helsinki: Ministry of Education.

————. (2007b) *Yliopistojen Taloudellisen ja Hallinnollisen Aseman Uudistaminen* [Reform of the financial and administrative status of universities]. Helsinki: Ministry of Education.

————. (2008a) Charter of Foundation for Aalto University to Be Signed in June. News release, May 29, www.minedu.fi/OPM/Tiedotteet/2008/05/saadekirja.html?lang=en.

————. (2008b) KOTA Database, https://kotaplus.csc.fi/online/Etusivu.do.

————. (2008c) "Structural Development of Higher Education, 2008–2011." Ministry of Education, Helsinki, www.cimo.fi/dman/Document.phx?documentId=mw291081145254 42&cmd=download.

Neave, G. and F. A. van Vught (eds.). (1991) *Prometheus Bound: The Changing Relationship between Government and Higher Education in Western Europe.* Oxford: Pergamon.

Nummi, J. (2006) University-Industry Collaboration in Medical Devices Development: A Case Study of the Oulu Region in Finland. MIT IPC Local Innovation Systems Working Paper 06-001, http://web.mit.edu/lis/papers/LIS06-001.pdf.

Oksanaen, T., A. Lehvo, and A. Nuutinen. (2003) *Scientific Research in Finland: A Review of Its Quality and Impact in the Early 2000s.* Helsinki: Academy of Finland.

Organisation for Economic Co-operation and Development (OECD). (1982) *Reviews of National Policies for Education: Finland.* Paris: OECD.

————. (2008) *OECD Factbook 2008: Economic, Environmental, and Social Statistics.* Paris: OECD.

Ormala, E. (2001) Science, Technology, and Innovation Policy in Finland. In P. Laredo and P. Mustar (eds.), *Research and Innovation Policies in the New Global Economy: An International Comparative Analysis,* 325–358. Cheltenham, UK: Edward Elgar.

Polt, W., C. Rammer, H. Gassler, A. Schibany, N. Valentinelli, and D. Schartinger. (2001) *Benchmarking Industry-Science Relations: The Role of Framework Conditions.* Vienna/Mannheim: European Commission, Enterprise DG.

Rantanen, J. (2004) *Yliopistojen ja Ammttikrkeakoulujen Tutkimuksen Rakenneselvitys* [Review of the structure of university and polytechnic research]. Helsinki: Ministry of Education.

Science and Technology Policy Council of Finland. (1987) *Science and Technology Policy Review 1987.* Helsinki: Science and Technology Policy Council.

————. (1990) *Review 1990: Guidelines for Science and Technology Policy in the 1990's.* Helsinki: Science and Technology Policy Council.

————. (2000) *Review 2000: The Challenge of Knowledge and Know-How.* Helsinki: Science and Technology Policy Council.

————. (2006) *Science, Technology, and Innovation.* Helsinki: Science and Technology Policy Council.

Singleton, F. B. and A. F. Upton. (1998) *A Short History of Finland.* Cambridge: Cambridge University Press.

Srinivas, S. and K. Viljamaa. (2003) BioTurku: "Newly" Innovative? The Rise of Bio-Pharmaceuticals and the Biotech Concentration in Southwest Finland. MIT IPC Local Innovation Systems Working Paper 03-001, http://web.mit.edu/lis/papers/LIS03-001.pdf.

———. (2008) Emergence of Economic Institutions: Analysing the Third Role of Universities in Turku, Finland. *Regional Studies* 42:323–341.

Statistics Finland. (2007) R&D Expenditure. Finnish Science and Technology Information Service, www.research.fi/en/resources/R_D_expenditure/R_D_expenditure_table.

———. (2008a) R&D Funding in the State Budget. Finnish Science and Technology Information Service, www.research.fi/en/resources/R-D_funding_in_the_state_budget.

———. (2008b) University Education. Statistics Finland, Helsinki, www.stat.fi/til/yop/index_en.html.

———. (2008c) University Students Numbered 176,300 in 2007. Statistics Finland, Helsinki, www.stat.fi/til/yop/2007/yop_2007_2008-04-17_tie_001_en.html.

TEKES. *See* Finnish Funding Agency for Technology and Innovation.

UNU-MERIT. *See* Maastricht Economic and Social Research and Training Centre on Innovation and Technology.

Vuorio, E., A. Halinen, and J. Ylikarjula. (2006) *Tutkijanuratyöryhmän Loppuraportti* [Report of the committee on the development of research careers]. Helsinki: Ministry of Education.

Wilén, H. (2008) Government Budget Appropriations or Outlays on R&D—GBAORD. In *Eurostat Statistics in Focus,* 29/2008. Luxembourg: Office for Official Publications of the European Communities.

World Economic Forum (WEF). (2004) *The Global Competitiveness Report, 2003–2004.* Geneva: World Economic Forum.

———. (2005) *The Global Competitiveness Report, 2004–2005.* Geneva: World Economic Forum.

———. (2008) *The Global Competitiveness Report, 2007–2008.* Geneva: World Economic Forum.

CHAPTER SEVEN

Germany

JÜRGEN ENDERS

From a bird's-eye view, Germany's innovation landscape may be characterized by a certain set of features. As a country with few natural resources, Germany has traditionally depended on its success in knowledge and technology creation. The contribution of R&D-intensive branches of industry to gross domestic product (GDP) is higher than in countries such as France, the United Kingdom, the United States, and Japan; and leading-edge technology contributes almost as much to the entire economy as in Japan and the United States (BMBF 2006). Germany relies to a considerable extent on export-oriented industry, in which machine construction, vehicle construction, mechanical engineering, and the chemical industry have the traditional competitive advantages. Certain disadvantages of a late-comer country can still be seen in areas such as information technology, biotechnology, and nano-technology. Military R&D was forbidden in West Germany by Allied law until 1955 and still plays a smaller role than in other countries such as the United States, the United Kingdom, and France.

The workforce includes a growing but still comparatively small proportion of higher education graduates compared with many other knowledge-based economies (OECD 2005). In 2002 about 35% of the 15-to-29-year-olds participated in higher education, an increase of 7% compared to 1998. Nevertheless, student enrollment in higher education in Germany is still considerably lower than in many other highly developed countries, which have participation rates of about 50% and above.

Compared to international standards, R&D expenditure in Germany is relatively high (amounting to 2.5% of GDP in 2005) but still below the Lisbon goal of 3% of GDP. As a percentage of GDP, about 0.4% of R&D expenditure is performed by academic research in higher education, about another 0.4% is performed in public research outside higher education, and 1.7% in R&D in the private sector.

The institutional infrastructure for R&D is thus broad and diversified. It includes both an important business sector for R&D (with a share of about two-thirds of the gross domestic expenditure on R&D) and an important sector of public research outside of higher education that is similar in size to the binary higher education system, including universities and universities of applied science (*Fachhochschulen*).

The governance and funding of public research are also highly differentiated and decentralized. Governments on the national level as well as on the subnational level of the *Länder* share responsibility in this area. The role of government in governance and funding is supplemented by quasi-nongovernmental organizations and intermediary bodies, such as the German Research Foundation (Deutsche Forschungsgemeinschaft [DFG]) and the Science Council (Wissenschaftsrat [WR]) as well as some large private foundations for public research (e.g., Volkswagen-Stiftung, Bertelsmann-Stiftung, and Thyssen-Stiftung).

Obviously, the public research system and the responsibility for its governance and funding are highly differentiated and decentralized. Internationally, this has been seen as a characteristic advantage of the German setting (Krull and Meyer-Krahmer 2001). At the same time, such a system must be closely networked and must exhibit sufficient flexibility and dynamism to promote ongoing innovation and adaptation. Since the 1990s many reviews and policy recommendations have stressed that these requirements have not been met (Internationale Kommission 1999; OECD 2002; VolkswagenStiftung 2005; Wissenschaftsrat 2006). Policy recommendations call for improvement in the directions of more cooperation and competition, more flexibility and responsiveness, and more self-organization and strategic planning within the system.

In this overall context, the role and functioning of academic research has come under closer scrutiny, and the debate on the "crisis of the German university" has been enriched by a relatively recent focus on its research function. An important consequence for universities (and also in part for public-sector research organizations) is the emergence of policies for change in organization, funding, accountability, communication, interaction, and motivation within the academic research world. There is also a shift in the regulatory philosophy and the policy instruments used for universities. While it will be argued that these shifts in governance have been introduced only recently and in incremental steps, they add up to a serious challenge for the academic research system. Traditional notions of strong state regulation, academic freedom and self-governance, the unity of teaching and research, and the egalitarian homogeneity among universities and their units are under reconsideration and are due for structural reforms. This process is accompanied by a remarkable shift in public attention and governmental concern. Today, higher education is more likely to be regarded as a private consumption good, while research is seen as key for the global competitiveness of the knowledge-based economy and therefore as a core strategic area for national policies. As universities are increasingly considered an important part of an overall innovation system, numerous questions emerge concerning the status and boundaries of the university and the type of knowledge produced as well as the processes by which this knowledge is generated. Currently, the governance of academic research is in limbo as some forces and actors support the status quo and others support a major change in the system.

Structure, Functions, and Financing of Academic Research within the Public Research Sector

The governance and funding of German public research take place in a variety of institutions, involving a mix of actors at varying levels in many subsystems, each of which has a unique history, function, and interrelationship with other subsectors (Krull and Meyer-Krahmer 1996; Schimank and Hohn 1990; Schmoch 1996). Academic research is part and parcel of a wider setting of public research organizations.

The West German public research system was established along federal lines, with centralized and decentralized ways of decision-making based on principles such as the freedom of teaching and research, the autonomy of higher education institutions and public-sector research organizations in their substantial affairs, and the strong impact of federal and regional government on their governance and funding (Schimank and Hohn 1990). Some observers suggest calling it a "multi-actor network" (Kuhlmann 1998), given the high degree of institutional differentiation, the mission-based division of work, and the autonomy of the major actors. The following can only provide a brief sketch highlighting certain structural features that are of specific importance to the understanding of the recent dynamics and reforms in German higher education and public research overall.[1]

STRUCTURE AND FUNCTIONS OF ACADEMIC RESEARCH WITHIN THE PUBLIC RESEARCH SECTOR

There are basically two sectors within the German research system that are institutionally financed by the state. First, state-financed research is undertaken by universities and universities of applied science. Second, in addition to universities, there are several important groups of state-financed research institutes outside of higher education.

The Higher Education System. The higher education landscape is comprised of about 170 institutions with university status, including about 84 universities and 13 technical universities. The remaining institutions are teacher-training colleges, colleges of music and fine arts, and theological colleges. Universities are subordinate to Länder ministries, from which they receive basic financing, while the federal government contributes to infrastructural funds as well as to research funding via its own programs and the German Research Foundation (the DFG). Nowadays, universities' legal framework is mainly affected by regional laws, although some general frameworks are set by federal law. Universities' main objectives are to carry out research-based teaching, to conduct basic as well as applied research, and to train postgraduate and postdoctoral researchers. The 13 technical universities possess the same legal and organizational framework as the other universities but with a somewhat different mission; the transfer of knowledge and technology to business enterprises is among their main objectives. Certainly the universities provide an important backbone of the German research system, and no other research sector covers such a broad range of functions and themes. Such diversity means that research at universities is carried out in relatively small units, a feature that is also supported by the strong position of the individual professor within the chair-based organization of German universities. Further, universities have played an important role in taking up growing student numbers in the wake of educational expansion. Germany established a nonuniversity sector to absorb part of the massification in higher education. This sector is, however, relatively small and does not really shield the universities and their research function from the consequences of massification.

Universities of applied sciences (*Fachhochschulen*) provide practice-oriented studies in engineering, business fields, and social sciences / social work, including one-year practical study periods at firms. The same legal and financial framework applies as that of universities, in that they are mainly subordinate to Länder governments. The amount of research activities in these teaching-oriented institutions is rather low. Many colleges have, however, developed strong ties to regional enterprises, especially SMEs (small and medium-sized enterprises), in the field of consulting and technical development. Such activities largely depend on external research funds and technology transfer units.

Research Institute System (outside of Higher Education). There is an important sector of public research undertaken outside of higher education with its own legal, financial, and organizational frameworks. Most of the public-sector research institutes are either members of loosely coupled associations or part of holding organizations, representing their interests on the national level. In the mid-1970s an overall political compromise was reached that established a functional differentiation with regard to the type of research conducted by the different groups of research organizations. Further, a consensus on the political competencies of the federal state and the Länder for each group of organizations has been achieved, while a precarious balance between their institutional autonomy and state regulation has been established. Table 7.1 provides an overview of these various research organizations.

The Max Planck Society[2] (MPG) consists of about seventy-five institutes and is financed jointly by the federal government and the Länder. Its mission is to carry out top-level basic research in selected areas according to international standards. The vast majority of research funds are provided via general public grants that are supposed to cover most of the costs needed to run an internationally competitive institute. External acquisition activities are restricted by formal as well as informal rules. The MPG, as a corporate actor, chooses its own portfolio of research areas and decides on the establishment of new institutes and the abolishment of old ones. The MPG allows the director of an established institute to set his or her own lines of research and provides a high level of autonomy for its researchers overall. All in all, these conditions have allowed the MPG to be the top runner in basic research within Germany as well as in the international public research system. However, there has been increasing latent pressure on the MPG to demonstrate its willingness to contribute to political priority-setting and societal relevance. The establishment of certain new areas of research has clearly been influenced by concessions to governmental research priorities. The amount of external research funding is still relatively low but is constantly growing. Finally, the MPG has recently started to highlight the activities of its technology transfer office (Garching Innovation), which has functioned as a mediator between research and industry since the 1970s.

The Helmholtz Research Centers[3] (HGF) is the largest public-sector group of research institutes, consisting of sixteen large, organizationally independent research labs. The centers cover a wide spectrum of interests, from basic research through strategic research and the development of industrial technology. These centers also run large-scale research infrastructure facilities (e.g., the German Electron Synchrotron [DESY] in Hamburg) that are to a large extent used by academic researchers. So far, research financing has been based mainly on general funds, of which 90% is provided by the federal government and 10% by the Länder. Basically, the HGF is a loosely coupled association with a weak central administration; the research centers interact directly with the federal ministry regarding research priorities and institutional funding.

The Fraunhofer Society[4] (FhG) consists of forty-eight research institutes that carry out a broad range of strategic and applied research for government and in-

TABLE 7.1.
Public research organizations in Germany

	A	B	C	D	E	F	G
	Budget 2001 (€M)[a]	Research personnel 2001 (FTE)[a]	A/B	Institutional funding 2003 (%)	Total CSI papers 2000–2002	Total science and nature papers 2000–2002	Total DPA, WPI patent applications 1999–2001
Universities	10.119 (56.9%)[b]	100,455 (71.3%)[b]	0.101	—[c]	145,847 (71.7%)	474 (51.4%)	6,394 (70.7%)
Helmholtz Research Centers (HGF)	2,288 (12.9%)	10,252 (7.3%)	0.223	78	15,352 (7.5%)	125 (13.5%)	1,206 (13.3%)
Max Planck Society (MPG)	938 (5.3%)	3,692 (2.6%)	0.254	80	20,414 (10.0%)	314 (34.0%)	245 (2.7%)
Leibniz Association (WGL)	568 (3.2%)	3,348 (2.4%)	0.170	70	8,558 (4.2%)	44 (4.8%)	188 (2.1%)
Fraunhofer Society (FhG)	947 (5.3%)	5,647 (4.0%)	0.168	39	1,988 (1.0%)	2 (0.2%)	1,011 (11.2%)

Source: SCI, WPINDEX, PATDPA (host: STN).

Notes: Not all German research institutes are covered; therefore columns A and B do not add up to 100%. Nonfractional counts in E, F, and G.

[a] Excluding social sciences and humanities.

[b] Including teaching.

[c] No figures available.

dustry. The institutes perform contract research and offer information and services related to new technologies, products, and processes. Of the basic financing, 90% is provided by the federal government and 10% by the Länder. The actual budget, however, depends to a considerable extent on contract research. Competition for user markets thus plays a strong role in research priority-setting for each institute, which is done relatively independently from government. Instead, central coordination by the FhG as a corporate actor has importance due to its capacity to cross-subsidize between institutes. This implies capacities for market analysis as well as responsiveness to societal demands on the central and institute level. It also means that leeway for self-driven research agendas is rather limited.

The Leibniz Association of Research Institutes[5] (WGL) comprises about eighty organizational units that are very heterogeneous in regard to size, research topics, and objectives. The WGL includes nonresearch institutions, such as museums, libraries, and thematic information centers as well as some of Germany's top-level research centers (e.g., in communication engineering and semiconductor research).

The main similarity among the institutes is that their basic financing is provided jointly by the federal government and the Länder. As with other centers that are jointly financed, the Länder are primarily interested in assuring the continued existence of "their" institutes and accept political intervention by the federal ministry as long as this remains untouched. In 2000, an evaluation of all institutes was finalized, and attempts to strengthen the profile of the WGL are currently under way.

In addition, most federal ministries run departmental research institutions. There are a total of fifty-two such institutes, most of which carry out applied research and provide research services. They include some large research centers as well as some high-level research institutes. Although their main objective is to provide research support for the federal government and public services in research-related areas, there is quite a high level of interaction with industry. A major system evaluation of all of these federal research institutes is currently being carried out by the Science Council (WR).

Both parts of public research, higher education and public research institutes, exhibit a firm and partly mission-based stability within a complex and multilayered institutional setting (Heinze and Kuhlmann 2007). But the situation is quite differently assessed by relevant actors. The institutional setting of public research institutes—including a basic research organization (MPG), an applied research organization (FhG), large research facilities (HGF), and the colorful remainder (WGL)—is not without problems and tensions. These seem to be more manageable by government than does the university sector, however, partly because of the public research institutes' better performance, greater determination to solve problems, and evaluation-based governance. The conclusions that emerge from various evaluations mainly concern (1) the negative effects of the segmentation of the system in regard to interorganizational cooperation between public research institutes as well as between them and universities and industry; (2) the introduction or intensification of internal competition for financing; (3) the need for more flexibility in the regulatory framework of management, financing, and staffing; and (4) whether there should be continuous monitoring of the functioning and performance of the public research sector (Internationale Kommission 1999; Wissenschaftsrat 2001a, 2001b).

In comparison, research conditions at universities have been and still are much more critically assessed (Wissenschaftsrat 2006). This is so for three interrelated reasons. First, the expansion of higher education has certainly led to a deterioration of academic research conditions. Since the 1970s, capacities for teaching have had to be extended without an appropriate amount of further financing. An unavoidable consequence has been further competition between teaching and research for scarce resources. Research is more and more often carried out in the "shadow of teaching"

(Schimank 1995). This problem has not yet come to an end, and special efforts will be needed to avoid a further decline of capacities of academic research in the near future due to dramatically increasing student numbers. According to recent prognoses, German higher education will have to host 700,000 more students in 2013 than in 2004 due to both the overall demographic trend and a reform in secondary education. This means an increase of 35% over the 2004 number of students (Kultusministerkonferenz 2004).

Second, quite a bit of the research carried out by public-sector research institutes was once carried out by universities. Since the end of the 1980s, universities and their representatives have again and again called for an end to the further extension of public research outside universities and asked that some capacities be returned to universities. Until very recently, however, political actors were reluctant to do so because of the perceived structural problems of the university system.

Third, the policy framework for higher education in general and academic research more specifically is structured in such a way that it is quite difficult to institute necessary reform across the system. It was only after the German reunification (during which the West German higher education and research system was implemented in East Germany)[6] that major steps for reform were undertaken. At that time, calls were also made for a change in the regulatory framework.

FINANCING OF ACADEMIC RESEARCH WITHIN THE PUBLIC RESEARCH SECTOR.

Currently, Germany spends about 2.5% of its GDP for R&D. During the 1990s R&D intensity declined, and in 1995 Germany spent only about 2.25% of its GDP for R&D. This development was partly due to the decline of private R&D in East Germany and partly due to cuts in public R&D financing. This trend has been reversed since the end of the 1990s. Nevertheless, Germany currently spends the same percentage of its GDP for R&D as it did at the beginning of the 1990s (table 7.2).

About two-thirds of R&D expenditures occur in the private sector. In the public sector, academic research and public research institutes outside higher education account for about equal shares of the remaining one-third of overall R&D expenditures. This distribution of R&D expenditures between the public and the private sectors as well as between academic research and other public research organizations has basically remained the same during the past two decades. Within higher education, universities hold a share of about 44% of the total public R&D financing, technical universities about 7%, and universities of applied sciences about 2% (BMBF 2004). Accordingly, the share of higher education researchers among all

TABLE 7.2.
Expenditure for higher education R&D as a percentage of GDP, 1995, 2000, and 2004

	1995[a]	2000[b]	2004
Germany	0.40	0.39	0.41
Australia	0.39	0.40	—
Austria	0.39	0.40	—
Belgium	0.38	0.40	0.43
Canada	0.46	0.55	0.70
Denmark	0.45	0.45	0.61
Finland	0.44	0.60	0.69
France	0.38	0.40	0.41
Greece	0.22	—	—
Iceland	0.43	0.44	—
Ireland	0.26	0.23	0.33
Italy	0.25	0.32	—
Japan	0.60	0.43	0.42
Korea	0.19	0.27	0.28
Netherlands	0.55	0.51	0.50
New Zealand	0.29	—	—
Norway	0.44	—	0.48
Portugal	0.21	0.30	—
Spain	0.25	0.27	0.32
Sweden	0.73	—	—
Switzerland	—	0.59	0.67
Turkey	0.26	0.39	—
UK	0.38	0.38	—
US	0.31	0.31	0.36
EU25	0.35	0.37	—
Total OECD	0.34	0.36	0.39

Source: OECD, Main Science and Technology Indicators, June 2006.
[a] 1993 for Austria.
[b] 1998 for Austria.

public and private researchers has been stable during the past two decades and amounts to about 28% (table 7.3).

Academic research is financed by basic funding provided by the Länder, by basic infrastructural funds provided by the federal government, by program- or project-based financing via the German Research Foundation, by various private research foundations, by the various governmental actors, and by the European Union as well as by industry. Academic research acquires about one-third of its research funds from such (competition-based) program and project financing. In the years 2001 to 2003, about 31% of the external research funding came from the German Research Foundation (mainly financing basic research at universities), 30% from other governmental and public sources (mainly financing strategic and applied research in

TABLE 7.3.
*Higher education researchers as a percentage of all
researchers, 1995, 2000, and 2004*

	1995	2000	2004
Germany	27.9	26.0	—
Australia	—	59.9	—
Austria	—	59.9	—
Belgium	42.1	28.6	—
Canada	34.5	27.2	—
Czech Republic	22.5	27.2	26.2
Denmark	34.6	—	—
Finland	38.4	31.6	—
France	35.5	35.8	—
Greece	62.5	—	—
Hungary	38.5	40.6	39.6
Iceland	35.3	—	—
Ireland	33.3	25.2	38.0
Italy	45.7	38.9	—
Japan	36.1	27.7	—
Luxembourg	—	1.3	—
Mexico	48.7	—	—
Netherlands	36.6	36.8	—
New Zealand	49.6	—	—
Norway	31.4	—	—
Poland	55.6	62.1	—
Portugal	50.4	51.3	—
Slovak Republic	40.7	50.3	60.7
Spain	58.4	54.9	—
Sweden	35.3	—	—
Switzerland	—	35.6	—
Turkey	74.3	73.2	—
UK	32.3	—	—
US	17.5	—	—
EU25	37.1	37.0	—

Sources: OECD, Main Science and Technology Indicators, June 2006;
OECD 2006, 33.

universities and technical universities), 27% from industry and business (mainly financing applied research in universities and technical universities), 7% from foundations, and 6% from international organizations (mixed portfolio) (DFG 2006).

Over time, the relative importance of external financing for academic research has increased. In 1991, external financing from second- and third-stream income was about 30% of all expenditures for R&D in higher education. In 2002, external financing made up about 36% of all expenditures for R&D in higher education (table 7.4).

Further, private financing of academic research via contract research has certainly increased over time as well. We may hypothesize various interrelated developments

TABLE 7.4.
Higher education R&D financed by industry,
as a percentage of total higher education R&D, 1995,
2000, and 2004

	1995	2000	2004
Germany	8.2	11.6	12.8
Australia	4.7	4.9	—
Belgium	13.1	11.8	—
Canada	8.0	9.5	8.4
Czech Republic	2.0	1.1	0.6
Denmark	1.8	2.0	3.0
Finland	5.7	5.6	5.8
France	3.3	2.7	—
Greece	5.6	—	—
Hungary	2.1	5.5	12.9
Iceland	5.4	—	—
Ireland	6.9	5.3	2.6
Japan	2.4	2.5	2.8
Korea	22.4	15.9	15.9
Mexico	1.4	2.0	—
Netherlands	4.0	7.0	—
New Zealand	9.4	—	—
Norway	5.3	—	—
Poland	11.4	7.8	5.6
Portugal	0.9	1.0	—
Slovak Republic	1.0	0.3	0.6
Spain	8.3	6.9	7.5
Sweden	4.6	—	—
Switzerland	—	5.1	8.7
Turkey	13.1	19.4	—
UK	6.3	7.1	—
US	6.8	7.1	5.0
EU25	5.9	6.6	—
Total OECD	6.2	6.6	—

Sources: OECD, Main Science and Technology Indicators, June 2006;
OECD 2006, 26.

supporting a considerable increase in academic contract research for industry. Certainly, the lack of basic funding for academic research and the growing competition for external funding have stimulated a more proactive approach by academic researchers in this area. Institutional barriers to private research financing have also been reduced over recent decades, including new rules and regulations for taxes and intellectual property rights. Studies suggest that such regulations may in themselves not be sufficient to stimulate further private investment, but redirection of federal financing in R&D may encourage further private investment (Krull and Meyer-Krahmer 1996).

Academic research is strongly oriented toward natural science, engineering, and medicine. About 75% of all research is performed in these fields. Research in the social sciences and the humanities accounts for about 20% of the overall financing. Relative figures—that is, external financing per professor—also show the differentiation in funding among the fields. In three years, a professor in engineering accounts for €1.1 million in research grants, in chemistry for €478,000, in the social sciences for €145,000, and in the humanities for €107,000.

A recent ranking of the capacities of 100 universities (including technical universities) in regard to external financing of research shows a very interesting picture (DFG 2006). First, external financing is quite concentrated in the top of the system. In regard to project funds from the German Research Foundation (the DFG), the top 10 universities account for about 33% of the total of external research grants, the top 20 universities for about 58%, and the top 30 universities for about 78%. The remaining 22% is distributed to the other 70 universities. The top 20 universities in the area of federal funding for large research infrastructure receive 82% of these funds; about 78% of federal financing in biomedical research, and about 82% of EU Framework money in biotechnology-related research goes to the top 20 universities in this area.

Second, size and disciplinary profile matter. It is especially big universities with a strong profile in the life sciences and engineering that are able to score across the board in obtaining different sources of external research financing, such as grants from the research council, from federal ministries, or from the European Union. There is a diversified landscape of smaller universities that score in specific niches in science and engineering or in the social sciences and humanities.

Third, there are, nevertheless, certain universities and regions that have established a top place across the board in fields of research and in sources of external financing available, usually characterized by cooperation with a critical mass of public research organizations outside higher education within the region.

Fourth, comparisons over the years between 1991 and 2004[7] show a certain stability as well as mobility of institutions with respect to success in external research financing. For example, four out of the top ten universities kept a stable position while the other six universities were replaced by others.

We may thus say that the system is currently characterized by an overall investment in R&D that is relatively high by international standards, serving a dual-track system of public research in higher education and other public research organizations. In the public research sector, there is a large variety of institutions and financial sources. Financing of academic research is predominantly based on basic funding, while the landscape of external financing provides quite broad and diversified opportunities serving different types of basic, strategic, and applied research.

The disciplinary structure and distribution of research financing and the concentration of external financing among top universities should also provide an attractive framework for academic research. Such a statement, however, paints an overly rosy picture, hiding certain structural problems of academic research financing.

Most importantly, there is no doubt that universities are structurally underfinanced. In relative terms, financing of universities is still based on formulas developed some twenty years ago, whereas student numbers are already about one-third higher than predicted.[8] Basic financing as the major source of income for universities is still jointly provided for teaching and research. Internal competition for scarce resources between teaching and research has increased. In many cases, academic research has been the "second winner" in this game, while relatively little has been done until recently to assure the universities' research function under conditions of the mass university (Schimank 1995). A comparison between research-intensive and prima facie well-equipped German universities on the one hand, and leading US universities and European universities on the other hand reveals, for example, the following picture: the annual budget per student at leading research universities in the United States is about six to ten times as high as for leading research universities in Germany. And the annual budget per student of Eidgenössische Technische Hochschule Zürich in Switzerland is about six times the budget per student of the Technische Universität München, which is one of the winners of the recent Excellence Initiative in Germany (Wissenschaftsrat 2006).

Further, academic research in Germany has to deal with a well-established competitor for resources within the public research sector. Currently, the gross expenditure for higher education in Germany (16.8%) is lower than the average of OECD countries (18.7%), while expenditure for public research organizations outside universities is higher in Germany (13.4%) than the OECD-average (10.9%) (OECD 2005).

Recent Developments in the Policy Framework

The traditional governance configuration of the German university system was once characterized by Clark (1983) as a combination of strong state regulation and professional self-control by an academic oligarchy. Despite the reorganization of the university system after 1945, the "democratization" of universities' decision-making processes in the early 1970s, and German reunification in the early 1990s, this historical compromise has persisted. In legal terms, this combination of state funding and regulation with constitutional guarantees of the freedom of teaching and research is expressed by the dual nature of universities as both public institutions and corporations of autonomous individual chair-holders. In effect, this means that institutional autonomy of the university and related capacities for organizational self-steering are low. In addition, the overall constellation of relevant political actors is complicated by the dual responsibilities of the federal state and the Länder for higher education and public research.

All together, these deeply institutionalized actor constellations created conditions that forestalled major reform of the system. They cumulated into a situation that may well be characterized by Scharpf's (1985) notion of a *Politikverflechtungsfalle*—a mutual blockade of actors in a wicked, multilayered political game. First, for a long time universities and government actors stressed quite different viewpoints of what needed to be done to cope with problems in higher education and research. While the universities wanted more money, vaguely promising "some reforms later on," government actors insisted on "big reforms now" and "some money afterward." Second, the dual responsibility of the federal state and the Länder tended to complicate the picture on the side of government. Basically, major reforms required agreement between the federal and the regional level as well as among the different Länder. The outcome of this constellation tended to be the smallest common denominator for reform that still had to be implemented later on by the Länder. Third, governmental capacities to achieve certain goals within universities were also quite limited. Their realization would have required a substantial capacity for self-regulation within the universities, which would have had to act as implementation agents. But it is this very capacity that universities were lacking. Fourth, actor constellations within universities were instead dominated by a strong and risk-averse professoriate that brought up mutual nonaggression pacts. Support for the status quo and the smallest common denominator for reform were likely outcomes. Thus, a mutual blockade between the federal and the regional government, between universities and government, and within universities has occurred. Only since the mid-1990s have initiatives been taken to break up the "reform blockade," as it was called.

REGULATION AND LEGAL FRAMEWORKS

Since the mid 1990s, several amendments to the Federal Framework Act for Higher Education have delegated certain responsibilities to the Länder and the universities. In 2005 and 2006 further changes in state regulation took place. They were the outcome of a recent and hot debate about the division of responsibility between the federal government and the Länder for higher education that nearly led to a constitutional crisis. The federal government agreed to give up its responsibility for setting legal framework conditions in higher education, except for rules and regulations regarding access to higher education and degree structures. Further responsibilities thus are delegated to the Länder, while the federal government and the Länder will continue to jointly fund public research infrastructure and research programs. The new division of responsibility has already stimulated competition between the different Länder in the governance and financing of higher education and research. Such a "competitive federalism" is, in fact, likely to break up the reform blockade of the past while it also creates new problems for the overall coordination of the academic system in Germany.

Tight regulation of universities' internal procedures by the state was replaced by output control mechanisms to be established. In the future, financing of universities has to be linked to performance. Within a university, resource allocation should be based on performance as well, while institutional leadership is expected to be strengthened. The present situation is that all Länder have implemented those aspects of deregulation expected to bring about efficiency gains. They have given universities and professors more room to maneuver with regard to financial resources by abandoning many features of the old budgeting system and introducing, instead, lump-sum budgeting. In five Länder, universities can choose their legal status. They may remain public institutions or can become foundations of civil law. These options open up additional possibilities for maneuver in financial and organizational matters, even though universities remain bound to the public-sector salary structure and its rigid employment categories. The approval of study programs has been delegated from the ministries to newly founded agencies of accreditation, in which academic peer assessment and quality criteria have a stronger role than before. However, it is still up to the ministry of a particular *Land* to decide whether a given program at a given university fits into the overall planning of that Land. State authorities are still reluctant to relax regulations relating to the structure and size of faculties and the appointment of professors. Only a few Länder have done away with the ministry's right of approval over the appointment of professors and have delegated this decision to rectors.

During the 1990s, the formal powers of rectors and deans increased in all Länder. Many decisions can now be made without a majority in the university senate or the faculty council. In six Länder deans now allocate financial and personnel resources on their own.[9] Terms of office for these positions have been extended. Deans who were traditionally elected for two years now serve four. In five Länder, deans now need dual approval—not only from their faculty but also from the rector. They are beginning to be seen as important "persons in the middle" who not only represent their faculty's interests to the rector but are also supposed to implement the rector's policies within their faculty—if necessary, against the will of the majority within their faculty council. All in all, the system is acquiring elements of hierarchy; the top of the hierarchy is no longer merely a symbolic figure.

At the moment most measures to build managerial self-governance, however, remain incomplete. Change needs time and is not without resistance. The consensus-oriented culture of the academic profession compels many in leadership positions to act as if they have no new powers. Thus, formal competencies remain unused, and consensus, at least among professors, is still sought by rectors and deans. One reason for this is that those in leadership positions know that one day they will return to the "rank and file," and they do not want to make enemies among those who may come into power after them. But the more important reason is that many have internalized the traditional organizational culture of consensus during their long academic socialization and lack capacities for professional management.

All in all, we observe, however, a potentially far-reaching transformation that is beginning to gather speed in German higher education and research. Its basic message: replace the old regime, dominated by a state-regulated profession, with a new regime in which the Länder and organizational leadership gain in power, and in which certain rigid rules of the game are relaxed. There are important differences between as well as within the various Länder. Some Länder and some universities want to be front-runners while others are more reluctant to reform. Still, the reform agenda is gathering momentum, and the general direction of change is the same everywhere.

FINANCING AND BUDGET ALLOCATION

There has always been an important element of competitive pressure among individual researchers at universities. The pressure to go for competitive external financing of research has become stronger with growing student numbers and declining intrauniversity support for research. Recently, two major competitive funding programs have been initiated. The Pact for Research and Innovation is expected to increase the annual budget of public research outside higher education by 3% per year (about €150 million). The flagship for academic research is the Excellence Initiative, with a five-year budget of €1.9 billion for the support of graduate schools, for clusters of research excellence, and most importantly for a select group of up to ten elite universities.

The Excellence Initiative was established to increase the worldwide competitiveness of the German university system. Universities as a whole have to set up overall strategic plans for the further development and sustainability of their research portfolio and performance in order to succeed in this highly selective program.

The program brings new money and a new element of peer-reviewed competitive financing into the system. But the total amount of money involved in the Excellence Initiative, to be spent by 2011, is less than the annual budget of one top American research university. Hopes are thus high that such initiatives will generate spill-over effects, attracting further private investment as well as further cooperation between universities and public research institutes.

The latter element may well be crucial for the future standing of academic research in Germany. The diverse German landscape of public research may be attractive when taken as a whole but calls for further cooperation and synergy if academic research is expected to enhance its overall standing. Up to this point, the policy framework has inhibited structural cooperation between academic research and public research outside of higher education beyond local and incremental means. Here a variety of issues come into play: different schemes of funding and responsibility on the federal and regional level for universities and for public-sector research institutes, different rules for staffing and salary, and the deeply institutionalized "pillarization" of the system (Heinze and Kuhlmann 2007). The Excellence Initiative has clearly stimulated further cooperation and strategic alliances between universities and the other public research organizations in their region. Such collaborations are supposed to contribute to the establishment of poles of critical mass that provide a competitive advantage within the program. The Technical University of Karlsruhe (annual budget €280 million) plans, for example, to merge with the Research Centre Karlsruhe (annual budget €320 million). The other two universities selected in the first round of the program—the University of Munich and the Technical University of Munich—capitalized on the extension of further cooperation with ten Max Planck institutes and other public research labs in their region.

Since the beginning of the 1990s, several of the major public programs for R&D have indirectly affected academic research as well. The two major, large-scale programs—nuclear/fusion energy and space—have changed places, with space research taking over a prominent role; biotechnology, information technology, and new materials have received a steadily increasing share of funds, as has environmental research and technologies. Further, new programs have been implemented to support entrepreneurial activities of universities, patenting and licensing, and public-private partnerships.

The German Research Foundation (DFG) has also changed its funding policies to some extent. The DFG still spends the main bulk of its budget (around 40%) on its very first program, the Individual Grants Program. Any German who holds a PhD and has been a resident researcher for a certain period at a German research institution can submit an application for the funding of a research project without any programmatic quotas or other limitations. The program is completely open to all kinds of research proposals.

However, several new programs addressing, for example, groups of researchers have been introduced. In recent years, about 28% of DFG funds have been used to fund "collaborative research projects" that are long-term, cross-disciplinary research programs carried out at universities. Another 5% of DFG funds went into "research training groups" that promote doctoral students through systematically structured and usually cross-disciplinary study programs. There is also programmatic funding in "coordinated programs" and programs providing funding for research infrastructure.

The introduction of tuition fees is a very recent phenomenon in Germany and was accompanied by major political conflict between the federal government and the Länder as well as political stakeholders (Kehm 2006). In 2005 a decision by the Constitutional Court opened a window of opportunity for the introduction of tuition fees. Currently, almost all German Länder have decided to introduce moderate tuition fees of about €500 per semester. They provide a very welcome, substantial additional income for the universities. Income from tuition fees is, however, legally required to be invested in the improvement of teaching. While the outcomes regarding access and equity for students cannot yet be examined, it is unlikely that this further source of income will provide a major opportunity for cross-subsidization of academic research in the near future.

Finally, the political framework for resource allocation to and within universities has seen major changes since the amendment of the Higher Education Framework Act in 1998. Tight regulation of universities' internal procedures by the state was replaced by output control mechanisms to be established. Intraorganizationally, resource allocation is supposed to be based on performance. The Länder have abandoned many features of traditional public budgeting by introducing lump-sum budgeting.

Overall, these changes are widely considered to be a clean break with the previous policy rationale underlying governmental regulation of higher education and research. The actual leeway is, however, likely to be restricted. Basically, academic research in Germany still suffers from low public support per capita compared to academic research in many other highly developed countries. Further, within universities there is still little "free money" floating around that may be used for priority-setting; most resources are obviously bound to costs in infrastructure and personnel.

INFORMATION AND EVALUATION

In the mid-1990s a whole series of evaluations of the role of public research were begun. Some of these evaluations and their related policy measures had an impact on the overall public research system. The most prominent examples include the work of a nonpermanent commission set up by the federal ministry that had a decisive influence on governmental R&D policies. For example, action was taken to invest in biotechnological research and to create a National Health Council that reflects priorities in medical research. Foresight exercises and long-term prospective studies were launched and contributed to a whole range of new programs, "lead projects," and support schemes that mark a shift from a mission-oriented science and technology policy to a diffusion-oriented policy.

Ironically, the evaluation of the public research system of East Germany by West German scientists and scholars undermined their traditional veto against any kind of evaluations in German universities and public research institutes. In 1996 the federal government and the Länder agreed to start a major system evaluation of those organizations that they jointly fund. First, some research fields at universities were evaluated to identify directions for their future research priorities. Subsequently, all public research organizations outside higher education and the German Research Foundation were evaluated by national and international commissions. These evaluations were mainly meant to assess the structures and procedures of these organizations and to make recommendations for organizational reform. Assessment of research performance played a secondary role, partly because systematic monitoring of performance was lacking. It is thus not surprising to note that these system evaluations recommended, among other things, the introduction of a regular system of organizational and systemic monitoring, including research performance indicators.

Thus far, there has been no federal evaluation scheme covering all German universities. This is partly due to the decentralized responsibility of the Länder and partly due to the reluctance of many actors (not only within universities) to implement such systems in a supposedly homogeneous university system. In general, performance measures have not been used to allocate basic funding, nor have there been evaluations for this purpose. Recently, some Länder as well as some individual universities have begun evaluations on their own. They are mainly intended to improve self-information, and thus self-learning, within the system, but some of them are also used for resource redistribution to the strong performers. Evaluation methods and criteria differ considerably. In most cases some kind of peer review is established, but there are also examples of indicator-based formulas.

Since the late 1990s, Länder have set up commissions to assess universities and their overall teaching and research performance. These commissions have initiated redirections in study programs and research priorities. Recently, "management by objectives" has become institutionalized in the form of mission-based contracts between ministries and universities. In theory, such contracts should not contain concrete recommendations, only goal statements; in practice, this flexibility is often not granted to universities, allowing ministries to revert to regulation under the guise of guidance. For example, instead of formulating the goal that the share of female researchers in certain areas shall be increased by x% over the next six years, leaving the actual pursuit of this goal to each university, ministries prescribe detailed and uniform procedures as well as organizational structures for gender mainstreaming.

Finally, two recent developments point in the direction of the establishment of a national research evaluation scheme. The German Research Foundation has set up the Center for Research Information and Quality Assurance (IFQ), which is supposed to provide evaluative measures for in-house information on the activities of the German Research Foundation and also on academic research in general. The Science Council (the WR) is currently undertaking a federal research evaluation in two pilot disciplines that is supposed to form the basis for an encompassing system-wide evaluation.

LINKING ACADEMIC RESEARCH TO ECONOMIC DEVELOPMENT

As in many other countries, linking public research in general and academic research more specifically to economic development has gained in importance in political priorities. Recent analyses of knowledge and technology transfer suggest a move from an early information and documentation model, via a cooperation stimulation model, to a blurring-of-boundaries model (Krücken, Meier, and Müller 2007).

Traditionally, the German knowledge-transfer system may be characterized as relatively stable, structured, and homogeneous. University-industry relationships rely heavily on personal contacts, organizations, and institutions that are national in scope, highly differentiated by industry and technology, and in many instances well interconnected. These structures and their relative stability support cross-institutional as well as incremental innovation and technology diffusion among a number of important industries. The German system is thus well organized to perform in the application of new technologies in major existing industries but less well organized to perform in new or fast-moving fields (Abramson et al. 1997).

Since the mid-1990s, many policy instruments have been enacted to improve links between universities, research institutes, and industries. First, financial incentives have been established in various ways. Since 1995, major programs have supported small and medium-sized enterprises (SMEs) in technologically mature industries as well as the establishment of new SMEs in emerging fields. Such programs also include components to strengthen science-industry relationships, especially in new fields. They have helped to establish hundreds of new SMEs, most successfully in certain regional poles in the south of the country with a high density of preexisting science-based industries and public research organizations. Direct research promotion within thematic programs is a major means to strengthen contract research of public organizations for industry as well as collaborative R&D projects. A special and prominent approach in this area is "lead projects" that follow a bottom-up network approach in setting research priorities for public support. So far, empirical evidence suggests that such financial stimuli have been mainly employed by public research organizations that have preexisting strong ties to industry (technical universities, engineering in universities, Fraunhofer institutes, universities of applied sciences) (European Commission 2001). Tax reduction and subsidies for industry investment in public research have also been established, but they seem to have only limited effects in creating new links or start-ups (Krull and Meyer-Krahmer 2001). Instead, they tend to reinforce existing ties in public-private cooperation.

Institutional arrangements have changed to some extent as well. The public not-for-profit status of universities complicates contract research for industry that has to become part of the regular activities and budgets of the universities. This has necessitated organizational innovations establishing contract research institutes outside universities that are partly funded by governmental money and research contracts (called "An-Institutes"). Many observers feel uneasy about their status at the boundaries of universities, where they still have to comply with a variety of public rules covering financing, personnel, and organization while relying to a considerable extent on demand from user markets. In effect, such institutes tend to share the strengths and weaknesses of the Fraunhofer institutes, which have, however, much more developed and coordinated structures to deal with such a situation.

In 2002 the universities became responsible for patenting activities in place of the individual academic, who traditionally held the title to all potentially patentable innovations. It was argued that individual academics may lack knowledge and capacity to judge the patentability of their research as well as the financial and legal resources to manage patent applications. Nevertheless, patent applications per researcher in science and engineering had already increased substantially before the enactment of this new legal framework (European Commission 2001). It remains to be seen whether and to what extent the current establishment of coordinated structures in universities for patenting and licensing will further encourage such activities as well as increase university income from such sources. Experience from the United States tends to show that these activities are costly and that only a small number of patents generate real financial benefit to universities.

Further, changes in rules and regulation regarding the mobility of personnel between universities and industry are currently under consideration. Traditionally, "enforced mobility" of younger academic researchers to other employment sectors has been very high (Enders and Bornmann 2001). A widespread system of temporary contracts for younger academics and lower wages in universities attract a considerable proportion of them to other public and private employers. In later career stages, job mobility between sectors is very low due to unfavorable framework conditions, such as different wage systems, contractual arrangements, and pension systems.

Finally, growing attention to public-private cooperation and knowledge diffusion has led to a change in the overall policy philosophy that can be described as a move from a top-down approach to a network approach. Science and technology policy now routinely postulates the efficiency and effectiveness of steering in and by heterogeneous networks (Krücken, Meier, and Müller 2007). Innovation networks, regional clusters, science poles, excellence networks, and competence networks (BMBF 2004) are spreading as ways to stimulate cooperation between heterogeneous partners as well as ways to encourage neo-corporatist policymaking in these areas.

In summary, several changes in the policy framework have been established in order to strengthen old ties and create new ties. Support for SMEs from federal programs both helps new companies become established and enhances new science-industry links. Spin-offs are fostered by different activities of the federal government and the Länder. Organizations working at the boundaries of universities and industries have been established, regulations of intellectual property rights have been changed, numerous programs to support contract research and cooperative research have been enacted, and transfer units and technology consultancy have been newly established. It is certainly too early to provide a fine-grained assessment of all these recently introduced measures. What can be said is that interactions between non-academic public research and academic research, on the one hand, and industry on the other hand, have increased and have reached a high level (European Commission 2001).

The instruments described above are, however, mostly elements of a transfer-push approach. Public-sector research in general and academic research more specifically are the areas being reconsidered (accompanied by a focus on SMEs). But demand is another important element in the equation of industrial innovation and its link to academic research. So far, instruments for further cooperation and diffusion of knowledge and technology seem to have strengthened traditional lines of cooperation significantly but have not helped emerging new companies very much or increased demand for innovative products. We find path dependencies of the traditional German innovation system not only on the side of the public research enterprise but also on the side of private industry, with its traditional base in "first innovator" strategies in specific branches. Small countries have shown that there are more flexible and more efficient strategies of "the intelligent fast imitator," especially in markets and products of new technologies and services (Krull and Meyer-Krahmer 2001; Pavitt 2001). From this perspective, the performance of the system is nowadays determined not so much by static competitive advantages as by its "dynamic efficiency."

Reform of Doctoral Training. At the crossroads of recent policies for higher education and research lies an important province: doctoral training. Sensitivity to the virtues and plights of research training in universities and its usefulness for the further careers of PhD graduates has obviously grown. In Germany, as in a considerable number of other European countries, doctoral training is shifting away from the traditional Humboldtian model and toward the professional model (Kivinen, Ahola, and Kaipainen 1999). The underlying rationale for these new policies is three-fold. Reformers hope first for efficiency gains in terms of the PhD factory's input and output, second for employability gains in terms of widening career perspectives of PhD graduates beyond the traditional labor markets in academe and science, and third for innovation gains in terms of increasing knowledge transfer during and after research training (Enders 2005).

Criticism of the German system of doctoral training and junior staff careers has a long tradition that dates back to Max Weber's famous lecture "Science as a Vocation." Certainly, there is ample evidence of structural and cultural deficiencies that impact not only the efficiency of the training and career system but also the innovative capacities of the system overall. Many stereotypical dichotomizations contrasting the old-fashioned "Humboldtian system" with the modern "professional system" fall short, however, because of a failure to understand the mix of strengths and weaknesses of the German tradition and its recent reform.

Research training and *career* are not independent commodities but highly dependent variables presupposing a well-functioning university system, adequate resources for teaching and research, and developed links with society and the marketplace. In regard to the latter, conventional wisdom acknowledges that Germany belongs to those countries where the PhD has not only a relatively high prestige within society but a relatively high value on the overall labor market.

Recent studies that have delved for the first time into the labor market–PhD connection (Enders and Bornmann 2001; Hartmann 2002) confirm such a strong link as well as a deeply embedded function of the PhD for the self-reproduction of elites within society. Half of the PhD graduates leave higher education and research for employment in government, private industry, and nonprofit organizations—all areas where the PhD responds to a wide range of different labor markets. Labor market links to the various sectors are, of course, clearly differentiated by discipline. Empirical evidence also shows that PhD graduates have a significant career advantage in comparison with holders of master's degrees or the equivalent. Furthermore, the PhD provides an entrance ticket, especially for elite positions—consider, for example, that 50% of the members of the boards of the two hundred biggest German companies hold a PhD. This nexus with the labor market serves also as a major channel for the dissemination of knowledge and know-how into other sectors of the labor market and society.

What does this mean for the science system? Most importantly, the science system's needs for self-reproduction and for PhD candidates as a cheap and willing workforce can rely on a relatively promising external labor market. External labor market prospects are an important stabilizer for the research sector in times when universities and other research producers face a decrease in resources and can offer only a few permanent positions to the young scientists and researchers. Such a training-career nexus certainly has an advantage in comparison with other systems that serve more exclusively the higher education and R&D labor market (Gläser 2001).

Currently, German universities graduate about twenty-six thousand PhDs per year. The number of PhD graduates increased by 35% between 1993 and 2002 (excluding medicine). Medicine and the sciences accounted for the bulk of PhD graduates (one-third each), followed by the social sciences (13%), and the humanities and engineering (10% each). Overall, one out of six of German holders of master's degrees or the equivalent finished a doctoral thesis. The percentage varied a lot across fields of study: in medicine, 80% of this group achieved a PhD; in the sciences the proportion was 34%, with peaks in chemistry (69%), biology (44%), and physics (42%). The proportion of holders of master's degrees or the equivalent achieving a PhD was much lower in other fields, including engineering (between 9% and 13%) (Wissenschaftsrat 2002).

Germany has, however, reached a stage of development at which an amalgam of latent problems has raised—and still raises—the questions of whether and how a system of research training could be designed more systematically (Enders 2002). Research training traditionally had—and to a considerable extent still has—an unclear status within the higher education system. It was not part of regular university studies, and there were no graduate schools with restricted attendance and course requirements. Research training was also only indirectly part of national science policy. There were some policies dealing with the various financial support systems for posts and fellowships run by universities, academies, foundations, and research councils. But a research candidate was either left by him- or herself or was accepted as an apprentice of a chair or research team. Rarely did a doctoral candidate register as such—a persisting situation that leads, among other things, to a lack of available basic statistics about the number of doctoral candidates in the system or about drop-out and success rates. There was hardly any national market for the selection and recruitment of research trainees. Most of them were locally recruited, thus solving the information problem of mutual selection of supervisors and candidates by lo-calized knowledge based on face-to-face interaction. Competition for more-or-less prestigious places among candidates, and competition for the talented among uni-versities, rarely took place. Some studies showed a considerable lack of guidance and supervision by mentors as well as a lack of integration of doctoral candidates into the broader scientific communities and the daily work routines of departments and institutes. The incredible length of this qualification period and the increasing age of PhD graduates were related to these problems as well (Wissenschaftsrat 2002).

Certainly, all of these factors create more of a problem in those disciplines in which research is carried out in a rather individualistic way. But science and engi-neering fields also experience problems. Various efforts have been undertaken to overcome the weaknesses of the system, most importantly by the introduction of *Graduiertenkollegs* (graduate colleges) since the 1990s (DFG 2003). In the graduate colleges, PhD candidates are embedded into the context of a larger group of aspiring students who are supervised by a number of professors who provide some further training courses as well. Graduate colleges are not established on a permanent basis but for three years, with the possibility to extend this period to a total of nine years after successful performance and positive review. PhD candidates are expected to be recruited nationwide on a competitive basis, and candidates are expected to fulfill special criteria, such as a relatively short length of first-degree studies, a relatively low age (maximum of 28 years), above-average marks, and a thesis proposal of outstand-ing quality.

Up to this point, the impact of the graduate colleges on the performance of research training has been assessed positively. Baldauf (1998, 176) concluded, for ex-ample, that the graduate colleges "introduced a flexible curricular structure; helped to overcome the isolation of doctoral candidates, especially in the humanities; and seemed to have reduced the duration of the doctoral degree, but not necessarily to the extent the Science Council was hoping." Other studies in this field (DFG 2003; Hein, Hovestadt, and Wildt 1998) have come to quite similar conclusions. But we have to keep in mind that the graduate colleges are currently a 10% phenomenon, while most doctoral candidates work under conditions of the traditional system, with its strength and weaknesses as detailed above. The establishment of further graduate colleges—not financed by the special program of the DFG but by other means—has been strongly supported by the Science Council and the German Rectors Conference (Hochschulrektorenkonferenz 1996). Recently, a number of universities have established some kind of graduate school. Further programs are in the making, supported by some of the German Länder and by the MPG, the DFG, and the Volkswagen Foundation. A very recent incentive toward such structural contexts for research training has been provided by the Excellence Initiative, which supports about twenty new graduate centers. Given the various aims, functions, and views of the PhD, it is premature to assume that such new role models will become standard for all situations and all disciplinary and trans-disciplinary contexts. A diversity of organizational and structural forms, as well as different validation criteria and procedures, will probably determine the future face of research training in Germany.

Autonomy and Accountability of Universities. Traditionally, institutional autonomy of the university and related capacities for organizational self-steering have been limited. As a consequence of the reshuffling of authority and responsibilities across the different levels in German higher education, universities as organizations have become important foci of attention in the system's coordination. The "upgrading" of the university suggests, however, that the university is transforming from a loosely coupled to a more tightly coupled system. As shown above, basically two types of measures have been undertaken in order to strengthen institutional autonomy as well as capacity for organizational self-steering: deregulation and managerialism.

Regarding deregulation, university reform offers a specific example of the fact that changes in the legal framework are a necessary step in bringing about changes in practice, and thus changing the significance of other governance mechanisms, such as academic self-governance or hierarchical self-steering in universities. It has already been shown that the federal government has withdrawn in a stepwise manner from the legal authority to set overall and detailed frameworks for higher education and research. Trute et al. (2007) show that through several interconnected decisions, the Federal Constitutional Court has substantially reduced the powers of the federal government to make framework laws, so that it has become impossible for it to set standardization de jure. The Federal Constitutional Court has also made basic revisions of earlier rulings on the guarantee of academic freedom. Whereas the court traditionally emphasized the individual freedom of academics as well as broad legal requirements for the organization of institutions of higher education, in a more recent decision it significantly changed this position and gave legislators broad freedom in regulating the organization of these institutions. This means that the previous harmonizing effect of the rulings of Germany's Federal Constitutional Court on organization of higher education in the individual states has largely ended. One effect of this development is a substantial broadening of the options available to the Länder. It is still an open question whether advantage will be taken of these options or whether practice will lead to convergence into one specific model. Trute et al. (2007) show, for example, that in both Brandenburg and Lower Saxony, the ministry's decision-making rights and ability to impose conditions before granting approval have been transformed into the right to lodge objections to proposals, and powers have been transferred from the ministry to the universities or to actors who provide stakeholder guidance. Many powers of the ministries have been converted into supervisory and cooperative participation, for instance through the introduction of jointly formulated performance targets or through contractual agreements with the universities.

Recently, "management by objectives" has become institutionalized, in the form of mission-based contracts between ministries of the Länder and universities. Such contracts have become widespread, partly on the initiative of the ministry and partly on the initiatives of universities that thought to establish some stability in framework conditions set by the ministry. The contracts take different forms, sometimes setting the basic funding and the overall structure of all universities in a Land, sometimes applying only to some universities, and sometimes including performance targets, sometimes not. Usually, such contracts are a mixed blessing to universities because attempts to reform the relationship between ministries and universities and to establish internal quality assurance measures have been overshadowed by budget cuts introduced via the same contracts. In theory, such contracts should also not contain concrete recommendations, only goal statements. In practice, this flexibility is often not granted to universities. Consultations between the ministries and the universities are carried out in the shadow of the hierarchy: "If goals are not agreed upon the ministry can revert to the governance mode of state regulation by specifying its own binding targets" (Trute et al. 2007, 14).

Regarding personnel matters, universities so far remain bound to the public-sector salary structure and its rigid employment categories. German professors are civil servants and in theory must be treated alike. A recent reform of the professorial salary structures was meant to change this situation. Salaries were split into two parts: a basic salary component that is lower than the traditional pay for professors and a performance-based component. Also agreements between the university and the professor about his or her support from the university have changed. Basic subsidies provided by the university are nowadays limited to five years, and their renewal is based on the outcomes of internal or external evaluations. First appointments of young professors are currently based on fixed-term contracts of five years. All these measures are obviously meant to push the professoriate toward more accountability and efficiency, but many observers believe that they will be detrimental to the attractiveness of an academic career.

Regarding "managerialism," it has been pointed out above that during the 1990s the formal powers of rectors and deans were strengthened in all Länder; in six Länder deans now allocate financial and personnel resources on their own. Terms of office for these positions have been extended; in five Länder, deans now need dual approval—not only from their faculty but also from the rector. Further new elements of managerial capacities within universities concern internal resource-allocation systems and contract management, full-cost systems, and the implementation of controlling and accountancy systems as well as the development of mission statements and midterm strategic development plans. A recent survey of the current state of the implementation of management measures in German universities provides a scattered picture of a work-in-progress (Wigger 2005). Overall, so far, much less has been implemented than has been discussed. There are huge differences between the Länder and the individual universities. Most universities describe their own situation as being in a state of "experimentation."

Academic Research Output and Potential Consequences of Recent Reforms

Obviously, any reflection on the consequences of the changes in the policy framework for academic research that have been addressed has to be cautious and preliminary. Many measures for reform in governance and funding of academic research, internal organization of universities, their accountability, and their links to the economy have been introduced recently or are currently under construction. We observe a bewildering variety of new rules and regulations, programs and initiatives, on all levels of the system. In addition, different regions in Germany not only have different capacities in public research but follow partly different methods in the instruments used and their implementation. Further, the traditional governance regime for higher education and research itself is under reconstruction and has not yet found another prestabilized equilibrium. In the German context, locking up traditional institutions and actor constellations is most likely to be good news given the mutual blockage for reform that had emerged over time as well as the traditional belief in top-down legal reform. Consequences for academic research, however, can only be hypothesized and are due for further analysis of ongoing change and reform.

What can firmly be said is that in private- and public-sector R&D in Germany, including academic research, publication output increased in the past decade in absolute as well as in relative terms (that is, per million inhabitants). Many other countries have increased their publication output as well, and Germany thus forms part and parcel of an overall race of "publish or perish" that has become increasingly fierce (table 7.5). In consequence, in terms of publication output, the midlevel positioning of German R&D has not really improved in cross-national comparative perspectives.

Because the total number of researchers, both those in the higher education sector and those in the other public research organizations outside of higher education, has increased much less than their publication output, we can also conclude that publication output per researcher has increased as well. In cross-national comparative perspectives, there is, however, quite a gap between the publication output per researcher in Germany and in a number of other countries (fig. 7.1). Basically the same holds true if we look at the citation impact scores of German research articles in the international scientific and technological journals. The normalized citation impact score for Germany (1.13) is well above the average score, which is fixed at

TABLE 7.5.
Scientific and engineering publications, 1991 and 2001
(published articles per million inhabitants)

	1991	2001		1991	2001
Australia	618	758	Netherlands	671	786
Austria	353	564	New Zealand	598	742
Belgium	416	582	Norway	564	721
Canada	817	727	Poland	102	147
Czech Republic	279	256	Portugal	65	208
Denmark	733	931	Slovak Republic	—	177
Finland	640	983	Spain	187	387
France	402	514	Sweden	945	1,159
Germany	412	530	Switzerland	886	1,117
Greece	153	304	Turkey	15	60
Hungary	175	243	UK	696	807
Iceland	403	610	US	766	705
Italy	243	385	Total OECD	454	468
Japan	—	451	EU25	—	485
Korea	31	233	EU15	416	556
Mexico	13	32			

Sources: National Science Board, Science and Engineering Indicators 2004, www.nsf.gov/statistics/seind04/; population data from OECD MSTI database, June 2004; OECD (2004, 33).

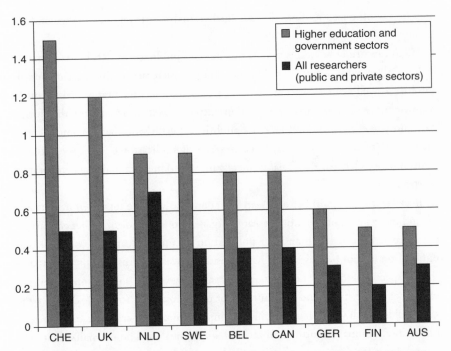

Fig. 7.1. Research articles per researcher, 2002–3. *Sources:* CWTS / Thomson Scientific, OESO/MSTI 2003/2; data treatments CWTS.

1.0 per field of science, and comparable to the scores of countries such as Sweden (1.14), Belgium (1.13), and Finland (1.16). Switzerland (1.41), the Netherlands (1.26), the United Kingdom (1.23), and Canada (1.21) score, however, much better than Germany (NOWT 2005).

Further, German R&D productivity in terms of patenting has increased in recent years (fig. 7.2). The number of triadic patent families per million inhabitants increased from about 80 in 1990 to 130 in 2002. In this area German R&D clearly performs above the EU average (81 per million inhabitants) and slightly above the United States (123 per million inhabitants). Some other European countries, such as Finland (269) and Sweden (241) as well as Japan (185) have improved their output in triadic patent families per million inhabitants much more than Germany.

Obviously, the main share of patenting results from the performance of the well-developed private R&D sector in Germany. But academic research in engineering and the applied sciences is well connected to the private sector and contributes to this performance in indirect and direct ways. Heinze and Kuhlmann (2007) analyzed German publications (SCI) and patent applications (DPA/WPI) (relative to

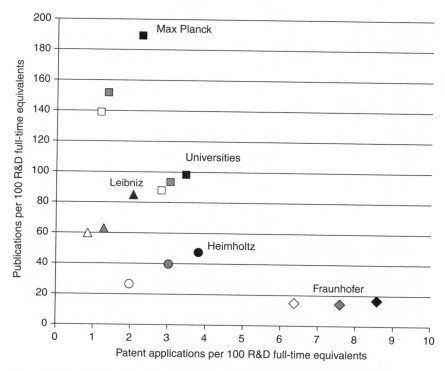

Fig. 7.2. Publications and patent applications of the public research organizations in Germany between 1990 and 2002. White shade = 1990–93; gray shade = 1995–98; black shade = 1999–2002. *Sources:* SCI, WPINDEX, PATDPA (host STN); Heinze and Kuhlmann 2007, 194. *Notes:* For universities the scaling factor is 50 R&D FTE (instead of 100) because the institutional mission of university faculty members embraces both teaching and research. Example: in the period 1999–2002, a university scientist published one article per year on average, while every thirtieth university scientist filed a patent. The Research Centre for Computational Sciences (GMD), which was transferred from Helmholtz to Fraunhofer in 2001, is not included in the data.

the number of research staff) between 1991 and 2002. Three trends emerged from this analysis. First, academic research (like other public research organizations) increased its productivity in both areas of publication output and patenting (fig. 7.2). "This is a clear indication of the high pressure on the research system to demonstrate higher output efficiency. Despite a substantial decrease in public sector research funding, scientists in 2002 produced significantly more research papers and patent applications than in 1991" (Heinze and Kuhlmann 2007, 193).

Second, increase in output in the direction of technological research (indicated by patent application) is more pronounced than shifts in scientific productivity (indicated by publications). "This development implies a decreasing institutional differentiation in the German research system in two ways. First, institutes that did not carry out technological research in the early 1990s apparently do so today. Second, institutes whose core competence has traditionally been in technology research have come under considerable pressure" (Heinze and Kuhlmann 2006, 193). Finally, important public-sector research organizations outside higher education (Max Planck institutes, Helmholtz research centers, institutes of the Leibniz Association) increased their output, especially in terms of publications, more than the universities. Within Germany, universities are thus losing ground within the internal competition of public R&D.

STRATIFICATION AND REGIONALIZATION

It seems likely that recent developments will contribute to a further stratification of universities. On the one hand, growing accountability, evaluative measures in implementation, and national and international rankings of different kinds highlight differences in quality and performance that have already existed across the system for a long time. The notion of "egalitarian homogeneity" has held basically true with respect to governmental policies for universities; however, it was always a myth when it came to the relative strengths of universities in research. It makes a difference, however, when strengths and weaknesses become publicly known and are used for competitive means beyond the traditional peer-review-based system of project funding. The recent Excellence Initiatives provide a most obvious example of attempts to break with the long-standing policy rationale. Usually, such policies follow the Matthew approach ("Those that have will be given more") and not the Robin Hood approach ("Take from the rich and give to the poor"). It is, however, unlikely that the policies will lead to a highly stratified system, at least not in the short run. The means provided to create some German elite universities may be sufficient to enhance their standing among continental European research universities, but certainly not to catapult them into the global league. Further, regional responsibility for universities and the diversified financing of academic research have created a patchwork of specific disciplinary strengths and weaknesses within individual institutions that is unlikely to disappear. True, the result is that many German universities rank at a (fairly) good middle level internationally. In addition, these changes in responsibility and financing provide a certain backbone for regional development of R&D and science-industry relationships across the board, instead of concentrating resources in a few world-class institutions. Nevertheless, attempts to strengthen the peaks of the system are accompanied by a regionalization of research: certain poles of excellence and relevance have already emerged and are likely to grow, such as in Munich, Karlsruhe, and Stuttgart. Some of them are intended to link up across international borders, such as in the Basel-Freiburg-Strasbourg corner of Switzerland, Germany, and France. While such developments are encouraged by national and EU policies and funding, they are not a consequence of these, but rather have much stronger historical roots.

SEPARATION AND COOPERATION

Intraorganizational separation of teaching and research is likely to be another outcome of change and reform in German higher education and research. The "unity of teaching and research" so far has had a rather detrimental effect on academic research in times of mass higher education. This is unlikely to mean that the idea of research-informed teaching and learning will be abandoned. Rather, ongoing separation of organization, financing, and evaluation tends to increase the centrifugal forces pulling high-level research away from mass higher education in Germany. An alternate possibility is that research capacities, on average, will erode in the future due to the dramatically increasing number of new students expected in the near future.

Further, a number of recent measures tend to stimulate additional cooperation and linkages between academic research and other public research institutes. This is likely to be of growing importance for German universities as well as for the innovative capacities of the public research sector overall. Strategic alliances of universities with the other public research organizations in their region are emerging and are contributing to the establishment of poles of critical mass that provide a competitive advantage. Certainly, many obstacles caused by different policy frameworks and institutionalized divisions of work still need to be overcome (Heinze and Kuhlmann 2007). Also, such alliances are in danger of falling short if they integrate public research institutes into the traditional structures of German universities. These alliances do, however, provide means to build focus and mass using existing resources rather than waiting for new resources, which are unlikely to emerge in the near future.

IN SEARCH OF MONEY AND EFFICIENCY

The indisputable underfunding of German universities—their heavy reliance on basic government funding and their limited capacities to raise money from other sources—is likely to be a major obstacle to further change and reform. Universities have been able to increase their income from private sources via contract research. They have also copied and pasted approaches from other countries and from each other, such as alumni networks, science and business parks, and collaborative research laboratories with industry. In addition, universities have introduced moderate tuition fees. All this is still very much in the making and does not fully compensate for the relative loss in public subsidies.

Finally, internal reorganization of universities, including strengthening their autonomy and capacity for self-steering, is also still in progress. Early empirical evidence has been gathered about the effects of this reorganization on institutional behavior and academic research (Minssen et al. 2003; Schröder 2003). The results are similar in that they identify only limited effects for the standing of academic research. It does not come as a surprise that partly autonomous institutions that have little financial and organizational leeway find it difficult to increase efficiency of resource allocation. Anyway, from the point of view of academic researchers, the success in the market of external research grants is seen as much more important than internal reorganizations (de Boer, Enders, and Schimank 2007).

Conclusions

Some fifteen years ago a major review of German higher education concluded as follows: "German higher education (and the public discussion of higher education) seems to include a few unquestioned assumptions, some of which represent the underlying components of the coordination mechanisms characterizing the German higher education system. . . . German higher education must face the challenges of a large and still growing system, and address the question of the extent to which these assumptions contribute to the strengths or weaknesses of the higher education system. This is necessary to make the system fit for the challenges of the 1990s and the next century" (Frackmann and de Weert 1994, 132).

The current situation of the academic research enterprise in Germany is probably best described by the notion of "shaking up a path-dependent system." The analysis of the strengths and weaknesses of the traditional system of academic research is relatively clear-cut and has been enriched by a broader perspective on its functioning within the overall system of innovation and public research:

- The success story of "Made in Germany" relies on a heavily export-oriented system in certain classical areas and some new areas of industrial production that are well connected to R&D-based innovation in general and academic research more specifically. Germany is a late-comer country in certain new areas, such as information technology, biotechnology, and nanotechnology, both in its industrial and its research infrastructure.

- R&D in Germany relies on a predominant private sector and an important public sector that is highly differentiated and segmented. The mission-based segmentation of public research and its highly differentiated funding base provide ample room for productive variety but call for further networking and synergy in order to overcome "pillarization."

- Academic research in Germany still performs on a good level according to international standards and has improved its output under difficult conditions. German academics are a welcome asset to other countries that are competitors in the "global academic arms race." In contrast, governance and financing of academic research in Germany have provided quite unfavorable conditions for academic research in the past.

- The traditional German governance regime for academic research tends to favor the status quo due to a mutual blockage of actors in a wicked, multilayered political game. Until recently, the mutual blockade between the federal government and the Länder, between universities and government, and within universities has been detrimental to the responsiveness and adaptability of higher education and research.

- Financing of academic research relies on a theoretically good mix of basic support and competitive funding via multiple sources. However, little has been done so far to assure proper financing of academic research under conditions of the mass university. Research has increasingly fallen under the shadow of teaching; some of universities' research capacities have been moved to other public research organizations; and overall financing has decreased.

- University-industry relationships rely heavily on personal contacts, organizations, and institutions that are national in scope, highly differentiated by industry and technology, and in many instances, well interconnected. The German system is well organized to perform in the application of new technologies in major existing industries but less well organized to perform in new or fast-moving fields.

- Output of PhD training is remarkable and relies on a well-developed and diversified funding base for PhD candidates as well as on attractive links to the external labor market that provide a major channel for the dissemination of knowledge and know-how into other sectors of the labor market. Recruitment, selection, and training of PhD candidates are highly personalized and lack competitive market elements or coherent structures.

- Individual freedom of academic researchers has been a major intrinsic motivation of the system overall and has contributed to the attractiveness of an academic career. Institutional autonomy and related capacities for self-steering within universities have been low, as well as capacities for systematic monitoring of the strengths and weaknesses of the system.

Since the mid-1990s, a bundle of reforms, programs, and initiatives have been enacted to overcome the well-known structural weaknesses of the system. It is easy to fault these efforts for their incremental and unsystematic character. However, in the German context they amount to a major challenge for the academic research system and its political coordination. Traditional notions of strong state regulation, academic freedom, and self-governance; the unity of teaching and research; the egalitarian homogeneity among universities and their units; and the separation of universities from other public research organizations are under reconsideration and are due for structural reforms.

Various measures have been implemented or are currently being implemented in regard to legal frameworks and the regulation of the institutional setting, financing and budget allocation, and instruments for information and evaluation:

- Regulation and legal frameworks: The regulatory framework of political coordination has been shaken up by the retreat of the federal government and the subordination of certain authority to the Länder and the universities. Measures to enact institutional autonomy and accountability are being implemented. Some Länder and universities use the "new freedom" of universities to establish a competitive advantage via internal reorganization, strategic planning, and interorganizational alliances.
- Financing and budget allocation: Competition for external research grants as well as for program-based financing of universities has increased. Academic research is attracting increased funds from contract research for industry. Additionally, modest tuition fees and new means of fund-raising for universities are being implemented. The Excellence Initiative brings in further support and a new element of competition to the system.
- Information and evaluation: System monitoring for public research has been established and is under preparation for academic research. Mechanisms to evaluate the existing stratification among universities and within fields of research as well as measures to support the top of the system have been established. Within universities, measures to build information systems and regular evaluations are under way.

Further specific measures have been put in place to support reform and to enhance system performance regarding technology transfer, PhD training, and the autonomy and accountability of the universities.

- Programs have been established to support R&D in the emerging new fields of interdisciplinary science and technology studies. Existing cooperation between academic research and public research institutes as well as with industry has been strengthened; incentives for new linkages have also been provided.
- New forms of recruitment, selection, and training of PhD candidates have been implemented that support a more transparent market for research trainees as well as more structured means of coordination, supervision, and peer review.
- Measures to enhance institutional autonomy and accountability have been implemented; institutional leadership and management are expected to be strengthened.

Any reflection on the consequences of these changes in the policy framework for the performance of academic research has to be cautious and preliminary. Many measures for reform in the governance and funding of academic research, the internal organization of universities, their accountability, and their links to economy have been introduced recently or are currently under construction.

So far, performance indicators do not show a "quantum leap" in academic performance but rather a consolidation of Germany as a good performer in absolute as well as in relative terms. Overall, productivity has increased, although the relative standing of the German system has not improved within the international competition. The recent policy measures may have had an impact in terms of assuring the German position in the increasingly fierce international race in science, technology, and innovation. A major step forward can be expected only if the recent measures truly unfold and unexpected negative consequences can be avoided.

It seems probable, however, that recent reforms will contribute to a further stratification of the German universities, though a steep stratification is unlikely, at least in the short run. The means provided to create some elite German universities may be sufficient to enhance their standing among continental European research universities but certainly not to catapult them into the global league. Attempts to strengthen the peak of the system are accompanied by the emergence or strengthening of certain regional poles of excellence and relevance.

Intraorganizational separation of teaching and research is likely to be another outcome of change and reform in higher education and research. The "unity of teaching and research" so far has had a rather detrimental effect on academic research in times of mass higher education. Ongoing separation of organization, financing, and evaluation increases the centrifugal forces pulling high-level research away from mass higher education in Germany. Further, a number of recent measures tend to stimulate cooperation and linkages between academic research and other public research institutes. Strategic alliances of universities with the other public research organizations in their region are emerging and are contributing to the establishment of poles of critical mass that provide competitive advantage.

Germany is a late-comer to the reform of the academic research enterprise. Many obstacles still need to be overcome, not only due to the traditional structures in the university system but also due to a certain set of strengths and weaknesses in the overall research and innovation system. In regard to academic research, three aspects call for further attention: the structural underfunding of universities in general and of their research function more specifically, the need for increased participation in higher education in general, and the extension of the nonuniversity sector in higher education more specifically. A lot will thus depend on the capacity of actors to give momentum to recent reform instead of searching for a way to lock in a new prestabilized institutional harmony, division of work, and responsibility. A lot will also depend on capacities to generate further sources of income for universities. The extent to which academic research will benefit from change and reform is due for further observation.

NOTES

1. For a more comprehensive overview, see Meyer-Krahmer 2001 and European Commission 2001. For an overview of the historical development of the German system of technological innovation, see Keck 1993.

2. Max Planck Gesellschaft, Web site: www.mpg.de.

3. Helmholtz Gemeinschaft Deutscher Forschungszentren, Web site: http://helmholtz .de.

4. Fraunhofer Gesellschaft zur Förderung der angewandten Forschung, Web site: http:// fhg.de.

5. Wissenschaftsgemeinschaft Gottfried Wilhelm Leibniz, Web site: www.wgl.de.

6. For an assessment of the transformation of the East German research system in the process of German reunification, see Mayntz 1990.

7. Longer-term monitoring of these indicators does not exist.

8. The student-professor ratio was 49:1 in 1980 and 63:1 in 2003.

9. Excluded are resources personally dedicated to individual professors.

REFERENCES

Abramson, H. N., J. Encarnacao, P. P. Reid, and U. Schmoch (eds.). (1997) *Technology Transfer Systems in the United States and Germany: Lessons and Perspectives.* Washington, DC: National Academy Press.

Baldauf, B. (1998) Doctoral Education and Research Training in Germany: Towards a More Structured and Efficient Approach? *European Journal of Higher Education* 33 (2): 161–182.

Bundesministerium für Bildung und Forschung (BMBF). (2004) *Bundesbericht Forschung 2004*. Berlin: BMBF.

———. (2006) *Forschung und Innovation in Deutschland 2005*. Berlin: BMBF.

Clark, B. (1983) *The Higher Education System: Academic Organization in Cross-National Perspective*. Berkeley: University of California Press.

de Boer, H., J. Enders, and U. Schimank. (2007) On the Way towards New Public Management? The Governance of University Systems in England, the Netherlands, Austria, and Germany. In D. Jansen (ed.) *New Forms of Governance in Research Organizations: Disciplinary Approaches, Interfaces, and Integration*, 137–154. Dordrecht: Springer.

Deutsche Forschungsgemeinschaft (DFG). (2003) *Entwicklung und Stand des Programms "Graduiertenkollegs." Erhebung 2003*. Bonn: DFG.

———. (2006) *Förder-Ranking 2006. Institutionen—Regionen—Netzwerke. DFG—Bewilligungen und Weitere Basisdaten öffentlich geförderter Forschung*. Bonn: DFG.

Enders, J. (2002) Serving Many Masters: The PhD on the Labour Market, the Everlasting Need of Inequality, and the Premature Death of Humboldt. *Higher Education* 44: 493–517.

———. (2005) Border Crossings: Research Training, Knowledge Dissemination, and the Transformation of Academic Work. *Higher Education* 49:119–133.

Enders, J. and L. Bornmann. (2001) *Karriere mit Doktortitel? Ausbildung, Berufsverlauf und Berufserfolg von Promovierten*. Frankfurt: Campus.

European Commission, Enterprise DG, and Federal Ministry of Economy and Labour, Austria (eds.). (2001) *Benchmarking Industry-Science Relationships: The Role of Framework Condition: Final Report*. Vienna/Mannheim: Institute of Technology and Regional Policy, Joanneum Research.

Frackmann, E. and E. de Weert. (1994) Higher Education Policy in Germany. In L. Goedegebuure, F. Kaiser, P. Maassen, V. Meek, F. van Vught, and E. de Weert (eds.), *Higher Education Policy: An International Comparative Perspective*, 132–161. Oxford: Pergamon Press.

Gläser, J. (2001) Macrostructures, Careers, and Knowledge Production: A Neoinstitutionalist Approach. *International Journal of Technology Management* 22 (7–8): 698–715.

Hartmann, M. (2002) *Der Mythos von den Leistungseliten: Spitzenkarrieren und Soziale Herkunft in Wirtschaft, Politik, Justiz und Wissenschaft*. Frankfurt am Main: Campus.

Hein, M., G. Hovestadt, and J. Wildt. (1998) *Forschen Lernen*. Gray material, new series, no. 141. Düsseldorf: Hans-Böckler-Stiftung.

Heinze, T. and S. Kuhlmann. (2007) Analysis of Heterogeneous Collaboration in the German Research System with a Focus on Nanotechnology. In D. Jansen (ed.), *New Forms of Governance in Research Organizations: Disciplinary Approaches, Interfaces, and Integration*, 189–212. Dordrecht: Springer.

Hochschulrektorenkonferenz. (1996) Zum Promotionsstudium. Entschliessung des 179. Plenums vom 9. Juli. Hochschulrektorenkonferenz, Bonn.

Internationale Kommission. (1999) *Forschungsförderung in Deutschland. Bericht der internationalen Kommission zur Systemevaluation der Deutschen Forschungsgemeinschaft und der Max-Planck-Gesellschaft*. Hannover: VolkswagenStiftung.

Keck, O. (1993) The National System for Technical Innovation in Germany. In R. Nelson (ed.), *National Innovation Systems: A Comparative Analysis*, 115–157. New York: Oxford University Press.

Kehm, B. (2006) Tuition Fee Reform in Germany. *International Higher Education* 45:2–3.

Kehm, B. M. and U. Lanzendorf. (2005) Ein neues Governance-Regime für die Hochschulen—mehr Markt und weniger Selbststeuerung? *Zeitschrift für Pädagogik*, Suppl. no. 51 (50) 41–55.

Kivinen, O., S. Ahola, and P. Kaipainen (eds.). (1999) *Towards the European Model of Postgraduate Training*. RUSE Research Report 50. Turku, Finland: University of Turku, Research Unit for the Sociology of Education.

Krücken, G., F. Meier, and A. Müller. (2007) Information, Cooperation, and the Blurring of Boundaries—Technology Transfer in German and American Discourses. *Higher Education* 53 (6): 675–696.

Krull, W. and F. Meyer-Krahmer (ed.). (1996) *Science and Technology in Germany*. London: Cartermill.

———. (2001) The German Innovation System. In P. Larédo and P. Mustar (eds.), *Research and Innovation Policies in the New Global Economy: An International Comparative Analysis*, 205–252. Cheltenham, UK: Edward Elgar.

Kuhlmann, S. (1998) Moderation of Policy-Making? Science and Technology Policy Evaluation beyond Impact Measurement—The Case of Germany. *Evaluation* 4 (2): 130–148.

Kultusministerkonferenz. (2004) *Prognose der Studienanfänger, Studierenden und Hochschulabsolventen bis 2020*. Documentation no. 176. Bonn: Statistische Veröffentlichungen der Kultusministerkonferenz.

Mayntz, R. (1990) Science in East Germany—Consequences of Unification. In F. Meyer-Krahmer (ed.), *Science and Technology in the Federal Republic of Germany*, 33–45. Harlow, UK: Longman.

Minssen, H., B. Molsich, U. Wilkesmann, and U. Andersen. (2003) *Kontextsteuerung von Hochschulen? Folgen der Indikatorischen Mittelzuweisung*. Berlin: Duncker und Humblot.

Netherlands Observatory of Science and Technology (NOWT). (2005) *Wetenschaps- en Technologie-Indicatoren*. Report. NOWT: Den Haag.

Organisation for Economic Co-operation and Development (OECD). (2002) *Steering and Funding of Research Institutions—Country Report: Germany*. Paris: OECD.

———. (2004) *Science, Technology, and Industry Outlook 2004*. Paris: OECD.

———. (2005) *Main Science and Technology Indicators 2005/2*. Paris: OECD.

———. (2006) *Science, Technology, and Industry Outlook 2006*. Paris: OECD.

Pavitt, K. (2001) Public Policies to Support Basic Research: What Can the Rest of the World Learn From US Theory and Practice? (And What They Should Not Learn). *Industrial and Corporate Change* 10 (3): 761–779.

Scharpf, F. W. (1985) Die Politikverflechtungs-Falle: Europäische Integration und Deutscher Föderalismus im Vergleich. *Politische Vierteljahresschrift* 26:323–356.

Schimank, U. (1995) *Hochschulforschung im Schatten der Lehre*. Frankfurt/Main: Campus.

Schimank, U. and H. W. Hohn. (1990) *Konflikte und Gleichgewichte im Deutschen Forschungssystem: Akteurkonstellationen und Entwicklungspfade der Staatlich Finanzierten Außeruniversitären Forschung in Deutschland*. Frankfurt/Main: Campus.

Schmoch, U. (ed.). (1996) *Technology Transfer in Germany.* Karlsruhe: Fraunhofer Institute for Systems and Innovation Research.

Schröder, T. (2003) *Leistungsorientierte Ressourcensteuerung und Anreizstrukturen im Deutschen Hochschulsystem—Ein nationaler Vergleich.* Berlin: Duncker und Humblot.

Trute, H. H., W. Denkhaus, B. Bastain, and K. Hoffmann. (2007) Governance Modes in University Reform in Germany—From the Perspective of Law. In D. Jansen (ed.), *New Forms of Governance in Research Organizations—Disciplinary Approaches, Interfaces, and Integration,* 155–176. Dordrecht: Springer.

VolkswagenStiftung. (2005) *Eckpunkte eines zukunftsfähigen deutschen Wissenschaftssystems, 12 Empfehlungen.* Hannover: VolkswagenStiftung.

Wigger, B. (2005) *Erfolge der Verwaltungsmodernisierung an Hochschulen.* www.foevspeyer .de/governance/intern/intranet/D%20Workshops/D1%20Workshops/D13%20Work shop%20III/Dateien/Verwaltungsmodernisierung%20in%20Hochschulen_Wigger.pdf (accessed June 7, 2006; link no longer active).

Wissenschaftsrat (WR). (2001a) *Systemevaluation der Blauen Liste—Stellungnahme des Wissen-schaftsrates zum Abschluß der Bewertung der Einrichtungen der Blauen Liste.* Berlin: WR.

———. (2001b) *Systemevaluation der HGF—Stellungnahme des Wissenschaftsrates zur Her-mann von Helmholtz-Gemeinschaft Deutscher Forschungszentren.* Berlin: WR.

———. (2002) *Empfehlungen zur Reform der Doktorandenausbildung.* Saarbrücken: WR.

———. (2006) *Empfehlungen zur Zukünftigen Rolle der Universitäten im Wissenschaftssystem.* Berlin: WR.

CHAPTER EIGHT

The Netherlands

BEN JONGBLOED

This chapter describes the policy framework for the academic research enterprise in the Netherlands and analyzes some of the impacts, both intended and unintended, of the policies. After describing the main actors in the Dutch academic research enterprise (ARE), I discuss the main government policies that shape the innovation system of the Netherlands, such as the funding policies, the regulations in place to assure the quality of academic research, and the policies for ensuring effective doctoral education. The final part of the section on the policy framework lays out the policies that are aimed at connecting academic research to economic development. After this description of the Dutch ARE, I offer an assessment of its performance, showing some statistics and discussing some of the main reform policies that have been implemented in recent years. The final section lists a number of conclusions that touch on important policy themes, such as financing, research commercialization, and research concentration.

The Academic Research Enterprise

This section describes the main elements and the principal actors in the Dutch academic research enterprise, starting with research universities, continuing with the research institutes that concentrate on fundamental research, and then completing the picture with the research institutes that are more occupied with applied research.

PRINCIPAL ACTORS

Dutch higher education is organized as a binary system, consisting of thirteen research universities[1] and some forty institutions providing vocational higher education (in Dutch, *hogescholen;* they are also known as *Hoger Beroeps Onderwijs*). There is also an Open University, and a number of other state-funded and non-state-funded institutions also provide higher education.[2] The main functions of the hogescholen and the universities are formulated in the national Higher Education and Research Act of 1993. Although the aims of the hogescholen mainly relate to the provision of vocationally oriented education programs geared to specific professions, the hogescholen are increasingly engaging in applied research activities. However, the hogescholen so far receive very little public funding for their applied research.

The mission of the universities is threefold: to perform academic research, to teach, and to transfer knowledge to society. They provide academic programs (bachelor's, master's, PhD programs) and carry out most of the basic research (fundamental and strategic) in the Netherlands. Of the fourteen public universities, three are geared to engineering programs and one to agriculture. Nine others are general universities, and the remaining one is the Open University. The universities' research covers all disciplines. Table 8.1 shows the distribution of the university research capacity across the disciplines.[3] For disciplines such as health, natural sciences, and engineering, research capacity is the highest.

The academic research enterprise of the Netherlands is a complex system, with many actors, funding mechanisms, and interrelations. Figure 8.1 shows the key actors in the system as situated on four levels:

— Level 1: Government
— Level 2: Ministries
— Level 3: Policy development/funding/intermediary organizations
— Level 4: R&D performers

The Ministry of Education, Culture and Science (OCW) has overall responsibility for the governance of the public universities and the numerous other public research organizations, the hogescholen, and the research institutes in the system. The research universities have a high level of autonomy; the government steering takes place mostly "at a distance" (see below). The Ministry of Economic Affairs has the responsibility for innovation policy instruments and all matters concerning applied and industry-oriented R&D. This ministry in its steering has more of a "hands-on" approach.

TABLE 8.1.
Total university research capacity by discipline, the Netherlands, 2003

	Researchers[a]	%
Health	4,631	28.7
Science	3,367	20.9
Engineering	2,965	18.4
Social sciences	1,715	10.6
Humanities	1,200	7.4
Agriculture	857	5.3
Economics	782	4.8
Law	582	3.6
Other	45	0.3
Total	16,144	100

Source: Association of Universities in the Netherlands (VSNU) Database, www.vsnu.nl.

[a] Full-time equivalents

Within the Ministry of OCW, higher education policies and science policies have long been treated as separate fields. Policy plans for academic research used to be published in a separate Science Budget (*Wetenschapsbudget*). Today, the Higher Education and Research Plan, which is the ministry's four-year strategic plan setting out its higher education policy, is a more integrated document, covering both the aims of higher education policies and science policies. For coordinating the various ministries' R&D-related policy agendas, a relatively new, high-level council (CWTI, the Committee on Science, Technology and Information Policy) has been created. At the Cabinet level, the RWTI (Council on Science, Technology and Information Policy) prepares the decisions to be taken by the plenary Cabinet.

Decisions on distributing the R&D budgets and on implementing the research and innovation policies are mediated through the various policy coordination bodies (situated on level 3 in fig. 8.1). Here, the Netherlands Organization for Scientific Research (NWO), which is the Dutch research council, acts as the most important intermediate funding organization in the field of fundamental and strategic research. NWO awards competitive research grants to researchers and research teams on the basis of proposals. In 2006 NWO funded 3,800 FTE (full-time equivalent) researchers working in the Dutch universities. In terms of subsidies, this amounted to more than €293 million (NOWT 2008). Apart from playing a role as a funding organization, NWO owns nine research institutes in the fields of astronomy, mathematics, computer science, physics, history, sea research, law, criminality, and space research. The total NWO budget allocated to its institutes was €160 million in 2003.

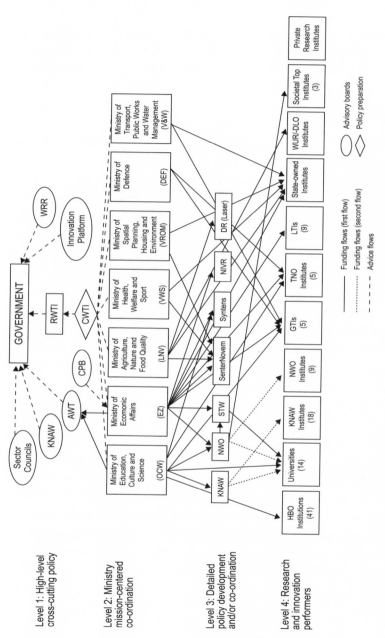

Level 1: High-level
cross-cutting policy

Level 2: Ministry
mission-centered
co-ordination

Level 3: Detailed
policy development
and/or co-ordination

Level 4: Research
and innovation
performers

Funding flows (first flow)
Funding flows (second flow)
Advice flows

Advisory boards
Policy preparation

Fig. 8.1. Overview of key actors in the Dutch academic research enterprise. *Source:* OECD 2005a, chap. 7, 211.

The KNAW is the Royal Netherlands Academy of Arts and Sciences. It primarily advises the government on matters of science and technology, especially in the field of basic research (e.g., codes of conduct, quality assurance, and research schools) and awards prizes and selective funds to excellent researchers. The KNAW owns eighteen research institutes that perform basic research, such as in the life sciences, the humanities, and scientific information. These are research institutes comparable to those of NWO. Some of the institutes also have a scientific service function that includes maintaining biological and documentary collections, providing information, and creating other facilities for research. In 2002 the KNAW had a budget of approximately €100 million. The Ministry of Education, Culture and Science contributed 77% of this budget.

Five Sector Councils advise the ministers focused on specific policy areas. They conduct foresight exercises and analyses for the programming and coordination of research and the organization of the knowledge infrastructure in their relevant sector. The members of these councils are from science and business in order to have both producers and users of science represented. Delegates from ministries act as observers in the Sector Councils. The Dutch system of Sector Councils is unique in the world, although it bears some similarities to organizations such as the Office for Science and Technology found in the United Kingdom and Canada. An important advisory body in R&D matters is the Advisory Council on Science and Technology Policy (AWT). The core of AWT's advisory task is focused on the knowledge and innovation process and its development. Some more general advisory bodies every now and then express their opinions on education and research matters: the Socio-Economic Council and the Advisory Council on Government Policy (WRR). The Netherlands Bureau for Economic Policy Analysis (CPB) and the Social and Cultural Planning Office also regularly produce reports on higher education and research matters.

The universities defend their common interests through their representative organization, the Association of Universities in the Netherlands (VSNU). VSNU is one of the partners in the national policy network with respect to higher education and research. *Hoger Beroeps Onderwijs* (HBO) institutions have organized themselves in the HBO council. Employers' organizations and trade unions as well as student unions and other interest groups at various times will also try to make their voices heard in consultative and policy development processes.

RESEARCH INSTITUTES

In addition to the research universities, the HBO institutions, and the NWO and KNAW institutes just described, the other publicly supported parts of the R&D

system (also part of level 4 in fig. 8.1) need to be mentioned. They include public laboratories and "para-university" institutes: five Large Technological Institutes (GTIs), the Leading Technology Institutes (LTIs), the Netherlands Organization for Applied Research (TNO), the DLO Agricultural Research institutes, a number of state-owned research and advisory centers, and several other institutes in the fields of health and the social sciences. Private research labs and institutes make up the remaining part of the R&D system. While they are outside the sphere of public authorities, they are affected by public policies and may receive public subsidies.

The group of Large Technological Institutes (GTIs) consists of five organizations conducting applied research and related activities, such as supporting industry and government in specific fields:

- Energy Research Centre of the Netherlands;
- GeoDelft, conducting research in highway and hydraulic engineering and soil (including pollution);
- The Maritime Research Institute Netherlands, conducting research in ship-building, offshore technology, and ocean engineering;
- The National Aerospace Laboratory, conducting research in aerospace engineering for both civil and military purposes; and
- WL Delft Hydraulics, focusing on ports, the coast, rivers, shipping, water management, and the environment.

The GTIs have two main functions: (1) they act as a center for technological knowledge to fulfill the knowledge needs of government and business enterprises, and (2) they develop technology and make it available to government and business enterprises. The total research budget of the GTIs was €276 million in 2006 (NOWT 2008). The two accountable ministries, Economic Affairs and Transport and Public Works and Water Management, together with the Ministry of Education, Culture and Science, allocate basic grants to the institutes. GTIs also obtain funds from targeted grants and, in particular, contract research. The three ministries provide basic funds that make up around 25% of the GTIs' total budget.

In the year 1923, TNO (the Netherlands Organization for Applied Research) was founded through the enactment of a law that was intended to promote applied scientific research in the Netherlands. The organization was specifically founded to perform research on concrete problems. Today, TNO is an independent contract research organization that provides expertise and research primarily to the small and medium-sized business sector. Until 2005 TNO organized its work in fourteen specialized TNO institutes, some of them working in cooperation with the Dutch universities. In 2005 TNO reorganized its activities around five strategic, application-oriented research areas (quality of life; defense, security, and safety; science and industry; built environment and geosciences; and information and communication technology). TNO currently employs over five thousand people, has a broad research agenda, and performs various activities, such as contract research, specialized consultancy, licensing of patents, development of software, and certification of products and services.

The institute had a turnover of €500 million in 2006. Of this turnover, about €200 million consists of two budgets from the national government: a basic allocation (€75 million) and an earmarked allocation (€125 million). The remainder (about €300 million) is acquired through research contracts, some with private clients (€230 million), others with public organizations (€75 million), that commission TNO to do work in specific strategic knowledge areas (NOWT 2008). For instance, the Ministry of Transport, Public Works and Water Management (V&W) benefits from TNO knowledge in the areas of transport systems, geological data, and soil and water management.

The Leading Technology Institutes (LTIs) were launched in 1997 as virtual organizations in which companies and knowledge institutes participate via public-private partnerships. The LTIs aim at stimulating R&D cooperation between public and private partners in areas of importance for the economy and society. Initially there were four such institutes (in the fields of nutrition, metals, polymers, and telematics) with TNO participating in all of them. Recently an additional five LTIs have started, which operate in the fields of pharmaceutics, molecular medicine, green genetics, water technology, and biomedical materials. The LTIs are qualitatively far more important than their modest share in the overall science and technology budget would suggest. (See below for further discussion of the LTIs.)

The DLO Agricultural Research Institutes used to be part of the Ministry of Agriculture, Nature and Food Quality. In the second half of the 1990s, the institutes became independent from the ministry (and the DLO Foundation was created) and merged with Wageningen University into a center for knowledge and research, the Wageningen University and Research Centre. The DLO Foundation and the university are separate entities but cooperate in expertise groups in five different areas. There are ten institutes within the DLO Foundation.

Finally, there are the state-owned research centers. A number of such institutes operate under a ministerial umbrella, although the number of these centers is decreasing. Some of them are directly connected to a ministry, such as the Research and Documentation Centre of the Ministry of Justice. Others are ministerial agen-

TABLE 8.2.
R&D spending as a percentage of GDP, 1995–2005

	Total R&D spending	Businesses	Research institutes	Universities
1995	1.97	1.03	0.37	0.57
2000	1.82	1.07	0.25	0.51
2001	1.80	1.05	0.26	0.49
2002	1.72	0.98	0.25	0.50
2003	1.76	1.01	0.25	0.49
2004	1.78	1.03	0.26	0.49
2005	1.72	1.00	0.24	0.48

Source: CBS 2007.

Note: Detail may not add to total because of rounding.

cies, such as the Royal Netherlands Meteorological Institute, which is part of the Ministry of Transport, Public Works and Water Management.

Summing up, Dutch R&D expenditure amounted to about €8.8 billion in 2005. This expenditure translates to 1.72% of the Dutch Gross Domestic Product (GDP). This fraction—also referred to as the R&D intensity—has decreased significantly since 1995 (table 8.2). The business sector accounts for 58% of the R&D spending (€5.1 billion). About €2.5 billion (28%) is spent by the universities. The research institute sector is responsible for the remaining 14% (€1.2 billion) (CBS 2007, table 3.1.1).

Policy Framework

The policy framework for the Dutch academic research enterprise revolves very much around funding and quality assurance mechanisms. In the most recent decade, a large emphasis was placed on policies that stimulate the transfer of knowledge from academia to industry, in particular through the promotion of public-private partnerships. The subsections below sketch out these policies, along with some of their developments over time. In a separate subsection the training of researchers and reforms in the system of doctoral education are discussed.

FUNDING FOR ACADEMIC RESEARCH

Public financing of research is an important mechanism through which the Dutch government tries to achieve its main research policy goals. This chapter concerns mainly the funding of university research, since universities are the centerpiece of the Dutch academic research enterprise. The funding that universities receive for research must be understood in the context of their overall funding framework, which is pictured in figure 8.2, which presents the main funding flows to the thirteen publicly funded Dutch universities (leaving out the Open University).

Universities carry out teaching and research. The recurrent funding they receive for these functions, provided as a block grant from the government, is called the first flow of funds. The university is free to spend the grant as it wishes (the "lump sum" principle), provided it is for teaching and research. The size of this grant depends on a formula that incorporates input and output criteria referring to teaching and research. The research part of the block grant is roughly two-thirds of the total grant. However, the universities are free to internally reallocate funds from research to teaching. Student fees, which are equal across all programs, are also seen as part of the first flow of funds.

The second flow of funds refers to the research grants made available to universities by the research council NWO, in the form of grants, and by the KNAW in the form of support for academy researchers and professors. NWO awards competitive grants for top-level research and research equipment and to coordinate research programs. The competition takes place on the basis of bottom-up research proposals prepared by researchers with tenure at Dutch universities. The selection is made on the basis of peer review. NWO encompasses all scientific fields. Apart from subsidies for individual researchers, NWO offers program subsidies (often in thematic areas), publication grants, and support for investment in instruments and facilities. Of the subsidies for individual researchers, the so-called Innovational Research Incentive Scheme (*Vernieuwingsimpuls*) is the most important. It is oriented at promising researchers in different phases of their career (new PhDs, postdocs, and senior researchers). Other subsidies for individuals are the prestigious SPINOZA prizes, four of which are awarded each year to a world-class researcher working in the Netherlands, and programs specifically oriented to women in science (Aspasia). In recent years the share of NWO funding for universities has been growing.

The third flow of funds consists of revenues from contract activities, consultancies, and research commercialization activities carried out by the universities. Over the past fifteen years, the third-stream share of university income has grown rapidly to around 12% of university revenues. In addition, figure 8.2 shows university revenues from interest, university bookstores, student restaurants, and so forth (9% of university income). A large part of contract activities involve knowledge transfer in the sense of research carried out for industry, government, and other public-sector organizations. A slightly smaller part involves contract teaching for companies (and individuals—such as in MBA programs), short courses, and lifelong learning. However, the largest part is contract research carried out for government and not-for

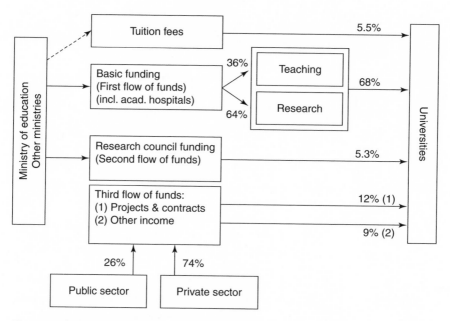

Fig. 8.2. Relative size of funding streams for universities, 2003.

profit organizations, such as the medical charities. Charities' funding of research is relatively low, but as this funding is concentrated on a few areas in the medical sciences, it may have a considerable impact in these fields. Funding is allocated on a competitive project base. Most charities comply with a protocol for "good research funding practices" that the charities have agreed upon. It includes the possibility for programmatic priorities, bottom-up project formulation, and selection of proposals by peer review and panels. The budgets of the charities depend on yearly contributions through donations and legacies and on returns from equities.

The funding of university research has changed quite drastically in recent decades. The biggest changes have been in research council funding and contract funding. More than half of all university researchers these days are paid through such competitive grants (table 8.3). More staff, therefore, are on short-term contracts. The number of researchers funded through recurrent funding is more or less stable. This is in line with the fact that for more than a decade the state's contribution to universities, corrected for inflation, has been relatively constant.

While recurrent funding may have been constant, the funding methodology for calculating the amount of core funding per university has changed over the years. Until 1983 the government did not allocate separate funding to universities for research. In effect, the teaching and research budget that a university received

TABLE 8.3.
University research capacity, 1997–2004

	Researchers, all funding sources combined[a]	% of research staff funded through		
		Recurrent funding	Research council	Contract research
1997	14,132	54	18	28
1998	14,245	53	18	29
1999	14,593	53	20	28
2000	15,000	52	21	27
2001	15,362	52	22	27
2002	15,638	50	24	27
2003	16,143	47	24	28
2004	16,150	48	24	27

Source: Association of Universities in the Netherlands (VSNU) Database, www.vsnu.nl.
[a] Full-time equivalents

was directly related to the number of students. In 1984 the government introduced a separate research compartment in the universities' core funding. This allowed for a university-specific research capacity and was originally allocated as earmarked, conditional funding. See further discussion below regarding this system of what was known as conditional funding. In 1993 a normative, partly formula-driven system was introduced for calculating the universities' teaching and research budget. Within this budget, the funding for research was partly based on the teaching budget and partly based on research performance as measured by the number of PhD degrees awarded by each university. On top of that, the universities continued to receive fixed research allocations based on the system of conditional funding.

Today, the allocation mechanism that distributes core funding across the universities is known as the BA-MA model (Bachelor-Master model). This name refers to the bachelor's and master's degrees. The BA-MA model is a distribution model that allocates a given amount of funding across the thirteen universities. Basically this means that the minister of education (or rather, Parliament) determines the budget for the university sector as a whole and subsequently distributes this budget across the individual universities according to a set of fixed rules (a formula). Apart from the BA-MA allocation, the universities receive allocations for academic teacher training, for academic hospitals, and for unemployment benefits paid to former university employees.

The BA-MA model is partly output-driven, which means that the allocation for each university depends on the degrees granted to students (BA as well as MA degrees), the number of new entrants per university, and the number of PhD degrees.

For the teaching function, a distinction is made between programs in social sciences, the humanities, and law, on the one hand, and programs in natural sciences, engineering, and agriculture on the other. The latter receive a greater weight in the funding formula to reflect the higher cost of programs in these laboratory-based fields. Medical programs receive an even greater weight. The ratios between the weights are 1:1.5:3. In addition, the weight for bachelor's degrees is twice the weight for master's degrees. Apart from these variable allocations for teaching, each university receives a fixed, historically based allocation for teaching that differs across universities.

In the BA-MA model, the allocation of research funds to universities within the first flow of funds consists of four separate components: (1) education-related research, (2) PhD programs, (3) research schools, and (4) a strategic considerations component.

The education-related research allocation is a formula-based allocation to each university that intends to express the fact that research is a prerequisite for university teaching. The connection with teaching is operationalized through the number of master's degrees and, to a lesser extent, bachelor's degrees. In the formula, degrees in engineering and natural sciences receive a higher weight (twice as high) than degrees in social sciences, the humanities, or law. For the university sector as a whole, a little over 15% of available research funds in 2005 were allocated under the heading of education-related research.

The second component in the recurrent funding for research is an allocation based on the number of PhD degrees and (for technical universities) the number of postgraduate "designer certificates" awarded. In other words, this part of research funding is performance-driven. It is meant to be a compensation (or premium) for doctoral work carried out by graduates in universities. Again, two rates apply: a rate for "inexpensive" doctoral theses (e.g., in the social sciences and humanities) and a rate that is twice as high for the "expensive" theses, such as in science, engineering, agricultural, and medical disciplines. Some 12% of the total research allocation is based on the number of doctorates.

The third component in the research funding is based on the number of research schools. This feature in the allocation model was introduced in the 1998 budget to strengthen the system of research schools in the Netherlands. In line with the "breadth and depth" strategy, the first part of this component encourages universities to continue on the road toward establishing research schools (the breadth strategy). As explained later in this section, more than one hundred research schools have been established so far, covering all disciplinary fields. The aim of research schools is, first, to have a structure in which researchers from different universities concentrate their research activities on certain (sub-)disciplinary fields and, second, to organize the training of new researchers (PhD students). This strategy, based on arguments of scale and synergy, seeks to strengthen the quality and profile of university research in general. The second part in the research school component (the depth strategy) is targeted at supporting those research schools (about six at present) that are considered to be among the best research teams in the world or show potential to gain that status. The underlying argument for this component is to reward excellence. The research schools component is relatively small: 3% of research funds are available for the breadth strategy and 3% for the depth strategy.

The fourth part of research funding is the most important, as it represents some two-thirds of research funds. It goes by the name of the "strategic considerations" component (SCC). The name derives from the fact that the Ministry of Education originally (in 1993) planned to use this component to fund research that would have strategic relevance, meaning relevance to society or the national economy. This implied that quality and relevance were supposed to affect a university's funding. Although the ministry and the universities agreed on the principle, the universities took the view that a reshuffling of research funds by the ministry would be a major intrusion into the universities' autonomy. The universities have managed to avert policies aimed at major reallocations in the strategic considerations budgets per university. Therefore, the SCC budgets turned out to be fixed allocations per university. Thus, the research function of universities, to a large extent, is funded on the basis of historical allocations. The distribution key for the SCC over the different universities has been discussed repeatedly over time but has changed very little in any substantial way. The largest (and oldest) universities continue to receive relatively larger SCC components than the younger universities. The younger universities have lobbied for a higher amount, arguing that their research budgets have not kept pace with rising student numbers.

Still, the government has continued its efforts to make part of the strategic considerations component more dynamic. In order to "shake up the Dutch innovation system" (OECD 2005a, 179), a policy known as the dynamization component was added to the SCC. This was part of the Smart Mix strategy, discussed below. In 2006, €50 million was taken out of the SCC and redistributed across the universities according to each university's success in winning research council grants (from NWO, the second stream of funding) and selected competitive research contracts in the third stream of funding. For five universities there were resulting budget reallocations of about €0.5 million (either plus or minus), while for some of the more entrepreneurial ones (e.g., Twente and Nijmegen) the additional budget was close to €1 million. In 2007, an amount of €100 million was taken out of the SCC for redistribution according to the Smart Mix. However, after a new Cabinet took

TABLE 8.4.
Composition of first-stream funding for research, 1992–2007

	Basic allocation (%)	PhD (%)	Strategic allocation (%)	Research schools (%)	Top research schools (%)	Dynamization / Smart Mix (%)	Total (€M)
1992	15	8	77				938
1993	15	9	76				976
1994	15	10	75				989
1995	15	11	74				1,004
1996	15	12	73				1,047
1997	15	12	73				1,070
1998	15	12	65	4	4		1,102
1999	15	12	65	4	4		1,166
2000	15	12	65	4	4		1,251
2001	15	12	66	4	4		1,294
2002	15	11	67	3	3		1,362
2003	15	12	66	3	3		1,404
2004	15	12	66	3	3		1,416
2005	16	12	65	3	3		1,417
2006	17	12	60	3	3	4	1,415
2007	19	13	55	3	3	7	1,442

Source: Baggen 2008.

office in 2007, the Smart Mix policy was abandoned, and the €100 million was to be redistributed by the research council for strengthening fundamental research in universities.

Table 8.4 illustrates that the funding model produced relatively stable allocations in the 1990s, although the national aggregates may hide effects on individual universities. The research schools component did not introduce a great deal of dynamics into the system. Only recently, through the introduction of the dynamization (Smart Mix) component, did the picture change. After 2006 the performance-related character of the BA-MA model increased quite substantially.

By combining the various components of the first-stream funds that are performance-driven (about €650 million in 2007) and adding the research council funds allocated by NWO (about €300 million) and the contract research income (about €600 million), one may conclude that, at present, some two-thirds of the universities' research revenues is based on performance. Only the "strategic allocation" is not based on performance. What the changes in the funding model and the increase in competitive funding have meant for the research performance of universities and the concentration (versus the dispersion) of research funds is discussed in the next section.

QUALITY ASSURANCE OF RESEARCH

The way in which universities render proof of their research performance is another important part of the research policy framework. Table 8.4 shows that 55% of the universities' core research funds for the year 2007 was allocated as a set of largely constant amounts ("strategic allocations") per university. This might give the impression that the Netherlands government, unlike its British counterpart, which translates the outcomes of the research assessment exercises into funding decisions, is not as occupied with the outcomes of research as it probably could be. Actually, the Dutch government, like that of many other countries, works under the assumption that research activities can hardly be captured in terms of volumes and prices. It feels that it should be left to the academic community itself to guarantee the quality of research and the sound use of research funds. However, the government frequently has tried to influence the research agenda of universities through the power of the purse. Targeted budgets have been made available and specific funding instruments have been employed to encourage the academic research community to meet specific goals. An important attempt was made in 1983, when the system of "conditional funding" was introduced. The goal of this system was to enhance quality and coherence in university research and to assess the relevance of research to society. To this end, university departments (faculties) had to draw up research programs that were to conform to the following conditions: (1) the program was to be of sizable scale (at least five full-time equivalents of researchers involved); (2) the program was to run over five years; and (3) the quality of the program was to be examined by independent, external peers, selected from the disciplinary fields.

One of the basic ideas was to translate differences in research quality into funding decisions. Although at the start of the system in 1983 a conditional research budget for each university was agreed upon, in practice from 1983 onward this budget was never actually adjusted. This was due to the opposition of the universities and their representative organization, the Association of Universities in the Netherlands (VSNU). However, the attempts to introduce more flexibility into the funding system did force the universities to focus much more on generating research output and revealing that output. In other words, the universities were encouraged to justify the public research funds granted to them. An unintended effect of the conditional funding system was that it contributed to the idea that, for academics, research is the most important part of a university's activities and the main determinant of an academic's career.

The 1993 introduction of a strategic considerations component of the universities' funding model (see above) did not change this. Reallocations between the universities' strategic research budgets have, so far, never been effectuated. The government had to resort to introducing additional components in the funding model and providing "new money" on top of the core funding stream, such as the special funding component for research schools and the Smart Mix mentioned above.

Ensuring accountability, therefore, has been mostly unconnected to the funding framework. Like all organizations that receive subsidies from the government, universities have to submit the usual annual reports. In the case of universities, however, a result of the government's "steering at a distance" philosophy is that, at present, universities are only expected to produce annual and financial reports that conform to rather general reporting standards. The annual reports do include information on research activities and research output. However, it is left to the universities themselves to determine to what extent they include detailed information on research inputs and outputs. All universities submit such information to a VSNU database that contains information on the scientific activity for each cluster of disciplines. This database contains information on the number of researchers (in the three flows of funds discussed above), the number of research publications (dissertations and scientific publications), and the inflow and outflow of PhD candidates (Jongbloed et al. 2005). The database replaced the paper versions of the scientific reports that previously were published by universities each year.

When it comes to policies that assure accountability on the area of research, the most important instrument is the research assessment, in place from 1993 on. The universities agreed to have their research programs examined in a system of peer review (i.e., by international committees of independent experts in the respective disciplines). Originally these research assessments were carried out every six years under the auspices of the VSNU. In 2003 the assessment system was modified to reduce the administrative workload for the universities and to place more emphasis on the management of research. It now is up to the university itself to arrange the assessment, to seek out the review committee, and to decide whether or not benchmark institutions are included in the peer review. Despite the increased autonomy for the Dutch universities in this respect, all universities agreed to adhere to a uniform evaluation approach, the Standard Evaluation Protocol for Public Research Organizations (SEP). The KNAW, NWO, and VSNU jointly developed the SEP protocol. The SEP-based research assessments take place every six years.[4]

The goal of the assessment continues to be to rate the quality of research programs in terms of four criteria: academic quality, scientific productivity, scientific relevance, and scientific long-term viability. The ratings make use of a five-point scale, where a rating of 5 represents excellent research and a rating of 1 indicates poor research. Table 8.5 shows the rating scale used in the SEP assessment protocol

TABLE 8.5.
Rating scales used in research assessments

SEP	Score	VSNU
Excellent	5	Excellent
Very good	4	Good
Good	3	Satisfactory
Satisfactory	2	Unsatisfactory
Unsatisfactory	1	Poor

as well as the scale that was used in the earlier VSNU-led research assessments. The "poor" rating was scrapped (in practice it was never assigned) and at the top end of the scale more differentiation was made possible.

The assessment produces ratings for each research group along all four dimensions. Because the research assessment reports are published, they perform an important accountability function. Moreover, the research assessments are used by the university's executive board, deans, and research directors in the elaboration of the university's internal research policy (Jongbloed and Van der Meulen 2006). However, these assessments are not used the same way as in the United Kingdom, where the research assessments reign; in the Netherlands, the results are not used as inputs in the ministry's decisions on research funding. In fact, the adoption of the SEP protocol has made such an option less likely because SEP assessments are not organized on a national scale and do not necessarily include all institutions operating in the same disciplinary area.

When it comes to the issue of the universities' internal research policy, a survey showed that the Dutch universities all explicitly strive for the highest ("excellent" or "very good") ratings in the research assessments (Jongbloed and Van der Meulen 2006). Universities regard academic quality as the most important criterion—more important than criteria like productivity, relevance, and viability. A quality rating of 3 (or less) these days will not do anymore. Scoring a 4 or a 5 is the goal. Research units that score a poor or unsatisfactory rating will—to put it mildly—"receive attention" from deans, program leaders, and often the central (executive) board of the university. In cases like that, restructuring, sometimes going hand-in-hand with some temporary extra funding, is the usual remedy. Sometimes underperforming research groups are merged with successful ones to "spread the success formula." Financial incentives are used to encourage better performance or to help an underperforming unit change course.

Many universities strive to create fewer, larger (often multidisciplinary) research units where the best researchers collaborate and produce high-quality research that has the potential to reap economic rewards as well. This goal nicely fits with the Dutch government's policy goal of "focus and mass" (see below). Universities try to distinguish themselves from other universities by carefully selecting their areas of excellence. They make use of research assessments as tools in choosing particular areas of strength and in building their research profiles (Salerno et al. 2005). However, when the central board of a university or a dean decides to eliminate, downsize, or restructure some research program, the quantitative ratings of a program in practice will not be the only criterion. Other goals and criteria also are taken into account. In practice, such criteria relate to issues like the desired profile of the university or a department, its education programs (master's education as well as PhD programs—increasingly organized in graduate schools these days), and the capacity of a research group to generate research income from competitive sources.

SUPPORTING EFFECTIVE DOCTORAL EDUCATION

The quality assessment protocol also requires universities to evaluate the training of doctoral students. This topic is quite real and pertinent, given the concern about the replacement of professors who are due to retire in the coming years. These days, the university's human resources policy is very much performance-based, and a tenure track system is in the making in many cases (Jongbloed and Van der Meulen 2006). By carefully organizing and structuring research groups, universities try to create the required financial room to appoint new (young) professors. The tenure track system is frequently supported by monitoring systems that keep an eye on the quantitative and qualitative research performance of research staff. Norms are being implemented for the number of articles in peer-reviewed journals that are to be produced by scientists—even for disciplines like law. The professionalization of the supervision of PhD candidates is also part of the human resources policy repertoire. On the same theme, one can see universities providing guidance to researchers who apply for research council subsidies. This is to the benefit of individual researchers as well as to the university that wishes to increase its success rate in competing for prestigious research grants.

Up until the 1980s, individuals pursuing a PhD in the Netherlands were usually employed as faculty staff, either in the position of research assistant or as regular (senior) staff. Traditionally, the process of writing a PhD thesis had the characteristics of the apprentice model. The doctoral candidate was working under the guidance of a professor. Gradually, a broad discussion started on the function and structure of research training. With the introduction of the Two-Tier Act in 1981, university education was structured in two tiers. The first tier covers the usual education up to the master's degree (or equivalent). This tier is accessible for many. Research training was positioned in a separate tier, in which entrance was subject to selection criteria. After some discussion, a law was passed in 1984 (the Beiaard Bill) to address the doctoral training that was to take place in the second tier. The AiO (in Dutch: *Assistent in Opleiding*) system was created. It meant that a doctoral candidate obtained a four-year appointment in the university as an "assistant in training" and actually earned a salary. The AiO is a distinct academic position. The objective of the AiO system is to provide advanced research training by way of active participation of the candidates in university research and, to a limited extent (less than 25% of working time), in teaching and administration. The idea is that an instruction and supervision plan is in place to guarantee guidance to the AiO.

Job openings for AiO positions mostly attract candidates who have recently finished their master's degree and wish to continue pursuing their university education toward the PhD. Table 8.6 shows the distribution of AiOs over the different discipline clusters. At the end of the year 2006, a total of 7,500 AiOs were employed in the thirteen universities (table 8.6). While the AiO contract officially lasts four years, the typical AiO required 5.2 years to complete his or her PhD (VSNU 2002). Of the male candidates, 58% finish within five years; for the female candidates the percentage is 44%. AiOs who take more than four years to obtain a PhD have to rely on social security benefits. On average, 70% of the doctoral candidates successfully finish their dissertation. The remaining percentage abandon their dissertations for various reasons.

Dutch doctoral candidates are basically supported through the three flows of funds (see above). From 1999 on, the AiOs that have been funded through the research council (NWO) have been appointed as part of the host university's academic staff. There are no precise data available on the proportions of doctoral candidates funded by the various sources. However, surveys carried out by Oost and Sonneveld (2004) on research schools indicate that almost half of the AiO positions are funded out of the university's first-stream (i.e., recurrent) funding, and almost 30% are funded by the research council. The remaining 20% are funded from revenues received through research contracts with government (including the European Union), nonprofit organizations, private companies, and charitable boards. The Philips Company, for instance, sponsors talented students who want to become researchers and who feel comfortable doing research located at the intersection between company and university. The company offers them a position in its Van der Pol program. In some cases, funding may be supplemented through the Casimir program of NWO or the Marie Curie program of the European Union. The Casimir program encourages the mobility of researchers between universities and the business sector.

TABLE 8.6.
PhD candidates in the AiO system, 2006

	Number of AiOs (FTEs)	%	Dissertations produced	%
Science	1,695	22.6	553	17.6
Engineering	2,115	28.2	499	15.9
Health	822	10.9	1,032	32.9
Social sciences	1,035	13.8	292	9.3
Humanities	561	7.5	255	8.1
Agriculture	472	6.3	224	7.1
Economics	434	5.8	160	5.1
Law	317	4.2	120	3.8
Other	56	0.7	5	0.2
Total	7,507	100	3,140	100

Source: Association of Universities in the Netherlands (VSNU) Database, www.vsnu.nl.

Partly as a result of the wish to improve doctoral training, and partly react-ing to statistics showing disappointing time-to-degree and success rates, the Dutch government in the early 1990s stimulated the development of a system of research schools. Research schools are expected to be centers of high-quality research in which structured training is offered to young researchers (KNAW 2005). The un-derlying rationale is that excellent training of researchers should be conducted in an environment of high-quality research. At present some 3,000 dissertations are produced every year (table 8.6). While most are the result of research by AiOs, some dissertations are written by academic staff appointed to other (non-AiO) positions. In the period 1993–2004, the number of PhDs was around 2,500 per year. In recent years the number has rapidly risen, with 3,140 in 2006. The number of AiOs rose from about 5,000 in 1993 to 7,508 in 2006 (VSNU 2008).

The system of research schools is supposed to give an impetus to high-quality research and education. This is why a broad yet selective system of research schools was set up, from which eventually a limited number of centers of excellence should develop. The government envisaged a diverse system of research schools sharing a number of common characteristics. A research school was to have a critical mass in terms of a minimum inflow of ten PhD students per year. This scale criterion is complemented with the condition that the school is to have autonomy with respect to financial and administrative responsibilities, meaning that sufficient funds should be allocated to the research school by the hosting university.

The Royal Academy of Sciences (KNAW) is responsible for the recognition (i.e., accreditation) of research schools. It has developed an accreditation procedure and established an independent committee to review the performance of research schools. The number of research schools has grown gradually since their first appearance (in 1991), but the pace of development differs across disciplines. Schools in the natural sciences were quick to take off, but schools in the social sciences developed somewhat later. Research schools in the humanities and law were even later. By 2005, 102 accredited research schools had been established, including virtually all disciplinary fields. Although the system of research schools was intended to include all doctoral candidates, participation rates differ among disciplinary fields.

Recently, the KNAW indicated that the system of research schools is not functioning as intended (KNAW 2005). The present system of research schools is entering a period of transition. Bartelse, Oost, and Sonneveld (2007) mention that, while the universities—through the VSNU—acknowledge the value of the system of research schools, they would like to see a more flexible organizational model for research training developed to host all doctoral candidates in the Netherlands. Universities wish to embed these schools in structures that align with their institutional strategy, and some are hesitant about participating in schools that are run by other universities. At several universities, doctoral training is currently organized under the internationally recognizable name graduate school. These graduate schools also host master's students that undertake a research master's program. Such a structure enables a smoother transition between master's programs and PhD training. It is expected that the (for the most part) locally organized graduate schools will gradually merge or closely cooperate with the (predominantly) nationally operated research schools (Bartelse, Oost, and Sonneveld 2007).

POLICIES CONNECTING ACADEMIC RESEARCH TO ECONOMIC DEVELOPMENT

Earlier in this section it was noted that the revenues that universities receive through contract research have increased rapidly in recent years. In the years 2003 and 2004, income from external sources covered 22% of total university revenue. This income from market-oriented activity is shown in figure 8.2. That percentage is up from 15% in the year 1990 (Baggen 2009) and illustrates the tendency for universities to be more responsive to demands that originate from their external environment. Partly this tendency is the result of government policies and can be seen as the universities' response to a decrease in their recurrent funding from the Ministry of Education. It is also partly the reflection of the universities' success in going after the public funds that have been made available through dedicated public budgets aimed at making universities interact more with business and industry. Some of these financial instruments are presented below. However, the fact should also be mentioned that disciplines like biotechnology, nanotechnology, and information technology have seen a dynamic of their own, independent of government policies, as a result of scientific discoveries made by "science entrepreneurs" working in a Mode 2 or Triple Helix–like context. This means that academic research is frequently taking place in collaborations between university, business, and government organizations (Etzkowitz 2000; Gibbons et al. 1994). Some of these developments have provided opportunities to universities in terms of research commercialization and revenue generation.

The first policy instrument to promote science-industry relationships has been in operation since 1981. It is the Innovation-Oriented Research Program (IOP), a collaborative research scheme that provides incentives to academic researchers to develop a more applied research agenda. The IOP program is a Ministry of Economic Affairs subsidy scheme that makes competitive grants available to innovative technological research projects at universities and other nonprofit research organizations within public-private research programs. At present, some fifteen programs are running. The IOP projects are supposed to be in line with the long-term research needs of the business community and support research networks between the research world and the business community in strategic areas. The IOP program should lead to the transfer of knowledge to the business sector. The program is run by SenterNovem, an agency of the Ministry of Economic Affairs (see fig. 8.1).

Another financial instrument aimed at encouraging universities to commercialize their discoveries is the Open Technology Program (OTP) of the Technology Foundation STW (in Dutch, *Stichting Technische Wetenschappen*) (see fig. 8.1). From 1990 onward, this research foundation has been part of the research council (NWO). It has a yearly budget of about €46 million. About 60% of its budget derives from NWO, and 40% comes from the Ministry of Economic Affairs. The OTP program stimulates high-quality university research projects with high user involvement and good prospects for utilization and research yield. It is a competitive program; a jury consisting of representatives from science and industry rewards some 40% of the applications for subsidies. Tenured university staff can apply for an OTP research grant. STW also awards so-called valorization grants. This is another subsidy scheme aimed at commercializing the research findings produced in universities.

The Leading Technology Institutes (LTIs), discussed previously, are public-private research partnerships created to encourage public-private collaboration and innovation in areas that have a strong counterpart in the country's manufacturing base. There are nine LTIs, cofunded by government and industry, initially for a period of ten years. Universities and (semi-)public research institutes participate in the LTIs. The first four LTIs were evaluated in 2002, and the outcome suggested that LTIs constituted a successful model that should be continued in the years ahead.

The public subsidies to LTIs, therefore, were continued after the first ten-year period, and funding for additional LTIs was made available from a number of ministries. In a 2004 OECD report, the LTI model was presented as a best practice to be considered by other countries. However, critics have suggested that the LTIs are operating in a rather closed (i.e., inward-looking) fashion, leaving little room for small and medium-sized companies to cooperate. They say that LTIs ought to be client-driven instead of technology-driven (Berkhout and Sistermans 2006; Van Beynum 2006).

The idea of creating additional LTIs that are oriented more toward the intersection between social sciences/humanities/law research and user groups in society (including government departments) was promoted by the Innovation Council. This advisory body launched the concept of Societal Top Institutes, the name reflecting the idea that this type of institute is to focus on social themes and social innovation, whereas LTIs focus on technological innovation. In 2005 the minister of education (through NWO) made funds available to support three Societal Top Institutes in the fields of pensions/aging, urban innovation, and international law.

The Economic Reinforcement Fund, later renamed Bsik (Knowledge and Research Capacity), is a government investment program set up in the 1990s. The Bsik program is a competitive fund out of which initiatives to strengthen the research infrastructure of the Netherlands are supported, using government revenues from natural gas exploitation. The Bsik program supports projects by public-private consortia. Bsik funding is partly a government subsidy and partly private funding from participants. The bulk of the Bsik funds are allocated to universities. The fund is aimed at strengthening research in the following priority fields that represent areas of strategic interest for Dutch industry and society at large: (1) genomics and life sciences, (2) information and communication technology, (3) microsystems and nanotechnology, (4) spatial planning, and (5) sustainable system innovations.

The Bsik program is in line with the government's wish to focus on areas of excellence. Bsik is now in its third round. Thus far, €802 million in public funds has been made available for thirty-seven selected programs that started in 2004 and 2005. The programs run over four to six years and are expected to produce user-oriented knowledge in multidisciplinary knowledge domains (or "clusters" in the Bsik jargon). The public-private networks that are supported consist of universities, private companies, research institutes, and other public organizations.

Over the years, the introduction of new policy instruments, such as the ones mentioned here, as well as the overall rise in external research funding, have led universities to respond increasingly to demands originating in society. External clients nowadays drive a large part of the universities' research agenda. In addition, research assessments also pay more attention to the user orientation of academic research. In other words, scientists are encouraged to become more entrepreneurial. As a result, many research groups have tried to increase their budgets through research council grants and externally funded research contracts. Often they are encouraged to do so through their university's internal resource allocation model that places a premium on every competitive research grant brought in. This "co-financing" of research is a common phenomenon in Dutch universities—and indeed in European universities (Salerno et al. 2005). External sponsors of research often are not prepared to pay the full cost of the research project; they may pay only labor costs and a small overhead rate. Research councils and charitable boards (many of the latter working in the medical fields) expect universities to augment research grants using income from the university's internal sources. So, even if the relative income from research councils and other nonprofit organizations may still be relatively low, research council funding and charitable funding will have a considerable impact on the research portfolio of a research group. For instance, in the medical field of chronic lung diseases (such as asthma and COPD), in which nonprofits fund about 15% of research, external clients will have a large impact. This is so in particular because the peer system that funders use is very similar to that of research councils. Therefore, winning a research contract provides prestige to the research group in question.

The practice of cofinancing (or *matching,* which is the term in the Netherlands) increasingly presents a problem for Dutch universities. In many universities, the increasing matching obligations (due to the rise in research council grants, Bsik programs, IOP grants, and European Framework Program projects) reduce the university's room to maneuver and to set their own research agenda. As mentioned above, half of the research capacity in Dutch universities is paid for by research councils or other external parties (see table 8.3). These external grants will not cover the total costs of the research that is commissioned. On average, research subsidies cover only 54% of the total costs (Ernst & Young Accountants 2004), implying that universities will have to add 85 eurocents to every euro of research subsidy they receive. To continue their involvement in externally driven research, universities therefore have to address the only other source of discretionary income that they have, which is the recurrent funding supplied primarily through the strategic considerations component (SCC; see above).

The practice of matching has some benefits. It offers research sponsors the certainty that universities will not "blindly" apply for subsidies. The fact that universities have to augment the grants implies that they will be selective in the research projects they apply for. However, in recent years the amount of money in research council funding and contract funding has risen substantially and has quickly increased the share of recurrent funding exhausted by matching. The SCC is acting more and more like a matching device instead of a device to sponsor research that is driven by the interests and curiosity of the researchers themselves.

Assessment

The overall performance of the academic research system is the topic of this section. The first two parts of the section examine the inputs and outputs of the system. The remaining parts focus on reform policies for the Dutch academic research enterprise: policies to strengthen public-private interaction in research and the funding policies to achieve selectivity and concentration in research.

INPUTS

The main actors in the Dutch academic research enterprise, as introduced above, are the research universities, the public research organizations (government labs), and the private sector. Table 8.7 presents some overall statistics on the organizations that fund R&D and the ones that carry out R&D. The Dutch R&D expenditures in 2003 (almost €8.4 billion, or 1.76% of GDP) were lower than the OECD average (2.2%) or the EU15 average (2.0%) and lower than the percentage of GDP spent by Germany, Belgium, France, Denmark, or Sweden (CBS 2007). The Dutch government often states that the Netherlands is at risk of falling behind many of its main competitors within the global knowledge economy. R&D expenditures fall short of the 2010 Barcelona target for the European Union, which is to achieve an R&D intensity of 3% of GDP.

As shown above in table 8.2, R&D spending by higher education institutions as a percentage of GDP declined over the period 1995–2005 (see also table 8.7, column 7). When corrected for inflation, the level of university R&D has declined in the previous five years by nearly 0.5% annually. This is a marked contrast with an almost 5% annual growth in other countries. However, compared to the OECD as a whole, R&D spending by Dutch universities is still relatively high, since the OECD average was 0.51% of GDP in 2003. Most large OECD countries, including the United States, Japan, Germany, France, Italy, and the United Kingdom, devote between 0.35% and 0.45% of GDP to R&D in higher education institutions. Businesses fund a growing share of the R&D performed in the higher education and research institutes sectors, but they still account for only 7% of the R&D income of Dutch universities (€161 million in the year 2003; see table 8.7). Once again, however, the OECD average is lower: 6.1% in 2003 (and 6.5% in the EU25). Most large OECD countries show even lower shares: 5% (United States), 6% (United Kingdom), and

TABLE 8.7.
Funding of R&D: Destination of R&D Funds, 2003 (€M)

| | Dutch Organizations (performing organizations) | | | | | R&D contracted out (6) = 1+2+3+4+5 | Expenditure on R&D carried out by own personnel (7) |
	Business enterprises (1)	Public research institutions (2)	Universities (3)	Private nonprofits (4)	Abroad (5)		
Origin of funds							
Total Dutch organizations	418	469	628	3	519	2,037	8,376
Business enterprises	398	196	161	—	500	1,255	4,804
Public research institutions	15	216	410	—	19	660	1,213
Universities	—	12	—	—	—	12	2,356
Private nonprofits	5	45	57	3	0	110	3
Abroad (excl. EU)	679	76	—	—	—	755	
Total (at home & abroad)	1,097	545	628	3	519	2,792	
Government support for R&D in the Netherlands	186	1,336	1,852	2			
Central government support	147	1,277	1,761	2			

Source: CBS 2007, table 4.1.1.

2.5% (Japan). Germany is an important exception, with almost 13% of higher education R&D financed by the business sector.

If the Dutch universities are relatively important to business R&D, how do companies perceive the public R&D sector? A study by the Observatory of Science and Technology (NOWT 2005) shows some survey results. The NOWT quotes the results of the latest of Europe's Community Innovation Surveys. Data for the years 2000–2002 indicate that some 22% of the "innovating companies" in the Netherlands—a subset of all Dutch companies—use the services of universities as a source of R&D-related information. About a third of those innovating companies (8% of all innovating companies) considered universities to be a "very important" source. As for the nonuniversity public research institutes, a 30% share of innovating companies declared that they have used input from such research institutes. The "intermediate" nonuniversity research institutes were appreciated to a significantly higher extent. R&D performance by Dutch research institutes is close to both the EU15 R&D intensity and the EU15 average growth performance. This sector has been one of the strongest links in the Dutch innovation system chain for many years.

The most important indicator of the input into the academic research enterprise is, of course, related to human resources, that is, the number of researchers. Table 8.3 presents time series for the number of researchers, indicating the growth in staff on research contracts. Figure 8.3 shows the shares of the three main categories for the research staff of the thirteen Dutch universities. The shares funded by means of recurrent funding (first flow of funds), research council funds (second), and other project funds (third) differ substantially. In 2005 the share of researchers on recurrent funding was the highest for Tilburg University (UvT) and the University of Groningen (RUG). In relative terms, the number of grant holders from the Dutch research council was the highest for the University of Twente (UT) and Radboud University Nijmegen (RUN). Wageningen University (WUR) had the highest number of research staff on projects in the third flow of funds.

OUTPUTS

According to international benchmarks, the quality and productivity of Dutch university research are high (NOWT 2008). The Netherlands is among the leading countries in terms of its focus on basic ("exploratory") longer-term research. The excellent research performance is reflected in the relatively high level of research output. Dutch researchers contribute some 2.5% of all scientific publications worldwide. The publication output rate of Dutch researchers puts the Netherlands among the most productive countries in the world: it is among the top three nations in terms of productivity per researcher. Some three-quarters of the publication output

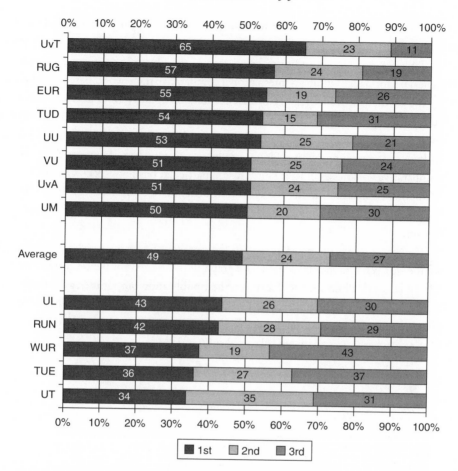

Fig. 8.3. University research staff by funding source (1st, 2nd, and 3rd flow of funds), 2005. *Source:* NOWT 2008, 91. *Note:* UvT, Tilburg University; RUG, University of Groningen; EUR, Erasmus University Rotterdam; TUD, Delft University of Technology; UU, Utrecht University; VU, Free University Amsterdam; UvA, University of Amsterdam; UM, Maastricht University; UL, Leiden University; RUN, Radboud University Nijmegen; WUR, Wageningen University and Research Centre; TUE, Eindhoven University of Technology; UT, University of Twente.

is produced by scientists and scholars employed at the fourteen universities in the Netherlands and their medical centers. Researchers in publicly funded, nonuniversity research institutes account for an additional 12%. The private sector accounts for about 5% of the total publication output. The remaining 6% is produced by general hospitals and governmental organizations. In the four-year period 1990–93, the Netherlands produced almost 60,000 scientific research articles, and in the period 2003–6 this output increased to almost 100,000, amounting to an annual output of about 25,000 articles.

Table 8.8 provides an indicator-based scan of the Dutch scientific and technological performance within an international comparative framework. Based on work by the Netherlands Observatory of Science and Technology (NOWT), the table summarizes performance indicators for the Netherlands and sixteen benchmarking countries. The overview shows that the Netherlands has an excellent record in knowledge creation, with a highly productive science base. However there are comparative weaknesses in R&D intensity, in particular business R&D. Overall, the Netherlands ranks around the average for the benchmark countries according to the International Competitiveness Index, which is compiled by the World Economic Forum.[5]

Figure 8.4 shows the development in the number of research articles produced by the thirteen Dutch universities. The graph illustrates that in the seventeen-year period covered, all universities have seen their publication output increase steadily. In the most recent decade, there was a 50% increase in scientific output. Because the number of researchers increased much less rapidly (see table 8.3), one may conclude that researcher productivity has increased substantially.

TABLE 8.8.
Performance indicator ranks for the Netherlands and benchmark countries, 2000–2007

	International competitive ranking	Productivity		% R&D expenditure			
		Scientific research	Technological development	Total	Private sector	Higher education	Research institutes
US	1	15	6	5	5	16	7
CHE	2	1	1	4	4	7	17
DNK	3	9	9	8	9	8	12
SWE	4	6	4	1	2	11	16
GER	5	13	5	7	8	14	5
FIN	6	14	2	2	6	13	10
JPN	8	17	3	3	3	15	11
UK	9	4	12	13	12	9	8
NLD	10	2	7	14	14	2	4
KOR	11	16	17	6	1	17	6
CAN	13	8	14	11	15	1	9
AUT	15	10	11	10	11	4	15
NOR	16	11	13	16	16	3	3
FRA	18	12	10	9	13	12	2
AUS	19	5	15	15	17	5	1
BEL	20	7	8	12	7	10	14
IRE	22	3	16	17	10	6	13

Source: NOWT 2008.

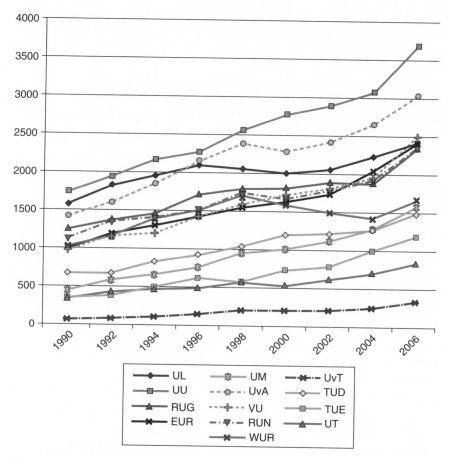

Fig. 8.4. Output of Research Publications for Universities, 1990–2006. *Source:* NOWT 2008, based on Thomson Scientific Web of Science Database. *Note:* For abbreviations and acronyms, see note to figure 8.3.

Bibliometric research carried out by the Centre for Science and Technology Studies of the University of Leiden (NOWT 2008) suggests that quantity and quality of research do go hand-in-hand for the Netherlands. Looking at the citation impact of the Dutch research from an international perspective, we may conclude that the Netherlands is among the world's most highly cited nations, showing a score that is 34% above the worldwide citation average for the period 2003–6 (table 8.9). The "field-normalized citation-impact-score" has a value of 1.34. This indicator, based on publications in 2003–6 and citations made during the same period, has shown an increase of 12 percentage points compared to its 1993–96 value. The normalized citation impact score indicates the extent to which a country's citation

TABLE 8.9.
Citation impact scores, 1993–96 versus 2003–6

	Impact 1993–96	Impact 2003–6
Switzerland	1.43	1.41
United States	1.44	1.37
Netherlands	1.19	1.34
Denmark	1.13	1.33
United Kingdom	1.19	1.24
Canada	1.14	1.22
Norway	0.93	1.22
Belgium	1.09	1.20
Sweden	1.10	1.20
Ireland	0.88	1.17
Finland	1.17	1.16
Germany	1.05	1.13
Austria	0.97	1.12
Australia	0.98	1.10
France	0.93	1.06
Japan	0.82	0.86
Korea	0.66	0.84

Source: NOWT 2008, 29, based on data by Thomson Scientific.

impact differs from the average impact score of all countries within the same field of science.

Naturally, behind the average score lies wide variation. The citation impact performance of the Dutch science base varies significantly across fields of science. Some fields have a much larger publication output when compared to output levels in the benchmark countries. Some of the fields that have a significant "overrepresentation" in the Netherlands include information and communication sciences, agricultural and food science, astronomy and astrophysics, and language and linguistics. Some of these fields enjoy citation impact rates that are more than 40% above the world average (information and communication sciences), while others are at the world average (such as language and linguistics). Conversely, some fields produce comparatively lower quantities of articles but manage to achieve high levels of citation impact within the international research literature. Fields such as chemistry and chemical engineering, physics and materials science, and electrical engineering and telecommunication are among the most highly cited in the Netherlands. Very few fields of Dutch science are cited below the worldwide average.

Many of the Dutch research universities feature strongly within the international ranking of universities published by the Times Higher Education Supplement (2007) and the ranking by the Institute of Higher Education of the Shanghai Jiao Tong University (Institute of Higher Education 2007). In fact, each Dutch university excels in at least one field of science in terms of producing a relatively large number of papers that are among the top 10% of most highly cited research publications worldwide. These include fields within the natural sciences, the medical and life sciences, and the technical sciences, but also the social sciences (including economics).

This evidence supports the conclusion that the growth in research output did not come at the cost of a lower quality. However, numbers of ISI publications and their impact are only one way of showing the quality of research. The research ratings in the recent research assessments (over the period 2003–5) may also reveal the quality of research carried out in the various disciplinary research groups. Figure 8.5 gives the quality scores for 146 research groups that were assessed in the period 2003–5; it indicates how many research programs received ratings of 1 (unsatisfactory) through 5 (excellent), using the SEP rating scale (see table 8.5). Overall, the quality of academic research in most university research programs in the Netherlands was "good" to "very good." More than 40% of the research groups received a rating indicating very good research, while almost 30% scored "excellent."

Another indicator of research output is the number of PhD degrees conferred. The distribution of PhD degrees over the various disciplines is shown in table 8.6.

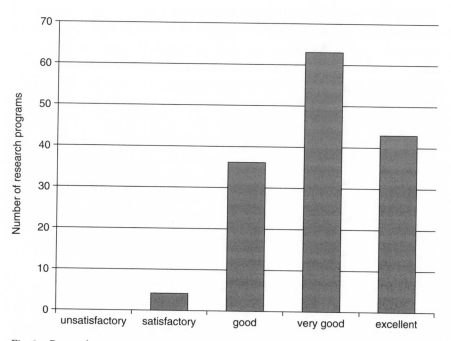

Fig. 8.5. Research assessment scores awarded to research programs over the period 2003–5. *Source:* OCW 2005.

The health-related sciences have the highest PhD production (33%). The number of PhD degrees in science, engineering, social science, the humanities, and law has been rather stable over the period 1997–2006. The number of PhDs in health-related subjects rose by 50%, while there was an increase of 30% for agriculture and an increase of about 40% for economics.

Other indicators of research output are rather difficult to find. There is, however, scattered material on the issues of technology transfer and research commercialization (table 8.10). Dutch universities and public-sector research institutions are gradually embracing patents as a means to protect their discoveries and inventions. Increasingly, they are exploiting their intellectual property rights. The number of academic patents has increased steadily during the past decade, both domestic patents and patents registered at the European Patents Office and the US Patent and Trademark Office. The commercial value of the university-owned patents has been limited so far. The Netherlands typically stands out favorably in international benchmarks of patenting activities (OECD 2005b). It performs at an average level within Europe in terms of external commercialization by means of launching R&D-based start-up companies. The number of university-based spin-offs is about 30% to 40% lower than the number in some important competitor countries (Van Tilburg 2003).

While the above indicators seem to paint a rather positive picture of the research output produced in the Dutch academic research enterprise, some problems have emerged recently. As an OECD report on the performance of the Dutch national innovation system observed, "there are features of the Dutch innovation system that contribute to a lack of dynamism observable for some time now. As a consequence, the Netherlands has been losing ground in innovative performance *vis-à-vis* a number of high-performing economies. At present, the Netherlands does not sufficiently succeed in translating the results of scientific research into economic performance. Furthermore there are new challenges ahead. One of these challenges to be addressed is an imminent shortage of skilled personnel, in particular in science and technology, but also the impact of globalisation of R&D" (OECD 2005b, 147).

The remainder of this section discusses some reform policies that have been implemented lately to address these concerns.

INCREASING SCIENCE-INDUSTRY INTERACTION— CREATING PEAKS IN THE DELTA

Overall, we have observed that the Dutch universities and semipublic research institutes offer a broad and high-quality organizational base in many fields of science. This structure and its scientific achievements have resulted from R&D investments

TABLE 8.10.
University spin-offs and patents

	Number of spin-offs 1999–2001	Patent applications 1990–1999	Research staff (in FTEs) 2001
Leiden University	3	24	1,610
Utrecht University	3	19	2,108
University of Groningen	6	26	1,470
Erasmus University Rotterdam	5	3	847
Maastricht University	10	3	783
University of Amsterdam	4	10	1,708
Free University Amsterdam	1.5	4	1,269
Radboud University Nijmegen	10	8	1,483
Tilburg University	0	0	274
Delft University of Technology	10	90	1,416
Eindhoven University of Technology	7	17	743
University of Twente	20	18	771
Wageningen University and Research Centre	10	2	839
Total	89.5	224	15,321

Sources: Spin-offs: Van Tilburg 2003; patent applications: Bekkers, Gilsing, and van der Steen 2006; research staff: Association of Universities in the Netherlands (VSNU, 2008).

made and policies in place in the past, often as early as the 1970s and 1980s. Although the level of investment in the public sector has declined since the early 1990s, the indicators shown in table 8.8 reveal a fairly efficient research system with relatively modest expenditure levels and comparatively high levels of productivity and international scientific impact (table 8.9). However, innovation activity, which entails the development and application of new knowledge in new products and/ or processes, appears to be only mediocre. According to the European Innovation Scoreboard Summary Innovation Index, which brings together twenty-two indicators considered to reflect innovation activity (OECD 2005c), innovation activity in the Netherlands ranks twelfth out of the twenty high-income countries for which the index has been calculated.[6] Although the Netherlands ranks slightly above the EU15 average, it is far below the leaders. Along with Norway, the Netherlands ranks six places lower on innovation activity than on the scientific publications score; the largest fall in ranking between these two indicators. This performance represents a paradox, because rankings for scientific publications and innovation activity are in general highly correlated. There appears to be a gap between knowledge creation and innovation activity. This, of course, is a diagnosis that applies not only to the Netherlands but to Europe as a whole (the "European Paradox") (OECD 2005b, 154).

The knowledge paradox nowadays is pervading all policymaking related to higher education, research policy, and technology policy in the Netherlands. In 2003 a high-level advisory council, the Innovation Platform (Innovatieplatform), was created, composed of high-level representatives from research, industry, and the policy arena and chaired by the prime minister (see figure 8.1). This council, now in its second round after a change in Cabinet following the 2006 elections, is developing policy recommendations for strengthening the innovative capacity of the Netherlands. In 2004 the council gave advice on the problems that researchers and knowledge workers from abroad face if they seek employment in the Netherlands. The council also gave advice on stimulating the dynamics in the Dutch innovation system and identified key areas for the Dutch economy (Innovatieplatform 2004).

The scientific cooperation among research teams from the public sector and between public research organizations and industry are regarded as critical areas. It is argued that universities are not given enough incentives to exploit their in-house knowledge and patents and have insufficient capabilities to commercialize their research (Carey et al. 2006). This problem was already identified in the 1990s, leading to the creation of university research schools and the Leading Technology Institutes (LTIs) discussed previously. These initiatives are specifically meant to enhance both economies-of-scale and focus within the Dutch science system. They serve to achieve critical mass within science fields by promoting more collaboration between the various relatively small populations of Dutch researchers, leading to a more attractive infrastructure for foreign companies and external firms who are considering new R&D investments in the Netherlands. The first four LTIs that were created seemed to perform well. Nevertheless, an OECD (2004) review also found that the national innovation system would benefit from making existing LTIs less dependent on subsidies as they mature. Furthermore, incentives should be improved for LTIs to strengthen their linkages with government-sponsored international cooperative ventures.

The LTIs have been an important exception to the traditional government policy of pursuing a generic rather than a specific approach in technology policy to strengthen innovation. A generic policy meant that it was not the government that should make the choices but industry. Industry should be able to use generic instruments and choose the technologies in which to invest. The most important generic instrument aimed at stimulating business R&D is the WBSO, an R&D promotion act. The WBSO research and development tax credit is a fiscal facility for companies, knowledge centers, and self-employed persons who perform R&D work. Under the act, a contribution is paid toward the wage costs of employees directly involved in R&D. Today, it amounts to some €0.4 billion, making it a relatively prominent part (over two-thirds) of the total innovation budget of the Ministry of Economic Affairs. A 2002 evaluation of the WBSO found that the scheme is cost-effective. According to the evaluation (Brouwer et al. 2003), the WBSO makes a significant contribution toward increasing the R&D intensity of the Dutch private sector.

When it comes to universities, we can observe the same generic instruments at play. While universities were given some incentives to achieve economies-of-scale-and-scope by joining departments from different universities, in principle the individual universities made the choices themselves. From the late 1990s on, this overall policy philosophy gradually changed. Observing that companies were less and less engaged in fundamental research and that research in universities at the same time was becoming fragmented (also due to constant budget cuts), the Netherlands was seen to lack sufficient critical mass to be among the international state-of-the-art economies (Boekholt and den Hertog 2004). In addition, relatively few high-tech spin-offs and new ventures came out of the Dutch universities and research centers. This resulted in a relatively weak Dutch position in technology areas that were considered the new growth potential areas: life sciences, e-commerce, and nanotechnology. The "hands-off" governance mechanisms and policy instruments proved insufficient to radically change the research agendas of the actors involved. Therefore, the Dutch government started moving more of its support for innovation toward specific instruments. The authorities felt that focusing public support on specific fields that were crucial for the economy would produce large external benefits. Such a policy would contribute to the creation of "focus and mass" in research, or "Peaks in the Delta." The latter phrase was the title of a government policy document that announced where the Ministry of Economic Affairs expected the future hot spots of technology-based economic activity to be located (Ministry of Economic Affairs 2004).

The policy to provide more specific support was evident in instruments like the Bsik subsidies (see above). Two policy documents were published to set out the main lines of science, technology, and innovation policy in October 2003. The Ministry of Economic Affairs (2003b) published its Innovation Letter, entitled *Action for*

Innovation. The Ministry of Education released its Science Budget 2004, entitled *Focus on Excellence and Greater Value* (Ministry of Education 2003). The Science Budget 2004 included these main elements:

- *Stimulating focus and mass in research* in order to concentrate research funds on national priorities (such as genomics, information and communication technology [ICT], and nanotechnology). Other areas of special importance from a societal point of view, such as traffic management, water control, and the vitality of major cities, were also mentioned.
- *Rewards for excellent research groups,* which should be based on simple, non-bureaucratic procedures, such as track records of individual researchers.
- *Promoting the utilization of research results* by strengthening the societal role of universities and stimulating a university patent policy.
- *Attention for human resources in science and technology,* with a particular focus on the impending shortage of knowledge workers, especially in science and engineering.
- *Raising public awareness of science and technology.* Science centers and museums as well as schools, universities, and industries play a vital role, and more can be done in pooling current regional initiatives.

The main elements of the Innovation Letter are

- *Strengthening the climate for innovation,* which includes intensifying the fiscal incentive for R&D, the introduction of a new (project-based) scheme for R&D collaboration and policy directed at tackling the impending shortage of knowledge workers.
- *Encouraging more companies to be innovative* by improving the climate for high-tech start-ups, by better exploiting the potential of SMEs (small and medium-sized enterprises) through policy actions aimed at the dissemination of knowledge to SMEs, and by attracting knowledge-intensive business activity to the Netherlands.
- *Taking advantage of opportunities for innovation by opting for focus and mass in strategic areas.* The aim here is to strengthen the knowledge base and stimulate public-private cooperation in key technology areas, such as ICT and life sciences.

Summing up, the Dutch policies on research and innovation are centered around three core elements: excellence, focus and mass, and interaction.

Primarily, the government wishes to improve science-business linkages by increasing incentives and improving the institutional frameworks. It established two coordination organizations (*regie-organen*) in the fields of genomics and ICT. Improving cooperation between companies and public research organizations was seen as a priority. In this regard, a new policy instrument, the Smart Mix, with a budget of €100 million for research initiatives proposed by consortia of business enterprises, nonprofit organizations, and knowledge institutes (primarily universities) was launched to encourage initiatives to create focus and mass in excellent research as well as to stimulate the social and economic payoff from this research. The keyword in all of this is *valorization*. Valorization encompasses all activities that contribute to ensuring that the outcomes of scientific knowledge add value beyond the scientific domain.

Policies were also implemented to strengthen science-business interactions through greater spin-off activity. The existing instruments to support the creation of technology-based start-ups were streamlined into one program in 2004, the TechnoPartner Action Program (OECD 2005b). The TechnoPartner Action Program aims at improving the start-up climate through the following means:

- a seed facility to support the bottom end of the Dutch venture capital market, thereby helping high-tech start-ups (including spin-offs from public institutes) to satisfy their capital requirements at an early stage;
- the Knowledge Exploitation Subsidy Agreement, which has the objective of speeding up the utilization of scientific knowledge by high-tech start-ups both inside and outside of knowledge institutes (universities, hogescholen, and publicly financed research institutes); and
- the TechnoPartner Platform, set up to promote an entrepreneurial culture through education and to identify and remove institutional barriers that negatively impact entrepreneurship or the creation of high-tech start-ups.

These are complemented by policies that increase student awareness of entrepreneurship, such as business plan contests, matchmaking events between universities and businesses, and the Dreamstart agency, which advises potential starters (Waasdorp 2002, 34–35).

The Casimir program, which is based on the European Marie Curie grants scheme and the French Cifre Scheme, was established by the Dutch Research Council, NWO, to foster mobility of researchers between public research organizations and the private sector and to make jobs in research more attractive. It awards subsidies (up to €160,000 per project) for projects having three partners: a company, a university, and an individual researcher.

Interaction between science and business is also strengthened through the distribution of "knowledge vouchers" to SMEs. These vouchers (representing a value of either €2,500 or €7,500) are an initiative of the Innovation Council and may be used by SMEs to purchase knowhow from universities or other research institutes. Since their small-scale introduction in 2004, these vouchers have proven to be a great success and each year have been quickly exhausted, pressing the government to make more vouchers available in subsequent years.

To strengthen innovation, a large number of instruments and organizations were created over the years, leading to concerns about the complexity and effectiveness of the system and its administrative costs. The Ministry of Economic Affairs therefore worked to integrate the various initiatives (e.g., one outcome was the TechnoPartner program) and merged its two agencies for implementing policies on innovation and sustainable development (*Senter* and *Novem*) into a single agency (*SenterNovem*) (see fig. 8.1) charged with innovation support to SMEs. This was done to streamline the public support schemes and to provide a single package of information, advice, and capital facilities for all entrepreneurs. In addition, the government increased expenditure on innovation. The already substantial WBSO tax breaks (discussed previously) were extended, especially for SMEs (an increase of €100 million), and an experiment with a small business innovation and research initiative (SBIR) was started.

To stimulate the transfer of knowledge further, a program was created to strengthen "knowledge circulation" between the vocational higher education institutions (the hogescholen) and companies, especially SMEs. An important element in this program is the so-called *lectorate,* a senior staff position in the *hogeschool* that was created in 2001. Government subsidies were made available to create around one hundred lector positions in the hogescholen. The lector is expected to build a "knowledge circle" of professionals from within the hogeschool and link up with the business sector (in particular, with SMEs), work on stimulating the external orientation of the hogeschool in his or her particular area of expertise, make innovative changes in the curriculum and staff development of lecturers, and encourage knowledge transfer between the hogeschool and its environment. Recent evaluations of the lectorates provided evidence of modest successes in these areas and contributed to the continuation and extension of the lectorate experiment. The government also made it clear that it saw the knowledge circulation function of hogescholen as an explicit task, next to the teaching mission. This showed that, more than before, hogescholen are regarded as an essential part of the innovation system. There still is debate on the research function of hogescholen as compared to the research mission of universities. Whereas university research is fundamental (developing new insights and in-depth knowledge), research at hogescholen is expected to focus on new applications of existing knowledge (AWT 2005). It should primarily have a "design and development" function and should be demand-led.

Another goal of the government is to attract scientists and engineers from abroad and to encourage those already in the Netherlands to stay. Complaints were made by foreign investors about the barriers to bringing in highly skilled workers from abroad. The Netherlands has not been very successful in attracting and retaining foreign human resources in science and technology (HRST). Not only are HRST immigration flows relatively low, but such immigrants also tend not to stay on in the Netherlands, seeing the country as a stepping-stone to other destinations. In order to increase entry of "knowledge migrants" from outside the European Union, the government recently took steps to facilitate their entry. It established a single point of contact, shorter procedures, and limited levies for knowledge immigrants. The government expressed in the *Deltaplan Bèta Techniek* its intention to consider further steps to facilitate such immigration (Ministry of Economic Affairs 2003a).

Part of the *Deltaplan* is to promote science- and technology-based studies and careers among young people. The attractiveness of the Netherlands for R&D activities is believed to depend on the number of science and engineering graduates in the country. University graduates and PhD graduates—in particular those who graduated in natural sciences and engineering (S&E)—constitute a pool of human resources out of which companies and research institutes can select their future researchers and technical personnel. The Netherlands has not had much success to date in maintaining its stock of S&E graduates. There is a low share of science and engineering students in total graduations (OECD 2005b). Poor career prospects in S&E jobs, relative to those in management, encourage students to pursue studies in economics, law, and business. In addition, S&E programs are perceived by students as risky and difficult. All of these factors appear to have a stronger effect on women than on men. Indeed, the increase in the share of females in total graduations has contributed to a reduction in the share of S&E graduates in total graduations in the past twenty-five years (CPB 2005).

The number of female researchers is still relatively low compared to that of male researchers. The fraction of female researchers is growing steadily, but women are still significantly underrepresented among university research staff, especially in the higher ranks. In view of the current problems in recruiting sufficiently large quantities of R&D staff from the higher-educated Dutch population, particularly females, it would seem that a significant segment of the human capital in the Netherlands is underutilized.

PERFORMANCE-BASED FUNDING

Reforms in the model for the funding of academic research have been undertaken because of concerns that university funding is still heavily based on historical considerations instead of research performance. A major government concern is to make academic research more dynamic. The government would like to see the universities' own decisions on the resourcing of their research activity made more dependent on measures of research excellence and the societal and economic relevance of research.

The 2006 introduction of the Smart Mix element in the universities' funding formula and the competitive Smart Mix budget, both mentioned above, aimed at strengthening selectivity and concentration in research, thus contributing—it was hoped—to excellence in research. They also aimed at removing the knowledge gap between the research enterprise and the business sector and strengthening cooperation between industry and research institutes. However, both Smart Mix instruments were short-lived, lasting only one round. The new Cabinet in 2007 decided to abolish the Smart Mix and make the funds available through the "ordinary" research council channels for investing in fundamental research. Dynamizing research through a more performance-based system of research funding, however, keeps reappearing on the government's policy agenda. For the time being, it appears that the way forward will be through the allocations of the research council (NWO) and not through formula-based recurrent funding. The government continues its course set out in the Science Budget to create more focus and mass in academic research and to strengthen the incentives for a more dynamic research landscape. In 2005 the education minister installed an independent commission, the Dynamics Commission, chaired by the director of the Dutch Research Council's FOM division, Hans Chang. This advisory commission released seventeen proposals for "dynamizing research" (Commissie Chang 2006). Attention was also paid to the issue of performance-based funding of research, although this was not formally included in the committee's mandate.

The Chang proposals may be grouped into two parts. The first set of measures does not require any extra resources on the part of the government. The second set of measures requires an extra commitment of €1 billion per year. In the first group are recommendations such as urging universities to make more explicit what results are produced from public research investments. Thus, universities are called upon to become more transparent and work harder on communicating their research performance more clearly to their stakeholders and funders. Related to this is the issue of whether and how the university rewards (or punishes) research groups on the basis of external research evaluations. To what extent does the university's executive board or its deans take into account the ratings assigned by external peers to universities' research programs? Do the ratings translate into the resourcing of research groups?

The background to the latter question lies in the fact that 55% of the universities' recurrent funding for research is independent of performance (see table 8.4). Should this percentage be brought down, thus increasing the performance element of the budget? Doing so would make the funding of university research more dynamic, that is, less dependent on allocations made on a historical basis and more dependent on measures of performance, such as the research ratings produced in the peer-review-driven research assessments. The use of such ratings in research funding might lead to a system that resembles the British funding model based on the Research Assessment Exercise (Venniker and Koelman 2001).

One may argue that the need to introduce performance-based funding is larger when there is little evidence of research ratings affecting the universities' internal research policies. This is why the Chang Commission found it useful to study this issue. They examined whether quality ratings affected decisions about the structure, funding, and human resources dedicated to research in a university's faculties and departments. It was found that rectors, deans, research directors, and individual researchers regarded the system of external research evaluations in the Netherlands as very effective (Jongbloed and Van der Meulen 2006).

In their decision-making, university leaders explicitly take into account the quality and productivity of the research groups' output. The evaluation of academic research has become an essential part of the apparatus for accountability, governance, and management in universities, although some researchers have complaints about the administrative burden. Management on the basis of performance and ratings of quality has become standard practice in Dutch universities. This is the case for the institution as a whole and at the departmental level as well as at the individual level. Strategic decisions about the continuation of research institutes or the setting up of new ones are made with a very keen eye upon the academic research performance. The same holds for human resource decisions (hiring of professors). To achieve a critical mass of researchers working in a selected number of strategic research areas, many universities have restructured departments and constructed larger research units that have the best chance of performing high-quality research and/or have a large income-generating potential (Jongbloed & Van der Meulen 2006). Thus the national "focus and mass" policy has been mirrored in university policies.

Universities try to distinguish themselves from other universities by carefully choosing their areas of excellence. They make use of research assessments in building their research profile (see also Salerno et al. 2005). The bundling of research groups around multidisciplinary themes and (sometimes) expensive research infrastructure is a common strategy in the Dutch research enterprise. If this is the case, one may argue that at the national (i.e., ministry or government) level, there is little need to add more teeth to the existing system of research assessments or indeed to connect a university's research allocation to its research ratings. Moreover, the universities would perceive the introduction of such a UK-type funding model as a sign of a lack of trust in the management of the university and the academic professionals.

However, what may seem good practice from the perspective of the individual university may not be optimal when the combined decisions of all Dutch universities are taken into account. At the system level there may be important gaps or indeed duplications in the research landscape. In such a case, the instrument of research council (NWO) grants may be more effective in creating focus and mass and encouraging universities to take a route that is more optimal from a national perspective. Whereas allocating recurrent funds creates stability and lays the foundation for the academic research landscape, the funds allocated by the research council can emphasize particular research areas. By nature, research council funding is performance-driven and dependent on the academic peer evaluation. The research council route, therefore, might be an effective as well as a legitimate choice for achieving focus, mass, excellence, and relevance. The current (2008) education minister has signaled that the cofunding (or "matching") requirement for research council grants will be abolished for part of the grants.

Most stakeholders (government, universities, research council, and advisory boards) support the idea that more focus and mass are needed in university research. The main reason is to be better prepared to face the global competition for research talents and resources. Constructing larger research units or strengthening the cooperation between disciplinary research units is seen to safeguard the quality and vitality of education and research, particularly in those areas where there is a danger of fragmentation and failure to achieve critical mass. "Cooperate to compete" seems to be the maxim these days. As science becomes increasingly international and the interaction between researchers and users becomes increasingly important, the need for research groups to be recognizable and visible increases, as well, helping to achieve more focus and mass.

Conclusion

Two things stand out when one looks back over a period of, say, twenty-five years and observes the development in the Dutch academic research enterprise. The first is the increased attention being paid over the years to research quality, beginning in the 1980s. Second, in the 1990s there was an increased emphasis on relevance and valorization. Related to the latter is the emergence of a "third task" for universities in addition to the traditional tasks of providing degree programs and carrying out scientific research. In short, *performance* has become the key goal in the academic research enterprise, although the meaning of this concept currently is vastly different from its meaning twenty-five years ago.

While we observed that many policymakers, advisory boards, and opinion leaders suggest that the interaction between universities and the outside world should be intensified, the topic remains a contentious one. It is probably good to step back a bit from this discussion and ask why there is a lack of interaction. As has been made clear in this chapter, the Dutch academic research enterprise consists of (semi-) public applied research institutes (e.g., TNO, GTIs, and LTIs) next to the universities and the HBO institutions. The applied research labs seem to work quite well in responding to the demands and research needs of the business sector. In other words, the universities have for a long time been "exempted" from having to do contract research with third parties, in particular with industry. They have mainly dealt with fundamental, "blue sky" research.[7] Therefore, the university sector has never developed a culture of conducting contract research for external customers and establishing bridges to industry. Moreover, on the national level, the governance of the universities and their research agenda has always been "at a distance." The universities have always resisted changes to their recurrent research-funding allocations.

The consequences of this university governance model affect the academic research enterprise. It has contributed to a certain degree of fragmentation and duplication in research domains and research efforts. However, one may also argue that this model has led to variety and competition in research. In any case, the hands-off approach has meant that few centers of excellence with critical mass have developed over the years. For a long time universities were only to a small extent reliant on contract research funding. Add to this history the existence of a network of public research labs, and one may understand why a "third mission" only recently emerged in (some) universities. As a result, the record of the Netherlands in university start-ups and research commercialization activities up to this point has not been very strong from a comparative European perspective.

For decades there were hardly any linkages between the "academic" and the "applied" subsystems in the Dutch innovation system (Boekholt and den Hertog 2004). In terms of policy design, funding, and research performers, there has always been a strong "division of labor" between science, on the one hand, and technology and innovation on the other. This situation has only recently begun to be addressed. "Corrective measures" were introduced: thematic research programs through NWO, public-private research institutes, LTIs, specific subsidies like the Bsik funds, Technostarter subsidies, and the Smart Mix instrument. These measures have contributed to a rather fragmented array of science and technology policy initiatives that are targeting a quite decentralized and fragmented science and research community consisting of universities, hogescholen, public research institutes, and public-private constructs.

The hands-off governance mechanisms have gradually been replaced by a more intrusive approach, in which policymakers have tried to support specific areas of research (and economic) potential, such as life sciences/genomics, nanotechnology, and social innovation. The selection of instruments and areas of strategic interest, however, has often involved a great deal of consultation and stakeholder involvement. "Committees of wise men" and intermediary institutions were called upon to select priority areas; this procedure has received criticism because of its lack of transparency and its high transaction costs.

Today, research and innovation policies still seek to address the challenge of how to achieve excellence, focus and mass, and public-private interaction. Actually, most of these challenges were already on the table in the year 1983, when the system of conditional funding was introduced (see above). The fact that the same policy challenges are still around does not mean that nothing has been achieved. It is very unlikely that these challenges will ever be solved. The mix of policy instruments for addressing the challenges will never be a stable one. The academic research enterprise, the actors involved, and in particular the environment they are part of is changing constantly. Therefore, policies need to be adapted, too.

The chapter ends with some policy challenges for the Netherlands innovation system. These are based on a reaction by the Advisory Council on Science and Technology Policy (AWT) (see above) to the strategic plans of the Dutch research council (NWO), the Academy of Sciences (KNAW), and the TNO research organization (AWT 2006). The challenges very much relate to issues already touched upon in the preceding text.

In the multilayered and very diverse academic research–innovation system, each different actor is supposed to have its own role and responsibility. The government is expected to keep an eye on the overall functioning of the system and to lay out the framework within which the system is expected to operate. It should be occupied with questions like these: Should the system be engaging more (or less) in international relations? Should it focus more on research commercialization? Should research priorities be set? Should research be more concentrated and its funding be made more selective? The government's Science Budget is the official medium for taking a position on these various questions and communicating its choices to the research community. The Dutch tradition so far has been to allow the various research organizations a relatively high level of autonomy to operate; they are not steered in any direct sense by the government. Universities, NWO, KNAW, and TNO each have their roles to play. However, the large amount of freedom that the various actors have in carrying out their tasks will not in itself lead to an innovation system or performance that, from a societal point of view, is optimal. Holes in the patchwork of institutions may appear, or duplications may emerge. In such cases, the government should intervene to make the various organizations coordinate their activities. Financing is one of the instruments it can use. Dialogue, consultation, and persuasion are complementary instruments.

The government—that is, the minister of education and his or her colleague from economic affairs—is expected to outline the broad goals of the research system in order to provide the required planning stability for the various publicly funded research organizations. Such a strategic framework would need to be based on an analysis of the trends and underlying forces in research. The Dutch government and the public research organizations are expected to express their ambitions based on the challenges and trends they have identified. While these ambitions, in any case, need to address important issues like guaranteeing the quality of research, intensifying internationalization in research, and making sure that sufficient numbers of researchers are trained, some important trade-offs that cannot be made by the individual organizations themselves also need to be addressed. Four main trade-offs are highlighted below:

- *The breadth versus width of research activity.* Given the current state of research activity and performance, should the Netherlands continue to cover all research areas, or should it specialize in particular areas or clusters of disciplines? Should it follow a broad strategy and thereby negatively affect the ability to concentrate resources and achieve excellence in selective clusters? Or should it choose a selective strategy that may harm the absorptive capacity in the remaining areas? Currently the minister of economic affairs opts for a selective approach, meaning that a number of strategic technologies are selected and receive the backing of the government (the "backing winners" approach). But what consequences does this policy have, and how has the selection come about? What should be done about vulnerable areas that also may be of strategic importance (such as the humanities, languages, and some of the natural sciences)? Is it better to "back the challengers" instead? When it comes to quality, will all areas of strategic importance have to meet the same criteria in terms of excellence? What should be done in the case of less-than-excellent research groups?

- *The differentiation of functions in the innovation system.* In the innovation system a number of different functions need to be fulfilled. Knowledge will need to be produced, validated, structured, transmitted, and used. Over the years, the system arrived at today's functional differentiation, showing different tasks for the various organizations. In contrast to some other countries, the Netherlands has a large public labs sector next to the university sector. However, in recent years, we can observe a development in which research organizations take on functions that they previously did not have. This is the case in particular when it comes to the valorization theme. Compared to the past, universities as well as the research council (NWO) are expected to pay more attention to the transfer and utilization of knowledge. Previously, this task was reserved for TNO and the applied, public research labs. At the same time, one can observe the emergence of new institutions and organizations alongside the already existing ones. These have come into existence as a result of an increase in program- and project-based forms of (extra) funding. Examples are the Leading Technology Institutes, the newly formed Societal Top Institutes, and the Bsik institutes—some of them virtual institutions. Many of these were created in order to channel the extra research funding made available by the Dutch government. This leads to the question of how the academic research enterprise–innovation system will develop in the future: What tasks will the various organizations be performing? What differentiation in tasks is desirable or feasible? And what position do the newly created institutes have vis-à-vis the already existing ones? Various options are available, each of which has its advantages and disadvantages. An integration of tasks can help reduce the transaction costs and raise efficiency in the research system. Sticking to the traditional tasks, however, may be beneficial in the sense that organizations will continue doing the things they can do best. Being clear on the functional differentiation also contributes to the formulation of an evaluation framework for the various organizations in the innovation system.

- *Competition versus collaboration.* A clear choice will need to be made when it comes to the question whether research organizations are supposed to compete or cooperate. This question touches on the relationship between the different types of institutions (e.g., universities versus TNO) as well as the relationship between institutions of the same type (e.g., university X versus university Y). The question is relevant within the Netherlands, but at the same time there is an international dimension to it: who is it we are competing with? Answers to these questions affect the way research will be financed: should one introduce policies and incentives to promote competition within the Netherlands, or should one encourage collaboration? The answer also has implications for the question of how many institutions of a similar type (e.g., universities) will be needed in the future. In the Netherlands, one can observe that the financial steering of research organizations is very much inspired by the wish to achieve competition, but at the same time policymakers also call for more cooperation between the various organizations. More clarity is needed on this issue. It also relates to the issue of creating focus and mass in the research landscape. Here, the question is, Who should be in the director's seat and at the same time protect the breadth of research activity?

- *To what extent is the government prepared to step back?* In the twenty-first century, it is no longer possible for the government to play a central role and plan the innovation system from the top. R&D organizations are increasingly becoming connected to organizations from outside the traditional academic circles. Such networks exert an important influence on the functioning of the public research organizations. In other words, the influence of the government is bound to become smaller: government is stepping back. How far the government is prepared to step back and in what way, is not an easy question to answer. Will new forms of horizontal governance be replacing the traditional, vertical-steering framework, or will they be complementary? On which areas and issues is the government prepared to protect the public interests and assume responsibility for the performance of the innovation system as a whole? What will be the role of other stakeholders (e.g., business) in giving direction to the research organizations? What does this imply for issues such as demand-driven steering and market-type coordination in the sector?

Answers to the above-mentioned questions and trade-offs will shape the further development of the academic research enterprise in the Netherlands.

NOTES

1. See the note to figure 8.3 in the "Assessment" section for the names of the universities.

2. There are a few small "designated institutions" that are part of the university sector: a university for business administration, four institutes for theological training, and a humanistic university, as well as several international education institutes catering primarily to

international students. The latter are also active in research and these days are increasingly cooperating with universities.

3. We have categorized the disciplines into the so-called HOOP sectors, which is the customary Dutch way of classifying disciplines. There are nine HOOP sectors, eight of which are covered by the universities (not counting the "other" category). The remaining cluster is the education cluster, which includes only teacher-training activities. Universities do not carry out research in this cluster. Pedagogy/didactics research is included in the social sciences cluster.

4. After three years, there is a so-called midterm review. This is a self-evaluation that does not necessarily involve external evaluators. It is mostly for internal purposes and is not made public.

5. World Economic Forum, www.weforum.org.

6. In the latest (2008) version of this scoreboard, the Netherlands ranked eleventh. See European Innovation Scoreboard, www.proinno-europe.eu/metrics.

7. The three technical universities (Delft, Twente, and Eindhoven) have a more applied research focus and work closely with industry.

REFERENCES

Advisory Council on Science and Technology Policy (AWT). (2005) *Design and Development: The Function and Place of Research Activities in Universities of Professional Education.* Advisory report no. 65. The Hague: AWT.

———. (2006) *De Strategische Plannen van TNO, NWO en KNAW. Briefadvies 20 Juli.* The Hague: AWT.

Association of Universities in the Netherlands (VSNU). (2002) *Digitaal Ontsloten Cijfers (DOC) Promovendi.* The Hague: VSNU.

———. (2008) *Wetenschappelijk Onderwijs Personeels Informatie (WOPI).* Utrecht: VSNU. Downloadable tables available at www.vsnu.nl.

AWT. *See* Advisory Council on Science and Technology Policy.

Baggen, P. (2009) De Financiering van de Onderzoekstaak. In P. Kwikkers (ed.), *Geldstromen en Beleidsruimte.* The Hague: SDU.

Bartelse, J., H. Oost, and H. Sonneveld. (2007) Doctoral Education in the Netherlands. In S. Powell and H. Green (eds.), *The Doctorate Worldwide,* 64–76. Maidenhead, UK: Open University Press.

Bekkers, R., V. Gilsing, and M. van der Steen. (2006) Determining Factors of the Effectiveness of IP-Based Spin-Offs: Comparing the Netherlands and the US. *Journal of Technology Transfer* 31 (5): 545–566.

Berkhout, G. and J. Sistermans. (2006) Geen TTI's; Innovatiesucces Vergt meer dan Technologie. *Het Financieele Dagblad,* March 24.

Boekholt, P. and P. den Hertog. (2004) *Governance of the Netherlands' Innovation Policy System.* Utrecht: Technopolis BV.

Brouwer, E., P. den Hertog, T. Poot, and J. Segers. (2003) *WBSO Nader Beschouwd: Onderzoek naar de Effectiviteit van de WBSO.* WBSO Evaluation Report. The Hague: Ministry of Economic Affairs.

Carey, D., E. Ernst, R. Oyomopito, and J. Theisens. (2006) Strengthening Innovation in the Netherlands: Making Better Use of Knowledge Creation in Innovation Activities. OECD Economics Dept. Working Paper No. 479, OECD, Paris.

Centraal Bureau voor Statistiek (CBS). (2007) *Kennis en Economie 2007.* Voorburg: CBS.

Commissie Chang (Commissie Dynamisering). (2006) *Investeren in Dynamiek. Eindrapport Commissie Dynamisering.* Amsterdam: Joh. Enschede.

CPB (Centraal Planbureau). *See* Netherlands Bureau for Economic Policy Analysis.

Ernst & Young Accountants. (2004) *The Scope of Matching: A Study of the Effects of Matching of Subsidies (Indirect Funding) and Contract Funding on the Room for Strategy Development in Dutch Public Knowledge Institutions.* AWT Background Study. The Hague: AWT.

Etzkowitz, H. (2000) The Dynamics of Innovation: From National Systems and "Mode 2" to a Triple Helix of University-Industry-Government Relations. *Research Policy* 29:109–123.

Gibbons, M., C. Limoges, H. Nowotny, S. Schwartzman, P. Scott, and M. Trow. (1994) *The New Production of Knowledge: The Dynamics of Science and Research in Contemporary Societies.* London: Sage.

Innovatieplatform. (2004) *Vitalisering van de Kenniseconomie.* The Hague: Innnovatieplatform.

Institute of Higher Education, Shanghai Jiao Tong University. (2007) *Academic Ranking of World Universities—2007,* http://ed.sjtu.edu.cn/ranking.htm.

Jongbloed, B., J. Huisman, C. Salerno, and H. Vossensteyn. (2005) *Research Prestatiemeting: Een Internationale Vergelijking.* Vol. 113 of *Beleidsgerichte Studies Hoger Onderwijs en Wetenschappelijk Onderzoek.* The Hague: DeltaHage.

Jongbloed, B. and B. Van der Meulen. (2006) De Follow-up van Oderzoeksvisitaties. Onderzoek in Opdracht van de Commissie Dynamisering. In Commissie Chang 2006, 6–65.

KNAW. *See* Royal Academy of Science.

Ministry of Economic Affairs. (2003a) *Delta Plan Bèta Techniek: Actieplan voor de Aanpak van Tekorten aan Bèta's en Technici.* The Hague: Ministry of Economic Affairs.

———. (2003b) *In Actie voor Innovatie. Aanpak van de Lissabon-ambitie.* The Hague: Ministry of Economic Affairs.

———. (2004) *Peaks in the Delta: Regional Economic Perspectives.* The Hague: Ministry of Economic Affairs.

Ministry of Education, Culture and Science (OCW). (2003) *Wetenschapsbudget 2004. Focus op Excellentie en Meerwaarde.* The Hague: Ministerie van Onderwijs, Cultuur en Wetenschap.

———. (2005) *Kennis in Kaart 2005.* The Hague: Ministerie van Onderwijs, Cultuur en Wetenschap.

Netherlands Bureau for Economic Policy Analysis (CPB). (2005) *Scarcity of Science and Engineering Students in the Netherlands.* CPB Document 92. The Hague: Centraal Planbureau.

NOWT. *See* Observatory of Science and Technology.

Observatory of Science and Technology (NOWT). (2005) *Wetenschaps- en Technologie Indicatoren Rapport 2005.* The Hague: Ministerie van Onderwijs, Cultuur en Wetenschap.

———. (2008) *Wetenschaps- en Technologie Indicatoren Rapport 2008.* The Hague: Ministerie van Onderwijs, Cultuur en Wetenschap.

OCW. *See* Ministry of Education, Culture and Science.

Oost, H. and H. Sonneveld. (2004) Rendement en duur van Promoties in de Nederlandse Onderzoekscholen. The Hague: Ministerie van OCW.

Organisation for Economic Co-operation and Development (OECD). (2004) *Public-Private Partnerships for Research and Innovation: An Evaluation of the Dutch Experience.* Paris: OECD.

————. (2005a) *Governance of Innovation Systems.* Vol. 2 of *Case Studies in Innovation Policy.* Paris: OECD.

————. (2005b) Innovation Policy and Performance: A Cross-Country Comparison. Chapter 5 of *Innovation Policy and Performance in the Netherlands.* Paris: OECD.

————. (2005c) *OECD Science, Technology and Industry: Scoreboard, European Innovation Scoreboard.* Paris: OECD.

Royal Academy of Science (KNAW). (2005) *Positie van Onderzoekscholen.* Amsterdam: Erkenningscommissie Onderzoekscholen.

Salerno, C., B. Jongbloed, S. Slipersaeter, and B. Lepori. (2005) Changes in European Higher Education Institutions' Research Income, Structures, and Strategies. Interim report for the project Changes in University Incomes: Their Impact on University-Based Research and Innovation (CHINC). NIFU STEP, Oslo.

Times Higher Education Supplement. (2007) *World University Rankings.* November 9. London: TSL Education.

Van Beynum, G. (2006) Innovatiecluster: De Markt Bepaalt Succes, Niet de Nieuwe Techniek. *Het Financieele Dagblad,* March 24.

Van Tilburg, J. J. (2003) *Researchers op Ondernemerspad, Internationale Benchmarkstudie naar Spin-offs uit Kennisinstellingen.* The Hague: Ministry of Economic Affairs.

Venniker, R. and J. Koelman. (2001) Public Funding of Academic Research: The Research Assessment Exercise of the UK. In CPB and CHEPS (eds.), *Higher Education Reform: Getting the Incentives Right,* 101–118. The Hague: SDU.

VSNU. *See* Association of Universities in the Netherlands.

Waasdorp, P. (2002) Innovative Entrepreneurship: A Dutch Policy Perspective. In Ministry of Economic Affairs (ed.), *Entrepreneurship in the Netherlands,* 27–42. The Hague: Ministry of Economic Affairs and EIM Business and Policy Research.

CHAPTER NINE

The United Kingdom

MARY HENKEL AND MAURICE KOGAN

For most of the twentieth century, "there was resistance to the idea that governmental programs in science and technology should be coordinated in any but the minimal sense of avoiding undue duplication, so that the idea of a deliberate `science policy' was explicitly rejected" (Gummett 1980, 1). The academic research enterprise was largely self-regulating, and universities in the United Kingdom enjoyed remarkable degrees of autonomy.

However, during the past three decades, the United Kingdom has been a site for radical policy and system change in higher education and research, and universities have had to come to terms with political vulnerability and financial exposure as neoliberalism has become a more dominant ideological component of policy regimes. In the 1980s economy, efficiency and value for money in the public sector were prime government concerns. Universities' own conceptions of the purposes of higher education and research were challenged. They came to be regarded as institutions for public service, as defined by others as well as by themselves, and were required to be publicly accountable in those terms for how they used public resources. There was an explosion of demand for higher education, the provision for which was massively expanded and molded into a national system in the 1980s and 1990s, but with significantly reduced units of resource. The resultant degree of underfunding had severe consequences for the academic research infrastructure.

Most UK universities have been transformed during the past twenty years in terms of their missions, modes of governance, and relationships with government, industry, and commerce. After a period of weakening resistance to the idea of a deliberate science policy, 1993 saw a decisive declaration of such a policy in the form of a white paper from the Conservative government (Chancellor of the Duchy of

Lancaster 1993). The central issue it addressed was in fact long familiar, going back to the Victorian era: the failure of the United Kingdom to exploit more effectively the potential of its strong science base to contribute to national wealth creation and quality of life. It sought to create significant culture change in academe and in industry and to shape science policy within a framework of national research priorities focused on economic competitiveness. The Technology Foresight Programme was seen as the key to such changes and industrial innovation as a primary goal. It was, however, left to the Labour government that came to power in 1997 to frame its policies explicitly in terms of the creation of a national innovation system (HM Treasury, DfES, and DTI 2004; see also DTI 1998).

The Academic Research Enterprise
INSTITUTIONAL BASE

In 2005, 25.6% of all R&D (research and development) in the United Kingdom was carried out in higher education institutions, 61.6% by the business enterprise sector, and 10.6% by government (OECD 2007, tables 17, 18, 19). Almost all research in the academic sector is based in the universities, which now number 118 in the United Kingdom, although that figure rises to 133 if the colleges of the universities of Wales and London are counted as separate "university institutions" (Universities UK 2008). However, the institutions that now hold the title of university are diverse in their research histories, missions, and current capacities.

The 1960s saw the first of two major increases in the number of universities in the second half of the twentieth century, together with the establishment of a short-lived binary system of higher education, comprising fifty-two universities and thirty polytechnics by the time of its abolition in 1992. Seven new universities were created, mainly on greenfield sites, institutions that would combine research and teaching in the sciences, engineering, the social sciences, and the arts and humanities. At the same time a number of existing technological institutions were given university charters, forming a new set of universities with a more restricted and specialized research and teaching mission. However, the United Kingdom continued to be an elite higher education system—in Trow's (1980) terms—into the 1970s, when the age participation index rose to 15%, and indeed, in 1992 it was still below 30%. Massification thus came late in the United Kingdom, but it happened over a very few years. Student numbers doubled in the last two decades of the twentieth century, but this increase was achieved with a drop of 40% in the unit of resource. It was therefore a major factor in the financial weakening of the universities and had severe consequences for the research infrastructure (Georghiou 2001).

The second expansion in university numbers came with the ending of the binary structure in higher education in 1992. At that point, polytechnics were allowed to take the title of university, which most did almost immediately. However, unlike the existing universities, they were not given charters.

Until 1992, while universities' infrastructure funding had been based on an assumption that research was an essential institutional function, the core funds of the polytechnics were almost entirely for teaching. Although the research project and program funding administered by the research councils was also open to applications from polytechnic staff, they were heavily disadvantaged as far as this source of financing was concerned. Research councils, for the most part, regarded themselves as guardians of curiosity-driven research, whereas polytechnics were expected to concentrate on applied research and development, although a number of them increasingly sought to extend their research and research-training functions. By 1988, thirteen polytechnics had powers to register students for PhDs, and twenty-four had established their own research degree committees.

Therefore, all pre-1992 universities, including those created in the 1960s, had a clear advantage in the building of research capacity, although there were great differences between them as to their research resources, reputations, and outputs (Kogan and Hanney 2000). Georghiou (2001) notes that at the end of the century the five leading universities, as measured by research income, received 25% of all such income and the top fifteen about 50%. The bottom 50% accounted for less than 10% of research funding.

Universities continue to share the academic research function with a variety of specialist bodies, known collectively as public-sector research establishments (PSREs). These include national laboratories and research facilities as well as research institutes and centers wholly or partly funded and managed by one or more of the research councils. Few are wholly owned by research councils, and in practice the boundaries and distinctions between universities and PSREs are blurred. Many of the research institutes and centers are hosted by and/or jointly managed with universities, although a few work under the aegis of independent or charitable research organizations.

In 2004, about 45% of the annual research council budget of approximately £2.2 billion was devoted to PSREs' research activities and resources. Current research grant allocations by individual research councils between PSREs and universities vary widely. At the extremes, the Medical Research Council (MRC), which invests heavily in its research institutes, allocates nearly two-thirds of its grant money to them, while for the Arts and Humanities Research Council (AHRC) the sum is nil (DIUS 2008).

DISTRIBUTION BETWEEN DIFFERENT TYPES OF RESEARCH

In 1993 the Conservative government explicitly accepted the continuation of its role as the main funder of basic research and declared "public patronage of the pursuit of knowledge for its own sake [to be] an essential activity at the heart of our culture" (Chancellor of the Duchy of Lancaster 1993, 25). However, at the same time it rejected the idea of setting up separate funding bodies for curiosity-driven and mission-oriented research. It determined that the research councils should be in a position to "identify areas of cross-fertilization and integration along the continuum of basic, strategic and applied research" and to promote all three of these categories.

Moreover, the emergence of the concept of strategic research in the 1970s (see, e.g., in the UK context, the 1971 Dainton Report)[1] had already blurred the distinction between basic and applied research. Strategic research is research not targeted at specific wealth creation or application but which lies in areas where the emergence of application might reasonably be expected (Henkel and Kogan 1993; Irvine and Martin 1984). It became an increasingly dominant concept in UK science policy over the next thirty years.

All of this, combined with the growing importance of interactive, as opposed to linear, models of the science-technology relationship, means it is difficult to make definitive statements about the allocation of funding between basic and applied forms of research. However, table 9.1 attempts to quantify it for the period 1997–98.

Among the different research fields in receipt of research funding, the biological and biomedical sciences have been the dominant beneficiaries. In 2004, the Medical Research Council (MRC) and the Biotechnology and Biological Sciences Research Council (BBSRC) together received over £750 million of the annual research councils' budget, as compared with the Engineering and Physical Sciences Research Council's share of £500 million (DTI 2005). On top of that there was substantial funding from charitable foundations, notably the Wellcome Trust, which spent 90% of a £2.4 billion budget on UK biomedical research between 2000 and 2005. At the other end of the scale, the allocation to the Arts and Humanities Research Council (AHRC) was about £68 million and to the Economic and Social Research Council about £105 million.[2]

Under the Comprehensive Spending Review for 2008-11, the MRC allocation was set to rise by 30% over the three-year period, as compared with the average increase for the research councils of 18% (DIUS 2007). Moreover, it was the only research council whose allocation went up in real terms, taking account of councils' commitments to new government bodies and to the policy of full economic cost-

TABLE 9.1.
UK government R&D expenditure by Frascati[a] type of research activity and sector of government, in percentages

Type of research activity	Civil government department	Science budget (mainly research councils)	Defense	Higher education funding councils
Basic/pure	2.9	20.9	—	100.0
Basic/oriented	2.4	38.9	—	—
Applied strategic	41.2	31.2	6.7	—
Applied specific	46.4	8.2	21.6	—
Experimental development	7.1	0.8	71.6	—
Total	100.0	100.0	99.9	100.0

Sources: SET Statistics 1999; Georghiou 2001, 257.

Note: Percentages may not add to 100 because of rounding.

[a] The OECD Frascati manual provides a standard methodology (including definitions) for R&D surveys worldwide. The manual, now in its sixth edition, originated in a meeting of OECD experts in Frascati, Italy, in 1963.

ing (House of Commons Innovation, Universities, Science and Skills Committee 2008).

RESEARCH PRODUCTIVITY

The United Kingdom's strong relative international performance in scientific research is well attested (Georghiou 2001; King 2004), although it has long been achieved with a lower-than-average investment compared to its competitors and with a relatively lower availability of people with research training and skills (Evidence Ltd, DTI and OST 2003). In 2001 it lost its place to Japan as second to the United States in the number of journal article publications produced. However, it has retained its second place to the United States in its share of cited publications (12%) and of highly cited publications (that share increased to 13.2% in 2005 and remains broadly the same). The United Kingdom's main strengths are in the biosciences (Patel 2003), environmental sciences, and clinical sciences, although its output in the social sciences has increased strongly in recent years (Crespi and Geuna 2006). In mathematics it lost its second place to China in 2004.

Doctoral students constituted a small minority (9%) of all those receiving postgraduate qualifications in 2002–3 (HESA 2004). The graduation rate for advanced research programs in the United Kingdom was 1.6, as compared with Sweden's 2.8 and Germany's 2.0. In the academic year ending in 2006, 16,515 PhDs were awarded (12,950 on the basis of full-time study) (HESA 2007). More than 30% were gained by students at the universities of Cambridge, Manchester, Oxford, and three London University colleges. In 2003, 39% of all PhD graduates were in the biological and medical sciences.

After a period of decline in research student recruitment (Sastry 2004), the number of students registered for the final year of the PhD grew by 31% between 1998 and 2003. However, non-UK students (over 65%) and part-time UK students (over 72%) accounted for most of that growth and non-UK-domiciled students at that time accounted for 38% of all registered PhD students. The number of full-time, UK-domiciled PhD students grew by 11% in the same period. In 2006–7 about 60% of doctorates awarded went to UK-domiciled candidates (HESA 2008).

Policy Framework
RESEARCH POLICY

The 1980s marked the beginning of a period in which UK governments assumed clear leadership of research policy, establishing national priorities for research and introducing more varied and powerful policy instruments to achieve their economic ends. They installed an increasingly strong regulatory regime, comprising new laws, new structures and modes of governance, and performance measurement. They increasingly exploited the "power of treasure" (Hood 1983; Van Vught 1994) through financial incentives, wider use of competitive funding, and new principles of exchange for services, mainly to persuade universities to increase their efficiency and redirect their work. More recently they have placed greater emphasis on the selective enhancement of capacity.

Structural Change. In 1987, the Conservative government began "a searching review of R and D priorities . . . with a view to increasing the contribution of government funded R and D to the efficiency, competitiveness and innovative capacity of the UK economy" (Becher, Henkel, and Kogan 1994, 19). The government set performance indicators for the research councils, requiring them to develop corporate plans and submit annual reports, thus demonstrating their commitment to enhanced efficiency and economic goals as well as research excellence.

However, research council policies were developed within an advisory structure, consisting of the Advisory Board for Research Councils and an advisory committee of the chairs of the research councils. Following the 1993 white paper, these committees were abandoned and the post of director general of the research councils (DGRC) was created, evidence of an intention to coordinate subject areas and to pay less attention to their idiosyncrasies. In 1994 the minister for science wrote to the director general offering guidance. He was to "help . . . the Councils to set their policies within the wider framework of Government support for S&T as set out in the Annual Forward Look" (Waldegrave 1994, v).

Since then an increasingly elaborate and more centralized government structure has been established for the development and coordination of science policy within an economic instrumentalist framework and for the funding, monitoring, and evaluation of the academic research enterprise. In 1995, overall government responsibility for R&D was given to the Office of Science and Technology (OST) within the Department of Trade and Industry (DTI).[3] Science lost its dedicated cabinet minister. Instead, the executive head of OST, the government's chief scientific adviser, provided advice directly to the prime minister and contributed at cabinet level on science matters (Georghiou 2001).

The DTI, through OST and supported by the DGRC, "exercise[d] statutory control" of the (now seven) research councils.[4] The year 2004 saw the creation of the Arts and Humanities Research Council (AHRC). Chartered organizations though they are, the councils work to objectives set by government. In 2002, pressures for greater cooperation and strategic coordination between them resulted in the establishment of Research Councils UK (RCUK). Within the RCUK is the Executive Group, in which the heads of the councils "collectively own, drive and monitor the strategic and operational activities that deliver the RCUK mission."[5]

However, a fault line remained in the structure for the national governance of research, and, if anything, it was widening. Overall responsibility for the research infrastructure in the universities continued to lie with the Department for Education (under its various nomenclatures) and the funding councils.[6] Moreover, with increasing emphasis on research infrastructure as well as on the educational foundations of innovation economies, these bodies were assuming a higher profile in research policy. In 2003, the department appointed its first director general of higher education, whose remit included the strengthening of university research.

Meanwhile, since 1997, the Treasury had been taking an increasingly strong role in the direction and design of science and technology policies.[7] When the chancellor of the exchequer became prime minister in 2007, he carried out a major government restructuring exercise in which the interdependence of innovation, research, and higher education was cemented in the creation of the Department of Innovation, Universities and Skills (DIUS). The Department for Education and Skills disappeared, as did the DTI. The Government Office of Science, successor to OST and OSI (the short-lived Office of Science and Innovation), was then established in the DIUS.

The Funding System. For most of the twentieth century, the public funding of academic research was administered within a dual support system under the effective control of co-opted academic elites. Infrastructural funding was determined and allocated by the (academic-dominated) University Grants Committee (UGC) under the aegis of the Treasury and, for a brief period, the Department of Education and Science. It worked on principles of trust, discretionary decision-making, informality, and the essential role of academic judgment in determining resource needs and allocation—principles that were tolerated, if not always shared, by civil servants (Kogan and Hanney 2000). It supplied resources for the infrastructure and the teaching functions of universities. It also, mainly through student-based funding of teaching, provided the staffing costs and much of the housing and equipment for research. One-third of the UGC grant to institutions was hypothecated as enabling academics to undertake research.

Project and program grants came from five research councils. As already indicated, these councils, to varying degrees (e.g., the Agricultural Research Council was strongly conditioned by the stated needs of the agriculture ministries), regarded themselves as the supporters of independent and curiosity-driven research both in universities and in some research units of their own. The British Academy acted as the research council for the arts and humanities.

The dual support system continues; but under the Education Reform Act of 1988, which abolished the UGC, and the Further and Higher Education Act of 1992, some important principles changed. First, funding bodies for higher education were created by law rather than by administrative arrangement. Second, the primary mechanism for funding allocations was to be formula-led and thus transparent. Formulas are substantially output- and outcome-based. Third, university funding, instead of being a subsidy to public institutions with the freedom to determine the nature of their work and spend the money accordingly, was now to be based on the principle of exchange for public services, teaching, and research, and conditional upon their delivery (Williams 1992; see also Bekhradnia 2004; Geuna 1999). Universities must submit annual research plans to their funding council. Allocations come in the form of block grants, but the secretary of state has power to define the terms and conditions for grant allocation in his or her annual letters of guidance. With their grants, councils issue a financial memorandum to each university that is effectively a contract and can be detailed, not only as to the services the university is expected to deliver with the money, but also in the regulation of financial decisions about, for example, the borrowing of money (Bekhradnia 2004; see also Fulton 2002).

Under the 1992 act, all higher education institutions were incorporated into a unified core funding system for research and teaching, administered by statutory bodies, national funding councils of England (and Northern Ireland), Scotland, and Wales. They were defined as nondepartmental public bodies, but the secretary of state for education appointed their members and, as indicated, was given the legal power to give them general directions in the administration of their functions. They also work within an increasingly clearly defined government policy and public management framework. Their statutory responsibilities include ensuring the maintenance of a system for the quality assessment of higher education (though not of research).

Under the new funding system, there were shifts in the balance of public funding for the academic research enterprise away from the block grants of the funding councils, to the project and program funding of the research councils (see fig. 9.1 on university research funding sources) (HM Treasury, DfES, and DTI 2004). Also a significant proportion of university research money now comes from third-party funding. In particular, the amount of support for university research contributed by UK-based charities is now well over half of that which comes from the research councils.

Meanwhile, during the past two decades research councils have been active in identifying priorities within their corporate plans and developing research programs, thus moving from an entirely responsive mode of allocation toward a more structured one. Research councils still insist on the prime importance of the responsive mode,[8] but strategies, priorities, programs, and themes undoubtedly provide strong indicative frameworks for academic choices and the construction of grant applications. Responsive-mode funding still predominates in the more recently created AHRC. However, this research council is reducing this commitment over the years 2008–11, placing increasing emphasis on strategic research and also on team-based and collaborative research (AHRC 2008; see also AHRC 2004).

As far as the two sides of the dual support system are concerned, Adams and Bekhradnia (2004, para. 6) argue that although they operate on quite separate criteria, they "allocate in a remarkably consistent way. There is a very close alignment between the money received . . . from Funding Councils and what they receive in grant and contract income from Research Councils, charities and other sources."

Outside this system are the funds made available by government departments commissioning research and development for their own policy needs. By 2003–4 nearly 20% of the grants and contracts for higher education research came from government departments (central and local), as compared with 30% from research councils and about 25% from UK-based charities (HESA 2005, table G). Government initiatives constitute a growing proportion of the funds offered by public authorities.

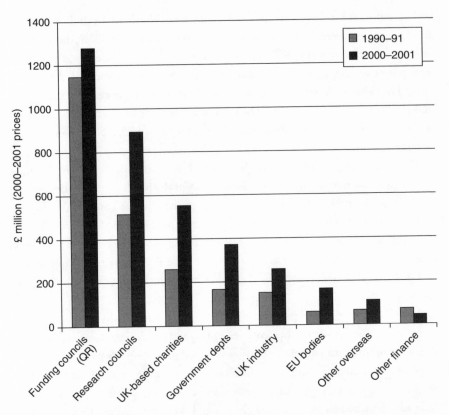

Fig. 9.1. University research funding sources. *Source:* HM Treasury, DfES, and DTI 2004, chap. 3.

Research Selectivity and Concentration. The moves toward a stronger central direction of R&D policies in the United Kingdom began well before 1993 and in some aspects had the support of the co-opted academic elite in the UGC. Government concerns with economy and efficiency in the public sector in the 1980s reinforced a growing conviction among elite scientists that the quality and reputation of British science depended on greater allocative efficiency of research resources in universities. Substantive policies for the selectivity and concentration of research, long discussed in the United Kingdom, emerged in the mid-1980s. These included, for example, a mooting of the idea of a functional classification of universities as R, T, and X (i.e., "research," "teaching," and a combination of these) and were accompanied by what became one of the most powerful and enduring instruments of change: the Research Assessment Exercise (RAE), initiated by the UGC in 1985 and institutionalized under the funding councils (Kogan and Hanney 2000).

Moves to strengthen competitive resource allocation gathered force with the ending of the binary system in 1992 and the resultant massive increase in the number of institutions that became eligible to apply for core research money from the new funding councils. Since then public research funding through the dual support system has become increasingly performance-based. The growing influence of selectivity and concentration as well as the idea that "critical mass" is key to research excellence is reflected in successive funding council allocations of quality-related (QR) money, based on the results of the RAE. In the RAE of 2001, departments or basic units of research in universities were graded on a seven-point scale.[9] Those awarded the four lowest grades attracted no QR funding; over 85% was allocated to those in the two top grades. Planned increases in the allocations based on the 2008 exercise will further concentrate funding in these two categories (HEFCE 2006a).

All subject centers wishing to receive public infrastructure funding for research had to submit themselves for this formal assessment, which took place at regular intervals of four or five years, until 2001. (The next and final exercise in this form was in 2008). The allocations were calculated by reference to individual subject center grades, but they were distributed to universities.

The units of assessment were subject groups or centers within individual institutions, defined within approximately seventy categories. The broad framework of the RAE was established by 1989, but it was modified for each exercise in light of review and consultation with stakeholders and, for that of 2008, after a commissioned review (Roberts 2003). A key change made in 1989 was that the work of each individual researcher in subject groups was assessed, rather than the best work of the group as a whole. In the 1996 exercise, the framework required a list of "research active" staff in the submitting unit. Thus, accountability became both an individual and a collective matter.

Research for the purposes of this exercise was most recently defined as "original investigation undertaken in order to gain knowledge and understanding. It includes work of direct relevance . . . ; scholarship; the invention and generation of ideas, images, performances, artifacts including design, where these lead to new or substantially improved insights; and the use of existing knowledge in experimental development to produce new or substantially improved materials, devices, products and processes, including design and construction" (HEFCs 2005, annex B).

Assessment was made by panels of mainly academic peer reviewers. In addition to assessing the quality of selected publications of each researcher, they were required to take into account quantitative performance indicators, such as the number of research students and studentships and external research income, as well as the research plans and the research culture of the unit.

The main changes applied to the 2008 exercise were as follows. The results were to be expressed as quality profiles of research in each unit. "These profiles will measure the different proportions of work in a submission that reach each of four defined levels of quality" (HEFCE 2004, 1). One purpose of this was to discriminate more finely between units so that the exercise could more effectively underpin the policy of selective funding (HEFCE 2004).

The exercise was explicitly focused on excellence, which, it is now argued in the context of an increasing emphasis on innovation policies, may be found in any form of research. In the framework for 2008, particular emphasis was laid on the recognition of applied research, new disciplines, interdisciplinary research, and also joint submissions, in response to criticisms that these dimensions had been disadvantaged in earlier exercises.

A new two-tiered panel structure was designed, in part, to ensure full recognition of interdisciplinary research. Panels included international scholars, whose role was to "validate the standing of research considered to be of the highest quality worldwide." "Where possible," users and commissioners of research were to be involved in the assessment process (HEFCs 2005, 2).

The RAE became an increasingly elaborate exercise. As the policies for the concentration of the research funding based upon it intensified, universities invested more resources, human and financial, in attempts to maximize their performance. Preparations for 2008 were under way in institutions from at least 2004.

The intensity of these efforts was due partly to a widely held assumption that the 2008 exercise would be the last of its kind. In 2006, the government announced its intention to replace the RAE with a system based on metrics. Since then the process of establishing a new Research Excellence Framework has been set in motion. Early intentions to make a sharp distinction between the methodology to be applied to the sciences (wholly metric) and that for other subject areas (the retention of a peer-review component) have been abandoned in favor of installing "a single unified framework for the funding and assessment of research across all subjects," with assessment based upon metrics and "light touch expert review" in different combinations according to subject areas (HEFCE 2008, para. 5).

Research Policy and the Economy: The Foresight Programme. Since 1993, UK science policy has been increasingly framed in terms of the contribution of science and technology to the national economy—wealth creation, economic competitiveness, and innovation (Chancellor of the Duchy of Lancaster 1993; DTI 1998; DTI, HM Treasury, and DfES 2002; HM Treasury, DfES, and DTI 2004). Users have been given a greater role in policy development and implementation. Emphasis on the need for prioritization in the areas of investigation and for selectivity and concentration of resources and activity has grown. Collaborative research production and sponsorship, on the one hand, and knowledge transfer between, in particular, universities and business, on the other, have been dominant themes. At the same time, strategic and financial coordination at all levels of research governance and between academe, industry, business, government, and charitable foundations has become more prominent.

In 1993, the Conservative government produced a white paper that was ambitious in both scope and title, "Realising Our Potential: A Strategy for Science, Engineering, and Technology" (Chancellor of the Duchy of Lancaster 1993). The government took the opportunity to introduce a new governing system for research and to set a new policy framework, the broad components of which were first sustained and then developed by their Labour successors. Central in this framework in 1993 was the Technology Foresight initiative.

The decision to adopt a Foresight approach originated from the activities of a comparatively small group of science-policy analysts and advisers who examined, or knew about, the adoption internationally of such policies (Irvine and Martin 1984). Many countries had experimented with Foresight before the opportunity to introduce it in the United Kingdom came with the appointment of William Waldegrave, the first British science minister of cabinet rank for many years.

In the 1980s, government had become increasingly persuaded of the need to direct science policy more firmly. The 1986 report of the Advisory Council on Applied Research and Development (ACARD), *Exploitable Areas of Science*, concluded that "a process is needed to prioritize and guide a substantial proportion of . . . the national scientific resource . . . and to stimulate its effective exploitation to the benefit of the United Kingdom. . . . We do not have a forum . . . where we can manage [this] process. . . . It is, we believe, a matter of national priority that such a forum be established . . . [to] gather, analyze priorities, and direct relevant information into the decision-making machinery. The decision-making process must recognize the three spheres of activity, scientific, technological and commercial" (ACARD 1986, 9). ACARD's report outlined a framework for the process of generating strategic research priorities, which included addressing, among others, the following questions: "Which areas of generic technology are supported by a particular area of strategic science? Has the UK the scientific resources to advance a particular area of strategic research? For a given generic technology, what new products or processes will become possible within 10–20 years? What are the likely costs of translating scientific knowledge into marketable products and processes?" (11).

According to Waldegrave, the most important aspect of the new policies in 1993 was getting a structure to achieve "a bridging of the cultural and organizational divides that sometimes exist in UK" (Henkel et al. 2000, 26). The government's intention, as set out in the white paper, was to use Foresight to achieve two means to its ends: to provide a robust basis on which government could set priorities for science funding, and "to achieve a key cultural change: better communication, interaction and mutual understanding between the scientific community, industry, and Government" (Chancellor of the Duchy of Lancaster 1993, 5).

Academics had long shown themselves willing to do work useful to the economy and society, but government-steered science policies had often been met with academic skepticism on the grounds that they conflicted with the norms of scientific activity and with well-established beliefs about how science works (Polanyi 1962). Despite some meeting of minds between scientists and government on the need for more effective collaboration between science, government, and industry (e.g., House of Lords 1987), academics were cautious about how far the trend toward perceiving science as part of a corporate wealth-creating enterprise should go (Higher Education in Trouble 1988).

Foresight, as articulated by Martin and Irvine (1989), comprised a complex set of ideas and carried with it a number of controversial assumptions. It required a comprehensive approach and strong strategic leadership from the top. Priorities had to be set for research funding.

Foresight must be distinguished from forecasting; Foresight assumes that it is possible to shape the direction of research without impairing scientific creativity or the degree of innovation required to sustain scientific and economic competitiveness. Strategic research is a key component of the program.

Partnership between academic researchers, industry, and government is central for the selection of priorities for strategic research and generic technologies, for their development, and for the dissemination and exploitation of results. These tasks require iterative interaction between scientists, engineers, and companies as well as active networks for collaboration and dissemination (Chancellor of the Duchy of Lancaster 1993; see also OST 1995). Linear and unidirectional assumptions about the relationship between fundamental, strategic, and applied research and between science and technology are increasingly untenable. Such assumptions have, in fact, long been challenged in the social studies of science (Layton 1977; Mulkay 1977; Ziman 1984; See also Brooks 1994; OECD 1995). Traditional boundaries between science and industry, the academy and the market, begin to shift. Multidisciplinary research is increasingly recognized as essential for bringing work on emerging technologies to fruition and for advancing knowledge in some scientific fields.

The approach envisaged wide communication with academics and an attempt to generate a sense of commitment to the results. Before the end of 1993, the high-level

Technology Foresight Steering Group had been established to oversee the Foresight Programme and to provide a general framework in which fifteen (later sixteen) sector panels—comprising members from academia, business, finance, and government—would begin to implement it with Foresight assessments of their sectors. In the main stage of the Foresight process, panels identified key issues and trends, developed scenarios for their sectors, and consulted widely through a Delphi survey and regional workshops (Georghiou 2001). Individual panel reports, together with the overarching Steering Group Report, were produced in 1995 and were followed by a process of dissemination and follow-up action.

In its 1995 report, the Foresight Steering Group, while making it clear that research councils "are important players in the implementation of Foresight findings within the science base," also stated that they "recognize that the research councils' flexibility for shifting priorities in the short-term is somewhat limited because of their existing commitments and because of the need, which we support, to promote innovative research irrespective of discipline. We nonetheless see scope for a sustained readjustment of priorities over a number of years" (OST 1995, 5.28 and 5.29).

In 1996, the Higher Education Funding Council for England (HEFCE) explored how Technology Foresight priorities might map onto the units of assessment used in the RAE, and in 1998 HEFCE considered whether a policy factor should be used in calculating the allocation of total research funds between subjects. Though the reaction from institutions was negative (HEFCE 1998), the issue has not gone away.

Under a new Labour government, commitment to Foresight was sustained, and it entered into a second and a third phase. The second round of Foresight (1999–2002) saw a change in structure to three thematic and ten sector panels. There was also a stronger emphasis on the potential social benefits and on implications of findings for education, training, and sustainable development. In the third phase, from 2002, the program reverted to a central focus on science, technology, and innovation and sought to strengthen future expertise in its operation, as a basis for informing new priorities and directions for research. The Horizon Scanning Center was established to support this work. Panels were dismantled and a rolling program of a maximum of four projects per year was set up. A simpler and more flexible structure was thus combined with strong government oversight of the program.

Research Policy and the Economy: A National Innovation System. As already indicated, the Labour government maintained substantial continuity in the science policy it inherited, notably in its emphasis on the contribution of science and technology to the economy, in its promotion of academic-business collaboration in a variety of forms, in its endorsement of "academic capitalism," and in its commitment to competitive, performance-based resource allocation.

However, it did undertake a major review of science policy and made some significant changes of emphasis. In particular, it incorporated science policy more decisively and explicitly into innovation policy and the building of a national research and innovation system. It also declared investment in the science base to be an essential means to its ends. The government's ambition is for the United Kingdom to be a "key knowledge hub in the global economy with a reputation not only for scientific and technological discovery but also as a world leader in turning that knowledge into new products and services" (HM Treasury, DfES, and DTI 2004, 1.1). The achievement of this ambition is set out in a science and innovation investment framework, incorporating six dimensions: world class excellence; sustainable and financially robust universities and PSREs; greater responsiveness of the research base to the needs of the economy and public services; business investment and engagement in innovation; enhancing the supply of skills in science, technology, engineering, and mathematics; and public understanding of, engagement with, and confidence in science. This framework, with its six dimensions, provided the basis for the goals and performance indicators developed by the DTI as part of a "robust management system" to ensure the fulfillment of the public service agreement it concluded with the Treasury on the implementation of the innovation policy (HM Treasury, DfES, and DTI 2006).

Gross domestic expenditure on R&D in the United Kingdom amounted to 1.88% of GDP in 2003 (as compared with the highest OECD performer, Sweden, at 3.98%). The aim, as declared in the framework document, is now to raise this to 2.5% "by around 2014" (HM Treasury, DfES, and DTI 2004, 1.4.), still well short of the Lisbon target for EU member countries of 3%. Achieving the UK target depends, however, on continuing financial investment from charitable funding bodies as well as on a significant growth in business R&D.

Central to the government's policy goals was a commitment to redress the problems of underinvestment in the science base and to ensure the sustainability of the academic research enterprise. In 1998, the government launched a public-private partnership between the DTI, the English funding council (HEFCE), and the Wellcome Trust: the Joint Infrastructure Fund. This initiative was later extended in the form of the (still temporary) DTI/funding council-based Science Research Investment Fund (SRIF) under the government's ten-year investment program. It was focused on research infrastructure and also incorporated incentives for universities to engage in strategic restructuring of science research, including mergers between universities. By 2006 it had amounted to over £2 billion (HM Treasury, DfES, and DTI 2006) and could be used to support humanities and arts research as well as the sciences, even if these forms of research figured little in the statement of ambitions. Under the 2007 comprehensive spending review, the SRIF's transition to the permanent Capital Investment Fund was announced.

Initially the infrastructure funding was allocated through a bidding process. Later it was distributed largely through a formula based on a combination of RAE rankings and the amounts of other research money awarded to institutions in grants and contracts.

As part of a new focus on the sustainability of the national research enterprise, government adopted a policy aimed at full and transparent economic costing of the research undertaken by universities and other public-sector research establishments. It expressed concern that much external funding of university research (including that from research councils) "provides only partial coverage of the full economic costs of research sponsored." Therefore, either it is subsidized by public infrastructure funding, or its long-term-cost impacts are left uncovered (HM Treasury, DfES, and DTI 2004, 3.11).

Universities are now required to base all of their research proposals to research councils on full economic costing, and government has allocated moneys to enable the research councils to pay 80% of these costs (moving to 100% by 2010).[10]

These policies represented a clear commitment on the part of government to strengthen the infrastructure of the academic research enterprise and to enhance its sustainability as well as to increase its annual allocations from the science budget (it received a 26% increase from 2004–5 to 2007–8). However, selectivity and concentration remain crucial components of these policies, now expressed in terms of the prime importance of retaining and building "world class centers of research excellence," which are seen as the way to secure "growth in [the nation's] share of internationally mobile R&D investment and highly skilled people" (HM Treasury, DfES, and DTI 2004, 6).

Equally clear is the government's desire to incorporate the whole range of contributors in a commitment to the research base and its exploitation. It established a "Funders Forum" in 2003, which includes business, charities, Regional Development Agencies, and government departments as well as the funding councils and research councils. Its aim is to take a strategic overview of the working of the science base and the collective impact of funders' strategies on its sustainability, health, and outputs (HM Treasury, DfES, and DTI 2006).

Research Careers and the Academic Research Enterprise. By the early 1990s, there was considerable anxiety in the universities, particularly among scientists, about their capacity to attract a sufficient number of talented people to academic research. Concern centered partly on the low value of student research stipends and increasingly on the growing proportion of academic researchers employed on fixed-term contracts. In 1994–95, the proportion of "research only" staff in universities was 29%, and by 2002–3 this figure had risen to 33%. The overwhelming majority of these staff were on fixed-term contracts derived from project or program grants.

One approach to these problems focused on career management and development for this group. In 1996, the Research Careers Initiative was launched by the Committee of Vice Chancellors and Principals (now Universities UK), the research councils, the British Academy, and the Royal Society. They agreed on a "Concordat" to provide a framework for the career management of contract research staff in universities and colleges and to monitor its implementation. One of the aims was "to change the culture in which contract research staff worked so that they were seen as central to the pursuit of good science research" (DTI 2003, 10). There has been some progress in the conditions and terms of employment of contract staff, probably due in part to the 2002 introduction of new EU regulations.

Although this initiative did not incorporate any intention to reduce the proportion of researchers employed on fixed-term contracts, and institutional action within it has been primarily in the provision of more career assessment and advice, some new sources of funding for early-career researchers have been provided by government (one thousand new academic fellowships over a period of five years and funding to raise the salaries of research council postdoctoral researchers by an average of £4,000 by 2006) (DTI 2003,18). The Wellcome Trust and the Royal Society also provide fellowships designed to attract the most promising and creative young researchers by giving them freedom to pursue their own research.

The year 2003 brought a broader, structural approach to the improvement of career opportunities in universities. A National Framework Agreement, designed to enhance equality of reward in a more diversified workforce, was concluded between the University and Colleges Employers' Association and the unions (UCEA 2003). It provides a more flexible structure for career progression that allows for career-long specialization in, for example, research or teaching, among other pathways, new and old. Its implementation is, however, a matter for local negotiation and decision.

The United Kingdom and the European Union. References to the European Union or the Lisbon agenda are largely conspicuous by their absence in UK science-policy documents. Despite this, the government has consistently given public endorsement to the development of the European Research Area and, in particular, to the Framework Programmes and the European Research Council, in terms of their potential to deepen the United Kingdom's international collaborative research connections as well as to offer not insignificant sources of research funding for British universities. It has also been quick to report the high level of participation by UK researchers in the Framework Programmes.

In fact, a glance at the policy instruments developed by the United Kingdom for the implementation of its science policy agenda demonstrates how closely aligned with EU developments it has been in recent years. There is no doubt that Britain has been an active participant in taking the Lisbon Strategy forward, although its advertisement of this in its Annual Report of 2007 on progress in implementing the innovation and investment program was unusual (HM Treasury and DIUS 2007).

DOCTORAL EDUCATION POLICY

Until the 1980s, policy for graduate education in the United Kingdom was characterized by benign neglect. Uncertainties about purposes and resources were reflected in a spate of uncoordinated but sometimes significant assaults on different aspects of policy. There was no national consideration of research training and no attempt to relate numbers of places to the staffing of higher education. There was some consideration of academic content. There was discussion, but at that stage resulting in negative conclusions, about the desirability of taught doctorates. The universities took on board the need to supervise and examine more carefully students reading for research degrees. Moves to control the distribution of research training were implied in the attempts to stratify institutions in terms of their research quality and in the research councils' attempts to reduce the number of outlets for research awards. In the late 1980s the research councils began to prescribe the patterns of graduate studies, particularly at the master's level; sanctions were applied to institutions to improve completion rates of doctoral degrees. Policies were introduced to broaden the purpose of doctoral education beyond simply the reproduction of the academic profession.

The funding structure for graduate studies followed in the shadow of that for research policy, with two funding streams. The funding councils provide for teaching and research staff and buildings and equipment; the research councils fund studentships for the payment of tuition fees and living costs for a selected minority of students. Research councils made 13,650 awards and UK charities just under 2,000 to doctoral students in 2003–4.[11]

Research councils receive bids from university departments for their awards, which they determine based on judgments of the quality of supervision, courses, and associated provision. Since 2001 the research councils have declined to give grants to students in departments with an RAE grading below the top three bands. Institutions may determine what doctoral or other graduate studies to offer, but these decisions are increasingly framed by the resources made available.

Postgraduate research students are concentrated in a small number of multi-faculty universities with established research reputations. These institutions also graduate most of those who become research students. This concentration conditions the ultimate recruitment to the academic profession. Oxbridge colleges offer studentships to meritorious students who fail to get research council grants. More important, the colleges award research fellowships, in effect, postdoctoral awards, with a tenure of three or four years, to the best of the cohorts. In October 1992, 304 research fellows were attached to Cambridge colleges. Of these 177 were fully or substantially supported by their colleges, and the rest received some kind of external support.[12] These figures compare with a total of 477 postdoctoral fellowships awarded from all public funding sources in 1991–92 (Becher, Henkel, and Kogan 1994). This Oxbridge source produces a major part of the academic elite.

There has been great variation in postgraduate recruitment to different subjects over time. About two-thirds of the 15,255 doctorates awarded in 2003–4 were in the natural sciences (led by the biological sciences). Awards in the social sciences and the humanities were in roughly equal numbers. In 2002–3, 38% of doctorate or research master's students were thirty years or over, and 45% were women (Sastry 2004).

There is no formal accreditation of research doctoral programs. But some informal and de facto recognition is afforded by the research gradings of departments, the decisions of research councils to award studentships to be held in selected departments, the external examiner system, and, probably most potent of all, the national quality-assurance procedures.

The graduate school is now the dominant mode of organization of graduate education in UK universities, but the form it takes varies. Although the single, institution-wide model is found in the greatest number of institutions, most pre-1992 universities have moved to a devolved model, in which graduate schools are based in faculties, departments, or groups of departments (Woodward and Denicolo 2004).

A longstanding scheme is that for Cooperative Awards in Science and Engineering. First developed by the (then) Science and Engineering Research Council in the early 1980s, it and its social science counterpart fund research students who are jointly supervised by academics and external sponsors, who might come from industry or from public-sector bodies.

KNOWLEDGE TRANSFER POLICY

University-Industry Collaboration.   Financial incentives for developing collaboration between universities and industry predated the Foresight Programme and have proliferated since. The most enduring was LINK, founded in 1986 and effectively incorporated into the Foresight Programme in 1995, focusing on priorities as defined in the Foresight process. LINK was the main mechanism for promoting collaboration between business and the research base in precommercial research. It focused on areas of strategic importance for the future of the national economy, encouraging innovative research well ahead of the market. Its prime aim was to significantly enhance the exploitation of such research by business in the development of innovative and commercially successful products, processes, and services (OST 2003). Government departments and research councils provided up to 50% of the total eligible costs for "core research projects," 75% of funding for "feasibility studies," and 25% for nearer-to-the-market "development projects," with, in each case, the balance of support coming from business. An OST-commissioned strategic review of LINK endorsed it and recommended its expansion (OST 2003).

The role of industrial liaison had been substantially developed in universities by the end of the 1990s. More recently it has become more professionalized, and the government has now invested £1 million to support training for the staff, now often part of technology or knowledge-transfer offices. In 2004, plans were announced for an Institute for Knowledge Transfer, to be developed by the Association for University Research and Industry Links (HM Treasury, DTI, and DfES 2004).

Since the late 1990s the government has launched a stream of initiatives aimed at enhancing the contribution made by universities to the economy and to society and strengthening the relationships among universities, industries, and other businesses. The *Lambert Review of Business-University Collaboration* (Lambert 2003) suggested that these initiatives had had significant impact on the culture of universities and had facilitated development in a range of activities: networking with business, including small and medium-sized enterprises (SMEs); marketing university research and teaching to business; establishing business or industrial liaison offices; establishing spin-off companies; providing entrepreneurial training for science and engineering graduates; and providing work placements for students in industry. However, *Lambert* was also critical: the developing infrastructure was too complex, and there were too many initiatives and too much hypothecated funding.

This line of policy has since been institutionalized within the concept of "third stream" funding, or funding for knowledge transfer (see HEFCE 2006b), and has been substantially consolidated in the Higher Education Innovation Fund (HEIF), which is jointly resourced by the Government Office for Science and HEFCE. Other activities supported under this heading include the commercialization of world-class research, collaboration between higher education institutions to exploit shared facilities, and the involvement of universities and colleges in the regeneration of their regions and beyond, through working with SMEs and noncommercial organizations and other partners. The HEIF has, inter alia, incorporated an earlier initiative, under which universities can bid for seed funds to assist them to manage the riskiest stage of the venture process and to bring research discoveries to a point where their commercial usefulness can be demonstrated and first steps taken to secure their utility.

Funds allocated under HEIF were more than doubled to £185 million for the years 2004–6. In the next phase of the program, 75% of the funds were allocated by formula. Both *Lambert* and a more recent paper by Hatakenaka (2005) were critical of the HEFCE approach to third-stream funding allocation, though it does have the support of the government (HM Treasury, DfES, and DTI 2004). Lambert and Hatakenaka argue that the adoption of a formulaic approach might discourage the essential diversity of these activities and prevent universities from responding to the particular external needs they perceive. Second, they support, too, the idea of light-touch qualitative evaluation rather than the use of metrics. The HEFCE strategy, Hatakenaka suggested, might induce universities to set up separate third-stream units whose primary purpose would be to secure funding rather than to embed a value commitment to contribute to the economy and society across whole institutions.

Meanwhile, the other half of the dual support system, the research councils, has been required to take a step further in the direction of involvement in impact. Long under pressure to take more account of the needs of research users, the councils received earmarked funding in 2005 "to boost their knowledge transfer capabilities." Each research council now has a "knowledge transfer" component of the "delivery plans" it is required to produce, together with a set of metrics to measure its performance on this dimension. According to the government's annual report of 2006 on progress within the science and innovation investment framework, "all councils agree that greater progress will be needed to deliver the step change in the economic impact of research council investments sought by the Government" (HM Treasury, DfES, and DTI 2006, 3.9).

Academic Capitalism. If the main emphasis in the Thatcher government's early policies affecting the academic research enterprise was on economy and efficiency in public expenditure, one initiative in 1985 seemed to represent a more decisive step toward neoliberalism: the promotion of "academic capitalism" and the reduction of universities' financial dependence on the state. The monopoly rights in the exploitation of research results of the National Research and Development Corporation, the precursor body of the British Technology Group, were abolished, opening the way for universities and academics to secure intellectual property rights themselves through patents, licensing agreements with private firms, and the creation of their own spin-off companies. From then onward, academic institutions became increasingly preoccupied with intellectual property rights and how to deal with the range of issues involved.

Since 1997, there have been further policy developments to facilitate the com-mercialization of research by universities and individual researchers. The Intellec-tual Property Working Group of representatives from business and universities was established in 2004, its purpose to develop a range of model collaborative contracts and a protocol for the ownership of intellectual property in research collaborations. The starting point for this was that universities should own any resulting intellec-tual property, with industry free to negotiate license terms to exploit it (Lambert 2003).

The *Lambert Report on Business-University Collaboration* (Lambert 2003) also recommended that government should set guidelines for third-stream funding to rebalance commercialization activities toward licensing industries to pursue com-mercial exploitation of university research and to give less emphasis to university spin-off companies, the long-term sustainability of which, it argued, is uncertain. Funding for "proof of concept" (the first stage toward the transfer of intellectual property to market) should be increased, the report claimed. The government, in its response, gave qualified support to the report's arguments on these points and noted that the bids already made for HEIF funding reflected strong demand for "proof of concept" funding, notably from prestigious research-intensive universities (HM Treasury, DfES, and DTI 2004, annex C44). However, the government made clear its commitment to include the encouragement of well-founded spin-off companies in its policies for research exploitation.

Regional Development Policies. The role of science and technology in facilitating regional growth and the importance of close collaboration with Regional Develop-ment Agencies (RDAs) have been strong themes in UK science policies in the past decade (see, e.g., DTI 2001; Lambert 2003). They were supported by Universities UK, which published a series of reports together with HEFCE that gave evidence of the "triple helix" (Etzkowitz and Leydesdorff 1997) phenomenon in a number of regions and of the range of contributions being made by universities to regions (Universities UK and HEFCE 2002). The white paper "Excellence and Opportu-nity—A Science and Innovation Policy for the 21st Century" underlined RDAs' role in supporting "the development of clusters, geographical concentrations of compa-nies, specialized suppliers and associated institutions such as universities, co-located for mutual competitive advantage" (DTI 2000, 38). This policy strand was further developed in the more recent Science Cities initiative. As devolution of government in the UK advances, it might be expected to become a more important feature of innovation policies in England.

However, a research report by Adams and Smith (2004) raises some important questions in connection with such policies. They identify the tension between government policies—notably selectivity and concentration—that reflect the idea of the research base as a national entity, and region-centered policies, such as that of regional clusters. They conclude that clusters involving higher education institutions and companies that are close spatially can be powerful promoters of innovation and have strong economic effects on their locale. However, they counsel caution, arguing that while research on the cluster phenomenon is usually precise as to what it means by "cluster," policymakers are often less so in the type and scale of interactions they support under this heading. They also show that regions have different potential to support the most productive collaborations. There are heavy "regional concentrations of public and private sector research and they are co-located around London and the South East" (Adams and Smith 2004, 7). Adams and Smith raise questions, too, about the importance of co-locality for collaboration (see also Poveda 2002). Sometimes, for example, sectoral affinity is more important than spatial affinity. Companies co-located with the concentrated research base in the Southeast are also collaborating with universities all over the country and outside of it. Finally, they point out that RDA cluster priorities are not, on the whole, distinctive but have a great deal of similarity to each other—a strong focus on biotechnology, pharmaceuticals, communications and IT, leisure and media, advanced engineering, and tourism (Adams and Smith 2004).

If problems remain in the relationship between national and regional policies, overall it is clear that knowledge transfer is assuming an increasingly high profile in science and research policies, within a broader policy framework committed to both economic instrumentalism and scientific excellence.

Assessment
POLICY INSTRUMENTS

During the past two decades, UK governments have been prepared to make their authority explicit in fields where it had previously been largely implicit. They developed and articulated policies based on stated government priorities for higher education and the academic research enterprise.

They have advanced their agenda through an increasingly strong regulatory regime, which has included new laws and new structures and modes of governance, along with substantial use of "the power of treasure" (Hood 1983; Van Vught 1994). A growing reliance upon competitive resource allocation, financial incentives, performance measurement, and principles of exchange has been central to the change agenda during a period in which universities have been coming to terms with new levels of financial need and political exposure. Only since 2002 (DTI, HM Treasury, and DfES 2002) has there been some shift of emphasis in the use of such instruments toward enhancing the capacity of the academic research enterprise as well as its efficiency and utility to the economy.

The role of new law was not insignificant in governments' assertion of authority and willingness to apply rules and restrictions. The Education Reform Act of 1988 and the Further and Higher Education Act of 1992 were important landmarks in a country where there had been little by way of specific national higher education law, as distinct from the individual institutional charters held by both the research councils and the pre-1992 universities. There is now a legal framework that allows the secretary of state the authority to impose upon the funding councils general terms and conditions for grant allocation to universities; those allocations, as we have seen, are now made in what are effectively contracts between the councils and individual institutions.

It remains possible within this framework for the councils' function to remain that of "buffer institution," a term undoubtedly applicable to the UGC (University Grants Committee). Indeed, the chief executive of the English funding council (HEFCE) was still claiming its function as such in 2003 (House of Commons Education and Skills Committee 2003). However, there have been indications of change in practice as well as principle in recent years. HEFCE's role is increasingly understood and described as one of cooperation with the ministry in the achievement of government policies (Sabri 2007), in other words as an "agency of public purpose" (Neave 2006; Trow 1980). Meanwhile, the government has shown itself prepared not simply to lay down general principles for the allocation of the block grant, but to specify precisely, in the case of research funding, the degree of selectivity to be achieved and the means by which that will happen as well as apparently to overrule the funding council's judgment in the process (Bekhradnia 2004; House of Commons Education and Skills Committee 2003).

There is little doubt that as far as the academic research enterprise is concerned, the funding council has been responsible for the most effective instrument of change in terms of, first, the achievement of a policy goal of growing importance to government: research selectivity and concentration; and second, its influence upon institutional and individual behavior. The funding council's research funding is distributed to universities by an increasingly competitive, formula-driven resource allocation system, based upon the Research Assessment Exercise (RAE). Its power lies in several factors; most important is the magnitude of the impact it has had on both universities' finances and their reputation. Second is the degree of the internal as well as external credibility of the RAE; third is its longevity. Up till now, it has been subject to adaptation rather than major change of purpose and process.

The dominant evaluative criteria feeding into the formula are research incomes and outputs, and the dominant influences are those of peer review and academic values. Peer review lessens transparency, but it strengthens credibility. There may be severe criticism among universities of the degrees of research selectivity and concentration the RAE has helped to bring about, but the RAE has reputational as well as financial value. It also constitutes an influential signal for potential customers of university-based research. It is therefore not surprising that it has generated a high degree of self-regulation by universities, as we shall see in a later section.

Between them, the two arms of the dual support system exercise significant steerage, much of it in the direction determined by government policy. Both have developed special funding programs that offer financial incentives for the implementation by higher education institutions of government policies: research council programs of awards conditional upon matched funding from industry and, more recently, the stream of initiatives for knowledge transfer, now substantially consolidated in the funding councils' Higher Education Innovation Fund (HEIF). Most of these initiatives have provided money at the margins for universities, but such money can be very important in a context of public expenditure constraint and escalating demand. On the face of it, the weaker the institution, the more powerful this kind of policy instrument is, but it is not a particularly reliable source of reward for mission change by such institutions. Well-established and well-resourced institutions are more likely to have the infrastructure to enable them to exploit such initiatives; weaker institutions may feel obliged to invest effort in applying for them, with limited success (Bekhradnia 2004). This aspect of the problem has been mitigated in the United Kingdom by a shift, particularly once programs become more embedded, from reliance upon bidding to allocation by formula.

Probably the most direct form of steerage used by the research councils consists of the strategic planning and programs defining priority research agendas, which they have evolved since the 1980s. The influence of these instruments on long-term individual and institutional strategies is substantial. Within this form of framing, distinctions between directive and responsive funding modes may become less important.

If there is evidence of more direction in the area of inputs into the research enterprise, there are also more stringent controls on institutions' usage of funding and on outputs. The greatly increased national investment in research embarked upon by the Labour government is accompanied by a full economic costing policy, designed, in part, to control cross-subsidies, particularly of private contracts by public funding. Meanwhile, research council grant contracts are increasingly explicit and comprehensive and underpinned with financial sanctions. The responsibilities of the university or research organization in which the researchers are based are laid out in detail. They include ensuring proper financial management of the project; project organization within a framework of best practice; submission of the project to bodies regulating research ethics, as appropriate; and compliance with health and safety regulations. They identify institutions' obligations to research contract staff in accordance with the national Concordat and require them to demonstrate involvement of and value to research users.

Principles of exchange inform both arms of the dual support system. They are further embedded in an overall form of public management designed to secure more systematic public accountability through public service agreements, for example, between the DTI (and from 2007 the DIUS) and the Treasury, incorporating the use of targets, delivery plans, and increasingly elaborate frameworks of progress and performance measurement (HM Treasury and DIUS 2007; HM Treasury, DfES, and DTI 2004, annex B; 2006).

Whereas in the past the United Kingdom has epitomized pluralism and the autonomy of institutions of different kinds in the governance of research—rationalized in terms of the logic of epistemologically diverse disciplines or maximizing the creative force of individualism in a "republic of science" (Polanyi 1962)—there is now a definite trend toward coordination, coherence, collectivism, and indeed, control at the level of national government.

POLICY OUTCOMES AND THE ACADEMIC
RESEARCH ENTERPRISE

Innovation Outcomes. Policies in the United Kingdom are increasingly predicated on the assumptions that innovation is the key to a successful national economy and that the academic research enterprise (ARE) can contribute to this in various ways: through academic research at the leading edge of knowledge, through the education of new generations of researchers, through research collaborations with business and the public sector, and through a range of other forms of knowledge transfer and distribution in society.

As yet there seems to be little change in the UK business sector's perception of the role of academic research in industrial innovation. In 2006, 57% of all industries in the United Kingdom were defined as innovation-active. The definition is broad and includes both "novel" and "follower" process and product innovators, that is, those who introduce processes and products new to them as well as those who bring in something new to the industry as a whole. It also incorporates involvement in innovation projects and current expenditure on training, on internal R&D, and on acquisition of new external knowledge or equipment linked with innovation (DTI 2006, 2).

However, of enterprises that had actually developed innovations, 71% did not use universities as sources of innovation, and only 2% reported making high use of them. Of "novel" and "follower" process and product innovators, novel innovators had collaborated most, proportionately, with universities. Firms engaged in international collaborations with universities were likely to attach most value to the effects of innovation on their businesses. The greatest effects on market share, range of goods, and quality were perceived as coming from collaboration with universities elsewhere in the European Union.

A comparative study of collaboration between industries and universities in the United Kingdom and the United States found that while UK firms reported more interaction of various kinds with industries than their counterparts in the United States, the latter more frequently rated their interactions as having high value for their innovative activities. They were also more likely to spend money on their innovation-related links with universities and to make much more use of internships than UK firms (Hughes 2006).

However, though investment in and use of research is traditionally low in UK industry as a whole, such studies may present a misleading picture. For example, they ignore the importance of the production of tacit knowledge through postgraduate research education. They may also occlude the historic concentration in the United Kingdom of research-based, innovation-oriented industry in large firms in a few sectors, such as pharmaceuticals and aeronautical industries.

Meanwhile, interest has been growing in more direct participation by the academic enterprise in technological innovation and the knowledge economy. Universities in the United Kingdom have had some involvement in the commercialization of their research since the 1970s. However, it was not until the mid-1990s that this became a significant ambition for more than a small minority. Though 1998 saw the beginning of a substantial rise in patent applications, licensing activities, and the formation of spin-off companies (HM Treasury, DfES, and DTI 2006), 30% of the higher education institutions engaged in these activities embarked upon them in or after 2000. As might be expected, therefore, there are great disparities between universities as to the levels of their activity around, expenditure on, and income from the protection and exploitation of intellectual property. In 2004, 21% of universities spent nothing on this, in contrast with the 6% that spent over £500,000 (UNICO 2005). Overall, expenditure on intellectual property fell in 2004 to £12 million from £12.6 million in 2003.

We have seen that government policies have encouraged the formation of spin-off companies, though, as Geuna and Nesta (2006) have observed, on the basis of little supporting evidence of their financial value and in the face of criticism from the Lambert report (Lambert 2003). In the year 2005–6, there was a significant reduction in the number of spin-off companies owned or partly owned by universities, although the number of new spin-offs rose. Again, there were wide institutional disparities. A full 43% of the institutions responding to a national survey had formed no spin-offs in 2004; 19% held no shares in any existing spin-off companies; 41% held shares in between 1 and 6; and less than 10% held shares in 30 or more companies. In total 850 companies had shares held in them by the institutions that created them (UNICO 2005).

Overall trends in the form and scale of academic capitalism are as yet unclear, although there have been sharp rises in licensing activities since 2001. Over 1,400 licensing agreements were executed by institutions during 2004, increasing the average number per institution from four to fourteen. However, this form of activity was undertaken by only 62% of all institutions, and one institution alone executed over 45% of the agreements (UNICO 2005).

In 2004, a total of 569 patents were granted to UK universities, 100 fewer than in the previous year but still twice the number recorded in 2001. There is also some evidence that the percentage of successful to unsuccessful applications is increasing (Geuna and Nesta 2006; UNICO 2005). Overall, some 66% of higher education institutions had existing licensing agreements that brought in income. However, while 56% earned either nothing or less than £50,000 from these, only 8% earned more than £1 million.

As yet, economic gain from exploitation of intellectual property rights is limited to a very few universities; indeed, a majority made a financial loss on their investment in technology transfer offices.

Impacts on the Academic Research Enterprise: System Level. As indicated earlier, the United Kingdom continues to perform well in world rankings of research outputs—notably in the quality and impacts of publications, as measured by citation levels—even if there are signs that its relative position is less secure. A study in 2006 of research in the fifty-two pre-1992 universities showed that there were significant increases in publications, citations, and research student awards for the natural, medical, and social sciences between 1990–2001 and 2000–2001 (Crespi and Geuna 2006). However, it also showed that overall, the growth in research productivity slowed during the second half of this period. The authors suggested that one possible reason for this was that the impetus provided by the RAE had lost its initial force.

Policies geared toward selectivity and the concentration of research have undoubtedly been highly effective. The most authoritative studies of their impacts are by Jonathan Adams and colleagues at the universities of Leeds and Manchester and at the London School of Economics and Political Science (Evidence Ltd. 2002, 2003, 2006).

Following the 2001 RAE, over 90% of research activity—measured by research staff share, research income other than quality-related allocations by the funding council, published outputs, and the production of PhD awards—was concentrated in units assessed as being in the top three bands (5*, 5, and 4) (Evidence Ltd 2003, 2006). There was then a small further gain by the units in the top two bands at the expense of units in the next lower two bands. Moreover, HEFCE funding allocations saw an increase on average of 6% for units in the top two bands (and a 4% average decrease for units in the next band) (Evidence Ltd. 2006).

These aggregate figures obscure differences between disciplines and between indicators, but the lowest concentrations by discipline of research active staff in the top three bands are still on the order of 50%. Concentration of research income from all sources ranges between 95% (in biological sciences) and 75% (in history, an outlier) (Evidence Ltd 2006). Overall, the increases in concentration are most marked in disciplinary areas where significant diversity has hitherto been maintained.

During the past two decades, the UK academic research system has become even more highly stratified than before. As is the case in the rest of the world, the high performance of the United Kingdom in world rankings can be attributed in large part to the small proportion of outstanding performers, that is, the 1.5–2% whose citation rate is eight times or more the world average and the 5.7% whose citation rate is at least four times the world average (Adams 2006). At a time when other nations are increasing their competitiveness, it may seem rational to sustain or indeed even harden the pursuit of steep stratification, a position apparently reflected in the government's approach to the distribution of the new infrastructure funding under the ten-year investment program. Allocations are made on a formula basis, but the aim is to strengthen centers of world-class excellence.

However, this remains a controversial policy. Adams's earlier work (Evidence Ltd. 2002) shows that, as measured by citation indexes, research units in the fourth RAE band (3A) were around a world rating for research quality, and that departments or research centers graded 4 in 2001 were better by a factor of 1.25 than the world ratings of all centers. The proportion of researchers who scored 4 or above in the RAE rose from 58% in 1996 to 79% in 2001.

But what Adams demonstrates so clearly is that the 5 and 5* peaks of excellence rest on a broader platform of important—if less spectacular—activity. Grade

4 units are a "platform" level of quality research that can develop into world-class 5 and 5*.

High stratification, critical mass, and concentration on excellence constitute one form of diversity reduction in the academic research base. Another is the growing disciplinary imbalance in the academic research enterprise as funding is more heavily concentrated on the biological and medical sciences (DIUS 2008). By contrast, there has been increasing concern in the academic community and beyond about the impacts of research policies, particularly the RAE, upon the disciplines of chemistry and physics, on which developments in interdisciplinary research in the biological sciences depend. In the past ten years, some thirty chemistry departments have closed, and by 2003 less than 40% of UK universities were offering undergraduate courses in physics and chemistry (Clery 2005; see also HEFCE 2005). In 2005, government and HEFCE, in response to this situation, established an advisory group under the chairmanship of Sir Gareth Roberts to identify subjects that were both vulnerable and strategic (important to the economy and/or society) and to recommend approaches for supporting them. Physics and chemistry figured strongly, although not exclusively, in their work. They also focused on minority foreign languages and modern foreign languages more generally. Interestingly, they included "cultural and creative disciplines" in their review of strategic subjects and found them to be flourishing (though there might be very specific subdisciplines within them needing support).

Impacts on the Academic Research Enterprise: University Level. In 1980, Becher and Kogan raised the question of whether the higher education institution "has a substantive existence and a distinct purpose" (1980, 63). By the mid-1990s, there was no doubt that it did. Universities and other higher education institutions by then constituted a distinct level of governance with a pivotal set of functions in the UK academic system.

Their collective acceptance of a functional transformation was symbolized by the conversion of the vice chancellors' national organization from the Committee of Vice Chancellors and Principals into Universities UK. Universities UK has gone out of its way to celebrate the central role that universities have played in the government's ambitions to achieve a successful knowledge economy and the implications for mission diversification, in terms not only of scientific and technological innovation, the growth of human capital through the massification of higher education, and increasingly varied forms and targets of knowledge transfer to business and the public sector, but also of direct engagement in for-profit exploitation of research (Universities UK 2002). Beyond this, Universities UK has supported and promoted the contribution made by universities to social and cultural as well as economic regeneration and development of their localities, cities, and regions, as government has placed more emphasis on regional policies (Universities UK and HEFCE 2002).

We will now consider the major development and change in the institutional capacity of universities and the modes of management and governance that have accompanied their new power and status. We will focus on the role in this process of two policies representing largely conflicting values: the Research Assessment Exercise (RAE) and the succession of measures to change the relationship between universities and business and eventually embed universities at the heart of a national innovation system. Finally, we examine a significant paradox entailed in these policies: that the autonomy of universities was to be both enhanced and reduced. The paradox would be managed, in part, by a redefinition of academic autonomy and in part by interinstitutional inequalities in policy impacts. It is widely considered that in the United Kingdom and elsewhere institutional autonomy has been enlarged at the expense of individual academic freedom (see, e.g., Tapper and Salter 1995). However, the issue is actually somewhat more complicated.

Several studies have examined the impact of recent higher education and research policies upon universities and academics. Here we depend, first, on two of our own: a study of the impact of the Foresight initiative (Henkel et al. 2000) and the English component of an international study of higher education reforms (Henkel 2000; Kogan and Hanney 2000; see also Kogan et al. 2006).[13]

As already indicated, the single policy that has had the most transparent and longest-lived impact was the institutionalization of the RAE. It represents a successful attempt to sustain academic values and control in a context where the state was making new demands on research and higher education and seeking to impose its own structures for quality assurance upon them. The RAE raised the profile of research in universities, reinforced the value of disciplinary excellence, and increased productivity, as measured by funded research projects and publications.

At the same time the RAE has been a force for change in institutional behavior in academe and in the balance of power between the university and its basic units. Increasingly strenuous efforts have been made by university leaders to maximize institutional success in the RAE. Crucially, it has encouraged universities to replicate national policies of prioritization, selectivity, critical mass, and concentration, in some cases on a very small number of "niche areas" (Talib 2003). The results have included departmental amalgamations, restructuring, and sometimes abolition, in cases of failure to achieve expected levels of performance in the RAE.

The RAE was the trigger for universities to introduce strategic reviews of institutional research agendas (Henkel et al. 2000; McNay 1997); internal monitoring and evaluation systems; structures and processes to enhance research activity, external funding, and performance; and new staffing strategies. From 1996, institutions became increasingly directive in the designation of staff as "research active" and thus eligible for inclusion in the RAE (McNay 1999). Basic units began to appoint research directors. Changes in ethos occurred when faculty were interrogated on research productivity and outcomes.

Impacts of the second set of policies, though increasingly powerful, developed more unevenly. We have seen that Foresight was the spearhead policy in a determined attempt by government in 1993, well backed by senior scientists, to cause cultural change in both universities and industry and in the application of science and technology to economic development. Overall, as far as the key external drivers of university research policies and academic behavior were concerned, Foresight itself was not thought to have had a major impact. Rather it was perceived in universities to have been part of a complex of external factors affecting their policies and structures.

However, three of the civic universities in the Brunel study (Henkel et al. 2000) saw Foresight as significant and influential in the research policy environment. It brought together a range of policies, including the development of interdisciplinary research; the demand to prioritize and maximize the use of research resources; the promotion of collaboration as well as competition between departments, universities, business organizations, and the public sector; and synergy between the interests of academic institutions and industries. All of these policies, along with the increasingly high profile accorded—at least by research-intensive universities—to various forms of academic capitalism, have had a continued and growing impact on the strategies and structures of universities.

Interdisciplinary research already had a prominent place in university policies by the 1990s (PREST 2000). In four of the universities in the Brunel study sample, financial incentives for interdisciplinary research and interdepartmental collaborations were explicitly mentioned. They primarily took the form of earmarked internal funding for which research partners, groups, or departments could apply. Most predated Foresight. One institution had been setting aside internal funding of well over £1 million per annum starting several years before Foresight, to encourage interdisciplinary collaboration. Another had a scheme for interdepartmental centers, allowing researchers to receive pump-priming funding for up to three years (Henkel et al. 2000).

The trend toward superimposing a web of research centers, most of them interdisciplinary, on departmental or faculty structures has intensified during recent years. Such centers figure strongly in universities' presentations of themselves as bases for innovative research and as resources for businesses and other potential external users. Indeed, prestigious research-led universities increasingly regard inter-disciplinary and transdisciplinary research institutes and centers, with their associated graduate schools, as their flagships (Geuna 1999).

The Brunel studies and others (e.g., Howells, Nedeva, and Georghiou 1998) confirmed that universities were rapidly developing industry relationships that required leadership and investment from the top of the institution. Many were of the kind advocated in the Foresight initiative. Some involved partnerships between universities. Some involved strategic alliances between individual universities and particular companies. The firms might have a site on campus; or they might fund chairs, buildings, or facilities from which long-term benefits were expected; or they might contribute money to the institution. Some informants noted a shift from a conception of exchange, where scientists were being paid by industry to do work for them, toward a more collaborative, open-ended enterprise, in which they were sharing costs for mutual benefits (Henkel et al. 2000).

It was clear that some institutions were creating a multifaceted research profile, enabling them to capitalize on a variety of public and private initiatives and to defend themselves against uncertainty. The most active were developing policies that would encourage links and partnerships at local, regional, national, and international levels with other universities, with small and medium sized enterprises (SMEs), with local and regional public authorities, and with national and multinational companies, some of which had local bases.

The developments described represented a transformation of universities and their research enterprise. The institutional importance of research increased, not only in reputational terms, but also as an indispensable source of income. Responses to government incentives for the opening up of university boundaries; for the incorporation of new values of utility, economic instrumentalism, and commercialization; for research collaborations with industries and other businesses; and for a variety of forms of knowledge transfer became increasingly powerful institutional imperatives.

These imperatives and the mission diversification they entail are reflected in the mission statements published by individual universities, but here it is noticeable that the primary emphasis is still upon their academic functions—the advancement of learning and research. There is still no evidence of sharp polarization between universities' conceptions of themselves, as envisaged by the 2003 white paper "The Future of Higher Education" (DfES 2003) (and apparently endorsed by the 2007 Sainsbury review of UK science and innovation policies). Some mission statements make explicit reference to the maintenance of long-cherished academic values, such as academic freedom and the pursuit of truth. Universities are seeking to combine research excellence with contribution to the economy, international research connections with local and regional commitments, and competitiveness with interinstitutional collaborations. While some universities do now focus primarily on their educational functions in their mission statements, the importance of research in the institutional profile extends well beyond the elite or research-intensive universities.

Nevertheless, universities have become different kinds of institutions, with new modes of governance. The strong boundaries and clear distinctions between them and other organizations have been blurred. The conception of university as "corporate enterprise" (Bleiklie 1994; CVCP 1985) has in many ways been realized in the United Kingdom.

In the 1980s and 1990s, the government enhanced the power of and made more explicit the responsibilities of the (predominantly lay) governing bodies of universities. It reinforced their accountability for the financial viability of their institutions and the delivery of value for public money. Less successful, however, at least as far as research-intensive universities were concerned, was its promotion of the view that university governance should conform more closely to best business practice (Shattock 2004).

This said, most universities have made major changes in how they are run. Vice chancellors have embraced the role of chief executive. Authority has been substantially centralized in small senior management teams under their leadership and these teams comprise key nonacademic directors as well as academics.

Strategic planning and management are major concerns of governing bodies and senior management teams, and, within these, research strategy and management are of overriding importance for most universities. Structures and strategies for research management include research committees; cross-institutional units for research development, support, and marketing; research institutes and centers; and research appointments. Universities often have at least two pro vice chancellors whose roles include research management: one for research, and another whose responsibility involves "innovation and economic development" or "enterprise and innovation" and/or external relations. The function of university-industry liaison has rapidly developed in UK universities since the 1980s and has been substantially expanded and incorporated into functions of research commercialization and technology or knowledge transfer.

The proliferation of cross-institutional or "interstitial" units (Rhoades 2006) to lead as well as support the functions of knowledge transfer, innovation, and external engagement has entailed the introduction of hybrid and still emergent roles. Some require new types of personnel, including individuals combining high academic and research qualifications with technical and business management and/or government experience, who are employed to help institutions manage novelty and uncertainty in their expanding worlds. Such individuals will shape as well as be shaped by their institutions; their relationships with academics are likely to involve a combination of collaboration, negotiation, education, and persuasion as well as support and advice.

Universities in the United Kingdom, and particularly research-intensive universities, have, as predicted by theorists of the knowledge society (Bell 1973), become "axial structures." They have been forced to accept the neoliberal ideologies of successive governments, to engage with rather than stand apart from other sectors of society, and to enlarge the scope of their activities. They now work within an increasingly strong external framework of regulation, accountability, and governance. The spaces within which they exercise choices are substantially structured by government and shaped by values and purposes that are political and economic rather than academic. While they maintain freedoms previously regarded as essential for institutional autonomy—freedoms to choose their own staff and students, to determine how to spend their money, and to decide on curriculum content and degree standards (Ashby 1966)— they do so within a new dirigiste policy regime. Further, institutional autonomy is now a relative concept and conditional upon financial status.

At the same time, universities' "spaces of action" are literally and metaphorically far more extensive and contain within them multiple and complex options. In the necessity to accept more financial responsibility—to engage in markets and so reduce their financial dependence on the state—lies the possibility of loosening state controls.

In fact, most universities remain substantially dependent on public funding and the reputational value attached to it. However, public policies have combined to sustain and, indeed, enhance large inequalities in the power and dependency relationships between the state and universities. As we have seen, they have reinforced the stratification of universities in terms of reputation and command of resources, whether public or private. It is not as yet clear what this will mean across the whole spectrum of universities. However, there is evidence that while larger and older universities are predominant among the narrowing elite of institutions, newer and financially weaker institutions are more restricted in their strategic choices, more dependent on contract research, and more likely to be forced to focus on less academically interesting forms of research (Geuna 1999; PREST 2000).

Policies may also loosen the ties between the strongest universities and the nation. We have noted that, like most European states, the United Kingdom has encouraged its universities to take advantage of the EU Framework funding programs, which reward the generation of international networks and other forms of international collaboration among universities and between universities and innovative industries. These funds, although relatively small as a proportion of total funding,[14] undoubtedly have a powerful leverage effect on universities' research planning. Advantageous in themselves, they may open up further international funding possibilities and orient institutions toward global markets. Similarly, national policies fueling some institutions' ambitions to join the ranks of newly defined "world-class" universities may encourage them to become more distanced from their national identity.

Impacts on the Academic Research Enterprise: The Academics. Studies examining how policy changes have impinged upon academic values and agendas are still relatively rare, and most have focused primarily on relatively high-performing, research-based universities. Our own research (Henkel 2000), carried out between 1994 and 1997, on the impacts of higher education reforms on values and working practices found little evidence of decisive shifts in attitudes or modes of work. Respondents, particularly scientists, were overwhelmingly preoccupied by research-funding problems. At the same time, they emphasized the importance of academic autonomy for sustaining motivation and the quality of research. For most academics their discipline continued to provide the central organizational, epistemological, and value structure for their career paths. More recent international and UK research suggests that many successful researchers see no contradiction between strong disciplinary identities and agendas and interdisciplinary collaborations (Di Napoli 2003; Marton 2005; Nowotny, Scott, and Gibbons 2001). Research also suggests that the latter generate new forms of integration and new epistemic communities, reflecting both changing internalist constructions of knowledge and externally defined problems or needs (Geuna 1999).

A combination of internalist and externalist influences has generated change in the organization of academic research in universities. Discipline-based departments have often been converted into larger and more loosely defined schools. Within these schools, divisions may be organized around domains or fields or combinations of disciplines, and distinct structures for research may be established. We have already noted how interdisciplinary and transdisciplinary research institutes and centers have become the flagship basic units in some universities. Some have cross-boundary objectives built in, such as the development of connections and collaborations with other academic institutions or with other sectors or organizations. They, in turn, imply new and more fluid working structures, networks, consortia, and "hybrid communities."

Academics largely now accept that research is a collective rather than an individual responsibility to the institution as well as to the discipline and to the immediate research group. Attitudes toward wider social and economic responsibilities are more uncertain.

Team membership, for social scientists and arts and humanities scholars, as well as for natural scientists, is increasingly important for obtaining research funding. All academics are under pressure to develop multiple research profiles and connections: basic, strategic, and commercialized or applied research; and networks and collaborations, national and international, across basic units, between institutions, and with sponsors and partners outside academe. These are among the growing pressures for the separation of research from teaching.

Initially, policy developments were seen as relevant primarily, if not exclusively, to the fields of science and technology. However, first the Economic and Social Research Council began promoting team-based research, the social and economic value of research, and the needs of research users, and now the Arts and Humanities Research Council (AHRC) has followed suit (though these are more obviously contested ideas in the AHRC).

Overall, in the United Kingdom as elsewhere, changing policies posed the least threat to strong researchers in the natural sciences in research-intensive institutions. For them there is, in practice, little conflict between the RAE and economic instrumentalist policies. Without seeking collaboration with industry, they are more able to attract it than some who do (Calvert and Patel 2002); they can accommodate it within their programs without disturbance to the control of their agenda.

Foresight (and the policy complex of which it was a part) was not, for this group, a threat to cherished scientific norms of creativity and self-motivation. It was not, in the words of one, about "turning scientists into short-term hack researchers" (Henkel et al. 2000, 34). Such research was likely to be on the periphery of the elite disciplinary structures around which they were organized. Foresight, "not least because it gave a new set of social and commercial priorities, opened up thinking in the research councils and made them potentially into more dynamic organizations" (Henkel et al. 2000, 35). Researchers might be involved or take the lead in institutional initiatives to facilitate access by firms to the university or department, to promote the exchange of ideas, or to develop joint programs of knowledge development.

This is not to say that academics had no qualms about the changing policy environment. While relatively few respondents thought that the RAE had affected their own research practices, more detected some general shifts. The strongest concerns among social scientists, particularly sociologists and economists, were that it encouraged concentration on publication in a limited range of high-impact-factor journals and reinforced trends toward ideological conformity (see, e.g., Harley and Lee 1995 on economics; see also Harley 2002). Other criticisms were that the exercise could undermine some disciplines (particularly the humanities). The emphasis on output over a short period of four years (extended to six for the humanities) might discourage involvement in long-term and possibly ground-breaking research[15] or penalize departments recruiting young staff. It might reduce the diversity of some disciplines through its bias against small departments; physics was most mentioned in this connection. Impacts could be severe for scientists in departments outside the top rankings. As we have seen, present policies put at risk research in many departments assessed as being of national or even international excellence (Evidence Ltd. 2002, 2006; see also Association of University Teachers 2003).

More general critiques were, first, that the RAE represented an insufficiently multidimensional approach to the definition of academically excellent research, although the widely held view that it disadvantaged multidisciplinary research was not borne out by the results (PREST 2000).

Attitudes toward science policies promoting collaboration between industry and academics were mixed. Several interviewees in the Brunel Foresight study remarked on the excitement of working with industry (which might have been partly induced by Foresight), challenging long-held assumptions about the distinction between pure and applied research. There was wider support for the principle of collaboration but uncertainty about the compatibility of the parties' expectations. Some felt that what industry most wanted from universities was a supply of well-trained scientists and that firms' requirements for academic research were short-term and tightly defined. Some were specifically critical of public research awards that were conditional on obtaining industrial funding. It was argued that partnerships with industry cannot be generated easily and that industry should be expected to take more of the initiative.[16]

Relationships between academics and industry predominantly reflected different forms of exchange. At one end of a continuum, they might be largely instrumental exchanges, in which researchers tested materials or clinicians carried out clinical trials of drugs for companies in exchange for money that could then be spent on various forms of research support. Researchers tried to minimize such forms of relationship as being of limited intellectual value. An alternative form of exchange was one in which researchers might carry out similar forms of work but in doing so get access to material or data or extended time that enabled them to address problems of interest to themselves; or academic clinicians might seek opportunities to evaluate devices, equipment, or drugs. Exchanges might be more speculative, in which university-based researchers took new basic knowledge or theories to firms, with a view to interesting them in technology transfer or product development. Alternatively, firms might fund individual scientists or groups over a period of time to undertake basic research in a field of interest to them, in exchange for privileged access to the knowledge that emerged.

Overall, most academics who were developing partnerships with industry, either under pressure from their institutions or on their own initiative, still did so only within the terms of their own agenda. Our findings here were confirmed by D'Este, Nesta, and Patel (2005),[17] who also suggested that, for the most part, the generation of new products, processes, patents, and spin-off companies are of little importance to scientists.

Conclusions

The UK story in the past two decades is one of transformation, in the areas of policy, of governance, and of funding. For most of the twentieth century, academic research in the United Kingdom was an almost uniquely autonomous enterprise, subject to a loosely structured, even disorganized system of governance and funding. Since the late 1980s, it has had a greatly enhanced policy salience and, in consequence, has been working within an increasingly directive, national policy framework. This framework has been based on principles of strong, central direction, coordination, regulation, and accountability, albeit within neoliberal conceptions of the state and management of the public sector.

POLICY GOALS AND VALUES

Policy for the academic research enterprise (ARE) is embedded within a major national policy agenda, to maximize national economic competitiveness in a global market. The organizing concepts are innovation and the knowledge economy, both of which have moved to center stage during the past decade. The ARE is one component of a national research and innovation system. It is assigned an essential, instrumental role in the pursuit of the main policy goal based on the following assumptions: research (more specifically, research excellence) constitutes a key source of innovation; national research reputation acts as a magnet for overseas investment and overseas talent as well as being a source of national prestige; knowledge or technology transfer in various forms between academe and other sectors is an important means of achieving a knowledge economy; and the academic sector can contribute directly to the national economy through engagement in markets and, not least, in the commercial exploitation of its intellectual property rights.

A perhaps more contentious underlying assumption is that Mode 2 conceptions of knowledge production should be predominant, though not universal. Structures and functions at different levels of the system reflect the belief that knowledge needs to be produced in a context of application or expectation of utility, with the involvement of different institutional actors and often within interdisciplinary or transdisciplinary frameworks.

GOVERNANCE AND FUNDING

The past two decades have seen a fairly steady buildup of governance for an innovation system and for the enhanced authority of central government vis-à-vis the ARE through the establishment of a stronger legal base, more tightly coupled government departmental structures and mechanisms for coordination, strong regulatory machinery (triggering high degrees of self-regulation by universities and individuals), and a determined use of the "power of treasure."

In the 1980s, the government focused on the need for economy, efficiency, and effectiveness on the part of universities and achieved compliance in major policy changes through the use of financial incentives for institutions that were under substantial pressure from reduced public funding and increased demand. The main principles for public funding established then have been sustained and in some respects strengthened: transparency, financial incentives for institutional behavioral change, competitive performance-based resource allocation, and higher-than-ever degrees of selectivity and concentration that have reinforced and enhanced a historic system of university stratification. The main change in the past decade has been a commitment to significant government investment in the science base, but the drive to diversify the sources of research funding and to encourage the ARE to maximize its market potential has been sustained.

In what is now a "multi-echelon system" (Van Vught, this volume, chapter 5), the university constitutes a significant level of governance, with increased responsibilities and power within the terms of the new policy framework. The promotion by government of the university as a "corporate enterprise" requiring a managerial regime has been largely embraced by Universities UK and the vice chancellors of individual institutions. They have centralized authority and enlarged their strategic organizational capacity in order to cope with their enhanced functions in the national innovation system.

This system has increasingly reflected a blurring of the functional distinction between institutions concerned with sustaining the national knowledge and education base and those concerned with their application or utility. There is at the time of writing no government department that includes education or science in its title; research councils, universities, and researchers have responsibilities for knowledge transfer and economic impact as well as for its production and quality.

STRENGTHS AND WEAKNESSES

A major strength, as far as the ARE is concerned, is the degree of importance it has acquired in the national policy agenda. It is now seen as a national economic asset rather than as primarily a consumer of public resources. In embedding the ARE in a national innovation system, government has highlighted the interconnections between excellent research, sustainable universities, the responsiveness of the research base, business investment in innovation, the enhancement of skills throughout the population, and public understanding of science and technology (HM Treasury, DfES, and DTI 2004).

The combination in the policy framework of strong direction, rigorous evaluation, and extensive use of the power of treasure has succeeded in raising the profile of research in universities as well as in government and industry. It has sustained the pressures and provided the incentives for universities to enhance the quality and productivity of their research and to invest in strategies and structures to strengthen their connections with industries and other sectors of society. Universities and the researchers within them are now more publicly accountable, have a broader conception of their research responsibilities, and are more actively engaged with other organizations. At the same time, the importance of excellent research for national innovation and competitiveness has been reinforced. It is clear that, as in the United States, a small number of research-intensive institutions and the researchers within them are learning how to incorporate strategic and for-profit research into their basic agendas in such a way as to advance their income, without loss to their academic reputation or their academic freedom.

Probably the most important question about the existing policy framework is how successfully it can sustain the strength of the ARE as a source of innovation in the medium and long terms. Does it provide the conditions to support the capacity and motivation of researchers to produce new knowledge that is not only immediately useful but also capable of opening up new paradigms, new depths of understanding, and new potential for problem-solving in society?

There are several causes for criticism in these terms. They largely concern the balance and the dynamics in the system between values that are not wholly reconcilable but widely regarded as needing to coexist: between academic freedom and accountability, between academic and economic priorities, and between the concentration of research and epistemic diversity.

First, while it can be argued that stronger governance and regulation were needed to achieve significant change in the relationship between the academic research enterprise and society, the degree of surveillance, performance measurement, and competition for resources is now so high in the United Kingdom as to raise serious doubts about its implications in the longer term for academic capacity and motivation. In particular, there are concerns that it leaves insufficient space for diversity and nonconformity in academic inquiry and that the balance between reward and punishment may inhibit the academic ambition, risk-taking, and tolerance of uncertainty needed to produce innovative work.

The second problem is the degree of research concentration in the United Kingdom and its implications for the nature of knowledge produced and for the institutional base of the ARE. While it is true that the high reputation of UK science is based primarily on a tiny minority of academic researchers (Adams 2006), research resources and activity are now so concentrated as to pose a serious threat to the platform on which these achievements are built (Evidence Ltd. 2006) and, again, to the range and diversity of new advanced knowledge that should be expected in economies and societies deemed to be increasingly dependent on their knowledge capacities. The erosion of sciences fundamentally important to the future of frontier research has already been noted. Meanwhile, there is a danger that humanities disciplines are not only being squeezed in universities but also valued in excessively narrow instrumental terms, such as their contribution to creative industries and to the prevention of social disintegration.

Third, government policies on academic capitalism seem to be substantially shaped by the experience and potential of an elite group of universities. Recent research on the impacts of such policies suggests that they may have exaggerated the financial rewards to most universities from this form of activity, particularly patent applications and the formation of spin-off companies (Geuna and Nesta 2006; UNICO 2005). High returns depend heavily on a few highly marketable discoveries.

Fourth, there is the Research Assessment Exercise (RAE). We have seen how effective an instrument for change the RAE has been but at the same time how successfully it has sustained academic values in the emerging innovation system. Government's proposals for change in the exercise are designed partly to shift the balance between these two phenomena. The reduction of the role of peer review in the new Research Excellence Framework and the greater reliance on metrics will generate a change in the definition and evaluative criteria of research excellence. As yet the details are undecided, but it seems likely that research income and economic impact will be more important, with a resulting rise in incentives to undertake research that will score highly on these dimensions and further disincentives to work in unfashionable areas (see Sastry and Bekhradnia 2006a, 2006b). Sastry and Bekhradnia also predict that competition for prestigious project and program grant awards will intensify and that this result, in turn, will increase the scrutiny of applications by researchers' universities, at the cost of time, money, and intellectual originality.

Long-term evaluation will be needed regarding the quality of research and research education, the national and institutional epistemic ecologies, and the research institutional structures resulting from the changes that have occurred in the United Kingdom. Much of the UK story seems to support Geuna's argument (1999) that the challenges of a new era are opening up an unbridgeable gap between universities; only a few elite research universities will fully adapt to the new demands and also manage to retain some of the assumed defining features of universities; many will be marginalized and little influenced by international changes in the production of knowledge. However, as yet we do not know enough about the variety of academic responses to new demands. While it is clear that the most active collaborators with industry and the most successful "academic capitalists" are the academic high flyers, a wider range of institutions and individuals have established research "niches" for themselves with a certain amount of success (Calvert and Patel 2002; Universities UK 2002; Universities UK and HEFCE 2002). We need a more systematic understanding of the success of such adaptive behavior. Meanwhile, academic values and motives remain robust in the face of change.

NOTES

1. This report of the Science Policy Working Group, under the chairmanship of Sir Frederick Dainton, was published in 1971 in the same document as the influential Rothschild Report on Government R&D. It, however, differed from Rothschild in that it rejected the distinction between pure and applied research and proposed instead a threefold classification of scientific research: tactical, strategic, and basic.

2. The amount allocated to "knowledge transfer" activities was about £79 million.

3. For a short time this became the Office for Science and Innovation (OSI), through a merger with the DTI's Innovation Unit in 2006.

4. Arts and Humanities Research Council; Biotechnology and Biological Sciences Research Council; Engineering and Physical Sciences Research Council; Economic and Social Research Council; Medical Research Council; Natural Environment Research Council; Science and Technology Facilities Council.

5. Research Councils UK, www.rcuk.ac.uk/aboutrcuk/executivegroup/default.htm, para. 1.

6. In the 1990s, this post lost its remit for science and gained one for, first, "Employment" and then "Skills."

7. Key policy documents began to be published under its name, together with those of the DTI and the Education Department.

8. The BBSRC, for example, pledged an annual increase of 3% in the proportion of responsive-mode funding in the new three-year spending review period (BBSRC 2008).

9. The seven grades were as follows: 5*, 5, 4, 3A, 3B, 2, 1.

10. The government insists that there is no intention to change the balance in the dual support system under this policy.

11. Basic support for research-council-funded PhD students had by then been significantly increased, notably for the sciences with the most recruitment problems (DTI, HM Treasury, and DfES 2002).

12. No comparable figures are available for Oxford, but an authoritative source has confirmed that the position would be much the same in the two universities.

13. Other works on which we have drawn include Howells, Nedeva, and Georghiou 1998; Geuna 1999; PREST 2000; Georghiou 2001; D'Este, Nesta, and Patel 2005; Reichert 2006.

14. The sixth Framework Programme, for example, represented only 5% of the total expenditure on research by EU member states.

15. See, e.g., Mynott quoted in Kogan and Hanney 2000; Henkel 2000, 139.

16. In a more recent study on the attitudes of scientists toward university-industry collaborations and their outputs, scientists reported that the most important barrier to collaborative activities was the difficulty of finding suitable industrial partners (D'Este, Nesta, and Patel 2005).

17. The disciplines incorporated in this quantitative study overlapped with and extended beyond our own two qualitative studies in some respects. They included the range of engineering sciences, chemistry, computer science, mathematics, metallurgy and materials, and physics. Ours included biological sciences (mainly biochemistry), chemistry, physics, clinical sciences, health informatics and computer science, and materials, as well as economics, sociology, health services research, social policy, history, and English.

REFERENCES

Adams, J. (2006) *How Good Is the UK Research Base?* Oxford: Higher Education Policy Institute (HEPI).

Adams, J. and B. Bekhradnia. (2004) *What Future for Dual Support?* Oxford: HEPI.

Adams, J. and D. Smith. (2004) *Research and Regions: An Overview of the Distribution of Research in UK Regions, Regional Research Capacity, and Links.* Oxford: HEPI.

Advisory Council on Applied Research and Development (ACARD). (1986) *Exploitable Areas of Science.* London: HMSO.

Arts and Humanities Research Council (AHRC). (2004) *AHRC Delivery Plan, 2004.* Bristol: AHRC.

———. (2008) *AHRC Delivery Plan, 2008.* Bristol: AHRC.

Ashby, E. (1966) *Universities, British, Indian, African: A Study in the Ecology of Higher Education.* Cambridge, MA: Harvard University Press.

Association of University Teachers (AUT). (2003) *The Risk to Research in Higher Education in England*. London: AUT.

Becher, T., M. Henkel, and M. Kogan. (1994) *Graduate Education in Britain*. London: Jessica Kingsley.

Becher, T. and M. Kogan. (1980) *Process and Structure in Higher Education*. London: Heinemann.

Bekhradnia, B. (2004) *Government, Funding Council, and Universities: How Should They Relate?* Oxford: Higher Education Policy Institute.

Bell, D. (1973) *The Coming of Post-Industrial Society*. London: Heinemann.

Biotechnology and Biological Sciences Research Council (BBSRC). (2008) *Delivery Plan, 2008*. Swindon: BBSRC.

Bleiklie, I. (1994) *The New Public Management and the Pursuit of Knowledge*. LOS Paper 9411. Bergen, Norway: LOS, University of Bergen.

Brooks, H. (1994) The Relationship between Science and Technology. *Research Policy* 23:477–486.

Calvert, J. and P. Patel. (2002) *University-Industry Collaborations in the UK*. Brighton: Science Policy Research Unit, University of Sussex.

Chancellor of the Duchy of Lancaster. (1993) *Realising Our Potential: A Strategy for Science, Engineering, and Technology*. Cm 2250. London: HMSO.

Clery, D. (2005) Darwinian Funding and the Demise of Physics and Chemistry. *Science* 307:668–669.

Committee of Vice Chancellors and Principals of the Universities of the United Kingdom (CVCP). (1985) *Report of the Steering Committee for Efficiency Studies in Universities* (Jarratt Report). London: CVCP.

Crespi, G. and A. Geuna. (2006) The Productivity of UK Universities. Science Policy Research Unit (SPRU) Electronic Working Paper Series (SEWPS), No. 147, University of Sussex, SPRU.

Dainton Report. (1971) *Report of a Study of the Support of Scientific Research in the Universities*. Cmnd. 4798. London: HMSO.

Department for Education and Skills (DfES). (2003) *The Future of Higher Education*. London: HMSO.

Department of Innovation, Universities and Skills (DIUS). (2007) *The Allocation of the Science Budget, 2008/09-2010/11*. London: DIUS.

———. (2008) *Research Council Tables: Analysis of Expenditure by Type*. London: DIUS, www.berr.gov.uk/files/file40969.pdf.

Department of Trade and Industry (DTI). (1998) *Our Competitive Future—Building the Knowledge Driven Economy*. London: HMSO.

———. (2000) *Excellence and Opportunity—A Science and Innovation Policy for the 21st Century*. Cm 4814. London: HMSO, www.berr.gov.uk/files/file11990.pdf.

———. (2001) *Science and Innovation Strategy, 2001*. London: HMSO, www.berr.gov.uk/files/file16263.pdf.

———. (2003) *The Research Careers Initiative, Final Report, 1997–2002* (Roberts Report). London: HMSO.

———. (2005) *Science Budget Allocations, 2005-06 to 2007-08*. London: HMSO.

———. (2006) *Innovation in the UK: Indicators and Insights.* DTI Occasional Paper No. 6. London: HMSO, www.berr.gov.uk/files/file31569.pdf.

D'Este, P., L. Nesta, and P. Patel. (2005) *Analysis of University-Industry Research Collaborations in the UK: Preliminary Results of a Survey of University Researchers.* Science Policy Research Unit Report. Sussex: SPRU, University of Sussex, www.sussex.ac.uk/spru/documents/deste_report.pdf.

Di Napoli, R. (2003) Modern Languages: Which Identities? Which Selves? PhD diss., University of London.

DTI, HM Treasury, and DfES. (2002) *Investing in Innovation: A Strategy for Science, Engineering, and Technology.* London: HMSO.

Etzkowitz, H. and L. Leydesdorff (eds.). (1997) *Universities and the Global Economy: A Triple Helix of University-Industry-Government Relations.* London: Pinter.

Evidence Ltd. (2002) *Maintaining Research Excellence and Volume.* Leeds: Evidence Ltd.

———. (2003) *Funding Research Diversity: The Impact of Further Concentration on University Research Performance and Regional Research Capacity.* London: Universities UK.

———. (2006) *Monitoring Research Diversity: Changes between 2000 and 2005.* Research Report. London: Universities UK.

Evidence Ltd., DTI, and OST. (2003) *PSA Target Metrics for the UK Research Base.* London: Department of Trade and Industry.

Fulton, O. (2002) Higher Education Governance in the UK: Change and Continuity. In A. Amaral, G. A. Jones, and B. Karseth (eds.), *Governing Higher Education,* 187–211. Dordrecht: Kluwer.

Georghiou, L. (2001) The United Kingdom National System of Research, Technology, and Innovation. In P. Laredo and P. Mustar (eds.), *Research and Innovation Policies in the New Global Economy: An International Comparative Analysis,* 253–296. Cheltenham, UK: Edward Elgar.

Geuna, A. (1999) *The Economics of Knowledge Production, Funding, and the Structure of University Research.* Cheltenham, UK: Edward Elgar.

Geuna, A. and L. Nesta. (2006) University Patenting and Its Effects on Academic Research: The Emerging European Evidence. *Research Policy* 35:790–807.

Gummett, P. (1980) *Scientists in Whitehall.* Manchester: Manchester University Press.

Harley, S. (2002) The Impact of Research Selectivity on Academic Work and Identity in UK Universities. *Studies in Higher Education* 27 (2): 187–205.

Harley, S and F. Lee. (1995) *The Academic Labour Process and the Research Assessment Exercise: Academic Diversity and the Future of Non-Main Stream Economics in UK Universities.* Leicester: Leicester Business School.

Hatakenaka, S. (2005) *Development of Third Stream Activity: Lessons from International Experience.* Oxford: Higher Education Policy Institute.

Henkel, M. (2000) *Academic Identities and Policy Change in Higher Education.* London: Jessica Kingsley.

Henkel, M., S. Hanney, M. Kogan, J. Vaux, and D. von Walden Laing. (2000) *Academic Responses to the UK Foresight Programme.* Uxbridge: Brunel University.

Henkel, M. and M. Kogan. (1993) Research Training and Graduate Education: The British Macro Structure. In B. R. Clark (ed.), *The Research Foundations of Graduate Education,* 71–114. Berkeley: University of California Press.

Higher Education Funding Council for England (HEFCE). (1998) *Consultation, 98/55, Appraising Investment Decisions: Consultation.* Bristol: HEFCE.

———. (2004) Quality Profile Will Provide Fuller and Fairer Assessment of Research. *News and Events,* February 11, www.hefce.ac.uk/news/hefce/2004/rae.asp.

———. (2005) *Strategically Important and Vulnerable Subjects: Final Report of the Advisory Group* (Roberts Report). June 2005/24, Policy Development. Bristol: HEFCE.

———. (2006a) *Recurrent Grants for 2006–07.* Bristol: HEFCE.

———. (2006b) *Strategic Plan, 2006–2011.* Bristol: HEFCE.

———. (2007) Future Framework for Research Assessment and Funding. Circular letter 06/2007, HEFCE, Bristol, www.hefce.ac.uk/pubs/circlets/2007/cl06_07/.

———. (2008) *Research Excellence Framework.* Bristol: HEFCE, www.hefce.ac.uk/research/ref/.

Higher Education Funding Councils (HEFCs). (2005) *RAE 2008: Guidance to Panels,* RAE 01/2005. Bristol: Higher Education Funding Council for England.

Higher Education in Trouble. (1988) *Nature,* January 14, 331:99–100.

Higher Education Statistics Agency (HESA). (2004) *Students in Higher Education Institutions, 2002/03.* Cheltenham: HESA.

———. (2005) *Finance Data, 2003/04.* Cheltenham: HESA.

———. (2007) *Students in Higher Education Institutions, 2005/06.* Cheltenham: HESA.

———. (2008) *Students in Higher Education Institutions, 2006/07.* Cheltenham: HESA.

HM Treasury, Department for Education and Skills (DfES), and Department of Trade and Industry (DTI). (2004) *Science and Innovation Investment Framework, 2004–2014.* London: HMSO.

———. (2006) *Science and Innovation Investment Framework, 2004–2014, Annual Report.* London: HMSO.

HM Treasury and Department for Innovation, Universities and Skills (DIUS). (2007) *Science and Innovation Investment Framework 2004–2014, Annual Report.* London: HMSO.

Hood, C. (1983) *The Tools of Government.* London: Macmillan.

House of Commons Education and Skills Committee. (2003) *Fifth Report of Session 2002–03.* HC 425-ii. London: HMSO.

House of Commons Innovation, Universities, Science and Skills Committee. (2008) *Science Budget Allocations. Fourth Report of Session, 2007-08.* HC215-1. London: HMSO.

House of Lords Select Committee on Science and Technology. (1987) *Civil Research and Development.* Vol. 2. London: HMSO.

Howells, J., M. Nedeva, and L. Georghiou. (1998) *Industry-Academic Links in the UK.* Manchester: PREST; Bristol: HEFCE.

Hughes, A. (2006) University-Industry Linkages and UK Science and Innovation Policy. Working Paper 326, Centre for Business Research, University of Cambridge.

Irvine, J. and B. Martin. (1984) *Foresight in Science: Picking the Winners.* London: Pinter.

King, D. A. (2004) The Scientific Impact of Nations. *Nature* 430 (6997): 311–316.

Kogan, M., M. Bauer, L. Bleiklie, and M. Henkel. (2006) *Transforming Higher Education: A Comparative Study.* 2nd ed. Dordrecht: Springer.

Kogan, M. and S. Hanney. (2000) *Reforming Higher Education.* London: Jessica Kingsley.

Lambert, R. (2003) *Lambert Review of Business-University Collaboration, Final Report* (December). London: HMSO.

Layton, E. (1977) Conditions of Technological Development. In I. Spiegel Rösing and D. de Solla Price (eds.), *Science, Technology and Society: A Cross Disciplinary Perspective*, 197–222. London: Sage.

Martin, B. and J. Irvine. (1989) *Research Foresight: Priority Setting in Science.* London: Pinter.

Marton, S. (2005) Academics and the Mode 2 Society: Shifts in Knowledge Production in the Humanities and Social Sciences. In I. Bleiklie and M. Henkel (eds.), *Governing Knowledge: A Study of Continuity and Change in Higher Education*, 169–188. Dordrecht: Springer.

McNay, I. (1997) *The Impact of the 1992 Research Assessment Exercise on Individual and Institutional Behavior in English Higher Education: Summary Report and Commentary.* Chelmsford, UK: Center for Higher Education Management, Anglia Polytechnic University.

———. (1999) The Paradoxes of Research Assessment and Funding. In M. Henkel and B. Little (eds.), *Changing Relationships between Higher Education and the State*, 191–203. London: Jessica Kingsley.

Mulkay, M. (1977) Sociology of the Scientific Community. In I. Spiegel-Rösing. and D. de Solla Price (eds.), *Science, Technology, and Society: A Cross-Disciplinary Perspective*, 93–148. London: Sage.

Neave, G. (2006) Times, Measures, and the Man: The Future of British Higher Education Treated Historically and Comparatively. *Higher Education Quarterly* 60:115–129.

Nowotny, H., P. Scott, and M. Gibbons. (2001) *Re-Thinking Science: Knowledge and the Public in an Age of Uncertainty.* Cambridge: Polity Press.

Office of Science and Technology (OST). (1995) *Report of the Steering Group of the Technology Foresight Programme.* London: HMSO.

———. (2003) *Strategic Review of LINK Collaborative Research: Report of the Independent Review Panel.* London: HMSO.

Organisation for Economic Co-operation and Development (OECD). (1995) *Educational Research and Trends, Issues, and Challenges.* Paris: OECD.

———. (2007) *Main Science and Technology Indicators.* Vol. 2007/2. Paris: OECD.

Patel, P. (2003) *UK Performance in Science Related to Biotechnology: An Analysis of Publications Data.* Sussex: SPRU, University of Sussex.

Polanyi, M. (1962) The Republic of Science: Its Political and Economic Theory. *Minerva* 1:54–85.

Policy Research in Engineering, Science and Technology (PREST). (2000) *Impact of the Research Assessment Exercise and the Future of Quality Assurance in the Light of Changes in the Research Landscape.* Final Report. Bristol: HEFCE.

Poveda, J. (2002) The Geography of Innovation: A New Model of Technology and Innovation Policies in a Decentralized Country. *Science and Public Policy* 29:385–396.

Reichert, S. (2006) *The Rise of Knowledge Regions: Emerging Opportunities and Challenges for Universities.* Brussels: European University Association.

Rhoades, G. (2006) The Higher Education We Choose: A Question of Balance. *Review of Higher Education* 29:381–404.

Roberts, G. (2003) *Review of Research Assessment: Report by Sir Gareth Roberts to the UK Funding Bodies.* Bristol: HEFCE.

Sabri, D. (2007) The Assumptive Worlds of Academics and Policy Makers in Relation to Teaching. PhD diss., University of Oxford.

Sainsbury, D. (2007) *The Race to the Top: A Review of Government's Science and Innovation Policies.* London: HMSO.

Sastry, T. (2004) *Postgraduate Education in the United Kingdom.* Oxford: Higher Education Policy Institute.

Sastry, T. and B. Bekhradnia. (2006a) *Using Metrics to Allocate Research Funds.* Oxford: Higher Education Policy Institute.

————. (2006b) *Using Metrics to Allocate Research Funds: Initial Response to the Government's Consultation Proposals.* Oxford: Higher Education Policy Institute.

Shattock, M. (2004) The Lambert Code: Can We Define Best Practice? *Higher Education Quarterly* 58:229–242.

Talib, A. (2003) Institutional Behavior Impact of the 1996 RAE. *Higher Education Review* 36 (1): 57–77.

Tapper, E. and B. Salter. (1995). The Changing Idea of University Autonomy. *Studies in Higher Education* 20:59–71.

Trow, M. (1980) Dilemmas of Higher Education in the 1980s and 1990s. Paper presented at the Conference of Learned Societies, Montreal, June 3.

Universities and Colleges Employers' Association (UCEA). (2003) Framework Agreement for the Modernisation of Pay Structures. www.ucea.ac.uk/en/New_JNCHES/Frame work_Agreement_.cfm.

Universities UK. (2002) *The University Culture of Enterprise: Knowledge Transfer across the Nation.* London: Universities UK.

————. (2008) *List of Institutions, 2007.* London: Universities UK.

Universities UK and HEFCE. (2002) *The Regional Mission: The Regional Contribution of Higher Education.* London: Universities UK.

University Companies Association (UNICO). (2005) *UK University Commercialization Survey, Financial Year 2004.* London: UNICO.

Van Vught, F. (1994) Policy Models and Policy Instruments in Higher Education: The Effects of Governmental Policy-Making on the Innovative Behavior of Higher Education Institutions. In J. C. Smart (ed.), *Higher Education: Handbook of Theory and Research,* 10:88–125. New York: Agathon Press.

Waldegrave, W. (1994) Operation of the Research Councils. Letter to the chairman of EPSRC. Annex A of *Priorities for the Science Base: Government Response to the Second Report of the House of Lords Select Committee on Science and Technology, 1993–1994.* Cm. 2636. London: HMSO.

Williams, G. (1992) *Changing Patterns of Finance in Higher Education.* Buckingham: SRHE and Open University Press.

Woodward, D. and P. Denicolo. (2004) *Review of Graduate Schools in the UK.* London: UK Council for Graduate Education.

Ziman, J. (1984) *An Introduction to Science Studies.* Cambridge: Cambridge University Press.

CHAPTER TEN

The United States

DAVID D. DILL

The growing belief among policymakers worldwide that economic growth and international competitiveness are related to national innovation has refocused global attention on US policy and particularly on the role that universities play in the American National Innovation System (NIS). Empirical research on the performance of national innovation systems consistently places the United States among the leading nations (Balzat 2006). International scorecards, such as the European Innovation Scorecard developed by the European Commission, use the US NIS as a primary benchmark (UNU-MERIT 2008).

The national research system of the United States, which as in all countries is a core component of its National Innovation System, would be noteworthy if only for its massive scale, contributing over 44% ($343,747 billion) of the total gross domestic expenditures for R&D (research and development) ($773,998 billion) among the OECD (Organisation for Economic Co-operation and Development) countries in 2006 (OECD 2007a). The US investment in R&D is 2.6 times that of Japan ($130,745 billion), the OECD country with the next highest expenditures, and more than 2.4 times that of China ($144,037 billion), the non-OECD country with the highest (and fastest-growing) expenditures. For the United States, as for most other OECD countries, the Academic Research Enterprise (ARE) represents a relatively small portion of the national research system. In 2006, industry performed 70.3% of the research, universities 14.3%, and government laboratories 11.1% (OECD 2007a). The proportion of national R&D performed by the US ARE was in fact significantly below the OECD average of 17.6%, while US industry's proportion was somewhat larger than the OECD average of 68.0%.

The growing importance of national innovation to economic growth has inspired

US policymakers as well, leading to increased national investments in knowledge. The OECD suggests that national expenditures on R&D, software, and higher education can serve as proxies for knowledge investment (OECD 2007b). Within the OECD countries, this investment averaged 4.9% of GDP in 2004. The United States expended the highest percentage (6.6%), followed by Sweden (6.4%), Finland (5.9%), and Japan (5.3%); the EU average was 3.6%. The OECD suggests the United States and Japan appear to be moving more rapidly to a knowledge-based economy than the European Union, as their respective expenditures in knowledge as a percentage of GDP have grown more rapidly than the average for the EU countries since 1994. In 2005 the United States invested 2.62% of its GDP on R&D expenditures, in comparison to the OECD average of 2.25% (OECD 2007a). With regard to the ARE, the apparent US advantage is somewhat less obvious. For example, US expenditures on higher education R&D (HERD) as a percentage of GDP were 3.7% in 2005, below the OECD average of 4.0% (OECD 2007a).

The growth in the American economy during the 1990s and the comparative advantage of the United States in a number of high-tech fields have been attributed in part to the strength of the US ARE. A comparative econometric analysis has also suggested that the US ARE represents the "efficiency frontier" in terms of numbers of publications and citations relative to national HERD investments (Crespi and Geuna 2004). Therefore, international policymakers seeking models for reforming their national innovation systems have been particularly attentive to the framework conditions for the ARE in the United States.

Many OECD countries are now closing the efficiency gap with the US ARE due to positive productivity gains in their academic science systems, but also due to observed negative productivity growth in the US system (Crespi and Geuna 2004; King 2004). Several nations in the European Union as well as Australia and Canada now produce more scientific articles per million population than the United States, although the United States trails only Switzerland in the relative prominence of cited scientific literature (OECD 2007b). The OECD suggests three additional proxy indicators of the output and impact of national investments in science and technology (OECD 2007b): patents, the technology balance of payments in highly R&D-intensive industries, and the technology balance of trade in the same industries. The United States leads on many of the OECD R&D performance measures, but not all, and when controlled for gross domestic expenditures on R&D, the US advantage over EU countries and Japan is much less obvious.

Within the United States these recent trends have led to growing concerns about possible inefficiencies in the American ARE policy framework (Adams and Clemmons 2006; Foltz et al. 2005) and to calls for policy changes in order to maintain the effectiveness of national capabilities for innovation and productivity growth across the economy (COSEPUP 2006; Galama and Hosek 2008). In his 2006 State of the Union Address, President George W. Bush responded to these concerns by announcing the "American Competitiveness Initiative," which featured substantially increased funding for innovation-enabling research in the physical sciences and engineering, tax credits to encourage additional private-sector investment in innovation, and related policy measures to foster increased innovation and productivity growth in the service sector.

In international comparisons the US ARE stands out not only because of the acknowledged reputation and performance of its research universities, but also because of its distinctive framework conditions: In contrast to the governments of other OECD countries, the US national government has never exercised direct control over higher education and has also played a very limited role in regulating the academic, research, and administrative policies of its universities. This laissez-faire framework encouraged the development of an extensive and diverse set of public and private universities, which for their funding were initially dependent upon local sources of political and financial support. As a consequence, many universities were motivated to perform research of local benefit and to forge links with local industry. The earliest federal intervention into this emerging system was in fact an attempt to foster innovation in the leading industry of the day. The Morrill Land-Grant Act of 1862 was intended to promote scientific advances in agriculture. Since World War II, US federal policy has played an increasingly influential role in the development of the ARE, utilizing a number of distinctive policies. These have included allocating the vast majority of national support for academic research competitively (Trow 2000), providing this support to both private and public universities, and more recently altering the laws governing intellectual property rights in higher education. Reviewing the long antipathy in the United States to developing a national industrial or technology policy, and noting the increasingly influential role American universities play in technical innovation, Crow and Tucker (2001) concluded that the framework conditions of the US ARE now serve as the nation's de facto technology policy.

The sections that follow examine the instruments of federal policy and their influence on the US ARE. After a brief description of the composition of the US ARE, I discuss the evolution over time of federal policy on academic research. Succeeding sections explore the primary policy instruments currently employed by the federal government to influence academic R&D, research-doctoral education, and the relationship between academic research and economic development. In the concluding section I assess the overall strengths and weaknesses of the framework conditions for the ARE in the United States.

The Academic Research Enterprise in the United States

Given the large number of institutions in the American higher education system, defining the US ARE is not a simple task. The use of the term *university* is not regulated in the United States. As a result there are many institutions with "university" in their title that do not provide doctoral education, including primarily undergraduate academic institutions that offer master's and/or professional degrees, as well as the McDonald's corporate training facility, which is named Hamburger University. In contrast, the provision of doctoral education is regulated by six regional accrediting organizations, which assure that institutions offering the doctoral degree meet threshold academic standards. In order for students at these accredited institutions to be eligible for federal financial aid, the six accrediting agencies must in turn meet standards set by the US Department of Education. Consequently, the number of doctoral-degree-granting universities in the United States is indirectly and modestly influenced by federal policy. Therefore, if we define universities as those engaged in both research and doctoral education, the US ARE consisted of 283 institutions as of 2004 (Carnegie Foundation for the Advancement of Teaching 2007).[1]

Within this overall group of institutions the academic research effort is quite concentrated. The National Science Foundation (NSF) reports that the top 100 US research institutions accounted for 80% of all R&D dollars expended in 2006 (NSF 2007). This NSF analysis, however, includes a number of specialized medical and health centers, as well as the Woods Hole Research Facility, that do not offer doctoral degrees. If we further confine the analysis to traditional research universities offering both undergraduate and doctoral education of respected quality (Geiger 2004), we can identify the core of the US ARE as 66 public and 33 private universities (table 10.1) that in 2006 collectively performed 74% of federally funded academic R&D and granted 71% of US doctoral degrees (Hoffer et al. 2007; NSF 2007).[2]

The performance of the overall US research system is influenced by both federal and state policies, including direct financial support for R&D activity, tax incentives, and related regulatory instruments. The performance of the "public" research universities is also influenced by the policies of the fifty states, which vary substantially both in funding for academic R&D and in their regulatory frameworks for higher education (McDaniel 1996). The influence of US state policy on the ARE is therefore separately analyzed in parallel chapters in this book by Roger Geiger (chapter 11) and William Zumeta (chapter 12) on Pennsylvania and California, respectively. However, all of the institutions in the US ARE, both public and private, are substantially affected by federal research funding and regulatory policies. While only 29.3% of the overall US R&D system is financed by the federal government (OECD 2007a), the ARE is much more dependent upon federal support than that percentage implies. Of the $47 billion expended on academic R&D in 2006, over $30 billion, or 62.9%, was provided by the federal government, a proportion of subsidy that has remained relatively constant over the past decade (fig. 10.1). In contrast, as also depicted in figure 10.1, state and local government provided 6.3% of the R&D expenditures in colleges and universities (which was obviously concentrated in public universities), industry provided 5.1%, and the institutions provided 19.0% (consisting primarily of fees, gifts, and endowment funds). The remaining 6.7% expended on academic R&D came from other sources; among these sources were nonprofit foundations, which historically have been highly influential on the US ARE (Geiger 1993).[3]

Historical Evolution of Federal Policy on the ARE

The US Constitution does not explicitly mention education, thereby delegating primary responsibility for education to the states. Nonetheless, soon after the new republic's founding, proposals were advanced for a national university that would include advanced scientific training and be supported by the federal government (Dupree 1957). The first US president, George Washington, discussed the idea of a national university with the Congress. In addition, during his presidency Washington proposed federal support for a national military academy, which was opposed by his secretary of state, Thomas Jefferson, on constitutional grounds. Nonetheless, when Jefferson subsequently became president, he signed into being the first federal initiative in higher education, the founding of the US Military Academy at West Point in 1802. Presidents Jefferson and John Quincy Adams also became advocates for a national university, but because of continuing constitutional questions in Congress, the idea never came to fruition. However, four more federally operated service academies were subsequently created, all oriented, as was West Point, to bachelor's, or first-level-degree, education.[4]

The more traditional starting point of federal higher education policy is usually accorded to the Morrill Land-Grant Act of 1862, signed into law by Abraham Lincoln. As the US Military Academy's focus was on engineering, so the federal initiative on land-grant colleges was intended to stimulate education in the "agri-

TABLE 10.1.
Ninety-nine selected US public and private research universities

Public	Public	Private
Arizona State University	University of Colorado, Boulder	Boston University
Auburn University	University of Connecticut	Brandeis University
Clemson University	University of Delaware	Brown University
Colorado State University	University of Florida	California Institute of Technology
Florida State University	University of Georgia	Carnegie Mellon University
Georgia Institute of Technology	University of Hawaii, Manoa	Case Western Reserve University
Indiana University, Bloomington	University of Illinois, Chicago	Columbia University
Iowa State University	University of Illinois, Urbana-Champaign	Cornell University
Kansas State University	University of Iowa	Dartmouth College
Louisiana State University, Baton Rouge	University of Kansas	Duke University
Michigan State University	University of Kentucky	Emory University
New Mexico State University	University of Maryland, College Park	George Washington University
North Carolina State University	University of Massachusetts, Amherst	Harvard University
Ohio State University	University of Michigan, Ann Arbor	Johns Hopkins University
Oregon State University	University of Minnesota, Twin Cities	Massachusetts Institute of Technology

Pennsylvania State University
Purdue University, West Lafayette, IN
Rutgers, the State University of NJ, New Brunswick
SUNY, Albany
SUNY, Buffalo
SUNY, Stony Brook
Texas A&M University
University of Alabama, Birmingham
University of Arizona
University of California, Berkeley
University of California, Davis
University of California, Irvine
University of California, Los Angeles
University of California, Riverside
University of California, San Diego
University of California, Santa Barbara
University of California, Santa Cruz
University of Cincinnati

University of Missouri, Columbia
University of Nebraska, Lincoln
University of New Mexico
University of North Carolina, Chapel Hill
University of Oklahoma, Norman
University of Pittsburgh
University of South Carolina, Columbia
University of South Florida
University of Tennessee, Knoxville
University of Texas, Austin
University of Utah
University of Virginia
University of Washington, Seattle
University of Wisconsin
Utah State University
Virginia Polytechnic Institute and State University
Washington State University, Pullman
Wayne State University

New York University
Northwestern University
Princeton University
Rensselaer Polytechnic University
Rice University
Stanford University
Syracuse University
Tufts University
Tulane University
University of Chicago
University of Miami
University of Pennsylvania
University of Rochester
University of Southern California
Vanderbilt University
Wake Forest University
Washington University in St. Louis
Yale University

Source: Geiger 2004.

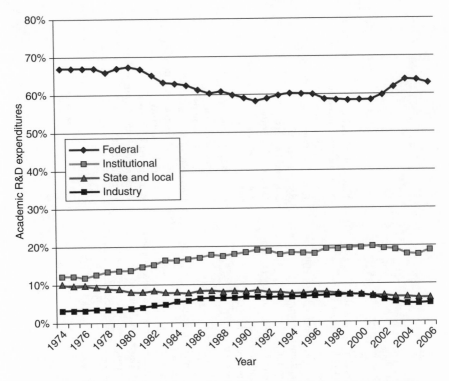

Fig. 10.1. Academic R&D expenditures by funding source, 1974–2006. *Source:* National Science Foundation, Division of Science Resources Statistics, Survey of Research and Development Expenditures at Universities and Colleges, WebCASPAR database, http://webcaspar.nsf.gov. *Notes:* WebCASPAR tables that include the "Highest Degree" classification variable may differ slightly from published tables because of institution branch aggregation that is done in WebCASPAR for ranking purposes.

cultural and mechanical arts" (Geiger 1986). These early federal initiatives in higher education thus reflected the public interest in practical and immediately useful education, in direct contrast to the focus on classical education then prevailing in both the private and emerging public colleges of the early nineteenth century. The first federal initiative explicitly addressing academic research was the Hatch Act of 1887, which established Agricultural Experiment Stations within the existing federally supported land-grant universities. These stations were intended "to aid in acquiring and diffusing among the people of the United States useful and practical information on subjects connected with agriculture, and to promote scientific investigation and experiment respecting the principles and applications of agricultural science."[5] This first federal effort to support academic science continued to reflect the public interest in utility but took a form that subsequently became unusual in federal policy—block grants to universities for research.[6]

Because of the US Constitution, the federal government took no active role in constraining the rapid proliferation of colleges and universities during the nineteenth century. The development of higher education in the United States was therefore shaped by four main factors: the Supreme Court decision in the Dartmouth College case of 1819, which established private colleges and universities as legal entities independent of the state; the generally permissive higher education licensing practices of the states; the provision of higher education by the states themselves; and the resulting competition among the many private and public institutions for students and resources. In this unique context, it was primarily those institutions with the size and resources to support graduate education that developed into research universities (Geiger 1986). Before the twentieth century, therefore, research prospered most at established private institutions, such as Yale and Harvard, or newly founded private universities, such as Johns Hopkins and Chicago, that had access to independent wealth.

The development of US federal policies supporting academic research in fields other than agriculture was significantly influenced by military requirements (Dupree 1957). In response to government's need for scientific advice during the Civil War, a small group of academic scientists quietly steered a bill through Congress in 1863 founding the National Academy of Sciences (NAS). The National Academy was a self-selected group of researchers who had "earned distinction by actual discoveries enlarging the field of human knowledge" (Dupree 1957, 147). The Academy was to be an advisory body to the federal government, providing research-based knowledge to requesting government departments. Members of the National Academy would receive no government compensation, but government agencies would pay the necessary expenses of commissioned research. However, with the end of the war the government's need for such advice declined, and the National Academy became largely an honorific body dependent for its survival on private support.

At the beginning of the twentieth century, federal funding for research had reached more than $11 million, which was larger than the total budgets of the top fifteen research universities of the day (Geiger 1986). Except for agricultural research, though, these federal funds supported applied research on government-defined problems carried out by a rapidly growing number of government bureaus and laboratories. During World War I the scientific challenges posed by submarine detection, poison gas, and explosives reinvigorated the US government's interest in harnessing academic science for war-related research (Geiger 1986). In response the NAS created the National Research Council (NRC) as an independent administrative entity supported by private foundations to coordinate research activities in the national defense. At that time no mechanism existed for directing federal funds for military research into civilian hands, so the NRC identified academic scientists who could be commissioned into the military to conduct relevant research in military labs. At the conclusion of the war in 1918, President Woodrow Wilson permanently recognized the NRC by executive order as the part of the NAS empowered to advise the government on military and related industrial problems.

During the early part of the twentieth century, large and influential private foundations, particularly those established by the fortunes of the Carnegie and Rockefeller families, altered their funding strategies for higher education to concentrate on graduate education and research. Given the large number of institutions of higher education, they adopted a competitive process of awarding grants in which proposals would be peer-reviewed by leading researchers from the established research universities.[7] During the interwar years, the foundations also supported the further development of the NRC by providing graduate fellowship funds for the NRC to allocate competitively for graduate education. As a consequence, the US ARE, as it rose to prominence prior to World War II, was essentially a privately funded system.[8]

This distinctive history influenced the fundamental norms of the US ARE and helped to shape the federal academic research policy that emerged during and after the war. In particular, the following characteristics were all visible before World War II: the US federal emphasis on supporting "best science" through competitive research grants; the tradition of federal (as well as private foundation) research priorities being defined primarily by civilian scientists; the reliance on peer review of proposals for federal research support by an elite of accomplished researchers; the resulting highly stratified nature of the US ARE; and the opposition of many academic scientists to proposals for better coordination of federal science policy, because of their belief that greater federal involvement would constrain scientific choice.

With the advent of World War II, the leaders of the NRC believed that scientific advances had rendered obsolete the tradition of academic researchers responding to research needs defined by the military and that a new system was needed in which the foremost civilian scientists would advise the military on potential scientific applications. Vannever Bush of the NRC's Committee on Policies outlined this idea to President Franklin D. Roosevelt in 1940. Soon thereafter the president appointed the National Defense Research Committee (NDRC), chaired by Bush and composed of academic researchers drawn from the NRC's Committee on Policies as well as liaison officers from relevant federal and military agencies. The research supported by the NDRC was thus coordinated by a group of civilian scientists who worked part-time, retaining their academic positions throughout the war. Because of the need to coordinate war-related medical research as well as research development and procurement activities, the NDRC was subsumed under the wartime Office of Scientific Research and Development (OSRD), also under the direction of Vannever Bush.[9]

The activities of the NDRC and the OSRD during World War II set the institutional framework for future federal science policy. These agencies established the tradition in which federal allocations in research were decided by an elite group of self-selected academic scientists. They also set the precedent of awarding the majority of federal research funds through contracts to the nation's best universities.[10] In contrast to the practice during World War I, a conscious effort was made by the civilian scientists governing these agencies to permit researchers to conduct war-related research at their home universities. Consequently the NDRC established new federal laboratories, such as the Radiation Lab at MIT, to be managed under contract by research universities. In the decentralized structure Bush and his colleagues designed for the NDRC and the OSRD, the influence, scientific choice, and autonomy of academic researchers was maximized.

The widely acknowledged contribution that academic scientists made to the US war effort, symbolized by the Manhattan Project, altered public as well as academic perspectives regarding federal involvement in academic research. But the participation of academic scientists in military research during World War I and the much more extensive effort in applied research during World War II had been purchased at some cost to the system of basic research in the universities. In the postwar world there was clearly a need to provide stable, long-term federal funding for basic research. In 1944, as the war effort wound down and the OSRD was to be closed, Vannever Bush arranged for President Roosevelt to request a report from OSRD outlining postwar science policy. Bush's response, *Science, the Endless Frontier* (1945), argued that continued federal support for basic research was essential for a strong economy and called for the creation of a national research foundation modeled on OSRD. The insulation of academic science from the pressures of national politics was to be ensured through a presidentially appointed board of civilian scientists, which would choose the foundation director. The foundation would be the primary means of supporting basic research in the sciences, including medical- and defense-related research.

This proposal to centralize funding for all federally supported basic research proved controversial, and over the next five years, while the Congress debated Bush's proposed foundation, it authorized the continuation of existing federal contracts for research and the development and initiation of new contracts under the Public Health Service (the National Institutes for Health); the army, the navy, and the air force; and the newly created Atomic Energy Commission. As a consequence, when a much diminished national research foundation, the National Science Foundation (NSF), was finally formed in 1950, it filled the remaining small space in an already established federal research matrix, one that provided support for basic scientific research and graduate education in the university sector. American postwar federal research policy therefore adopted a pluralistic, uncoordinated framework, in which mission-oriented agencies were the primary supporters of university research.

The birth of the NSF did not herald the expected new age of federal support for disinterested basic research. In 1953 over 87% of federal funds for academic R&D, which included the federal contract research centers operated by universities, were related to the military needs of the Atomic Energy Commission and the Pentagon (Geiger 1993). During this same period, presidential budget requests for the fledgling NSF were continually underfunded by the Congress. The Russian launch of Sputnik in 1957, however, presented a challenge to the United States that resulted in a radical alteration of federal science policy. A first response was the creation of the National Aeronautics and Space Administration (NASA) in 1958, which added another large mission-oriented agency supporting academic research in the basic sciences to the existing pluralistic federal funding system. More significantly, the Russian space exploits fostered a national insecurity about US educational achievement that led to the adoption of the unprecedented National Defense Education Act (1958). This legislation provided substantial new federal monies for graduate fellowships as well as for research in specialized academic fields, such as area studies and languages, that were deemed essential to the national defense (Geiger 1993). The NDEA legislation, which also subsequently provided support for higher education facilities, was therefore particularly helpful to the research universities.

Over the next decade, the cold war–inspired federal expenditures for the NDEA legislation and related increases in support of basic research by the NSF and the mission-oriented agencies transformed federal policy on academic research.[11] In 1953, prior to Sputnik, academic R&D represented .07% of US GDP; universities conducted 25% of national basic research; and 43% of higher education research was supported by the federal government. By 1968 academic R&D had more than tripled to .25% of US GDP; universities conducted 50% of national basic research; and 77% of higher education research was supported by the federal government, a proportion of government funding for academic research not since matched (Geiger 1993). The 1960s thereby represented the "golden age of academic science" (Geiger 1993) for the US ARE and shifted federal academic research policy from its postwar overwhelming emphasis on applied research to a predominant focus on basic research.

Paralleling this growing emphasis on federal funding for research and graduate education were attempts to provide greater coordination to federal science policy. In 1957 President Dwight D. Eisenhower appointed the first Presidential Science Advisor, a position that has been continued in succeeding administrations. This advisory domain within the executive branch was expanded in 1959 to include the Federal Council for Science and Technology (currently titled the President's Council of Advisors on Science and Technology) and in 1962 the Office of Science and Technology (currently titled the Office of Science and Technology Policy). Both branches of Congress created matching committees on science, and the Library of Congress, which serves as the research arm of the Congress, later added the Science Policy Research Division (currently titled the Resources, Science, and Industry Division).

Despite, or perhaps because of, these and preceding efforts to reorganize federal research policymaking, US science policy—and consequently the US ARE—is generally regarded as poorly coordinated, "composed of a fragmented matrix of science policy institutions, each with limited roles" (Kleinman 1995, 22). In cross-national comparative studies of science and technology policy, the US R&D system and organization is usually placed at the pluralistic, less centralized, market-oriented end of the spectrum (Lederman, Lehming, and Bond 1986). While most OECD countries have more than one government agency responsible for science and technology policy, the US framework is much more fragmented than that of the other leading industrial countries, and this is particularly true with regard to policies affecting the US ARE. The NSF has been assigned some responsibility for coordinating US national science policy, beyond its data-gathering activities, but it has had limited influence on overall federal science policy and has much less capacity for coordination and planning than comparable agencies in, for example, France and Japan (Lederman, Lehming, and Bond 1986). Some have suggested (Kleinman 1995) that this lack of coordination in federal science policy was an outcome of the political battle over Vannever Bush's proposal for a national research foundation and that if this policy proposal had been better managed, US federal science policy would have taken a different, more coordinated form. Appealing as this argument may be, there is ample evidence in comparative studies of US industrial, labor market, income, and social policies that poorly coordinated federal policies are more the norm than the exception (Wilensky and Turner 1987). The unique division of powers between the executive branch and the Congress, the complexities of federal-state relations, and the nonprogrammatic and non-discipline-enforcing character of US political parties create an institutional framework in which coherent national policies have been rare. From this perspective it may be debatable what can be learned from an analysis of US federal policy on the ARE. However, because of the high visibility of its research universities, the United States inevitably serves as an influential model for other national policymakers, and this is currently reflected in the implementation of competitive research funding as well as doctoral education reforms in many other countries. For this reason a systematic overview of current US federal policies on the ARE and their perceived strengths and weaknesses is likely to be of value.

The US Federal Policy Framework

Given its constitutional constraints and national cultural traditions, federal policymaking on the ARE in the United States has followed what Van Vught (1994) has characterized as the "state supervisory model" rather than the "state control model" typical of Europe and much of the rest of the world. Consistent with this general approach, US federal policy has relied primarily on the instruments of financial "incentives" (i.e., contracts and grants) and "market mechanisms," such as the redefinition of intellectual property rights and the provision of information on research doctoral education, with very limited recourse to the instrument of "rules" (i.e., regulation) and almost total avoidance of the instrument of "non-market supply" (i.e., direct government provision) (Weimer and Vining 2005).

The primary policy instrument affecting research has been federal funding of research contracts and grants as reflected in the overall investment in this activity, the divisions between basic and applied research, and the relative emphasis on different subject fields. A distinctive characteristic of US policy design on academic research has been the almost total federal reliance on allocating its support on a competitive basis and reimbursing indirect costs rather than allocating support directly in the form of institutional grants. Similarly, federal influence over doctoral education has been primarily through financial incentives in the form of competitively awarded scholarships and grants, although changes in the flow of foreign-born students after the terrorist attacks on the United States have revealed the potential influence of visa-governing regulations on PhD programs. The federal government has also subsidized the provision of information on the quality of research-doctoral education by the National Academy of Sciences. To better connect academic research to economic development, the federal government provides competitively awarded subsidies and grants for universities and industry. The related innovative change in framework laws governing intellectual property rights in universities in the early 1980s has also received a great deal of international attention.

The following sections outline the nature of these policy instruments and review related research on their impacts and limitations.

In the US higher education system, federal policies on academic research support are extremely influential on the conduct and performance of the ARE. As discussed above, the pluralistic structure of federal science or R&D policy means that US national policy on research support for the ARE is not formally stated but must be interpreted from the collective actions of various government funding entities. US federal policy can be inferred to a certain extent from the overall level of funding for academic research, the distribution of support between basic and applied research, the distribution of support among both social priorities and research fields, and the degree of concentration in academic research funding within the overall ARE. However, an important aspect of US federal policy on academic research, and one that does reflect a conscious national policy choice, is the nature of the academic research allocation process. From a comparative perspective, US academic research policy is distinctive in its emphasis on the competitive allocation of federal research funds, in its reliance on peer review, and in its emphasis on indirect cost recovery. These important aspects of US federal policy will also be discussed.

In March 2000, the European Council set out a ten-year strategy to make the European Union the most competitive and dynamic knowledge-based economy in the world. A key element of this "Lisbon Strategy" was a target for investments in R&D equal to 3% of EU GDP. This target was set to close the gap in R&D investment between the European Union and the United States. While US expenditures on R&D as a percentage of GDP have averaged 2.63% over the past decade, the federal share of R&D had been in continuing decline since reaching a high of 1.92% of GDP in 1964. Federal support for R&D as a percentage of GDP rose slightly after 2000, but to a level still below that of 1994 (fig. 10.2).

Federal support for university R&D as a percentage of GDP averaged 0.20% over the same period, similarly declining during the 1990s before rising to 0.24% in 2006. However, federal monies have a much more critical influence on academic R&D, and therefore on basic research, than on overall R&D. Between the early 1970s and early 1980s, the academic sector's share of basic research declined from slightly more to slightly less than one-half of the national total. In the early 1990s, however, its share of the national total began to increase once again and was an estimated 56.5% in FY2006 (figure 10.3). Federal funds supported 63% of overall academic R&D and 64% of academic basic research in FY2006 (NSB 2008).[12]

Of the $30 billion in federal academic research support in FY2006, 74%, or $22.3

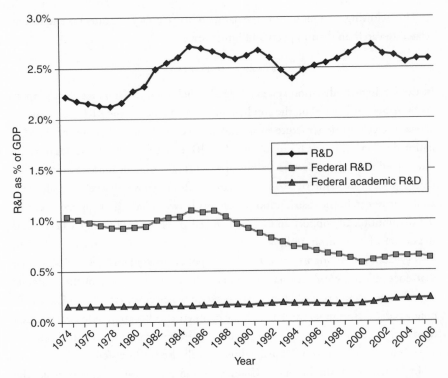

Fig. 10.2. R&D expenditures as percentage of GDP, 1974–2006. *Source:* National Science Foundation, Division of Science Resources Statistics, Survey of Federal Science and Engineering Support to Universities, Colleges, and Nonprofit Institutions, WebCASPAR database, http://webcaspar.nsf.gov.

billion, was defined as basic research (NSB 2008). However, this federal contribution is the combined result of discrete funding decisions by a number of R&D-supporting agencies with specific applied missions. Most of the federal government's R&D is mission-oriented; that is, it is intended to serve the goals and objectives of the agency that provides the funds (e.g., agricultural research for the US Department of Agriculture). As Harvard president James Conant once observed (Geiger 1993), the vast majority of federal academic research funding does not support "disinterested" basic research, but instead supports "programmatic" research that the funding agencies believe would eventually have utility for the sponsor. In contrast, as indicated in table 10.2, many other national governments have traditionally included as part of their support for R&D large block grants (i.e., General University Funds [GUF]) that are used at the discretion of individual higher education institu-

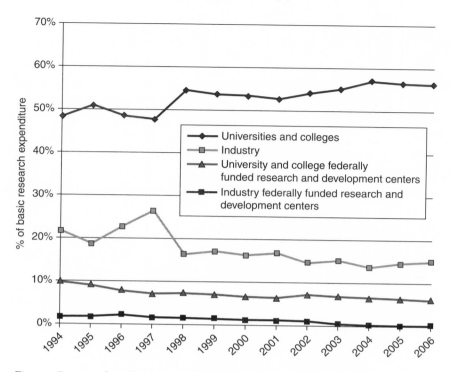

Fig. 10.3. Percent of total basic research expenditures, by selected performing sectors, 1994–2006. *Source:* National Patterns of R&D Resources: 2006 Data Update, National Science Foundation, Arlington, VA, www.nsf.gov/statistics/natlpatterns/.

tions to cover administrative, teaching, and research costs. In each of the European G-7 countries, these GUF account for 50% or more of total government academic R&D to universities and in Canada for roughly 45% of total government academic R&D support (NSB 2008). Consequently, by design US federal academic research policy involves the national government much more directly in defining the social purposes of academic research than do the national policies of governments in other countries.

As table 10.3 shows, the goal of improving human health (National Institutes of Health) motivated an estimated 63.2% of total US federal financing of academic R&D in FY2007. An additional 7.9% was funded for defense (Department of Defense), 4.8% for space (National Aeronautics and Space Administration), 3.2% for energy (Department of Energy), and 2.3% for agriculture (US Department of Agriculture). The National Science Foundation, whose mission is clearly basic research,

TABLE 10.2.

Government R&D support for selected OECD countries, by socioeconomic objectives, 2005 or 2006, in percentages

	Defense	Human health	General university funds	Nonoriented research	Space	Industrial production	Energy	Agriculture	Other
United States (2006)	57.9	21.8	0.0	5.5	7.6	0.3	0.9	2.0	4.0
Japan (2005)	4.0	3.9	33.9	16.3	6.7	7.3	16.8	3.3	7.8
Germany (2005)	5.8	4.3	40.6	16.3	4.9	12.6	2.8	1.8	11.6
United Kingdom (2005)	31.0	14.7	21.7	16.0	2.0	1.7	0.4	3.3	9.1
Canada (2006)	3.6	15.0	32.6	8.0	4.2	11.4	4.8	6.7	13.6

Source: NSB 2008, fig. 4-28, www.nsf.gov/statistics/seind08/start.htm.

Notes: Countries listed in descending order by amount of total government R&D. Data are for years in parentheses. R&D is classified according to its primary government objective, although it may support several complementary goals, e.g., defense R&D with commercial spin-offs is classified as supporting defense, not industrial development.

TABLE 10.3.
US federal obligations for academic R&D, by agency, 1997–2007, in percentages

	Department of Agriculture	Department of Defense	Department of Energy	National Aeronautics and Space Administration	National Institutes of Health	National Science Foundation	All other agencies
1997	3.5	10.7	4.6	5.7	56.2	14.5	4.7
1998	3.1	10.6	4.7	5.9	56.5	14.0	5.2
1999	3.3	9.9	4.2	5.3	58.6	13.9	4.9
2000	3.3	9.5	4.0	5.0	60.0	13.5	4.7
2001	3.3	11.7	3.8	4.8	58.9	12.8	4.8
2002	2.8	10.4	3.6	5.0	61.4	12.7	4.2
2003	2.8	8.5	3.5	4.8	63.3	13.1	4.0
2004	2.7	8.4	3.5	4.8	62.8	12.9	4.8
2005	2.9	8.8	3.3	4.4	63.0	12.4	5.2
2006 (est.)	3.2	9.5	2.9	4.6	62.0	12.3	5.7
2007 (est.)	2.3	7.9	3.2	4.8	63.2	13.3	5.2

Source: NSB 2008, appendix table 5-6, www.nsf.gov/statistics/seind08/start.htm.

provided 13.3% of academic R&D funding, although, as is noted below, these funds are concentrated in the fields of science and engineering and arguably are heavily influenced by the socioeconomic goal of industrial production. In comparative perspective, federal allocations reflect the significant and long-term national emphasis given to both human health and defense in academic research funded by the US government in comparison to the social goals of other countries (as suggestive evidence, see table 10.2).

These national policy differences are also reflected in the relative emphasis given to research in particular academic subjects by the US federal government. Table 10.4 provides the most recent analysis of federally financed R&D expenditures by academic field.[13] Corresponding to the agency expenditures on academic R&D discussed above, federal academic R&D support in 2006 was overwhelmingly concentrated in the life sciences, which represented 69.7% of federally supported expenditures in the sciences and 59.3% of total academic R&D (table 10.4). The more detailed analysis of life sciences expenditures available from the National Science Board (2008) indicates that 20.3% of total federal support for academic R&D was invested in the biological sciences and 33.9% was invested in the medical sciences. Engineering received 13.0% of total academic R&D funds, followed by the physical sciences at 10.3%. In contrast, the social sciences received 2.1% of total funding (psychology, which is separately categorized in NSF reporting, received 1.9%.), while the humanities, including the visual and performing arts, received .2% of total federal funding for academic R&D. Overall academic R&D expenditures in selected OECD countries (table 10.5) suggest that US federal policy places a higher priority on the biomedical sciences and a much lower priority on engineering, the social sciences, and the humanities than do other developed countries.

As previously noted, a widely acknowledged weakness of US federal academic R&D policy is its decentralized character. Because different federal agencies make budgetary decisions on federal academic R&D, and because these decisions are driven by the missions of each agency rather than some overall conception of the needs for academic basic research (or doctoral education), the results may not be socially optimal. A 2001 National Academy of Sciences (NAS) study (Board on Science, Technology, and Economic Policy 2001) reviewed national trends in federal support of graduate education and research and concluded that the existing policy apparatus was creating serious underfunding in the physical sciences, engineering, and mathematics. While there were real, significant increases in federal academic R&D in the first five years of the twenty-first century, the vast majority of those funds were designated for a doubling of the NIH budget. This growth of biomedical research reflected federal priorities, but the decreases in many S&E (science and

TABLE 10.4.
Federally financed R&D expenditures at universities and colleges, by field, FY2006 (current $M)

Field	Expenditure	%
Science	26,221	80.5
Computer sciences	1,015	3.9
Environmental sciences	1,763	6.7
Life sciences[a]	18,268	69.7
Mathematical sciences	373	1.4
Physical sciences	2,705	10.3
Other sciences	2,097	8.0
Psychology	629	1.9
Social sciences[b]	711	2.1
Engineering	4,236	13.0
All non-S&E fields[c]	773	2.3

Source: NSB 2008, table 5-1 and appendix table 5-3, www.nsf.gov/statistics/seind08/start.htm.

Note: Detail may not add to total due to rounding.

[a] Includes biological sciences, $6,240 million, and medical sciences, $10,434 million.

[b] Includes economics, political sciences, sociology.

[c] Includes humanities and visual and performing arts (which received a combined $60 million), as well as professional fields such as business and management, communications, education, journalism, library science, law, and social work. Education received the most federal support at $435 million.

TABLE 10.5.
Share of academic R&D expenditures, by country and S&E field, 2002 or 2003, in percentages

Field	United States (2003)	Japan (2003)	Germany (2002)	Australia (2002)	Netherlands (2002)
Natural sciences and engineering	91.0	67.8	77.0	73.2	72.8
Natural sciences	39.5	12.1	28.5	29.7	17.9
Engineering	14.5	24.7	19.8	11.5	21.0
Medical sciences	30.9	26.7	24.6	25.2	28.3
Agricultural sciences	6.2	4.3	4.0	6.9	5.5
Social sciences and humanities	7.3	32.2	20.2	26.8	24.8
Social sciences	6.2	NA	8.2	20.6	NA
Humanities	0.4	NA	12.1	6.2	NA
Academic R&D	100.0	100.0	100.0	100.0	100.0

Source: NSB 2008, table 4-15, www.nsf.gov/statistics/seind08/start.htm.

Notes: Detail may not add to total because of rounding or because some R&D could not be allocated to specific fields. For United States, $0.7 billion could not be allocated between NS&E and social sciences.

NA = detail not available but included in totals.

engineering) fields observed by the NAS were the result of independent decisions by federal agencies confronting a constrained budget environment, rather than any serious consideration of national S&E priorities. Because such a large proportion of doctoral students are supported by assistantships on federally funded research grants, these agency decisions also affect the future supply of S&E doctorates in critical fields. The NAS called for the president to develop a means for evaluating the overall federal research portfolio in light of national needs and to adjust budget allocations accordingly.

Several actions by the Bush administration appeared to be responsive to these concerns (Intersociety Working Group 2008). The president's Office of Management and Budget introduced the Federal Science and Technology budget. This budget, a collection of selected R&D and non-R&D programs that emphasize basic and applied research, was designed to provide an alternative measure of the federal investment in science and technology and to help track federal S&T investments in the budget process. The president's National Science and Technology Council also implemented a number of interagency R&D initiatives in global change research, information technology, and nanotechnology. Finally, President Bush announced in his 2006 State of the Union address the new American Competitiveness Initiative (ACI) to boost federal investments in physical sciences research. Federal budgets since that announcement have reflected these priorities, with increases for key physical science funding agencies, including the NSF.

However, as an American Association for the Advancement of Science (Intersociety Working Group 2008) analysis of the FY 2009 budget notes, increased funding for physical sciences and engineering research in the three federal agencies participating in the ACI has been offset by cuts in related research funding by other federal agencies. Consequently, the federal R&D portfolio remained unbalanced. In addition, while other countries are making significant investments in research as a means of enhancing their economic competitiveness, total US federal support of basic research in real terms decreased for the fifth year in a row, down 9.1% from 2004.

THE FEDERAL ACADEMIC RESEARCH ALLOCATION PROCESS

US federal academic R&D policy is also distinguished by its reliance on peer review of grant proposals, on competitive allocation of research funds, and on indirect cost recovery. While peer review[14] by researchers to help judge the merit of proposals for research funding plays an important part in US academic research funding, there is more variance in this policy than is often acknowledged (Foltz 2000). For example, the NIH uses exclusively panel reviews in which peers meet in one location to collectively review research proposals. NSF relies to a greater extent on the judgments of program managers using the advice of peers who review proposals by mail, either alone or in conjunction with a panel format. In contrast, many Department of Defense research allocation decisions are made exclusively by program managers. Criticisms have been raised as to the integrity of the peer-review process, particularly following the "big crunch" of the 1990s, when federal R&D funding ceased to match the growth in the academic research system (Foltz 2000).[15] In this more competitive context, it has been alleged that the peer-review system has become particularly vulnerable to politicization, favoritism, and strategic maneuvering, but there is limited empirical evidence to support these allegations (Chubin and Hackett 1990; Foltz 2000). However, it has been suggested that the US peer-review system for assessing grant proposals could be strengthened by focusing more on research output measures (Chubin and Hackett 1990). A more critical question may be the economic efficiency of an allocation system based largely on the review of individual grant proposals in a time of declining probability of obtaining federal research funds. Critics suggest that this new environment entails substantial opportunity costs in researcher time spent preparing and writings proposals rather than conducting research (Foltz 2000).

As already noted, the US federal academic R&D process is unusual in the high proportion of funds allocated competitively. This federal policy appears to provide the opportunity for a more efficient allocation of research funds than national policies that restrict government research funding to certain institutions or sectors of higher education. Historically R&D (and doctoral education) has been highly concentrated within the overall US higher education system. However, within the US ARE the concentration of academic R&D funds has been declining since the mid-1980s (Geiger and Feller 1995). The share of federal academic R&D received by the 99 universities (66 public and 33 private) previously described and listed in table 10.1 was 80% in 1979–80, 79% in 1989–90, 74% in 2000, and 74% of the $30 billion allocated in 2006 (Geiger 2004; NSF 2007). A recent analysis (NSB 2008) similarly notes that the top ten university recipients obtained about 20% of the nation's total academic R&D expenditures in 1986, compared with 17% in 2006. There was less change in the shares of the universities ranked 11–20 and 21–100 during this period. The decline in the top universities' share was offset by an increase in the share of those universities outside the top one hundred. This latter group's share increased from 17% to 20% of total academic R&D funds, signifying a broadening of the base of university performers.[16] The composition of the universities in any particular group is not necessarily the same over time, as mobility occurs within groups. Three of the top ten universities in 1986 were not in the top ten in 2006.

This dispersion of research funds could be an indicator of the effectiveness of a competitive research allocation policy. However, this argument must be interpreted with some caution. The state share of federal R&D funds has been a divisive issue of long standing in US national politics. The distribution of agricultural research funds on a formula basis to the states was the first federal academic R&D policy, and a similar approach was a favored congressional alternative to Vannever Bush's postwar proposal for a national research foundation. In response to increasing congressional concern about the "undue concentration" of R&D funds among the states, the NSF in 1978 initiated the Experimental Program to Stimulate Competitive Research (EPSCoR). The program's goal was to improve the competitive ability of the five states traditionally receiving low percentages of federal R&D support (Feller 2007). The EPSCoR program did not simply set aside funds for the selected states. Each eligible state submitted a proposal designed to develop and use the science and technology resources resident in the state's major research universities to improve the state's capacity to compete for subsequent R&D funding. Proposals were required to include state and industry matching funds and were subject to peer review. Although "experimental," the program continues and has now been extended to five other federal agencies granting R&D funds as well as to twenty-two additional states. Over the past decade the program has experienced a fourfold increase from $79.1 million in 1996 to $353.4 million in 2006 (NSB 2008). Studies of the impact of the NSF EPSCoR program discovered a modest increase in the share of competitively awarded federal R&D for states targeted by the program (Yin and Feller 1999) and a decrease in numbers of publications as well as an increase in citation impact (Payne 2006). Payne (2006) interpreted this latter divergent result as evidence that the design of the EPSCOR program, which included peer review of EPSCoR grants, motivated participating scientists to improve the quality of their research.

In contrast, another important influence on the dispersion of federal academic R&D—"earmarking"—provides less evidence of benefiting the public good and may be contributing to the detected inefficiency of the US ARE. Academic earmarking is the congressional practice of allocating federal R&D funds for research facilities or projects directly to colleges and universities through appropriations bills, thereby bypassing merit-based peer review. The practice began in the 1980s when federal funds for academic research facilities began to decline and has grown rapidly over the past decade. Between 1980 and 2003 the cumulative annual growth rate (CAGR) for academic earmarks was 19.4%, accelerating to 31% since 1996, while the CAGR for federal funding of academic research between 1980 and 2003 was less than 4% (de Figueiredo and Silverman 2007). Academic earmarks set a record of over $2.4 billion in FY2006, leading to a congressional moratorium on most domestic earmarks for that year (Intersociety Working Group 2008). But Congress resumed its earmarking prerogative in FY2008.

Academic earmarks are estimated to represent between 5% and 6% of federal academic R&D allocations (NSB 2004) and, as such, further diminish the actual amount of federal R&D competitively awarded to the ARE. Since these earmarks are allocated to a wide range of colleges and universities, including institutions that are not part of the ARE, they likely account for a measurable amount of the reduction in the academic R&D share of top-ranked universities observed over the past decades (de Figueiredo and Silverman 2007). Some Congress members have defended academic earmarks, arguing that these allocations, somewhat similar to the EPSCoR programs, represent "public values" that correct for biases in the federal R&D peer-review process (Savage 1999).

An analysis of the academic earmarking process, however, reveals that the allocations are influenced less by public values than by the location and influence of particular Congress members on congressional appropriations committees (de Figueiredo and Silverman 2007). Academic earmarks are not, consequently, allocated to the best research universities or the neediest states and therefore do not lessen research concentration so much as intensify it, based upon different variables. For example, the top state recipient of academic earmarks from 1988 to 1996 was Pennsylvania, which was also ranked in the top five recipients of federal R&D grants.[17] Academic earmarks also do not appear to enhance the research capability of universities. Receiving academic earmarks does not guarantee an improvement in an institution's share of competitively awarded grants (Savage 1999), and the citation rates of publications resulting from earmarked funds are statistically and substantially lower than those resulting from peer-reviewed projects (Payne 2006). Academic earmarks have appeared primarily within the budgets of federal agencies, such as the departments of defense and agriculture. To date the budgets of the National Institutes of Health and the National Science Foundation, which are the primary sources of federal academic R&D, have been relatively free of earmarked funds. But as de Figueiredo and Silverman (2007) note, lobbying for earmarks is increasing even among the leading US research universities. Left unchecked, this rent-seeking behavior can spiral upward to a "tipping point," at which the peer-review system will unravel.

A final distinguishing trait of the US academic R&D allocation process is its reliance on indirect cost reimbursement rather than on direct grants to universities for research infrastructure and facilities.[18] The reliance on indirect costs means that federal subsidies for research facilities and infrastructure are to a great extent allocated on a competitive basis and therefore are arguably more effectively concentrated in those universities that have proven their capability to perform high-quality research. Federal academic R&D grants include both the direct costs of research (such as materials and labor) and a percentage of indirect costs (such as facilities maintenance and renewal, heating and cooling, and research administrative staff). Reflecting this breakdown, indirect costs are now termed facilities and administrative (F&A) costs in the United States. Indirect cost rates are negotiated by each university with a lead federal agency and therefore vary by institution. Following a scandal at Stanford University in the early 1990s, the president's Office of Management and Budget placed a cap on the percentage of allowable administrative costs. Overall, F&A costs are estimated to be about 25% of federal academic R&D expenditures (Goldman et al. 2000).

Indirect cost recovery supplements, to some extent, the limited federal allocations for research facilities (table 10.6). The federal government's share of construction funding for research space experienced a decline and reached its smallest proportion (4.7%) since 1986–87 in FY2002–3, before rising to 7.4% in 2004–5 (NSB 2008).[19] Not coincidentally, the growth of academic earmarks, most of which are for facilities, corresponded with the decline in direct federal research facilities grants as well as with the rise in capital costs of conducting scientific research (Ehrenberg, Jakubson, and Rizzo 2003; Feller 2007; Savage 1999).

While indirect cost reimbursement potentially offers a more efficient allocation process for supporting research infrastructure and facilities than direct grants, the US policy has a number of weaknesses (Noll and Rogerson 1998). Negotiated indirect cost rates provide few incentives for universities to become more efficient in their use of research facilities and research infrastructure, as increases in these activities raise overhead reimbursements. Further, the accounting and auditing requirements necessary to justify negotiated rates are quite costly. By contrast, prospective reimbursement of research overhead based upon a benchmark of peer research universities would eliminate most of these costs and provide incentives for research universities to economize on their F&A expenses. Studies of research university overhead costs suggest that general variables, such as university region, research mix, control, and research quality could be used to identify peer groups for setting such benchmark F&A rates (Noll and Rogerson 1998).

DOCTORAL EDUCATION POLICY

Federal policy on doctoral education is exercised primarily through the provision of doctoral and postdoctoral fellowships, traineeships, and research assistantships,

TABLE 10.6.

Federal academic S&E obligations, by activity, 1971–2005 (constant 2000 $M)

	Federal obligations for S&E	R&D (%)	R&D plant (%)	S&E facilities and equipment (%)	Fellowships, traineeships, and training grants (%)	Federal obligations for general support of S&E (%)	Other federal obligations for S&E (%)
1971	8,256.27	5,466.49 (66.21)	105.50 (1.28)	101.23 (1.23)	1,483.54 (17.97)	351.20 (4.25)	748.31 (9.06)
1975	7,639.20	6,115.39 (80.05)	121.94 (1.60)	13.68 (0.18)	547.98 (7.17)	126.20 (1.65)	714.01 (9.35)
1980	9,022.55	7,835.30 (86.84)	71.15 (0.79)	7.11 (0.08)	395.71 (4.39)	172.39 (1.91)	540.89 (5.99)
1985	10,423.47	8,970.53 (86.06)	163.63 (1.57)	7.11 (0.07)	363.47 (3.49)	171.15 (1.64)	747.59 (7.17)
1990	12,887.10	11,097.39 (86.11)	174.34 (1.35)	23.68 (0.18)	484.70 (3.76)	177.13 (1.37)	929.87 (7.22)
1995	15,687.64	13,214.01 (84.23)	370.30 (2.36)	55.96 (0.36)	731.68 (4.66)	286.39 (1.83)	1,029.30 (6.56)
2000	19,857.23	17,269.81 (86.98)	239.62 (1.21)	59.70 (0.30)	782.84 (3.94)	314.18 (1.58)	1,191.08 (5.99)
2005	25,173.10	22,192.32 (88.16)	374.24 (1.49)	35.46 (0.14)	924.84 (3.67)	353.01 (1.40)	1,293.21 (5.14)

Source: NSF Survey of Federal S&E Support to Universities, Colleges, and Nonprofit Institutions, WebCASPAR Integrated Science and Engineering Resource System, http://webcaspar.nsf.gov.

Note: Detail may not add to total because of rounding.

the latter supported by the previously discussed academic R&D grants. US federal policy, in contrast to that of some other countries, exerts little or no direct influence over the conduct of doctoral education, but there are several federal policies that have an indirect effect. As already noted, federal policy supports institutional accreditation, but this process has a negligible impact on doctoral-level education in the United States. In contrast, changes in visa and immigration policies following the September 11, 2001, attacks on the United States have revealed both the relevance and the important influence that these framework laws have on doctoral education. Finally, the National Research Council (NRC) has been conducting publicly available evaluations of US research doctoral education programs for over twenty years, and there is some evidence that these assessments are affecting research university behavior (Dill 2006). Although the NRC is a private, nonprofit institution, it was originally formed to provide science, technology, and health policy advice to the federal government under a congressional charter. In addition, the NRC research doctoral assessments have been regularly subsidized by grants from the NSF and the NIH.

As in other countries, the most significant federal policy on doctoral education is financial support for students. In 2005 federal agencies provided over $1 billion in support of 82,448 full-time graduate students in doctorate-granting universities, almost double the 42,175 students supported twenty years earlier (NSF 1994, 2008). While federal S&E support for fellowships, traineeships, and training grants has almost tripled since 1985, federal graduate student support controlled for inflation still has not regained the levels of the early 1970s, although the US population increased from 200 million to 300 million over the same period (table 10.6). Research assistantships associated with federal academic R&D provided the largest proportion of federal graduate support (80%), with much smaller numbers of students receiving federal fellowships (10.9%), and traineeships (8.8%), proportions that have remained relatively constant over the past twenty years (NSF 1994, 2008).[20] Overall, federal agencies supported 21% of full-time S&E graduate students in doctorate-granting universities, with over 58% of those receiving their support from the NSF and the NIH (NSF 2008).

While the percentage of graduate students supported by federal funds is a relatively small share of total graduate enrollment, the percentage of students receiving federal support is particularly significant in the biological sciences (38.7%), the physical sciences (34.6%), and chemical engineering (34.0%) (NSF 2008). In addition, an NRC analysis of federal funding and doctoral enrollment trends in the 1990s discovered a high correlation between declining federal support in a field and declining graduate enrollment and PhD production (Board on Science, Technology, and Economic Policy 2001). National policy is even more significant in S&E postdoctoral training, where 70% of the 48,653 postdocs in 2005 received federal support (NSB 2008). Postdocs who are federally supported, like federally supported graduate students, receive the largest proportion of their government support from research grants. Over 56% of postdocs are supported by research grants. The influence of the national emphasis on health R&D is very evident in postdoctoral training, where over two-thirds of postdocs are working in the biological, medical, and life sciences.

The substantial increases in federal R&D and support for S&E doctoral education in the last decades of the twentieth century stimulated US PhD production in the natural sciences and engineering, peaking in 1996 with over 20,000 doctorates. But following the leveling-off of real federal S&E support in the 1990s, US natural sciences and engineering PhDs declined to 19,000 in 2002 before increasing between 2003 and 2006 to over 24,000 PhDs, primarily due to growth in the life sciences and engineering (Hoffer et al. 2007; NSB 2008). The proportion of PhDs produced in science and engineering has also increased, from 37.5% in 1976 to 53.4% in 2006, but this growth is also explained by substantial increases in the shares of the life sciences and engineering. While the 45,596 PhDs produced in 2006 represents a new and positive-appearing high for US universities, the overall figure is somewhat misleading. As discussed below, this growth is due almost entirely to foreign residents. The number of US citizens receiving PhDs actually declined from 1996 to 2006, despite the substantial growth in PhD production, and while the number of US citizens receiving PhDs in the life sciences increased over the period, the numbers in both the physical sciences and engineering declined.

As in the case of academic R&D funds, research doctoral education has historically been quite concentrated in the US ARE. The ninety-nine universities previously described (see table 10.1) produced 71% of the 45,596 research doctorates granted in 2006 (Hoffer et al. 2007). However, also parallel to R&D funding, the growth in the number of PhDs over the past thirty years has been accompanied by an increased dispersion of doctoral production. Because the most highly ranked universities tend to limit the growth of their doctoral programs, the vast majority of the increase in PhD production has occurred in smaller universities that receive less federal R&D money and have fewer highly ranked research doctoral programs (Freeman, Jin, and Shen 2004).

As noted, foreign students on temporary visas have accounted for virtually all of the overall growth in the number of US S&E doctoral graduates since 1976 (Hoffer et al. 2007). Temporary residents received 11% of newly awarded S&E doctorates in 1976 and 31% in 2006. Foreign citizens now account for an even larger percentage of individuals in postdoctoral training, with the proportion of temporary residents increasing from 40% in 1985 to 55% in 2005 (NSB 2008). Temporary residents earn a significant proportion of their doctoral degrees in particular fields. In 2006 foreign students on temporary visas earned more than half of doctoral degrees awarded in mathematics and physics and almost two-thirds of the PhDs awarded in computer science, engineering, and economics (Hoffer et al. 2007). Until the 1990s, about 50% of temporary residents receiving an S&E doctoral degree planned to stay in the United States, but this percentage increased during the 1990s, and by 2006, 76% of foreign-born recipients of S&E doctoral degrees reported plans to stay in the United States. As the Committee on Science, Engineering, and Public Policy noted in 2005, US R&D as well as the US ARE have accordingly become highly dependent upon foreign nationals who received their doctoral degrees in the United States. For example, S&E occupations data from the 2000 census indicated that about 38% of doctorate-level employees were foreign-born and nearly half of the doctoral-level staff and 58% of the postdocs and fellows at the NIH were foreign nationals. Furthermore, by 2003, 33% of full-time S&E doctoral faculty in research universities were foreign born, and in the physical sciences and engineering the proportion was 47% (NSB 2008).

As a consequence, the more restrictive visa regulations designed to safeguard the United States following the attacks of September 11, 2001, quickly revealed the significance of national immigration policies for the ARE. An NRC study of the impact of these new regulations (COSEPUP 2005) discovered that following 9/11 there was a measurable increase in the visa refusal rate and consequently a decline in the number of student visas issued. The number of GRE exams taken by foreign nationals also substantially declined. In 2003–4 there were significant drops in international graduate student applications (down 28%), admissions (down 18%), and enrollments (down 6%) at US universities.

If US leadership in S&E was to be maintained, the study's authors argued, changes in US immigration policy were required. The committee recommended extending the duration of visa clearances for students and scholars from all countries and ensuring that foreign students and scholars could attend scientific meetings outside the United States without serious delays in reentering. The committee also called for the creation of new nonimmigrant-visa categories for doctoral-level graduate students and postdoctoral scholars, whether they are coming to the United States for formal educational or training programs or for short-term research collaborations or scientific meetings.

The second significant policy instrument influencing US doctoral education is the Assessment of Research Doctoral Programs carried out by the National Research

Council. The NRC conducted its first assessment in 1982, repeated the assessment in 1993, and has now completed its third assessment.[21] The organization of the NRC rankings is quite different from that of other league tables (Dill and Soo 2005). The rankings are subsidized by federal agencies, including the National Institutes of Health and the National Science Foundation, and the assessments are designed and carried out by some of the leading social scientists in the United States. While the NRC rankings include reputational peer judgments, they also include objective data on measures that research and experience have indicated are important determinants of academic quality in research-doctoral programs. These include inputs, such as the number of faculty members and doctoral students in each program, and crucial process measures, such as student time to degree. Also included are objective output measures like the number of doctoral graduates each year and the number of faculty publications, as well as significant outcomes, such as the number of times faculty publications were cited and the number of distinguished awards received by the faculty.

Following the assessment conducted in 1993, the NRC commissioned a study by leading social scientists of the methodology used in that assessment. The committee's report (Ostriker and Kuh 2003) concluded that valid academic rankings can assist funders and university administrators in program evaluations and are useful to students for graduate program selection. However, rankings would be harmful if they gave a distorted view of the graduate enterprise that encouraged behavior inimical to improving its quality. To guard against this, the committee recommended the following improvements:

- presenting ratings as ranges rather than rankings to diminish the focus of some administrators on hiring decisions designed purely to "move up in the rankings";
- expanding the quantitative measures used in the rankings to include institutional characteristics, doctoral program characteristics, and faculty characteristics that research has shown contribute to a reputation for quality;
- surveying a sample of advanced graduate students in selected fields regarding their assessment of their educational experience, their research productivity, program practices, and their institutional and program environment in order to encourage a greater focus by programs on education in addition to research; and
- determining whether programs collect and publish employment outcomes of graduates for the benefit of prospective students, in order to encourage programs to pay more attention to improving those outcomes.

Of particular interest was the committee's analysis of the reputational indicators used in the NRC rankings. The reputational measures had traditionally included two questions, one on the scholarly quality of the program faculty and a second on the effectiveness of the doctoral program in training scholars. The reputational survey had been limited to members of the discipline being rated. Nonetheless, the committee concluded that the strong correlation between the two reputational measures in past NRC assessments "suggests that raters have little knowledge of educational programs independent from faculty lists" (Ostriker and Kuh 2003, 36). Therefore, although the reputational measure will be continued, it will be limited to scholarly reputation of the program faculty alone. Furthermore, the NRC committee determined that because more highly ranked programs were the most visible, some measure of the rater's familiarity with the program should be included. Finally, the NRC assessment, unlike most available league tables, presents all its data in an unweighted form. Thus, users of the assessment can apply their own preferences to the data and make their own comparative judgments—impossible with weighted measures.

There is some evidence that the impact of the NRC rankings has been positive for US doctoral education. In a series of well-known articles in the higher education literature, Martin Trow (1983, 1999) described the extensive changes made in the departmental structures of the biological sciences at the University of California, Berkeley, over a period of twenty years. Berkeley radically changed the means of appointing and promoting faculty members in the university's biological community and redesigned the nature of facilities for the biological sciences. Trow argued that the impetus for these dramatic changes, which markedly strengthened biological research and education at Berkeley, came in part from the decline in the rankings of several of the biological sciences departments revealed in the NRC assessment of 1982. Similarly, Ron Ehrenberg (2002; Ehrenberg and Hurst 1996) has described how administrators at Cornell were able to use the objective measures of the NRC rankings of 1995 to develop a causal model of quality in research doctoral programs that helped to guide the university's strategic research decisions. In sociology the analysis revealed that the department's low ranking was due to its small size, not to its faculty's productivity; therefore the university decided to continue the department and increase its number of faculty. In biology the assessment led Cornell to devote resources to particular areas in which the university had special strengths and which would likely be important in the coming years.

KNOWLEDGE TRANSFER POLICY

Policymakers are now directing increasing attention to the role of research universities in innovation policy, national strategies designed to enhance economic development and international competitiveness. A recent international assessment (Polt et al. 2001) placed the United States among the countries with high performance on industry-science relationships. Several indicators suggest the recent growth of these relationships in the United States. The share of university research funding provided by private industry tripled between 1970 and 2000, although it has been noted that this support, in fact, restored the strong links that had existed earlier in the twentieth century (Hall 2004). The number of joint research ventures with at least one university partner registered with the federal government doubled in the 1990s, and papers jointly authored by university and industry scientists increased from 6,000 in 1980 to almost 9,000 in 1990 (Polt et al. 2001).

Industry-science interactions have long been a characteristic of the US system. The framework conditions of US higher education, featuring limited federal control and a nationally competitive market composed of private and state-supported universities, served to encourage entrepreneurial behavior by research universities that included organized efforts to transfer basic research to local business and industry. The initial US federal policy on R&D was motivated by a desire to connect academic research to agricultural development. Since World War II the programmatic nature of federal R&D funding—for example, in defense, space, and biomedical sciences—likely further facilitated the linkages between many university researchers and private industry.

In the 1980s, US federal policy on technical innovation became more formalized with the passage of the Patent and Trademark Laws Amendment Act (Bayh-Dole Act of 1980) and the National Co-operative Research Act of 1984, which encouraged the formation of university-industry joint research ventures. A large number of federal programs have been initiated to promote industry-science links, such as the Advanced Technology Program (ATP) of the Department of Commerce and the collaborative research infrastructure programs of the National Science Foundation, which include the Industry/University Cooperative Research Centers, Science and Technology Centers, and Engineering Research Centers.

The Bayh-Dole Act of 1980 provided blanket permission to recipients of federal R&D funds, including universities, to file for patents on the results of federally sponsored research and to exclusively license these patents to others. Under the act, universities are required to share any resulting income with the relevant academic researchers and to use the remaining portion of revenues for scientific research or education. Prior to the act, the US government held the rights to inventions based on federally funded research and granted them on a nonexclusive basis to anyone. Various federal agencies had in the past negotiated Institutional Patenting Agreements (IPAs) with individual universities, but no uniform policy existed.

The passage of Bayh-Dole has often been hailed as a landmark policy that helped initiate university patenting and licensing activity with subsequent significant benefits for the US economy. An empirical analysis of the impacts of the policy, however, suggests a more complex picture (Mowery et al. 2004). The Bayh-Dole Act did encourage universities to be more directly involved in and expand their licensing activities. Prior to 1980 a number of universities, particularly public universities with incentives to be connected to the local economy, were engaged in patenting faculty research. But many of these universities sought to insulate themselves from the commercial activity by relying upon third parties or legally separate foundations to manage their patents.

The federal endorsement of university licensing reflected in the Bayh-Dole Act persuaded both public and private universities to establish their own technology transfer and licensing offices (TTOs) and to more systematically pursue the commercialization of university inventions (Dill 1995). Following the act, the number of research universities with technology transfer offices that manage and protect intellectual property increased eightfold to over two hundred, and the number of university patents registered increased fourfold (Phan and Siegel 2006). A 2000 survey of TTOs indicated that university income from licensing and patenting actives represented 4.7% of their research expenditures (Thursby and Thursby 2003). The average income per active license was $66,465, but only 43% of licenses earned royalties.[22] The average income per university was $8 million, but 79% of the universities earned less than $5 million, and half reported income less than $824,000. These data support the view that at many research universities the cost of running a TTO likely exceeds the revenue from licensing. In addition, a recent analysis of TTOs and research outputs reveals that universities with established offices prior to Bayh-Dole (all of which are high federal R&D performers) are much more efficient producers of articles and patents than the many universities that created offices following the act (Foltz et al. 2005). However, the purpose of the Bayh-Dole Act was not to add to the revenue of universities but to increase commercial innovation, and the evidence that the growth in university patenting and licensing has accomplished this is still unclear.

Indeed, the supposed impact of the Bayh-Dole Act on universities has been brought into question (Mowery et al. 2004). Universities have increased their share of US patenting after the act, but the rate of growth in this activity accelerated in the early 1970s, prior to Bayh-Dole, and remained fairly constant through the 1980s and 1990s.[23] Factors other than the Bayh-Dole Act have created incentives for the growth in university patenting and licensing activity, including the drop in federal funding per full-time researcher in the 1980s and 1990s and the rise of new science-based industries in biotechnology and microelectronics.

A central assumption of the Bayh-Dole legislation was that patenting and exclusionary licensing were necessary conditions for the transfer and commercial development of university inventions. Exclusive patents can provide incentives for commercial firms to invest in costly development of an embryonic invention. There is some evidence that at the time of patenting, most university inventions seem to fit this profile, the majority being filed as "proof of concept" or "prototypes" rather than "ready for practical use." But whether this means the inventions require exclusive licenses to facilitate commercialization is still unclear (Colyvas et al. 2002; Thursby and Thursby 2003). Field research on commercial innovations suggests that the primary channels by which university research influences commercial innovation are publications, conferences, and consulting, with patents and exclusive licensing less significant, and only then in certain fields such as biotechnology (Cohen, Nelson, and Walsh 2002).

One frequently voiced concern is that the increased incentives for patenting and exclusive licensing have diverted universities from basic to applied research. The available research, while limited, provides little evidence to date of a shift toward more applied work. NSF annual surveys of academic R&D expenditures reveal no decline in the basic research share since the late 1980s (NSB 2006). Case studies of major research universities and surveys of faculty members provide little evidence of shifts in research direction or changes in the proportion of basic research (Hall 2004; NSB 2006; Thursby and Thursby 2003).

A more significant concern is that the increasing incentives for patenting and licensing may affect researchers commitments to "open science" by encouraging publication delays, the withholding of data, and increased secrecy in research. There is some evidence, for example, that patenting and restrictive licensing are being applied to theoretical research results and research tools, such as software and databases that are critical inputs to future research (Hall 2004). While Bayh-Dole permits exclusive licenses, it does not require them, and nonexclusive licenses for research tools may better serve the public interest. The possible negative impact of university patenting and licensing on open science has led to calls for revisions in the Bayh-Dole Act and further suggests that a policy focus on property rights may be too narrow a conception for facilitating effective knowledge transfer (Nelson 2004).

In addition to redefining property rights, US policymakers have utilized other instruments to stimulate closer linkages between universities and industry. These have included legislation designed to encourage collaborative research among industry, universities, and federal laboratories (the National Cooperative Research Act of 1984); subsidies for joint research between industries and universities (the US Commerce Department's Advanced Technology Program); and grants for shared use of expertise and laboratory facilities (NSF Engineering Research Centers, Science and Technology Centers, and Industry-University Cooperative Research Centers). In the complex quilt of the US ARE, where the fifty states as well as private industry are also supporting similar initiatives, including research parks, the specific influence and impacts of these federal policies are difficult to identify. Research on university technology transfer activities, however, provides some insight into the validity of the assumptions underlying the design of these policies.

The early literature on university-industry cooperation often focused on a clash of cultures between academic researchers and industry researchers (Cohen et al. 1998). While there had been a long tradition of university-industry links in agriculture and engineering research, the increase in industry-funded research and the development of many new cooperative research centers in the 1980s and 1990s generated debate within universities. Surveys of those involved in cooperative university-industry research centers often noted the shift from the norms of open science to those of industrial innovation—research was more applied, the communication of research findings was more restricted, information was more likely to be omitted from published papers, and publication itself was more likely to be delayed (Cohen et al. 1998).

Nonetheless, universities actively sought to be involved in cooperative research centers, in part because of federal government support and the incentive they provided for industrial funds for academic research. The cooperative centers also evolved into a significant part of the US ARE. By 1990 these centers were receiving 25% of federal academic R&D, and the NSF centers were producing 20% of university patents (Cohen, Nelson, and Walsh 2002; Polt et al. 2001). Research on the cooperative research centers' impact on industry also suggested that they resulted in increases in the number of patents and R&D budgets of the participating industrial laboratories (Adams, Chiang, and Starkey 2001). The positive impacts on industry were observed to be greater for NSF centers and for centers at universities that received larger amounts of federal R&D, both indicators of high research quality. Studies of successful university incubators associated with the ATP program stressed the value of substantial knowledge flows from the university to the firm, noting that start-ups that had university licenses and university faculty on their senior management teams were less likely to fail (Phan and Siegel 2006).[24] Again, an important predictor of successful university-related start-ups was the amount of federal R&D received by the university.

However, studies of the benefits of the cooperative research centers to industry continually downplay the importance of identifying new technologies compared to the enhancement of scientific knowledge and the improvement of human capital (Bozeman 2000; Kim, Lee, and Marschke 2005; Phan and Siegel 2006). Industrial participants in NSF Engineering Centers reported that the single greatest benefit was the ability to recruit students and graduates (Bozeman 2000). Surveys of industrial participants in the cooperative research centers rarely cited direct or tangible outcomes, such as patents or licenses, as an important benefit of cooperative activity (Bozeman 2000; Phan and Siegel 2006). Instead, frequent mention was made of the benefits of knowledge transfer through the channels of open science: publications, consulting, personnel exchanges, and the informal exchanges among bench scientists.[25]

This research suggesting the important role of the channels of open science to university-industry cooperative research centers is consistent with the results of the major US empirical study of the influence of public research on industrial R&D (Cohen, Nelson, and Walsh 2002). University R&D was discovered to be an important influence on industrial R&D, but more influential in the development of projects than in their initiation.[26] Only in certain industries, notably pharmaceuticals, cars and trucks, and aerospace, was university R&D important to new projects. The contribution of university research to industrial R&D was discovered to arise primarily through research findings rather than through prototypes, a finding that is consistent with the previously noted emphasis on the positive benefits of scientific information in cooperative research centers. Research in engineering was revealed to be the most valuable contributor to industrial R&D, followed by materials and computer science research. Medical and health science research were, of course, influential on R&D in the pharmaceuticals industry. With the exception of chemistry, the basic sciences were perceived to be less influential on industrial R&D, a result consistent with the strong focus on engineering and applied sciences in the NSF cooperative research centers. The most important source of information on university research for industrial R&D by far was reported to be publications and reports, followed by meetings and conferences, informal interaction, and consulting. Recent hires were observed to be a less important source of information but of equivalent weight with patents and joint or cooperative centers. Licenses were given a low weight, equivalent to personnel exchanges. Cohen, Nelson, and Walsh (2002) observed variations in these weightings across industries. For example, patents and licenses were a key source of university research in pharmaceuticals, partly because patents are more effective in protecting inventions in drugs than in any other industry, but even in pharmaceuticals the channels of open science were rated higher. Overall, the study suggests that it is the public expressions of university research through publications, meetings, conferences, and informal interactions that convey the content of university research to industry, rather than the private channels of patents and licenses. Start-up firms were more likely to use university research in their R&D, although again this pattern was reported to be strongest in the pharmaceuticals industry.

In sum, an important contribution of the various federal programs designed to stimulate linkages between industry and universities was that the interactions they provided over time helped to change the norms and behavior within both universities and industrial R&D labs in ways that facilitated technology transfer. These programs also developed networks between university researchers and industrial scientists that facilitated informal and formal communication on university research and reduced the transaction costs of university-industry relations (Polt et al. 2001). Studies have found these types of networks to be important to successful industrial innovation in numerous industries.

US Federal Policy: An Assessment of Strengths and Weaknesses

Many OECD indicators suggest the strength of the US ARE. The numbers of publications, citations, patents, and Nobel prizes; the attractiveness of its doctoral and postdoctoral programs to the best students around the globe; and the various rankings of its universities in international league tables provide additional evidence of the quality of American research universities. There is a tendency to associate this strength with national policy, to assume that the quality of the ARE reflects steps taken by the federal government. To some extent this is certainly true. The United States is among the world leaders in the amount and proportion of support by the federal government for R&D, and the allocation of these funds to the ARE primarily through competitive processes is a distinctive and clearly influential national policy that has permitted a significant number of public and private universities to develop a world-class critical mass in research.

In comparative perspective, however, other significant framework conditions of the US ARE include its limited federal control and the nationally competitive market composed of private and state-supported universities, which encourages entrepreneurial research behavior on the part of the research universities (Polt et al.

2001; Trow 2000). As a consequence, US research universities have a high degree of autonomy in how they organize their research activities, doctoral education, and university-industry relationships. Furthermore, faculty members are not civil servants but have individually negotiated employment contracts, which permit them a great deal of mobility and flexibility. The competitive allocation of academic R&D, the competitive admission of doctoral students, and the multiple sources of university funding all create incentives for the autonomous universities to innovate new forms of research management (e.g., centers and institutes), new structures for doctoral education (e.g., graduate schools), and new approaches to university-industry relationships (e.g., technology transfer offices) that have proven effective. Federal policies over the past twenty years have also been influential in encouraging further productive linkages between universities and industry. While the decentralized nature of federal R&D funding has a number of obvious weaknesses, several international observers have suggested that the reliance on overlapping research funding agencies with different missions creates incentives for innovative interdisciplinary research and may have contributed to US leadership in new fields, such as biotechnology (Polt et al. 2001).

Whether the existing US framework conditions will be as effective in the new internationally competitive world of research universities is less clear. In a widely discussed report (COSEPUP 2006) titled *Rising above the Gathering Storm,* the National Academy of Sciences argued that US strengths in sciences and engineering are eroding and thereby threatening the nation's economic development. With regard to national policy on the ARE, the report called for new initiatives that reflect weaknesses already documented above. The National Academy recommended growth in federal support for basic research in the physical sciences, engineering, math, and computer sciences and substantial federal five-year research grants for the best early-career researchers. The report also recommended a major new centralized, competitive fund for facilities and equipment to improve research infrastructure; changes in federal visa policies affecting international students, particularly doctoral students in S&E fields; and adjustments in patent laws designed to ease researchers' use of patented inventions. The report focused heavily on human capital development in the sciences and engineering, including support for K-12 teachers and programs to encourage more undergraduate majors and doctoral students in these fields.

Despite the continued real growth in federal support for academic R&D in the United States, the productivity of American research universities, as measured by research publications and citations, has not been keeping pace (NSB 2008). The US share of world S&E articles dropped from 38.1% in 1988 to 30.3% in 2003, and the American share of world academic literature cited in S&E articles dropped from 51.7% in 1992 to 42.4% in 2003 (NSB 2006). While one would expect the US share to decline as other nations expand their pool of S&E researchers, there is also evidence that the number of US S&E articles plateaued during this period and in several science and engineering fields actually declined (Freeman 2005; NSB 2008). The costs of producing these academic publications also suggest waning productivity. In 1995 the United States produced 6.75 academic articles per $1 million of academic R&D; by 2000 this ratio had declined to 4.87, and by 2005 to 3.73 (NSB 2008).

Recent econometric studies confirm this growing inefficiency in the US ARE and suggest several reasons for it (Adams and Clemmons 2006; Foltz et al. 2005). Despite the significant increases in earmarking previously noted, the researchers suggest that federal academic R&D allocations are still associated with greater research productivity. However, the research share of total R&D expenditures of the less productive research universities is growing because of their ability to attract nonfederal support for their research and doctoral activities. These nonfederal sources include state and local funds, industrial support, and the institutions' own support of R&D, which has grown rapidly over the past thirty years (fig. 10.1). An econometric study of research universities that included both research publications and patents as outputs also discovered falling productivity (Foltz et al. 2005). While increases in federal R&D funding again positively affected research output, increases in industrial support for research positively influenced output only when federal R&D support remained dominant.

A recent qualitative national study of US universities (Brewer, Gates, and Goldman 2002) provides further insights into the possible sources of this inefficiency. The researchers detected evidence of an increasingly costly "arms race" for prestige among US universities. In the United States, research is a revenue market because of the competitive allocation of federal research funds and the growing funding of university research by business and industry. But the amount of external research funding received by a university and the number of publications by its faculty have also become important indicators of prestige in various university rankings. Many lower-ranked universities, therefore, seek to increase their prestige by investing in PhD programs, in laboratories, libraries, computer facilities, and research management as well as by attracting more research-oriented faculty. There is also increasing evidence that some universities subsidize their federal research activity through increased investment in grant matching funds and/or by attempting to lower their indirect cost rate (Feller 2000). Since the funds to support these research investments are derived from revenue markets (e.g., public financing, student tuition, private giving, and industrial support) that are not effectively tied to research quality, the inputs to the US ARE continue to rise but are not matched by equivalent outputs. One implication of this analysis is that the public interest might be served better by an even more rigorous link between federal R&D funding and university research quality than now exists. This could include larger amounts of competitively allocated federal funds for research facilities and equipment, as suggested by the NAS (COSEPUP 2006) and/or an increase in indirect costs as a means of better supporting the most productive research universities.

The most serious problem for US federal ARE policy, however, appears to be the human capital problem stressed in the recent NAS report (COSEPUP 2006). The United States, more than any other developed country, has historically been heavily reliant on immigration as a source of its strength in S&E research, and its reliance has become greater as the number of foreign-born doctoral recipients in key S&E fields has rapidly expanded over the past thirty years. As one observer noted, the US national strategy for the development of human capital in critical S&E fields has been to rely on the "idiocy" of other countries. However, as developing countries become more open, invest more in their ARE, and provide challenging economic opportunities for their most able citizens, the US strategy of relying on immigrants for its strength in S&E fields will become less feasible (Freeman 2005).

The NSF is specifically charged with developing basic research and human capital in S&E fields, but the overall federal approach to this task has a number of serious flaws. First, S&E doctoral production is close to a pure public good that must be heavily subsidized by federal policy. While the fifty states should have an obvious economic interest in S&E doctoral education, the most able doctoral students are highly mobile and less likely than other levels of students to stay in the state in which they are trained (Ehrenberg 2005). Therefore, support for doctoral education is often a low state priority compared to support for first-level degrees. US state-supported universities are consequently highly dependent upon tuition, private gifts, industry support, and most particularly federal R&D to maintain the quality of their S&E doctoral programs. Second, federal support for doctoral education and postdoctoral training is provided overwhelmingly through R&D grant funding, with several important consequences. As noted previously, the reliance on research grant support for doctoral education distorts the supply of graduates because the fields supported by federal grants reflect the research needs of mission-oriented agencies rather than the national needs in S&E education. Doctoral production in biomedical fields in the United States is therefore extremely large, while doctoral production in other critical S&E fields may not reflect national priorities.

In addition, while eligibility for federal graduate fellowships and traineeships is limited to US citizens, foreign-born students are eligible for assistantships on federal R&D grants. The emphasis on R&D grants to support US doctoral education therefore has the effect of attracting a very large number of foreign doctoral students to the United States. These foreign students are very able, and the majority currently stay and make significant contributions to US society and to the ARE. But the dependence on foreign immigration suppresses the enrollments of able, native-born students in S&E doctoral programs. An analysis of the US doctoral market confirms the general economic theory that increases in the supply of a particular skill group depress the earnings and employment opportunities of that skill group (Borjas 2006). A 10% increase of foreign-born doctoral students in a particular field or at a particular time reduces the earnings of that cohort of doctoral students by 3% to 4%. Further, US postdocs, who are most prevalent in the basic sciences, earn about 50% less than the wages doctoral recipients would receive in a regular job—the wage equivalent to a comparably aged baccalaureate recipient. This low wage is significantly influenced by the willingness of foreign-born doctorates to accept postdoctoral positions in the United States. The heavy reliance on foreign-born students in US doctoral and postdoctoral training thereby creates a "vicious cycle" in which the low wages for US S&E research assistants and postdocs do little to discourage foreign-born students who have limited economic opportunities in their home countries. However, these low wages have a substantial impact on the career decisions of native-born students, who have many attractive alternatives to S&E doctoral programs.

With the predictable exception of the biological sciences, this impact is clearly visible in the actual career choices of US permanent residents who received S&E baccalaureates. The number of undergraduate S&E graduates planning advanced study in any field declined steadily from 48% in 1984 to 28% in 1998 (Zumeta and Raveling 2002). Among the most able students, as measured by GRE scores, those planning graduate study in S&E fields declined by 8% between 1992 and 2000.[27] The largest decreases occurred in engineering and mathematics. Reflecting the career alternative available to able students in a developed country, students forsaking S&E graduate study chose professional schools in business and non-MD health professions, careers with good income prospects and less need for extended graduate education. Finally, even those citizens continuing on for S&E doctoral education are disadvantaged by current federal policy. The previously noted dispersion of PhD degrees among the US ARE particularly affected the quality of doctoral education obtained by US citizens (Freeman, Jin, and Shen 2004). As discussed, because top-ranked research universities resist expanding the size of their doctoral programs, the majority of the increase in doctoral education between 1973 and 2000 occurred at smaller universities of lower academic quality as measured by NRC research-doctoral rankings and federal R&D grants. During this period foreign students, whose numbers also dramatically expanded over these years, disproportionately enrolled in the high-quality doctoral programs, while native-born students, particularly women, disproportionately enrolled in lower-quality programs.

Federal policy indirectly influenced these results because of its very heavy emphasis on graduate research assistantships as a means of supporting doctoral education rather than doctoral fellowships and traineeships. As previously noted, federal support for scholarships and traineeships in real terms still has not regained the levels of the 1970s, even though the significance of S&E doctoral education to the economy and society has increased substantially over that period and the US population has increased by 50%. In a recent analysis of the NSF Graduate Fellowship Program, the authors discovered that the pool of applicants for the fellowships varies with the relative value of the stipend, the number of S&E baccalaureate graduates, and the number of awards granted (Freeman, Chang, and Chiang 2005). For every 10% increase in the size of the stipend, the number of applicants goes up 8%–10%, and their measured skills increase. Between 1999 and 2005 the NSF increased the value of its graduate fellowship stipends from $15,000 to $30,000. However, the number of awards has not changed over the years. The number of S&E fellowships granted per S&E baccalaureate in 2000 was one-third of those granted in the 1950s and 1970s. Freeman, Chang, and Chiang (2005) concluded that the supply of native-born doctorates in S&E could be substantially increased by providing more lucrative and larger numbers of federally funded doctoral fellowships.

Conclusion

An evaluation of the framework conditions for the US ARE suggests a number of important general points for public policy. First, institutional autonomy, which permits universities to respond to changing social demands, has been an important strength of the US ARE, providing opportunities for universities to develop innovative organizational forms that better meet the needs of the larger society. Second, the competitive allocation of academic R&D support, using overlapping agencies or research councils, encourages greater productivity and possibly generates greater originality in academic research. If anything, the recent US experience provides additional support for the importance of allocating national academic R&D on the basis of research merit. Third, the contribution of university research and knowledge to economic development can be enhanced by national policies encouraging stronger links between universities and industry. The redefinition of property rights to university research may play some role in encouraging technology transfer, but the US experience suggests that as important, if not more important, are policies designed to strengthen traditional channels of scientific communication between university and industry, including publications, conferences, academic consulting, and regional networks. Finally, high-quality S&E doctoral education plays an increasingly important role in developing the ARE. Since research doctoral education can truly be described as an internationally competitive market, government provision of valid and reliable information on the quality of doctoral programs can be a particularly influential instrument for academic improvement. In addition, the US experience suggests that national policies, while offering opportunities for the best international graduate students, also need to emphasize strong incentives for native-born students to pursue doctoral education in critical S&E fields.

NOTES

I wish to express my appreciation to Ms. Jennifer Miller, Ms. Stephanie Schmitt, and Dr. Maarja Beerkens for their research assistance in the development of this chapter.

1. The Carnegie Classification includes institutions as doctoral-granting universities if they awarded at least twenty doctorates in 2003–4; therefore this figure does not include doctoral-granting institutions below this level of activity. Doctoral-level degrees that qualify recipients to enter professional practice (e.g., JD, MD, PharmD, DPT) are also not included in this definition. By Carnegie Classification convention, this count also does not include Special Focus Institutions, such as free-standing medical schools and centers.

2. There are other comparable rankings of the leading US research universities. See, for example, the Carnegie Foundation for the Advancement of Teaching (2007), which identifies 96 universities of very high research activity, the Center for Measuring University Performance (http://mup.asu.edu/index.html), which identifies the top 100 research universities, and the analysis of Graham and Diamond (1997). Geiger's (2004) listing of 99 universities has the advantage of tracking research university R&D performance over several decades, as well as attempting to control for the quality of doctoral education. The institutions named on these various lists overlap significantly. For purposes of the present discussion, the key point is that each of these analyses confirm that US academic R&D activity is "highly concentrated" in the top one-third of those institutions comprising the overall ARE.

3. State support represents direct support for academic R&D, which significantly underestimates state expenditures for academic research in public universities. Within US universities, expenditures for faculty salaries are traditionally listed in an accounting category termed "instruction and departmental research." Getz and Siegfried (1991) note that if half of these salary expenditures were to be assigned to university-supported research, which is consistent with national surveys on faculty time allocations, then they estimate that the top ninety-eight public and private research universities in the United States alone spent an additional $5.9 billion of general university funds on research in 1987–88. This figure represents over 10% of the total federal expenditures on research and development for the same period.

4. The federal government is responsible for the operation of five service academies or

colleges: the US Military Academy ("West Point," founded in 1802), the US Naval Academy ("Annapolis," 1845), the US Coast Guard Academy (1876), the US Merchant Marine Academy (1943), and the US Air Force Academy (1954).

5. Hatch Act. (1887), *U. S. Statutes at Large,* 314, 440.

6. Federal funding of agricultural research stations within US land-grant universities continues to this day under the Department of Agriculture. In contrast to the majority of federal support for academic research, it is not competitively awarded but, since 1935, has been allocated on a formula basis to eligible universities.

7. Geiger (1986) notes that the National Cancer Institute, established as part of the National Institutes of Health in 1937 by Congress, also competitively awards research grants and fellowships to medical schools with the advice of an external advisory committee of academic scientists.

8. It was also, just prior to World War II, a system already dominated by a small group of research universities. The first comprehensive federal study of the national science effort, published in 1940, revealed that some sixteen universities then graduated 58% of the nation's PhDs and were responsible for 50% of research expenditures (Geiger 1986).

9. Kleinman (1995) argues that the National Advisory Committee for Aeronautics (NACA), established by Congress in 1915 to coordinate and support aeronautical research in anticipation of World War I, served as a model for Bush in establishing the NDRC and then the OSRD. NACA was appointed by the president, dominated by part-time civilian scientists, and during the interwar years initiated the practice of federal contracts with universities to conduct military research.

10. By 1941 the NDRC had signed 207 contracts, which were awarded to 41 universities and 22 companies (Kleinman 1995, 62). The magnitude of the federal research effort is suggested by the number of university contracts let by OSRD in 1943–44, which was three times the level of all prewar university research (Geiger 1986, 264). The NDRC adopted NACA guidelines stipulating that no one should profit from this research and that contracts should support direct and indirect costs.

11. Geiger (1993, 174) notes that between 1958 and 1966 the NSF budget alone rose from $40 million to $480 million. However, this growth was dwarfed by federal appropriations to the National Institutes of Health (NIH), which grew from $98 million in 1956 to $1,413 million in 1967. Though a mission-oriented agency, the NIH expended most of its resources on basic research grants and graduate fellowships in the biological sciences, which were allocated to university medical schools and health research centers. By 1960 NIH had surpassed the Department of Defense as the primary supporter of academic research and by 1965 was larger than the next two smaller sources combined (Geiger 1993).

12. As noted in figure 10.3, universities and industrial firms also conduct basic research through Federally Funded R&D Centers (FFRDCs), such as the Lawrence Livermore National Laboratory, research that the universities and firms administer under their respective contracts with the federal government.

13. Table 10.4 is derived from the NSF FY2006 Survey of Research and Development Expenditures at Universities and Colleges. Because not all institutions responded and those that did respond did not fill in every category, these data are incomplete and do not correspond with data from other sources. However, the response rate was over 96%, and this survey

currently provides the most complete information on overall federal funding of academic R&D by field.

14. The NSF and other federal agencies prefer to describe their R&D allocation systems as "merit review with peer evaluation" (Hackett and Chubin 2003).

15. For example, it has been asserted that over the past decade the probability of being funded for a federal research submission has declined from 50% to 10% (Foltz 2000; Hackett and Chubin 2003).

16. The explanation for this shift is subject to debate. As pointed out below, it may not be due to the competitive allocation of federal academic R&D funds.

17. Almost 30% of academic earmarks went to five states and 47% went to ten states (Savage 1999).

18. As a measure of overall federal support for research universities, federal academic R&D is slightly misleading. The federal government also provides support for research facilities, scientific equipment, and doctoral fellowships, which is more clearly reflected in the NSF S&E budget (table 10.6).

19. Note that federal policy also indirectly subsidizes research facilities by making university bonds tax-exempt and university gifts and grants, which may be used for renovation and/or new construction, tax-deductible.

20. Research assistantships are the primary form of federal support in all fields except the medical sciences, in which fellowships and traineeships are dominant (COSEPUP 2005).

21. These assessments follow a long tradition in the United States of multidisciplinary rankings of graduate degree programs based upon reputational surveys (Webster 1992). Raymond Hughes initiated the first rankings in 1925.

22. In keeping with the Bayh-Dole Act, universities share licensing income with faculty inventors. On average, universities return 40% to the faculty inventor and 16% to the relevant department or school, although it is not uncommon for a department to return its share to the inventor's lab or to permit the inventor to direct its share (Thursby and Thursby 2003).

23. The United States leads the world in the number of patents generated by universities (Polt et al. 2001).

24. Thursby and Thursby (2003) note the importance of university inventors being willing to provide time to licensees in order to increase the potential for successful transfer of technology.

25. Adams, Chiang, and Starkey's (2001) frequently cited empirical study of the impacts of cooperative research centers on industrial laboratories also notes that the most important channels of cooperation were joint publications and faculty consulting.

26. The study focused on "public research," that is, research from both universities and federal labs. However, Cohen, Nelson, and Walsh's (2002) analysis of US patents revealed that university research was 5.7 times more likely to be cited than research from federal labs, although the difference was less pronounced in biomedical fields.

27. These data likely understate the decline, since the number of top students electing to take the GRE also decreased over this period, reflecting students' desire to take professional school admissions tests instead (Zumeta and Raveling 2002).

REFERENCES

Adams, J. D., E. P. Chiang, and K. Starkey. (2001) Industry-University Cooperative Research Centers. *Journal of Technology Transfer* 26 (1–2): 73–86.

Adams, J. D. and R. Clemmons. (2006) Growing Allocative Inefficiency of the US Higher Education Sector. NBER Working Paper No. W12683, NBER, Cambridge, MA.

Balzat, M. (2006) *An Economic Analysis of Innovation: Extending the Concept of National Innovation Systems.* Cheltenham, UK: Edward Elgar.

Board on Science, Technology, and Economic Policy. (2001) *Trends in Federal Support of Research and Graduate Education.* Washington, DC: National Academies Press, www.nap .edu/catalog/10162.html#orgs.

Borjas, G. J. (2006) Immigration in High-Skill Labor Markets: The Impact of Foreign Students on the Earnings of Doctorates. NBER Working Paper No. 12085, NBER, Cambridge, MA.

Bozeman, B. (2000) Technology Transfer and Public Policy: A Review of Research and Theory. *Research Policy* 29:627–655.

Brewer, D., S. M. Gates, and C. A. Goldman. (2002) *In Pursuit of Prestige: Strategy and Competition in US Higher Education.* New Brunswick, NJ: Transaction Press.

Bush, V. (1945) *Science: The Endless Frontier.* Washington DC: Government Printing Office.

Carnegie Foundation for the Advancement of Teaching. (2007) The Carnegie Classification of Institutions of Higher Education. www.carnegiefoundation.org/classifications/index .asp.

Chubin D. E. and E. Hackett. (1990) *Peerless Science: Peer Review and U.S. Science Policy.* Albany, NY: SUNY Press.

Cohen, W. M., R. Florida, L. Randazzese, and J. Walsh. (1998) Industry and the Academy: Uneasy Partners in the Cause of Technological Advance. In R. G. Noll (ed.), *Challenge to the Research University,* 171–200. Washington, DC: Brookings Institution Press.

Cohen, W. M., R. R. Nelson, and J. P. Walsh. (2002) Links and Impacts: The Influence of Public Research on Industrial R&D. *Management Science* 48 (1): 1–23.

Colyvas, J., M. Crow, A. Gelijns, R. Mazzoleni, R. R. Nelson, N. Rosenberg, and B. N. Sampat. (2002) How Do University Inventions Get into Practice? *Management Science* 48 (1): 61–72.

Committee on Science, Engineering, and Public Policy (COSEPUP). (2005) *Policy Implications of International Graduate Students and Postdoctoral Scholars in the United States.* National Academies of Science. Washington, DC: National Academies Press, www.nap .edu/catalog/11289.html#orgs.

———. (2006) *Rising above the Gathering Storm: Energizing and Employing America for a Brighter Economic Future.* Washington, DC: National Academies Press.

Crespi, G. and A. Geuna. (2004) *The Productivity of Science.* Sussex, UK: Science and Technology Policy Research, University of Sussex, www.sussex.ac.uk/spru/documents/cres piost2.pdf.

Crow, M. M. and C. Tucker. (2001) The American Research University System as America's *de facto* Technology Policy. *Science and Public Policy* 28 (1): 2–10.

de Figueiredo, J. M. and B. S. Silverman. (2007) How Does Government (Want to) Fund

Science? Politics, Lobbying, and Academic Earmarks. In P. A. Stephan and R. G. Ehrenberg (eds.), *Science and the University,* 36–51. Madison: University of Wisconsin Press.

Dill, D. D. (1995) University-Industry Entrepreneurship: The Organization and Management of American University Technology Transfer Units. *Higher Education* 29 (4): 369–384.

———. (2006) Convergence and Diversity: The Role and Influence of University Rankings. Paper presented at the Consortium of Higher Education Researchers (CHER) Annual Research Conference, University of Kassel, Germany, September 9.

Dill, D. D. and M. Soo. (2005) Academic Quality, League Tables, and Public Policy: A Cross-National Analysis of University Ranking Systems. *Higher Education* 49 (4): 495–533.

Dupree, A. H. (1957) *Science in the Federal Government: A History of Policies and Activities to 1940.* Cambridge, MA: Belknap Press of Harvard University Press.

Ehrenberg, R. G. (2002) *Tuition Rising: Why College Costs so Much.* Cambridge, MA: Harvard University Press.

———. (2005) Graduate Education, Innovation, and Federal Responsibility. *Council of Graduate Schools' Communicator* 38 (6), www.cgsnet.org/portals/o/pdf/Ehrenberg%20 Articles.pdf.

Ehrenberg, R. G. and P. J. Hurst. (1996) A Hedonic Model. *Change* 28 (3): 46–51.

Ehrenberg, R. G., G. Jakubson, and M. J. Rizzo. (2003) Who Bears the Growing Cost of Science at Universities? NBER Working Paper No. W9627, NBER, Cambridge, MA.

Feller, I. (2000) Social Contracts and the Impact of Matching Fund Requirements on American Research Universities. *Educational Evaluation and Policy Analysis* 22 (1): 91–98.

———. (2007) Who Races with Whom; Who Is Likely to Win (Or Survive); Why. In R. L. Geiger, C. Colbeck, R. L. Williams, and C. Anderson (eds.), *Future of Public Research Universities,* 71–90. Rotterdam, Netherlands: SensePublishers.

Foltz, F. A. (2000) The Ups and Downs of Peer Review: Making Funding Choices for Science. *Bulletin of Science, Technology, and Society* 20 (6): 27–440.

Foltz, J. D., B. L. Barham, J. Chavas, and K. Kim. (2005) Efficiency and Technological Change at US Research Universities. Agricultural and Applied Economics Staff Paper Series, No. 486, University of Wisconsin.

Freeman, R. B. (2005) Does Globalization of the Scientific/Engineering Workforce Threaten US Economic Leadership? NBER Working Paper No. 11457, NBER, Cambridge, MA.

Freeman, R. B., T. Chang, and H. Chiang. (2005) Supporting "The Best and Brightest" in Science and Engineering: NSF Graduate Research Fellowships. NBER Working Paper No. 11623, NBER, Cambridge, MA.

Freeman, R. B., E. Jin, and C.Y. Shen. (2004) Where Do New US-Trained Science-Engineering PhDs Come From? NBER Working Paper No. 10554, NBER, Cambridge, MA.

Galama, T. and J. Hosek. (2008) *US Competitiveness in Science and Technology.* Santa Monica, CA: RAND.

Geiger, R. L. (1986) *To Advance Knowledge.* New York: Oxford University Press.

———. (1993) *Research and Relevant Knowledge.* New York: Oxford University Press.

———. (2004) *Knowledge and Money: Research Universities and the Paradox of the Marketplace.* Stanford, CA: Stanford University Press.

Geiger, R. and I. Feller. (1995) The Dispersion of Academic Research in the 1980s. *Journal of Higher Education* 66 (3): 336–360.

Getz, M., and J. J. Siegfried. (1991) Costs and Productivity in American Colleges and Universities. In C. T. Clotfelter, R. G. Ehrenberg, M. Getz, and J. J. Siegfried (eds.), *Economic Challenges in Higher Education*, 261–392. Chicago: University of Chicago Press.

Goldman, C. A., T. Williams, D. Adamson, and K. Rosenblatt. (2000) *Paying for University Research Facilities and Administration*. Santa Monica, CA: RAND.

Graham, H. D. and N. Diamond. (1997) *The Rise of American Research Universities: Elites and Challengers in the Post-War Era*. Baltimore: Johns Hopkins University Press.

Hackett, E. J. and D. E. Chubin. (2003) Peer Review for the 21st Century: Applications to Education Research. Paper presented at a National Research Council Workshop, Washington, DC, February 25.

Hall, B. H. (2004) University-Industry Research Partnerships in the United States. In J. Contzen, D. Gibson, and M. V. Heitor (eds.), *Rethinking Science Systems and Innovation Policies*. West Lafayette, IN: Purdue University Press.

Hoffer, T. B., M. Hess, V. Welch Jr., and K. Williams. (2007) *Doctorate Recipients from United States Universities: Summary Report 2006*. Chicago: National Opinion Research Center.

Intersociety Working Group. (2008) *AAAS Report XXXIII: Research and Development FY 2009*. Washington, DC: American Association for the Advancement of Science (AAAS).

Kim, J., S. J. Lee, and G. Marschke. (2005) The Influence of University Research on Industrial Innovation. NBER Working Paper No. 11447, NBER, Cambridge, MA.

King, D. A. (2004) The Scientific Impact of Nations. *Nature* 430 (6997): 311–316.

Kleinman, D. L. (1995) *Politics on the Endless Frontier: Postwar Research Policy in the United States*. Durham, NC: Duke University Press.

Lederman, L. L., R. Lehming, and J. S. Bond. (1986) Research Policies and Strategies in Six Countries: A Comparative Analysis. *Science and Public Policy* 13 (2): 67–76.

Maastricht Economic and Social Research and Training Centre on Innovation and Technology (UNU-MERIT). (2008) *European Innovation Scoreboard 2007: Comparative Analysis of Innovation Performance*. Maastricht, Netherlands: UNU-MERIT.

McDaniel, O. C. (1996) The Paradigms of Governance in Higher Education Systems. *Higher Education Policy* 9 (2): 137–158.

Mowery, D. C., R. R. Nelson, B. N. Sampat, and A. A. Ziedonis. (2004) *Ivory Tower and Industrial Innovation: University-Industry Technology Transfer before and after the Bayh-Dole Act*. Stanford, CA: Stanford University Press.

National Science Board (NSB). (2004) *Science and Engineering Indicators, 2004*. Arlington, VA: NSB, www.nsf.gov/statistics/seind04/.

———. (2006) *Science and Engineering Indicators, 2006*. Arlington, VA: NSB, www.nsf.gov/statistics/seind06/

———. (2008) *Science and Engineering Indicators, 2008*. Arlington, VA: National Science Board, NSF, www.nsf.gov/statistics/seind08/.

National Science Foundation (NSF). (1994) *Graduate Students and Postdoctorates in Science and Engineering: Fall 1994*. NSF 06-325. Arlington, VA: NSF.

———. (2007) *Academic Research and Development Expenditures: Fiscal Year 2006*. NSF 08-300. Arlington, VA: NSF.

———. (2008) *Graduate Students and Postdoctorates in Science and Engineering: Fall 2006*. NSF 08-306. Arlington, VA: NSF, www.nsf.gov/statistics/nsf08306/.

Nelson, R. R. (2004) The Market Economy, and the Scientific Commons. *Research Policy* 33 (3): 455–471.

Noll, R. G. and W. P. Rogerson. (1998) The Economics of University Indirect Cost Reimbursement in Federal Research Grants. In R. G. Noll (ed.), *Challenges to Research Universities*, 105–146. Washington, DC: Brookings Institution Press.

Organization for Economic Co-operation and Development (OECD). (2007a) *Main Science and Technology Indicators.* Vol. 2007/2. Paris: OECD.

———. (2007b) *OECD Science, Technology, and Industry Scoreboard.* Paris: OECD.

Ostriker, J. P. and C. Kuh. (2003) *Assessing Research-Doctorate Programs: A Methodological Study.* Washington, DC: National Academies Press.

Pavitt, K. (2001) Public Policies to Support Basic Research: What Can the Rest of the World Learn from US Theory and Practice? (And What They Should Not Learn). *Industrial and Corporate Change* 10 (3): 761–779.

Payne, A. A. (2006) Earmarks and EPSCOR: Shaping the Distribution, Quality, and Quantity of University Research. In D. H. Guston and D. Sarewitz (eds.), *Shaping Science and Technology Policy: The Next Generation of Research*, 149–172. Madison: University of Wisconsin Press.

Phan, P. H. and D. S. Siegel. (2006) The Effectiveness of University Technology Transfer: Lessons Learned from Qualitative and Quantitative Research in the U.S. and U.K. Rensselaer Working Papers in Economics 0609, Rensselaer Polytechnic Institute, Department of Economics, www.economics.rpi.edu/workingpapers/rpi0609.pdf.

Polt, W., C. Rammer, H. Gassler, A. Schibany, N. Valentinelli, and D. Schartinger. (2001) *Benchmarking Industry-Science Relations: The Role of Framework Conditions.* Vienna/ Mannheim: European Commission, Enterprise DG.

Savage, J. D. (1999) *Funding Science in America: Congress, Universities, and the Politics of the Academic Pork Barrel.* Cambridge: Cambridge University Press.

Thursby, J. G. and M. C. Thursby. (2003) University Licensing and the Bayh-Dole Act. *Science* 301 (5636): 1052.

Trow, M. A. (1983) Organizing the Biological Sciences at Berkeley. *Change* 15 (8), 44–53.

———. (1999) Biology at Berkeley: A Case Study of Reorganization and Its Costs and Benefits. Research and Occasional Papers Series, CSHE.1.99, Center for Studies in Higher Education, University of California, Berkeley.

———. (2000) From Mass Higher Education to Universal Access: The American Advantage. *Minerva* 37 (4): 303–328.

UNU-MERIT. *See* Maastricht Economic and Social Research and Training Centre on Innovation and Technology.

Van Vught, F. A. (1994) Policy Models and Policy Instruments in Higher Education: The Effects of Government Policy-Making on the Innovative Behavior of Higher Education Institutions. In J. Smart (ed.), *Higher Education: Handbook of Theory and Research*, 10:88–126. New York: Agathon Press.

Webster, D. S. (1992) Reputational Rankings of Colleges, Universities, and Individual Disciplines and Fields of Study: From Their Beginnings to the Present. In J. C. Smart (ed.), *Higher Education: Handbook of Theory and Research*, 8:234–304. New York: Agathon Press.

Weimer, D. L. and A. R. Vining. (2005) *Policy Analysis: Concepts and Practice.* 4th ed. Upper Saddle River, NJ: Pearson Prentice Hall.

Wilensky, H. L. and L. Turner. (1987) *Democratic Corporatism and Policy Linkages.* Berkeley: University of California.

Yin, R. K. and I. Feller. (1999) *A Report on the Evaluation of the National Science Foundation's Experimental Program to Stimulate Competitive Research.* Prepared for the NSF. Bethesda, MD: COSMOS Corporation.

Zumeta, W. and J. S. Raveling. (2002) Attracting the Best and the Brightest. *Issues in Science and Technology: Perspectives* (winter), www.issues.org/19.2/p_zumeta.htm.

CHAPTER ELEVEN

Pennsylvania

ROGER L. GEIGER

Since World War II, funding for scientific research in the United States has come predominantly from agencies of the federal government. As explained by David Dill (this volume, chapter 10) the national consensus upholding this federal responsibility is reflected in the federal agencies responsible for advancing basic scientific knowledge (the National Science Foundation [NSF]) and furthering knowledge in the specific areas of human health (the National Institutes of Health [NIH]), space (the National Aeronautics and Space Administration), energy, agriculture, and especially defense. This arrangement has been characterized as science federalism (Hallacher 2005, 22). The preponderant role of the federal government in funding science has shaped the respective roles of other stakeholders in American science. This is particularly true for universities, which have consistently performed roughly one-half of the nation's basic research. And it also applies to the states, which have long played a critical role in sustaining the universities and their capacity to perform research.

The states have been decidedly junior partners under science federalism and are chiefly responsible for the institutional base for academic research. In recent years, however, states have become increasingly active, formulating and implementing policies to further science and technology (S&T). This chapter provides an overview and analysis of those efforts in one state, the Commonwealth of Pennsylvania. Pennsylvania's problems typify those of a large industrial state that has long wrestled with the challenge of a declining manufacturing base. Pennsylvania is sixth in the country in population and in gross state product (GSP). It is fifth in public expenditures on higher education and fourth in performance of academic research and boasts four of the nation's top fifty research universities. Since the early 1980s, Pennsylvania has sought policies that could mobilize this strong academic and scientific base to assist in generating new, technology-based industries.

Before detailing the Pennsylvania case, however, I will explore some basic features of S&T policy in the states. The following sections trace the evolution of state S&T policies, the different ways in which they seek to affect the academic research enterprise, and the pitfalls of making science policy at the state level.

Evolution of the State Role in S&T

American states have chiefly supported science through their public universities. The Morrill Land-Grant Act of 1862 determined the subsequent path in two respects. It established an official connection between universities and the teaching of agriculture and engineering; and, by providing an inducement for states to establish universities, it helped create a public system of higher education. These two areas of state interest developed quite irregularly across the various states.

Some states began to support their universities with tax revenues in the late nineteenth century, but very few considered scientific excellence to be an objective. Before World War II, the flagship universities of California, Illinois, Michigan, Minnesota, and Wisconsin had become significant participants in academic research, and their counterparts in Texas, North Carolina, Ohio, and Iowa were following their path (Geiger 2004a; Richardson 2002). With the emergence of science federalism after 1945, states readily accepted the role of supplying the university vessels into which federal research dollars would flow. Still, this was by no means a universal process. States provided the funds to enlarge their flagship universities chiefly as an educational policy to cope with burgeoning student demand. Whether or not these resources were used to erect an effective research university depended on established traditions of research, campus leadership, and the level of financial support from state legislatures. In addition, a number of federal programs of the 1960s were designed to increase university research capacity; but again, capitalizing on these programs required campus leadership and state resources. The state universities best positioned to advance in the postwar environment were the ones mentioned above and some other midwestern and western schools. By the 1970s, Sunbelt states awoke to the benefits of first-class universities. Funds were committed to build or enhance research capacity in Texas, Arizona, North Carolina, Georgia, and Florida. States ultimately played the dominant role in developing research capacity through investments in their major universities.

Since 1990, state legislators in all regions have grown increasingly parsimonious in supporting the educational mission of universities with state appropriations. The resulting shift in the costs of public higher education onto students has been widely remarked. In this context it represents a secular shift away from the historic policy of supporting education and research as joint products. Instead of a stealth companion of state higher education policy, research has become the focus of state policies linking S&T explicitly with economic development.

State technology policies were first prompted by the 1887 Hatch Act, which provided regular federal support for state agricultural experiment stations. The act resulted in ongoing agricultural research in every state, and, once such research was established, agricultural interests prodded state governments to expand this work. Soon, enthusiasm developed for creating similar arrangements for engineering research. This movement foundered at the national level, but a few states went ahead with their own initiatives. In both of these cases, agriculture and engineering, state support for technology development has often been a substitute for research performed by industry. Decentralized industries tend to lack the capacity to conduct research internally, yet technological innovation may significantly improve performance and economic impact. After the precedent was set in agriculture, states found it expedient to support university-based research on mining in Minnesota, textiles in North Carolina, locomotives at Purdue (in Indiana), small-scale engineering at Georgia Tech, transportation at Penn State, and forest products at numerous campuses (Rosenberg and Nelson 1994). Thus, state S&T policies are not an anomaly, but rather a characteristic political reaction to the perceived special needs of local industries. Nevertheless, recent state policies have been oriented toward research-intensive or high-tech industries and have become far more ambitious as well.

The environment for universities and science policy began a dramatic change around 1980—the beginning of the current era for university research (Geiger 2004b). The economic malaise that persisted through the 1970s (stagflation) was attributed to the declining competitiveness of US industries. Since it corresponded with a relative decline in research performed by US industry, a popular remedy called for greater realization of the economic potential of research, particularly university research. The changing zeitgeist set in motion several trains of events. Intellectual property laws were liberalized to encourage increased patenting of research results. Universities embraced the role of technology transfer by seeking collaboration with industry, building research parks, and opening offices for patenting and licensing. Government policies to foster tech transfer were another aspect of this trend. Active federal programs soon followed. At the federal level, the 1980 Bayh-Dole Act was followed by a program of grants for Small Business Innovation Research (SBIR) in 1982; NSF Engineering Research Centers (ERCs) were launched in 1985; technology transfer from government laboratories was encouraged in 1986; Defense Department support for Sematech was pledged in 1987; and the Advanced Technology Program was born in 1988 (Hallacher 2005). In the meantime, a number of states had begun programs of their own.

It is unusual for the states to seize leadership in dealing with a consensus of national need. In this case, however, the Republican administration was strongly opposed to a national "industrial policy" (selectively strengthening targeted industries for global competition). States, in contrast, willingly accepted the idea of facilitating tech transfer for the special needs of local industry. The specific concern was the decline of US manufacturing, and the typical state response consisted of programs that supported collaborative research between local universities and industry partners. Unlike the ERCs, which were open to proposals affecting any industry, state programs tended to target local technologies. They typically established research centers at state universities for academic and industrial investigators to work collaboratively. Industrial partners were generally large corporations, and they were asked to contribute part of the funding.

Only a few states invested in building research capacity in basic science fields with the expectation that fundamental discoveries and highly educated, specialized workers would enhance local industries. Texas's Advanced Technology Program provided such support for areas where Texas universities were particularly strong. More consequentially, Texas provided special financing to bring the Microelectronics and Computer Technology Corporation to Austin in 1983, followed by Sematech in 1988. New York in 1983 created centers for advanced technology at public and private universities, and New Jersey made similar investments in the early 1980s. These steps generally responded to the needs of specific corporations, aiming to bolster their competitiveness through provisions for enriching local sources of generic or precompetitive research (Feller 1992a; Leslie 2001).

A somewhat different tack was taken by Pennsylvania when it created the Ben Franklin programs in 1982 (discussed further below).[1] These programs provided a variety of services to small firms to help them commercialize innovations; the services included matching grants and networking assistance to contract for university research. The first state programs of their kind, the Ben Franklin programs were widely admired and imitated. The appeal of short-term economic stimulation through aid for commercialization carried an obvious political attraction.

By 1990, some 144 university-industry research centers had been established in twenty-two states. However, in the new decade the enthusiasm and funding for proactive state science policies waned (Plosila 2004). State leaders, interested above all in job creation, became disillusioned with universities, which they felt used these programs largely to enhance their own research. The national mood became less supportive as well. Many Republicans remained adamant against "industrial policy" and "corporate welfare." When the GOP gained control of Congress in the 1994 midterm election, there was serious talk of reducing the federal science budget by one-third. American science, it seems, was no longer key to industrial competitiveness.[2] This depreciation of science also registered in the states. The growth of state contributions to academic research stagnated. In the midnineties, state policies toward science still reflected their traditional roles of supporting universities and of being substitutes for industrial research (Coburn 1995, 50; Jankowski 1999).

The economic environment changed in the second half of the 1990s. Steady economic expansion accelerated into the "dot-com bubble" at the end of the century. Technology once again was hailed as the savior of US industries and the best hope for the economy. State interest in promoting technology-based economic development quickened, especially after 2000. The National Governor's Association boosted the cause with published reports having titles such as "Building State Economies by Promoting University-Industry Technology Transfer" and "Using Research and Development to Grow State Economies."[3] The twenty-first century spawned a new generation of state policies for technology-based economic development, or TBED (pronounced "tee-bed" in economic development circles). Many of these policies, particularly in the larger states, emphasized investments in building capacity for basic research (Geiger and Sá 2005). They also tended to emphasize the creation and commercialization of intellectual property. Most important, university research in strategic fields became yoked to the cause of economic development.

State S&T Policies for the Twenty-first Century

State S&T programs—the constituent elements of policies—run the gamut from technical assistance for local shops to support for the most blue-sky basic research.[4] In general, "downstream programs" are intended to provide assistance directly to small companies or entrepreneurs trying to launch them. "Upstream programs" aim to assist the creation of knowledge underpinning innovation through basic or generic research, and they are more often found in the larger states having major research universities. However, the recent trend has been toward omnibus S&T legislation that promises multiyear funding for a set of programs with targets up and down the streambed. Such legislation usually comprises a mixture of new and old (often renamed) commitments. Rather than trying to describe or classify these multidimensional efforts, the discussion that follows identifies the principal themes that characterize state S&T policies of the twenty-first century.

From Endogenous Growth to Cluster Building. The dominant strategy of state policies like the Ben Franklin programs has been endogenous growth—the aim to achieve economic development internally by assisting local industries to develop and expand. The focus here falls on smaller firms, and assistance is typically offered in the form of technology development through access to research or loans for developing new products. Recently, policies of this genre have been given a new twist. States now profess to be eager above all to develop "clusters." Writings by Michael Porter (1998) and others have popularized the notion that the most robust and enduring form of economic advancement occurs through the concentration of firms in the same high-tech industry.

The logic of building clusters encourages two distinct approaches to TBED: the creation and nurture of new firms and the intensification of knowledge assets. Clusters by definition require both numbers and density, but beyond this consideration, start-up and early-stage firms are presumed to be the natural vehicles for innovation and job creation (Braunerhjelm and Feldman 2006; Kirchhoff et al. n.d.). Endogenous growth strategies of the new century have been consciously targeted on such neophyte firms through the creation of "innovation zones," the establishment of business incubators or "entrepreneurial development centers," and arrangements for the provision of venture or seed capital. Universities are expected to play a key role in forming or anchoring clusters, at least those formed around technology-based industries. New York, New Jersey, and Pennsylvania, for example, created their respective innovation zones around universities. Universities are an important source of the innovations underlying the genesis of new firms. They also enrich the labor pool, which is a vital factor for endogenous growth (Romer 1990). Finally, through their research base, they provide a continual source of the new knowledge and expertise needed to sustain competitiveness. This last consideration, however, requires a different kind of S&T policy.

Centers of Excellence. Since 2000 states have become increasingly willing to make major investments in the scientific prowess of universities in areas strategic to science-based technologies. Such policies have been motivated by recognition of the competitive nature of these fields and the crucial role of cutting-edge research as the basis for innovation, intellectual property, and industrial advantage. Individual programs typically support the hiring of "star" professors and the creation of vital research infrastructure. The rubric "centers of excellence" covers a variety of arrangements, many antedating 2000. However, states such as New York and Georgia have intensified these investments since that date. The most audacious commitment was made by California in 2000 in launching the four California Institutes for Science and Innovation (discussed by Zumeta, this volume, chapter 12). The California institutes, like most of these arrangements, required matching funds from industry participants, thus ensuring the relevance of the prospective research (Douglass 2006).

The policy of building scientific quality in selected areas has been a boon to universities, particularly financially strapped state institutions. It promises to pay for itself in part through enhanced federal research grants, regardless of economic impact. This strategy is particularly appropriate for biomedical fields that can tap NIH support.

Tobacco Road. In 1998, states negotiated the Master Tobacco Settlement Agreement with the principal cigarette manufacturers, an arrangement that amounted to a large, judicially imposed tax on smokers.[5] The settlement produced an enormous annual windfall of approximately $6 billion. Many states chose to invest some portion of these funds in biological and medical research. States largely followed the usual approaches. Tobacco funds were used extensively to expand research capacity in the biosciences, a move that was particularly timely given the rapid advances in genomics and biotechnology as well as the 1997 federal commitment to double the budget of the National Institutes of Health. In some cases funds for research were given directly to university institutes. In addition, these investments were typically linked with commercialization of medical therapeutics as part of a TBED strategy. Thus, the overriding aim of these investments was economic development rather than alleviating the evils of smoking or its alleged costs to the states (the pretext for the lawsuit).

Tobacco settlement funds served as an opening wedge for state support for bioscience programs. By 2006, thirty states were providing over $400 million in annual research funds for the life sciences, mostly through tax revenues. Additional programs invested in infrastructure for biomedical research and the hiring of star faculty. These upstream investments were accompanied by just as vigorous expansion of downstream commitments to commercialization funds, seed or venture capital, and business incubators for biotech firms (Battelle 2008). Thus, states invested in both the quantity and the quality of their bioscience capacity with explicit expectations for economic returns.

Upstream and downstream policies are not mutually exclusive but complementary. Both approaches are also consistent with science federalism. However, in one area state policies broke new ground by rushing to fill a lacuna in the federal system of science support.

Stem Cells. In a dubious effort to reach a compromise acceptable to Christian fundamentalists and the scientific community, President George W. Bush limited federally funded research on stem cells to a small number of preexisting cell lines. As scientific interest in stem cell research grew, along with popular expectations for miraculous regenerative therapies, these approved cell lines proved entirely inadequate for research needs. The resulting gap created an opportunity for the states to fill not only a critical void in human health research, but also to gain advantage in the subsequent commercialization of a potentially lucrative medical technology.

California was the first to act in 2004, when voters approved Proposition 71 to sell $3 billion in bonds to support stem cell research for ten years ($300 million/year). This initiative established the California Institute for Regenerative Medicine to coordinate and lead this huge effort (see Zumeta, this volume, chapter 12). Other states feared Californian dominance and reacted with projects of their own. New Jersey, with a pharmaceutical industry as large as California's, established a research grant program in 2005 and became the first state to actually award funding for research projects. By 2008 New Jersey and California had been joined by seven more states (Battelle 2008, 51–52).

By establishing research grant programs, the states have unwittingly usurped one of the critical roles of federal science agencies—judging the relative merits of competing proposals and supporting the "best science." These federal practices were developed in the 1940s and 1950s with considerable difficulty and have never been without controversy (Chubin and Hackett 1990; Cole 1992). But as long as states could rely on federal agencies to recognize best science, their own policies could focus on strengthening universities—assembling the star professors and centers of excellence that could win such support. With the federal abdication of responsibility for stem cell research, state programs had to assume the burden of evaluating research proposals. If they flubbed this task, of course, the research they supported would be worthless.

New Jersey confronted this problem by establishing an independent panel of experts to score applications according, it claimed, to science, not politics. However, these good intentions do not obviate factors like who chooses the experts, what specialties are represented, or what are the rules of engagement (scientific vs. nonscientific criteria). In California, stem cell research was stalled until 2007 over just such issues. Opponents challenged Proposition 71 precisely over the issue of who will decide on the allocation of research grants, arguing that the scientific experts who review grant applications represent the likely recipient institutions.[6] The stem cell situation is the most tangible manifestation to date of the cracks in the system of science federalism. However, there is good reason to question the ability of states to step into this role.

States as Agents of Science Policy

For more than a half century, the National Science Foundation, despite occasional criticism, has been a highly effective agent for disbursing funds for scientific research and for shaping science policy. Three indispensable qualities have made this possible. First, rigorous adherence to peer review has enabled it to support the best science. Second, relative insulation from the political process has allowed the foundation to avoid compromising the best-science standard, even if bending to political pressures when necessary. And third, internal subject-matter expertise has allowed foundation officers to shape and advance the development of scientific fields. Of course, states have not felt the need to build large and costly science bureaucracies precisely because the NSF and other federal agencies accomplish scientific choice for them. But for that same reason, they have difficulty implementing coherent and effective S&T policies that depend on these same qualities—identification of best science, insulation from politics, and internal expertise.

The new state S&T policies have imposed this responsibility on the states. In addition to stem cell research, Arizona has established its own Science Foundation, and Florida created the Florida Technology Research and Scholarship Board to vet proposals for centers of excellence grants. In order to identify best science, states often rely on the judgments of their universities. For special projects, they sometimes appoint independent panels of outside experts.[7] But there are nevertheless persistent pitfalls for S&T decisions.

States have no constituency supporting the advancement of knowledge. Traditionally, they have furthered it indirectly, at best, by supporting their flagship universities. When they have formulated separate S&T policies, motives have become mixed. As Irwin Feller (1992a, 293) noted, "these programs are economic development programs, not science and technology programs per se." This fact introduces a bias at all levels of policy formation. For policymakers, the programs must be sold with promises of economic impact, which, if not necessarily dishonest, are impossible to estimate. At the campus level, the existence of such programs encourages policy entrepreneurs to shape their agendas around state economic goals. The results, in the case of upstream policies, may produce some good science, but perhaps with less than optimal effectiveness and efficiency. As for economic results, these are generally portrayed to match the original promises that motivated the programs. Feller (1992a), in fact, has identified evaluation as the principal weakness of state S&T programs. Usually carried out by pliant consultants rather than independent scientific panels, evaluations typically invoke anecdotes, successful exemplars, and disembodied metrics to justify the original programs (as will be seen for Pennsylvania).

Lacking insulation from politics, state S&T policies are subject to gubernatorial hubris and the legislative pork barrel. Governors generally possess the authority to dominate both the formulation and the execution of S&T policies. They seem to crave credit above all, as state Web sites readily attest. Legislators naturally enjoy recognition too, but it principally comes to them by bringing state money to their home districts. If there is a pattern, it would seem that governors have capitalized on the popularity of TBED to formulate new, comprehensive programs; to induce legislators to authorize appropriations for them, the spending is scattered widely across the state.[8] Universities may have benefited substantially from these TBED initiatives, but that does not necessarily make them good science policy.

What does constitute good science policy is support for peer-reviewed, cutting-edge research, a focus on strategic fields, an emphasis on comparative advantage, and especially concentration of talent (clusters work for science, too). As for economic development, articulation between academic specialties and the absorptive capacity of the ambient economy is a crucial factor. The nature of state policymaking renders the former criteria more difficult to achieve than the latter. Political influence is conducive to what is referred to disdainfully in the foundation world as "scatteration"—both functional and geographic. Scatteration across the state occurs as political interests exert their power. Typically, the homes of flagship campuses are economically vibrant, while less populated locales of smaller institutions may badly need economic stimulation. The aim to stimulate the gamut of activities, up and down the streambed, can produce impressive-sounding programs with few dollars behind them. Governors like to consolidate these programs under catchy titles and project them over many years in order to announce nine- or ten-figure investments, but the available spending for a particular goal may be far less.

Almost every state has some form of commission to oversee its S&T and TBED programs. The commissions are intended to bring professional legitimacy to these operations and typically consist of blue-ribbon appointees. They consequently provide little in the way of scientific expertise. In a number of states, S&T policies are handled by economic development commissions or boards.[9] Where S&T programs are bundled into comprehensive initiatives, those entities tend to have their own advisory structure, as in Ohio's Third Frontier Commission and Board or NYSTAR's Advisory Council. The New Jersey Commission on Science and Technology is one of the few such organizations with "science" in its purview, but its members are mainly CEOs from high-tech firms and representatives of the governor. Ostensibly responsible for development and oversight of S&T policies, in fact it can hardly function as an operational unit. As in most states, the programs are run by the staff of executive agencies, which are not likely to contain research scientists. No matter how named or constituted, these commissions tend to remain under the heavy hand of politics. Florida provides a discouraging example. After its board recommended centers of excellence based on "independent peer review of the science as well as independent review of the economic viability of the proposals," the Florida Legislature in 2008 overrode the recommendations to make its own choice of centers. "The result [was] to award money to several proposals which have only mediocre scientific merit and minimum economic value while leaving several more promising proposals unfunded."[10]

State S&T Policies in Perspective

To summarize, whereas federalism in most government programs has diminished by having other entities assume significant roles, it still dominates science policy. Science federalism has nevertheless been inadequate for state aspirations to pursue an industrial policy, due to lack of consensus at the national level. Lately, and most likely temporarily, it has also failed to adequately support an important area of medical science, stem cell research. For the most part, science federalism plays by a well-known set of rules, exemplified by the NSF. (There is also a side game called "earmarks" with few if any rules and, to the discredit of Congress, rising stakes.) States have thus devised strategies to enhance their competitiveness in accordance with the federal playbook. Increasingly since 2000, they have sought to enhance the competitiveness of technology-based industries by bolstering the state research base. In doing so, states have involved themselves more directly in the formulation of science policy. However, these science policies have been subordinate components to overriding goals for economic development (TBED).

State S&T policies (science plus TBED) can, for the purposes of discussion, be separated into three levels: upstream policies aimed at strengthening research, economic strategies that hypothesize linkages between university research and economic activity, and downstream policies that assist small business in commercializing innovations.

States have, for the most part, adopted conservative approaches to their upstream policies. They have invested mainly in infrastructure, physical and human, as a means to stimulate top-quality research and thus complement science federalism. The distinctive feature of recent policies has been the targeting of these investments on specific science-based technologies. States have also included provisions for collaborative research with industry to ensure the relevance and the transmissibility of these activities. The stampede into stem cell research has explicitly raised the latent problem of passing scientific judgment on research proposals.

States have largely bought into the Schumpeter thesis that innovation is the key to economic growth. Never mind the current booms in hydrocarbons, agriculture, gambling, or fast food—the operating belief is that innovation is the key to enduring growth and a competitive edge in the global economy. In fact, the wellsprings of economic growth have been endlessly debated by economists, and such wisdom as has been shed on this subject is probably too complex and nuanced to be translated into public policy.[11] Yet, states must place bets on strategies that will not only stimulate economic growth but also retain the benefits within state boundaries.

The current infatuation with clusters represents a best guess for an effective approach amid these uncertainties. The examples of existing clusters are certainly alluring, but how to replicate these dynamics is still elusive (see Geiger 2008). Literature can be cited to corroborate the belief that university research gives rise to the formation of tech-based firms and increased employment (e.g., Kirchoff et al. n.d.). But it is also a fact that most innovations arise in industry, and such entrepreneurship is the most crucial factor in cluster growth (Braunerhjelm and Feldman 2006, 12). Nevertheless, most cluster strategies emphasize the founding and nurturing of new firms spawned by university research.

At least two examples can be cited of university-based state policies that spawned significant economic clusters: investments made by Georgia in broadband technologies, largely at Georgia Tech; and New York's investments in the Albany Nanotech complex centered on the Albany College of Nanoscale Science and Engineering (Geiger and Sá 2008). The former was almost purely an academic initiative, aimed at generating new knowledge and highly skilled workers for this field. The latter included old-fashioned subsidies to industry in order to bring IBM and other chip makers into the Nanotech complex. What both policies have in common is, above all, a concentration of resources. In addition, they utilize their respective universities to optimal effect—expanding the knowledge base, creating human capital, and providing a space for the interaction of academic and industry scientists. In this way, knowledge from university research will serve to enrich entrepreneurship in industry, the key factor in cluster dynamics, rather than relying on university spin-offs alone.

Other states have attempted to implement more-or-less concentrated policies in the life sciences. Probably Arizona's commitment to a focused Biosciences Roadmap is the most ambitious. But the temptation to scatteration is strong. Michigan's Life Sciences Corridor aspires to build a cluster over two hundred miles. The large state investments that have flowed into biotechnology, broadly defined, nevertheless represent a fairly safe strategy. Biotechnology, when seen as encompassing humans, animals, and plants, affects enormous industries and decentralized ones as well, Big Pharma notwithstanding. State investments are almost certain to yield some commercial spin-offs and thus to demonstrate the success of these policies.

Enhancing innovation in large corporations, especially in mature industries, is an alternative economic strategy (Geiger and Sá 2008; Lester 2005). Essentially, policies that call for co-investment and research collaboration aim toward this goal. Only fairly large corporations with internal R&D labs can make the kind of investments required by the California Institutes or the initiatives at Albany and Georgia Tech. Here the uncertainty is less in the nature of the investment—presumably the firms recognize value—but in capturing the economic benefit of resulting innovations. The research may be performed locally, but payoffs for multinational corporations can be widely diffused.

Downstream forms of state support for commercialization of innovations represent an economic strategy as well. Since these programs are focused on small tech and start-up companies, they complement the endogenous growth strategies mentioned above. Many of these programs address commercial risk—the perils of bringing innovations successfully to the market (Tassey 2001). These include provisions for seed or venture capital (or loans), business incubators, and assistance with management and marketing. Other provisions are aimed at helping to overcome technical risk, usually through access to technical assistance or university research. These last arrangements properly belong with S&T policies. In some ways, state programs supplement federal provisions for providing research grants for small businesses. Although scarcely new, state programs have become far more proactive in recent years, as will be seen for Pennsylvania. However, they generally contain relatively small outlays of state expenditures.

Finally, as the effectiveness of state S&T policies has come to depend increasingly on the identification and encouragement of top-quality science, the limitations of state policymaking have become more salient. Judgments of quality are best left to universities, for enhancing programs or choosing star professors, or to independent bodies like the Georgia Research Alliance. States have a limited capacity to make good judgments here, and other factors frequently intrude. These factors include the political agendas of governors, the power of legislators to direct state spending to their home districts, the vested interests of economic development organizations, and the ability of local policy entrepreneurs to work the system to achieve their own goals. When viewed solely as technology policy, these same factors may prove less distorting. The political process in this case can direct resources to places where they may be needed. But the efficiency or effectiveness will still be difficult to judge. The recurrent weakness running through state S&T policies, and TBED policies more generally, is lack of ability and lack of will to make critical evaluations. All of these difficulties are evident in the S&T programs of the Commonwealth of Pennsylvania.

The Academic Research Enterprise in Pennsylvania

The Commonwealth of Pennsylvania is an overachiever in academic research (table 11.1). Its universities perform about one-twentieth of the national total, and it trails only California, New York, and Texas in this measure (excepting the federal Applied Physics Lab associated with Johns Hopkins University). It possesses four of the nation's top fifty universities, each a national leader in selected fields and each engaged in healthy expansion of its research enterprise. All of these universities are, relatively, much greater contributors today than they were a generation ago. It is not clear how much credit the commonwealth deserves for this last development. Pennsylvania has led the country in what some call "privatization" and others call "fiscal neglect"—the shifting of the burden of public university finance from state taxpayers to students and parents. At the same time, it has endeavored to harness university research to counter the steady decline of its manufacturing economy.

Pennsylvania's research universities have diverse origins. The University of Pennsylvania, pieced together in 1791 from the wreckage of earlier ventures, is the nation's only private university to bear a flagship moniker. It still receives a state appropriation for its medical school but is independent in all respects. Before 1980, it was one of the weaker members of the Ivy League, plagued by the ills of its location in urban Philadelphia (O'Mara 2005).

The commonwealth's land-grant institution is now called Pennsylvania State University. Originally founded by state agricultural societies, it has never been embraced as *the* state university. Squarely in the center of a state divided politically by the rivalries of the Philadelphia and Pittsburgh regions, Penn State has often been described as "equally remote from all parts of the state." Its development was somewhat fitful; it did not even claim the title of university until the 1950s. It did little to capitalize on the richness of research possibilities in the 1960s, focusing its energies instead on establishing branch campuses in every part of the state and aspiring to be the "Beast of the East"—the best of the eastern schools that played big-time football.

The University of Pittsburgh (Pitt) was a fairly typical urban, private university with a local clientele—a streetcar campus with a well-respected medical school, a strange gothic tower called the Cathedral of Learning, and a football team to chal-

TABLE 11.1.
Science and engineering profile: Pennsylvania

Characteristic	PA	US	Rank
Doctoral scientists, 2003	26,940	566,330	6
Doctoral engineers, 2003	5,030	118,540	6
S&E doctorates awarded, 2005	1,397	27,974	5
Engineering (%)	27	23	n/a
Life sciences (%)	21	26	n/a
Social sciences (%)	15	15	n/a
S&E and health postdoctorates in doctorate-granting institutions, 2005	2,406	48,601	5
S&E and health graduate students in doctorate-granting institutions, 2005	24,085	527,767	7
Population, 2005 (thousands)	12,430	300,322	6
Civilian labor force, 2005 (thousands)	6,292	150,717	6
Personal income per capita, 2005 ($)	34,848	34,495	19
Federal spending			
Total expenditures, 2004 ($M)	94,900	2,136,440	5
R&D obligations, 2004 ($M)	3,282	98,936	7
Total R&D performance, 2004 ($M)	10,813	283,439	10
Industry R&D, 2003 ($M)	7,091	198,244	9
Academic R&D, 2005 ($M)	2,354	45,725	5
Life sciences (%)	60	60	n/a
Engineering (%)	17	15	n/a
Math and computer sciences (%)	8	4	n/a
Number of SBIR awards, 2000–2005	1,224	33,289	9
Utility patents issued to state residents, 2005	2,298	74,630	10
Gross state/national product, 2005 ($ billions)	489	12,492	6

Source: National Science Foundation Database, www.nsf.gov/statistics/.

Notes: S&E = science and engineering; SBIR = small business innovation research.

lenge and embarrass its detested rival, Penn State. Beginning in 1960, however, an ambitious new president attempted a vault into the ranks of research universities. He made some progress but quickly bankrupted the institution in the process (Geiger 2004b). In 1966 the state stepped in to rescue the sinking ship and in the process made it a public university. However, Philadelphia politicians insisted that the state do the same for private Temple University, the city's commuter campus with nearly as dismal financial prospects. Suddenly the commonwealth had three "state-related" universities (not counting historically black Lincoln University), and a modus operandi was soon reached in Harrisburg: neither Pitt, nor Temple, nor Penn State could receive more favorable treatment than the others. These three quite different universities became Siamese triplets, joined by comparable state appropriations. As for Pitt, its subsequent academic development was long cramped by a financial hangover and uninspired leadership.

Carnegie Institute of Technology was the product of its namesake's desire to found an anticollege for practical training. It developed instead into a well-regarded school for undergraduate engineers and after World War II pioneered innovative graduate programs. In 1967 it merged with its next-door neighbor, the Mellon Institute, at the latter's behest and was transformed into a small engineering university with a large research base. Carnegie Mellon made a virtue of its small size, developing a distinctive cross-disciplinary culture, considerably abetted by having polymath Herbert Simon on the faculty and by an early push into computer science (Resnick and Scott 2004).

Since 1980, these four universities have made notable academic progress. Penn State registered the largest gains on the National Research Council ratings (table 11.2). It broadened its strengths in geosciences and engineering and became the largest performer of research in materials science. Penn has evolved into one of the most selective universities in the country and a far wealthier institution as well. It is now the third-largest recipient of NIH research support, and its medical school was ranked third by *U.S. News & World Report.* The University of Pittsburgh has doubled its research share since 1980, almost exclusively on the strength of its medical school (ranked sixteenth by *U.S. News & World Report*). Carnegie Mellon expanded its S&E graduate programs (against the national trend). Among the best in the country in computer science, it is considered to be the leader in robotics (America's Best Graduate Schools 2006).

Pennsylvania's other research performers have varied institutional identities. Temple University has been an urban service university where most funded research has been confined to the medical school. Only quite recently has it begun to register a detectable academic pulse. Lehigh University has been a hybrid engineering school and liberal arts college. It publicly promotes its research role in just a few areas (e.g., materials and optics), but these commitments appear to be contracting even as (perhaps because) the undergraduate college flourishes. Drexel University, on the other hand, is one of the fastest-growing—and most entrepreneurial—universities in the country. Formerly known as a local engineering school, in 2002 Drexel took over two Philadelphia medical schools. It now claims to have the largest private enrollments in both medicine and engineering. Research, too, is emphasized. Expenditures doubled from 2002 (table 11.2) to 2004 to exceed $100 million.

These seven universities enroll about one-quarter of postsecondary students in the commonwealth. Otherwise, Pennsylvania is noted for an abundance (approximately 85) of private or (their preferred term) independent colleges. They help Pennsylvania to attract the largest number of nonstate students and make the state's private enrollments twice the national average. Enrollments in two-year colleges, in

TABLE II.2.

Universities performing research, 1992 and 2002, and NRC rating, 1982 versus 1995

		1992 & 2002 statistics				1982 vs. 1995
	R&D expenditures ($M)	% of US total academic R&D expenditures	% life sciences	% engineering & computer science	Full-time grad students in science, engineering & health	Average NRC rating
University of Pennsylvania 1992	222	1.2	73	9.5	2,299	3.76
2002	522	1.4	81	5.5	2,029	3.89
Pennsylvania State University 1992	278	1.5	32	44	3,201	2.93
2002	493	1.4	33	40	3,744	3.49
University of Pittsburgh 1992	157	0.8	81	5.7	2,508	2.83
2002	400	1.1	86	5.3	2,641	3.04
Carnegie Mellon University 1992	111	0.6	12	68	1,276	3.45
2002	188	0.5	4.2	79	1,924	3.81
Temple University 1992	54	0.3	63	2.8	1,226	2.30
2002	64	0.2	63	5.4	975	2.53
Drexel University 1992	23	0.1	8.2	60	941	n/a
2002	44	0.1	1.6	92	960	n/a
Lehigh University 1992	32	0.2	2.5	75	726	2.40
2002	23	0.06	4.1	78	601	2.68

Source: National Science Foundation Database, www.nsf.gov/statistics/, data adapted by author.

contrast, are about half the national average. In addition to the state-related universities, the State System of Higher Education consists of fourteen former teacher's colleges with barely more students than Penn State's twenty-two campuses.

The vitality of the academic research enterprise in Pennsylvania would seem to owe little to the state. If state policies have contributed at all, they could hardly explain this result. Geographically, Pennsylvania spans two cultures: Philadelphia, belonging to the northeast corridor, and Pittsburgh to the Midwest. However, the state's basic approach to higher education mirrors that of the Northeast in the historic dominance of the private sector. It is telling that none of Pennsylvania's public universities were founded by the state, and the state has accordingly accepted a limited degree of responsibility toward them. Characteristically, when Milton S. Hershey donated a hospital and medical school to Penn State in 1966, the commonwealth explicitly refused to accept any fiscal responsibility; and to this day it has provided almost no funding. However, state programs for higher education tend to be open to public and private institutions alike. In addition, Pennsylvania has one of the most generous programs for student financial aid, much appreciated by its private colleges and universities. All told, 15% of state spending on higher education is directed to the independent sector.

The prosperity of the private universities can be largely attributed to national trends favoring selectivity, high tuition, and voluntary support. Carnegie Mellon doubled its real income per student from 1980 to 2000, and the University of Pennsylvania tripled its figure.[12] Only Drexel has taken a singular approach, using enrollment and program growth for university expansion. All of these universities have been involved in different forms of community outreach and economic development. These latter programs have brought some funding from state programs.

Pennsylvania's state-related universities have also reflected national trends; in fact, they have led the trend of declining state support. This is a case in which the most fundamental state role in higher education, that of sustaining public universities, has been substantially eroded (Heller 2006). Since Pennsylvania treats its state-related universities alike, the case of Penn State can be used to exemplify this development.

Figure 11.1 depicts what is well-known locally as the "Penn State cross"—the remarkable inversion of university funding from state to students. However, this striking pattern represents two distinct developments and thus calls for some interpretation. From 1960 to 1975 Penn State grew from 21,000 students to 61,000, with half of that growth occurring between 1965 and 1970. At the outset, 80% of students were at the main, University Park campus; by 1975 the university had almost reached its present fifty-fifty division between the main and the "commonwealth" campuses.

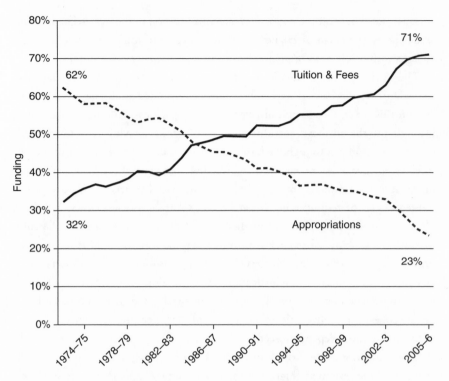

Fig. 11.1. Percentage of tuition and state appropriation revenues, Pennsylvania State University.

During these years, state appropriations grew steadily, but not as rapidly as enrollments. Thus, tuition revenues increased due to the rise in student numbers while the real per-student appropriation declined. One might justify the latter development by the fact that most of the new enrollments were in lower-cost, two-year programs at the branch campuses.

For the next decade (1975–85)—a period of severe inflation—real appropriations to Penn State first shrank and then recovered. Enrollments were stable, so Penn State ended the period roughly where it had started in terms of real and per-student appropriations. The university then began another period of expansion that added twenty thousand students. Unlike the earlier expansion, growth now included a substantial number of high-cost graduate and professional students. State appropriations kept pace during the most rapid growth (1985–90) but then stagnated. Since 1990, Penn State's appropriation in constant dollars has risen slightly and then ebbed. Today, it is about where it was in 1990. On a per-student basis, state support is at its lowest point in the modern era.[13]

Several conclusions can be drawn from this brief sketch. First, since 1970 the Commonwealth of Pennsylvania has never been generous in providing appropriations for Penn State, but it has been parsimonious since 1990, especially so since 2000. Developments within the university, like enrollment growth, have seemingly had no bearing on state appropriations, linked as they were with slow-growing Temple and Pittsburgh. Penn State, for its part, has compensated by generating tuition income, both by enrolling more paying customers and by increasing prices. Moreover, Penn State has not fared badly in spite of fiscal neglect. At its main campus, real income per student rose by more than 70% from 1980 to 2000. As noted, the ratings of its academic departments rose impressively. And in total research expenditures it advanced from seventeenth to tenth in those decades (Geiger 2004a). Certainly, such progress should not be taken as an endorsement of fiscal neglect; rather, it raises the question of how public universities can advance in spite of it. A combination of institutional and state policies seems responsible.

ECONOMIES OF SCALE AND SCOPE

Penn State clearly gained some capacity for internal maneuver by growing to gargantuan proportions. The basic model of the American public university is to educate large numbers of students at a relatively low average cost, and comparative studies suggest that very large size allows universities to achieve differentiation of work and specialization of function (Brint 2007; Geiger 2004a). These last two processes are by no means automatic, but rather require the expertise to manage a multipurpose, multiproduct organization.

RELATIVE AUTONOMY

As a "state-related" university, Penn State has been relatively free from the kinds of governmental interference that have hamstrung other public universities (performance-based funding, mandatory student assessment, etc.). The legislature has at times tied tuition increases to appropriations and threatened far worse; but no administrative body has authority to dictate to the university's Board of Trustees. Of course, autonomy does not guarantee effective governance either. As Penn State has become less reliant on the state, it has increasingly seized the initiative to plan and shape its own future.

TUITION-DRIVEN PUBLIC UNIVERSITY

In particular, Penn State has accepted its fate as a tuition-driven public university and has accordingly raised tuition aggressively when the opportunity presented itself. It now charges state residents the highest tuition in the United States ($13,400 per year for lower-division students and $14,400 for upper-division students: tuition at Temple and Pitt is comparable). The beneficial side of the Penn State cross is that the university no longer depends on state appropriations. Sudden cuts or rescissions can be more than made up with tuition increases. However, this dependence on students inclines the university toward a model of behavior described as the "elite public research university" (Geiger 2007), which resembles the prestige model of private universities. It also has some of the same regressive social implications, favoring students whose parents can afford high costs or students willing to incur substantial debt.

STATE STUDENT FINANCIAL AID

The Pennsylvania Higher Education Assistance Authority operates one of the most generous state programs of student financial aid, with maximum grants of $3,300 in 2005–6. Although the program is strongly supported by the private sector, the largest single recipient of these funds is Penn State (13% of total grants). The $47 million of grants to Penn State students represented 20% of the university's general funds appropriation. Thus, one of the state's basic programs supporting higher education distributes public monies widely to public and private institutions, allowing them to charge higher prices to those who can afford to pay them (price discrimination).

Pennsylvania's S&T Policies: Drifting Downstream

Pennsylvania has consistently lacked an overall vision for state investments in science, technology, and technology-based economic development. In its absence, a variety of programs and initiatives have been created, some from the governor's office and others from policy entrepreneurs in the field. The original programs evolved a decidedly downstream emphasis. Around 2000, the growing need for upstream investments resulted in separate initiatives in nanotechnology and biomedical science. These developments have been accompanied by important ad hoc investments in academic facilities made through the governor's capital budget, which, however, are difficult to characterize as S&T policy. Only since 2005 has the Department of Community and Economic Development sought to rationalize these developments into several coherent policies encompassing downstream economic development, nanotechnology, and the life sciences.

Pennsylvania's initial TBED program originated from a real sense of economic crisis. In the midst of economic recession and high inflation, the state's manufacturing industry, particularly steel, was in rapid decline, and unemployment had mounted in some areas to depression levels. The idea that universities could play a role in rejuvenating the economy became a theme in the national dialogue but could not be easily acted upon. Prevailing policies for economic development sought to attract industry by lowering operating costs through cheap space, subsidies for infrastructure, and tax abatements. Approaches to S&T were hobbled by a truncated federal initiative that induced states to establish science and technology foundations to guide and coordinate their efforts.[14] Instead, Pennsylvania governor Richard Thornburg had staff hold open forums at universities to ask how academic scientists and engineers might help the state's economy. Their suggestions helped to shape the program that was named, at the governor's suggestion, the Ben Franklin Partnership Programs.

The Ben Franklin programs represented a shift in strategy toward nurturing and developing small firms. Their chief feature was a matching-grant program that would allow firms to commission research at a participating university. The programs also sought to encourage complementary education and training programs, although universities were unenthusiastic about these possibilities. A third component was management assistance for new firms and business incubators. The state solicited proposals from state universities to host the program, ultimately covering the entire state through four regional centers based at Penn, Lehigh, Penn State, and Pitt. Thus, Ben Franklin programs were from the outset regionally organized, which made sense for so diverse a state, and they were university-based, since their meager initial resources ($1 million) would hardly cover separate administrative costs.

Compared with 1980s programs in other states, Pennsylvania's approach was decidedly "downstream" (Feller 1992b). It soon moved further in that direction. The centerpiece was the program of matching grants to small firms for university research projects. Pennsylvania was the first state to implement such an arrangement. This feature was widely admired and imitated, and its architect was invited to speak at numerous university-industry forums. However, the new Democratic administration of Governor Robert Casey attacked the program for being too cozy with universities. Universities assumed they were fulfilling their role by performing research relevant to industry and training a technological workforce, but the state measured results by jobs created. When the program was reauthorized in 1987, language requiring university participation was removed, and the regional centers became autonomous, nonprofit entities. Subsequently, university connections largely dissolved, and the program's chief activity became loaning seed capital for fledgling enterprises.

This last description can scarcely do justice to four autonomous centers, since they also became the vehicle for administrating other state S&T initiatives. However, one theme of this history is clear: a program that had been created to harness university research for technology transfer migrated under political pressure into one focused on the commercialization of new technologies, with only incidental links to universities. It thus constitutes a classic example of the evolution of state S&T: one governor's pet program is eviscerated by the successor, and the policy drifts from a longer-term strategy (research) to shorter-term instruments (seed capital). Thus, Pennsylvania was the leader both in creating a program to channel university research for TBED and in abandoning university ties as disillusionment set in.

The Ben Franklin programs were reorganized in 2002, but the basic orientation was not changed (table 11.3). In 2005 the Ben Franklin centers received $50 million of the $80 million the state appropriated for TBED. Some insight into their activities can be gleaned from a commissioned "in-depth evaluation" of the programs (1989–2001) (Nexus Associates 2003, 3). It defined the Ben Franklin activities as "financial investments . . . to serve as a bridge between entrepreneurial assets and third party capital"; "business and technical assistance"; and "regional infrastructure building." The evaluation reassured readers that "the majority of funds have gone directly to entrepreneurs, startups and established companies." No information was disclosed about the fates of the 1,850 companies in which the programs have invested or what proportion of loans were repaid; but the report states that Ben Franklin programs generated 93,105 direct and indirect "job-years," each at a cost of $3,342 ($8,741 for directly created job-years). Thus, *jobs* are the ultimate standard for public and political consumption. For this purpose, each job is assumed to be permanent, so that a job created in 1989 equaled thirteen job-years by 2001 and presumably had a total cost of $43,446 (or $113,633 for direct job-years). These last calculations are mine, and nothing like them appears in the report; but it did claim that Ben Franklin had increased state GSP by $8 billion (over thirteen years), or about double the payroll for those 93,105 job-years (Nexus Associates 2003).[15] This figure represented about 2% of GSP (my calculation). A classic case of disembodied metrics, this "evaluation" typically provides plausible figures for uncritical readers.

In 2004 Governor Ed Rendell recreated something like the original Ben Franklin programs. Named Keystone Innovation Zones, this initiative sought to stimulate economic development and also rekindle the involvement of universities. Each zone had an institution of higher education as a partner, and the funds were intended to facilitate technology transfer solely to new or fledgling firms within a small area. The program consisted of relatively small grants (starting at $250,000 and decreasing to phase out in 2008) for community development agencies and their university partners, plus tax credits for high-tech start-up firms. The program was not confined to research universities; in fact the first award was given to Franklin and Marshall College. Since the program strategy was to have a narrow geographic target, this form

TABLE 11.3.
Programs for technology-based economic development

Program	Year	Aim
Ben Franklin Partnership (BFP)	1982	Use university science and technology to strengthen industry competitiveness and job creation.
Industrial resources centers	1986	Assist small manufacturing firms to adopt advanced process technologies (precursor to the federal Manufacturing Extension Partnership).
Pennsylvania Technology Investment Authority (PTIA)	1999	Support university research and business development in technology areas.
Life sciences greenhouses	2001	Use tobacco settlement funds to support commercialization of discoveries arising from university biotechnology research.
Pittsburgh Digital Greenhouse	2001	Commercialize robotic and digital technologies.
Ben Franklin Technology Development Authority (merger of BFP and PTIA)	2002	Use university science and technology to strengthen industry competitiveness and job creation.
Keystone Innovation Zone	2004	Build alliances between regional economic development organizations and universities.
Jonas Salk Legacy Fund (proposed)	2006– ?	Securitize tobacco settlement fund to create $500 billion fund for biomedical capital.

of scatteration may be justified. However, given the plethora of existing programs for nurturing high-tech firms, one might question the rationale for this impressively named but diminutive program.[16]

Indeed, the number of downstream programs and regional development organizations would suggest that Pennsylvania should be nirvana for entrepreneurs. A 2005 status report on TBED listed no less than thirty-five such entities, all apparently eager to lend support for private companies to do something in Pennsylvania (DCED 2005b).[17] A breakdown of their activities showed concentration on the initial phases of creating innovation and founding firms. However, the evidence in the report showed rather meager accomplishments. Although the state was fourth in academic research (not mentioned), it was ninth in patents. Although sixth in population and GSP (not mentioned), it was ninth in Small Business Innovation Research (SBIR) awards and tenth in Small Business Technology Transfer (STTR) awards. Pennsylvania was seventh in venture capital invested and eighth in total R&D expenditures. The report concludes, laconically, "that Pennsylvania is doing well in building its technology-based economy, but other evidence shows there is room for improvement." But there are no suggestions for improvement. Pennsylvania not only is entirely uncritical in its self-serving "evaluations," but sells itself short by trumpeting its mediocre performance.

A different perspective on Pennsylvania's endogenous growth strategy may be inferred from the *2007 State New Economy Index*. The state was ranked twenty-first overall on the basis of some twenty-six indicators—the same rank it had in 2002 and slightly better than its 1999 rank (Information Technology and Innovation Foundation 2007). Most indicators fell near that same range. On only one indicator did Pennsylvania rank near the bottom—forty-eighth among fifty states—and that was "entrepreneurial activity." Unfortunately, entrepreneurial activity was considered most important for propelling economic clusters (Braunerhjelm and Feldman 2006). Could it be that the state's bureaucratic approach to economic stimulation through state-supported economic development organizations dampens rather than encourages entrepreneurship?

Pennsylvania S&T initiatives are lodged within the sprawling Department of Community and Economic Development (DCED). Perhaps one-tenth of the DCED's $630 million budget is directed toward S&T and commercialization. Some items represent successful past efforts of policy entrepreneurs to obtain state funding, like the supercomputer center in Pittsburgh ($1.5 million); and others are supported through the $52 million transferred to the Ben Franklin authorities (2007–8 budget). The DCED took steps in 2006 to obtain a better understanding of its efforts, commissioning a study by Battelle of the economic opportunities for Ben Franklin investments. It also engaged an economist to suggest measures of economic development, an effort that proved "very difficult" (Bagley 2007; Battelle 2007).

What should be obvious is that the Pennsylvania downstream policies have shown little evidence of invigorating a technology-based economy and at best sustain the status quo. They consist of numerous, decentralized programs that support and animate local economic development entities with small amounts of funding. These programs are able to assist only small businesses because they have only small sums to administer. If the reported figures are to be trusted, Ben Franklin has possibly boosted the state GSP by 2% over the past thirteen years. At that rate, how long will it take for the technology-based economy to emerge? Irwin Feller's (1992a, 300) critical judgment on Ben Franklin, circa 1990, seems apt for Pennsylvania's collection of downstream programs: "The overall program[s may] contribute to economic development, even though large components of its efforts are ineffective or inefficient."

Rationalizing a Policy for Nanotechnology

Investments in nanoscale science and engineering (NSE), along with initiatives in the life sciences, represent the science policy of the commonwealth.[18] The first NSE investment by Pennsylvania state government occurred in 1998, during the first term of Republican governor Tom Ridge. This was one of the earliest commitments to nanotechnology by any state, although it was scarcely planned as such. The apparent serendipity of these developments reflects the scope for policy entrepreneurs in the relatively open space of Pennsylvania's policy environment (Sá, Geiger, and Hallacher 2008). Unlike commercialization programs, nanotechnology policy was shaped by a pattern of opportunistic moves from the bottom up, ultimately producing six initiatives (table 11.4) (Geiger and Hallacher 2005).

  1. In 1997, the Semiconductor Industry Association announced plans to establish a network of university-based research centers. Pennsylvania made a be-

TABLE II.4.
Main state initiatives to support nanotechnology

1997	Allocates $2 million for Penn State's bid for a Semiconductor Industry Association research center on interconnects.
1998	Allows Penn State to use the funding for nanotechnology education and training; allocation becomes annual, supporting NMT Partnership.
2000	Matches NSF funding to establish a nanotechnology center in southeastern Pennsylvania.
2001	Funds the Nanotechnology Institute at $3.5 million/year in Philadelphia. Funds nanotechnology research & commercialization at Lehigh University at $1 million. Starts funding the Pittsburgh Digital Greenhouse at $3.5 million/year.
2002	Scales up investment at Penn State to $3.5 million/year to include research & commercialization. Increases funding for research & commercialization at Lehigh University to $3 million/year. First statewide conference, "Pennsylvania Nanotechnology 2002."
2003	Augments annual funding for Penn State's NMT Partnership ($2.3 million) and research commercialization ($3.5 million). State secretary commissions a study of nanotechnology in Pennsylvania by ANGLE Technology.
2004	Statewide conference, "The Business of Nano 2004."
2005	DCED releases the white paper "Pennsylvania Initiative for Nanotechnology."
2007	Statewide conference, "The Business of Nano 2005."
2008	PIN considered state nanotechnology policy. State spending for PIN = $9.8 million.

lated, unsuccessful effort to compete for this facility. When the proposal failed, Penn State requested and received authorization to use the $2 million allocated by the state to create a nanofabrication-technician-education program called the Pennsylvania Nanofabrication Manufacturing Technology (NMT) Partnership. The NMT Partnership has continued to receive annual state funding of $2–3 million since 1998, which has made it possible to create an associate degree program in nanofabrication for Pennsylvania community colleges (the nation's first such degree program), subsequently expanded to additional training and outreach programs.

2. In Philadelphia, the Ben Franklin Center of Southeastern Pennsylvania responded to interest in nanotechnology among local companies by organizing several industry forums in 1999 and 2000. The outcome of this regional effort was a successful proposal to the NSF, with matching state funds provided from the Ben Franklin Center, to establish a nanotechnology center. The center united a group of universities, economic development organizations, and industry partners in the region. Led by Penn, Drexel, and the Ben Franklin Center, this group submitted a proposal to the DCED in 2001 for establishment of the Nanotechnology Institute (NTI). The NTI proposal requested $10 million over three years to support nanotechnology research and commercialization efforts. The NTI proposal was particularly aimed at exploiting opportunities in the emerging field of nanobiotechnology, reflecting the heavy concentration of biotechnology and pharmaceutical companies in the Philadelphia area and the biomedical research strengths of Penn. The proposal was approved in 2001, and the group received $3.5 million of annual state funding.

3. In 2002, Penn State submitted to the state a proposal modeled on the NTI proposal, requesting $3.5 million, which was approved in 2002. The university sought to continue its workforce development program but also to receive support for research and commercialization at a level comparable to what Penn and Drexel had received. The university was successful in persuading the state of the merit of its request in 2003 and 2004 and received separate grants for the NMT Partnership ($2.3 million) and research and commercialization ($3.5 million).

4. During these same years, the commonwealth supported other initiatives that have funded nanotechnology in part. For example, since 2001 the DCED has supported the Center for Optical Technologies at Lehigh (with Penn State as a partner). The Lehigh center performs nanotechnology research in optics and photonics but is not entirely dedicated to nano as such. State funding for this center has risen from an initial $1 million grant in 2001 to subsequent alloca-

tions of $3 million through 2005. Roughly one-third of this funding is devoted to research at the nanoscale.

5. In addition, the DCED has provided annual funding of approximately $3 million to support participation by Lehigh in the NSF Materials Science and Engineering Center at Carnegie Mellon. Like the Center for Optical Technologies, these state-funded efforts impinge on nanotechnology, although they are not dedicated to nanotechnology per se.[19]

6. Pittsburgh belatedly joined this parade in 2005 by organizing the Pennsylvania NanoMaterials Commercialization Center under the auspices of the Pittsburgh Technology Council. Initial seed funding from the DCED was matched by support from the region's major materials companies, Alcoa, Bayer MaterialScience, PPG, and U.S. Steel. The center seeks to coordinate and influence university research at Pitt, Carnegie Mellon, and Penn State to complement the needs of these Pittsburgh-based companies.

When Governor Rendell took office in January 2003, the new secretary of DCED expressed some skepticism about the economic development potential of nanotechnology, especially in the near term. To gain some perspective (or leverage), he commissioned a study of nanotechnology in Pennsylvania by the ANGLE Technology Group, which had previously been associated with the creation of the NTI. The stated purpose of this effort was to examine the ongoing state investment in nanotechnology and to develop recommendations for these and other proposed investments to be made in the context of a state strategy for economic development.

The ANGLE report (2004) asserted that Pennsylvania had the capability to be the national leader in nanotechnology commercialization. The report also generally endorsed building and sustaining the academic research base, leveraging existing industry to drive commercialization, promoting statewide collaboration, developing industry clusters, supporting workforce development, and establishing national leadership. However, the report said nothing about how these goals might be achieved, nor did it evaluate the existing state initiatives.

During 2004 and 2005, the commonwealth undertook a series of conferences and workshops involving industry and universities intended to develop a response to the ANGLE report. This effort included a statewide conference in spring 2005 entitled "The Business of Nano," which followed up on similar events in 2004 and 2002. While no concrete initiative resulted from these deliberations, the DCED subsequently released a white paper (2005a) entitled "Pennsylvania Initiative for Nanotechnology" (PIN). Despite the workshops, conferences, and discussions that had taken place, the white paper largely echoed the ANGLE report. The PIN proposed a broad and ambitious array of objectives. These included maintaining academic research excellence, promoting commercialization in bioscience and electronics, promoting increased collaboration among universities and industry within the state, developing "value chains" from applied research through production, continuing to develop the nanotechnology workforce, and establishing Pennsylvania as the national leader in nanotechnology commercialization. Again, there was no indication of how these goals would be achieved.

It took until 2007 for the PIN to be officially adopted as state policy. It then consisted of the six initiatives described above (table 11.4). State funding appeared to be approximately what it had been before the efforts to formulate a new policy. The only new feature was the formal trappings of a policy. This consisted of three PIN strategies that could be related to the six initiatives: (1) university-based resources and technology transfer; (2) commercialization of nanotechnology applications; and (3) educational and workforce programs (Armstrong n.d.). In addition, recipients now reported semiannually twenty-two "PIN Initiative Metrics," including jobs created and jobs retained. The official stance of the DCED currently seems to be that the nanotechnology industry has a trillion-dollar destiny and that PIN will allow the commonwealth to claim its share.

Indeed, the PIN Metrics would seem to show a robust presence of NSE research at Pennsylvania universities as well as strong links with industry. However, they do not show that these qualities are connected with the state's minuscule contributions. Penn State is the nation's largest academic performer of materials research, and much of that activity touches nanotechnology. On a combined measure of NSF funding and publications, Penn State ranked fifth among all US universities in NSE (Prabhu 2006). Table 11.5 shows a breakdown by university of NSF funding for nanoresearch at Pennsylvania universities as well as numbers of nano-focused publications, patents, and copublications. Six major laboratories and the NMT Partnership are focused on nanotechnology, and all encourage industrial participants. In Philadelphia, the Nanotechnology Institute is predicated on the NSE expertise at Penn and Drexel. In 2004 Penn was awarded an NSF Nanoscale Science and Engineering Center in Nano-Bio Interface, complementing its national leadership in biotechnology. Carnegie Mellon boasts at least three centers that engage in nanoscale research in the course of developing somewhat larger devices. It has long been a national leader in micro electromechanical systems, working closely with industry and statewide consortia. Carnegie Mellon produces almost as many nanopublications as the much larger Penn. The University of Pittsburgh established the

TABLE II.5.
Nano-data for universities

	2002–4			1986–2003
	NSF funds ($M)	Publications	Patents	Copublications
Penn State	20.1	349	30	63
U Penn	19.1	169	32	29
Carnegie Mellon	8.6	166	12	40
U Pittsburgh	4.5	113	18	30
Drexel	5.4	80	6	5
Lehigh	3.8	40	1	15
Temple	0.3	18	1	7

Source: Prabhu 2006.

Institute of NanoScience and Engineering in 2002 and opened the NanoScale Fabrication and Characterization Facility late in 2005. Lehigh University may owe the greatest debt to the state in developing its research capacity in nanotechnology and photonics, now consolidated in the Nanophotonics Technologies Initiative.

All the academic units in Pennsylvania dedicated to nanotechnology have arrangements and expectations for collaborating with industry. For its part, industry has made considerable use of these channels, as demonstrated by formal affiliations and copublications. As far as state policies for TBED are concerned, there is no discernible difference between public and private universities. All participate in the multiple economic development agencies and form links with each other when appropriate.

The development of Pennsylvania's nanotechnology policy could be described as follows. First, state commercialization programs encouraged endogenous economic development through various forms of assistance to small and developing technology firms, and this institutional infrastructure readily incorporated initiatives in nanotechnology. Second, "policy entrepreneurs" at different universities and at different times couched nanotechnology initiatives in terms that resonated with the established endogenous growth policy and a growing national consciousness of TBED. These efforts found a favorable "window" at the DCED and, ultimately, with the governor. The result, third, was a number of discrete initiatives supporting nanotechnology in Philadelphia, Pittsburgh, at Penn State, and at Lehigh, an outcome that suggested successful promotion of regional interests rather than policy design. Fourth, the state sought some measure of control by rationalizing a policy after the fact with the ANGLE report and subsequent deliberations. Conspicuously lacking, however, was any evaluation of *existing* nanotechnology initiatives, any recognition of the challenges posed by the competitive national environment, or any articulation of specific future goals.

The constituent elements for the formation of a nanotechnology policy in Pennsylvania are apparent in retrospect. A favorable climate was created by the heady atmosphere surrounding the high-tech boom at the end of the twentieth century and the salience of nanotechnology as the "next new thing." In such a climate, investments in nanotechnology appeared attractive to the DCED and ultimately to the governor. These policy initiatives seem to have been confined to the executive branch, with little involvement of the legislature.[20] Pennsylvania already possessed ready-made conduits for these kinds of investments in the Ben Franklin Partnerships. Policy entrepreneurs at different universities and economic development agencies, at different times, were able to mold these elements into state initiatives. Penn State initially and serendipitously turned a failed industrial research center proposal into a program for education and workforce development. The Philadelphia group mimicked this educational component, at least on paper, but stressed research and commercialization. Penn State then asked for the same treatment for its own efforts in education, research, and commercialization. Lehigh combined these elements with a well-honed formula for regional economic cooperation. Pittsburgh, a latecomer to the nano-game, focused on harnessing nano-research for local corporations. All these initiatives dovetailed with the charter of the Ben Franklin program and advanced the agenda of the DCED. These policy entrepreneurs exploited this window of opportunity to achieve state commitments to nanotechnology at their respective universities.

The Pennsylvania policy for nanotechnology essentially provided supplemental support for the basic NSE activities of the principal universities, based on their own initiatives and linked with other educational and economic development organizations. Only the initiatives at Lehigh and, to some extent, Penn State enhanced basic research. Other funds have been devoted to proprietary research or have underwritten organizations that coordinate and facilitate the acquisition of technology by industry. Although evaluation of these activities has been perfunctory at best, technology transfer, especially to small and midsize enterprises is now coordinated and subsidized—and probably is as effective as conditions in industry allow it to be.

The policy reports recently produced under the guise of developing a coherent strategy to guide state investments in nanotechnology advocated more of the same. After the merest nod at strengthening the research base, the ANGLE report recommended undertaking more research for industry and—perhaps to please its constituency—adding more units to the organizational superstructure: "Regional

Collaboration and Focus Centers," a statewide advisory council, a NanoForum, a database, and a governor's advisory group as well as various promotional activities (ANGLE Technology Group 2004, 51–55). The proposed "Pennsylvania Initiative for Nanotechnology" echoed these recommendations, but they seem to have disappeared from the current PIN.

A different argument might be made, based on the literature on TBED, that the competition for a leading position in nanotechnology will depend more heavily on advances in basic science than on nurturing small companies and, furthermore, that only advances in NSE yield significant commercial development. The California experience, in particular, seems to show that scientific leadership produces the most consequential economic gains (Zucker and Darby 2005). Of the foremost states in nanotechnology, California, New York, Texas, and Illinois have all implemented upstream strategies; and Massachusetts relies on its private universities, whose basic outlooks are decidedly upstream (National Nanotechnology Workshop 2003). Pennsylvania has declared a commitment to be a national leader in "nanotechnology commercialization" but not to lead in the science and engineering that underlie this new field.

Opportunities and Opportunism in the Life Sciences

Pennsylvania has a competitive advantage in biomedical sciences. Home to major pharmaceutical companies, the commonwealth ranked fourth in academic research expenditures in the biosciences. Penn was the second-largest recipient of NIH research grants, and Pitt was ninth. Still, the state had no explicit policies to build upon this strength until it received the windfall tobacco settlement funds. Even then, it took three years (1998–2001) to devise a plan. Funds for the biosciences were allocated in three ways. An initial $100 million was used to establish three regional life sciences greenhouses—for Philadelphia, Pittsburgh, and everything in between.[21] Although they were organized somewhat differently, the greenhouses were intended to promote the commercialization of academic research in the life sciences and linkages with mature and emerging companies in this industry. Another portion was invested with four local venture capital funds with the intention of leveraging investments in Pennsylvania start-ups. Finally, 19% of the tobacco funds was allocated to research.

The three life sciences greenhouses are focused on early-stage companies emerging from research-based discovery. Their mandate is to provide all forms of assistance to make these ventures viable. They offer funding up to $250,000 for precommercial research on university-owned intellectual property. They also invest seed capital of up to $500,000 in new companies arising from academic research. The greenhouses thus are specifically oriented toward the commercialization of academic inventions. In this respect, they are consistent with the Pennsylvania predilection for downstream interventions. In fact, they often partner with the Ben Franklin programs that have similar programs and objectives.

The tobacco settlement research funds administered by the Department of Health (Pennsylvania Department of Health 2007) constitute an exceptional upstream program.[22] Since it began awarding funds in 2001, the program has disbursed roughly $70 million per year. Most of that is parceled out by a complex formula based heavily on NIH grants. Hence, these funds reflect quality as deemed by NIH peer review. The nonformula grants are targeted, multimillion dollar awards to the same institutions. As a result, Penn, Pitt, and Penn State (with a somewhat lesser share) receive the majority of these funds each year. Another substantial portion is given to the cluster of health sciences institutions in Philadelphia. In 2007, thirty-eight institutions received some award. The program of research funding thus follows the pattern of something for everyone, even though the bulk of these funds goes to those who presumably best know how to use them. Except for designating target areas for the few nonformula awards, these funds have no policy objective beyond sweetening the research pot. Hence, it might be possible to imagine strategies that would give the funds greater impact.

In 2006, the Rendell administration proposed redirecting the research funds into a new program, dubbed the Jonas Salk Legacy Fund (JSLF). Instead of annual allocations for research over the next twenty years, one-half of these funds would be securitized as a $500 million bond offering, and the proceeds would be expended over the next two years for capital expenditures for biomedical research infrastructure and faculty start-up packages. Two kinds of grants have been proposed: $100 million for "starter kits" for specialized equipment for newly recruited faculty; and $400 million for facilities and research infrastructure for universities, research institutions, hospitals, and economic development organizations. The grants would require a 1:1 match, so that the total investment would be $1 billion.[23] Other research funds would be redirected to venture capital and the three greenhouses. The governor pushed hard to get this proposal enacted in 2007 and again in 2008, but without success. Legislative approval seems unlikely on the governor's terms.

Unlike the cases of policy entrepreneurship just described, this policy was hatched in Harrisburg and exhibits telltale features of the capitol culture.

1. An absence of planning or analysis: American medical colleges have been expanding research space at an alarming rate despite a contraction of real funding

by the NIH.[24] This ominous pattern might at least have been considered. The biggest medical schools would probably prefer to continue to receive the bonus research funding currently in place. From the university perspective, planning major facilities or hires and raising the necessary matching funds, all within a two-year window, could be difficult. In addition, these institutions would be on their own for supporting the new faculty and buildings.

2. Hyperbolic advocacy: The governor in various pronouncements promised that JSLF would produce 12,000 (later 13,000) jobs, "provide the resources to save thousands, if not millions, of lives," and form "dozens of new compan[ies]."[25]

3. And politics: the JSLF would allow Governor Rendell to spend the future revenues of the tobacco settlement, leaving debt instead of income for his successors. Moreover, the capital grants could be awarded to many possible recipients, including economic development organizations. Philadelphia, where the governor used to be mayor, would be certain to claim a large share of these funds, including support for expansions already planned at Penn and Drexel. Thus, what looks on the surface to be a welcome addition to the state's research infrastructure, in fact may convert medical research funds into political pork.

The notion of securitizing tobacco revenues into a large capital fund for biomedical research might well have been a promising policy initiative. But such ventures require careful planning and execution. The contrast with the California Institutes is instructive (see Zumeta, this volume, chapter 12). The size of the potential investment rivaled that of California, but a process of working closely with universities and using peer-reviewed proposals to select carefully targeted research institutes is inconceivable in the political culture of Pennsylvania.

Conclusion

Without discounting the usefulness of downstream policies for small and start-up firms, the literature on science and economic development supports the efficacy of those activities encouraged by upstream investments (Salter and Martin 2001). Economic studies have generally found a very high rate of return to basic research, although large differences exist across industries. However, those fields in which industry derives the greatest value from research—materials, electronics, computers, and pharmaceuticals—encompass the strengths of Pennsylvania universities. University research is also associated with greater research in industry. Especially important, the economic impacts of basic research appear to remain localized to a large degree. Firms located near centers of advanced research have been found to possess major advantages, which would be consistent with findings of the importance of personal contacts and social interaction. University research has a particularly beneficial effect on utilization of advanced instrumentation, which is a critical factor in science-based technologies. Overall, the presence of high-quality, basic university research enhances the capacity for scientific and technological problem-solving in industry. Thus, "a state that improves its university research system will increase local innovation both by attracting industrial R&D and augmenting its productivity" (Jaffe 1989, 958).

On the other hand, this same body of literature somewhat dampens the case for downstream policies. Small firms are much less likely to be in a position to benefit from university research, and interventions on their behalf affect only one phase of the technology cycle. Basic research tends to generate university start-ups in a limited number of fields, which would certainly include nanotechnology and biotechnology. But university start-ups in general have low growth rates, even with the assistance of university incubators. The creation of start-up companies is symptomatic of a productive environment for technology transfer—the presence of economically relevant research and an entrepreneurial culture. However, given such an environment, basic research will translate to economic development in multiple ways.

The policy issue for Pennsylvania should not be whether to pursue an upstream strategy or a downstream strategy, but rather finding the proper balance of these two approaches. The basic questions underlying state policy are these: What are the ultimate public benefits? To what extent are they retained within the state? And, how much expenditure is justified? As we have seen, consideration of this issue in Pennsylvania has focused on endogenous growth of nanotechnology and other high-tech industries. However, such a development would be far from the only benefit to the commonwealth. Besides fostering the growth of new firms, scientifically robust Pennsylvania universities generate wealth from research and education by themselves and, additionally, anchor the presence of technology-based multinational corporations. Thus, three rationales exist for state investment in the quality of the academic research enterprise.

First, research itself is a valuable economic asset. It brings external dollars into the state and generates additional economic activity. A study of the economic impact of Penn State estimated that each $10 million of external research funds produced nearly $1 million in additional state revenues (Tripp Umbach & Associates 2004). Since research funds are obtained largely through a competitive process of peer review, research funding depends above all on the relative quality of academic science. In addition, high-quality research has spillovers in institutional prestige, voluntary support, and educational enhancement. The point to stress here is that investments in the quality or quantity of academic research create value irrespective of downstream commercialization. Moreover, they strengthen universities whose total contributions to their regions go well beyond potential economic gains in nascent industries.

Second, investments in the academic research base should contribute to anchoring the research and administrative organizations of multinational corporations. Given the inexorable trend toward globalization, these behemoths will manufacture their products wherever it is most efficient; however, their intellectual operations are best situated among the knowledge flows of vibrant innovation systems.[26] This effect cannot be easily quantified, but the evidence from nanotechnology in Pennsylvania showed major materials, pharmaceutical, and electronics firms actively associated with the state's universities. In at least one case, a major corporate lab chose to locate in the commonwealth solely for that reason.[27] These activities sustain a high proportion of professional jobs of the kind that Pennsylvania explicitly seeks. Corporations, for their part, seek cooperative partners with whom they can work, and the evidence cited here indicates that Pennsylvania universities have adapted well to that role. But corporations also seek cutting-edge research of the highest possible quality, and they can pursue it anywhere on the globe. University-industry partnerships must provide such science if they are to endure.

Third, endogenous growth has been the long-standing policy approach of the commonwealth. It now possesses a large institutional infrastructure and a stream of regular appropriations. Small companies presenting at the "Business of Nano" conferences, for example, frequently reported receiving assistance from these programs. Thus, one could question the need for additional "downstream" measures for more coordination or assistance to industry.

Pennsylvania's ad hoc policies for nanotechnology fulfilled some of these objectives, at least in part. Policy entrepreneurs all sought to bolster NSE research at their respective institutions, and to varying degrees they succeeded. But the price was always a basket of downstream commitments. This pattern typifies the relationship of the state to the academic research enterprise. Policy entrepreneurs in the universities can receive a sympathetic hearing in Harrisburg for schemes to promote research as long these projects also promise economic contributions—that is, jobs—in politically influential regions. In this way, Pennsylvania universities receive significant packages of state support for research infrastructure. However, policy entrepreneurship is not a state S&T policy, but rather a substitute for one.

Competition for scientific leadership may well determine where future clusters of nanotechnology and other science-based technologies will be located. Pennsylvania's annual investment of almost $10 million in nanotechnology might be adequate to compete against other leading states in that field if it were leveraged and focused on enhancing the competitiveness of the state's NSE research. Other states have found policies and practices to accomplish this by employing competitive proposals (California), creating star professorships (Georgia, New York), and utilizing external peer review for objective evaluation. Pennsylvania possesses an exceptional academic base in many science-based technologies and all the necessary machinery to facilitate the process of science-based innovation. Its S&T policy should focus aggressively on maintaining and enhancing the wellsprings of scientific discovery.

In this respect, Pennsylvania might do better by imitating the trends found in other states, mentioned at the outset of this chapter. Even contemplating how it might establish clusters or centers of excellence in strategic science-based technologies would force the state's leadership to extend their vision beyond their current myopia. It might prompt them to compare the state's modest level of investment with that of more aggressive states. And such comparisons might shake their complacent acceptance of low relative rankings of the nation's sixth-largest state. Finally—and a precondition for any progress on this front—a critical examination of efforts by other states to stimulate TBED might undermine the ostensible credulity toward its own self-serving propaganda. Above all, Pennsylvania should aim to supplement its current short-term approach to TBED, which at best sustains the status quo, with a long-term plan that would aspire to kindle a significant boost to the state's economy.

Postscript: In the fall of 2005, the Commonwealth of Pennsylvania committed $40 million to pay for one-half of a new materials science building at Penn State. As science policy, this would appear to be an enlightened investment. By bolstering the infrastructure of a leading academic performer of materials research, this leveraged commitment should strengthen an area in which the university possessed a comparative advantage. Negotiations for this award involved university commitments to economic development programs in politically powerful but economically depressed regions. Thus, concentration and augmentation of research resources was a direct result of policy entrepreneurship. For Governor Rendell, however, the issue was simple: "This is an investment . . . that will create a ton of jobs," he explained (Rendell Gives Funding 2005, 5A). If the foregoing analysis holds true, he could be right.

NOTES

1. These programs have changed names over the years while keeping the "Ben Franklin" appellation. They are called the Ben Franklin programs in this chapter.

2. According to Plosila (2004, 120), federal policy in the 1990s "focused on process improvement using off-the-shelf technology to improve the competitiveness of small and medium-sized manufacturing firms," as exemplified in the Manufacturing Extension Partnership Program (Hallacher 2005, 53–74).

3. The National Governor's Association (www.nga.org) published the following reports: "Building State Economies by Promoting University-Industry Technology Transfer" (April 28, 2000), "Using Research and Development to Grow State Economies" (May 1, 2000), "A Governor's Guide to Building State Science and Technology Capacity" (July 15, 2002), and "Science, Technology and Economic Growth: A Practicum for States" (March 25, 2004).

4. However, states generally do not sponsor programs explicitly for doctoral education. Research doctoral programs are national to a substantial degree in recruitment and placement of graduates, and hence of low or negative political salience for states.

5. By the terms of the agreement, the states were supposed to spend 25% of their revenues on antismoking campaigns. However, this amount (ca. $1.5 billion in 2006) would have produced a cloying volume of antismoking propaganda. Ironically, the states are now partners with cigarette companies in sharing the profits from smoking. See www.tobaccofreekids.org/reports/settlements/, accessed May 10, 2006.

6. CIRM was designed to operate as a scientific agency, not unlike the NSF. However, it is also a state agency and has thus far been harassed by the intrusion of religious fundamentalists, self-seeking politicians, and special-interest groups (Zumeta, this volume, chapter 12; Douglass 2006).

7. An example is California's choice of CISI projects, described in Douglass 2006.

8. *SSTI Weekly Digest* has a regular feature called *Tech Talkin' Govs.*

9. E.g., the North Carolina Economic Development Board, HiTechAlaska, the Wisconsin Technology Collaborative, the South Carolina Council on Competitiveness, and the Arizona Council on Innovation and Technology.

10. Joseph P. Lacher, chair of the Florida Technology Research and Scholarship Board, to Governor Charlie Crist, May 15, 2008.

11. Innovation and state policy are discussed in Geiger and Sá 2008, 63–108.

12. See Geiger 2004a, chapters 2 and 3 for trends in private higher education, and appendix A for university revenues, not including research or health centers.

13. The state also provides regular capital appropriations. Since the mid-1990s, Penn State has received $40 million annually for capital expenditures, with additional grants for specially authorized projects.

14. Congress in 1977 authorized the NSF to establish the State Science and Technology Program, which provided planning grants to establish S&T foundations. However, no sustaining appropriations followed, and these initiatives were gradually terminated. Only those of New York and North Carolina persisted (Plosila 2004, 114–115; Carnegie Commission on Science 1992, 25–26).

15. The payroll for the 93,105 job-years would approximate $4 billion, so business profits and "other income" (?) apparently doubled that figure. Assuming this figure is cumulative over the thirteen years, Ben Franklin's cumulative expenditures were $311 million.

16. Keystone Innovation Zone Grants were budgeted at $3.3 million in 2005 and $2 million per year thereafter. The money available hardly made them worth the effort for colleges

or universities, but participation appeared to please the governor and local legislators. By April 2006, sixteen innovation zones had been created, with thirty participating colleges and universities, most of which perform little research.

17. A handbook, *Business Assistance Programs,* lists forty-four such programs, one of which is "University Research Funding" (DCED 2005b, 16).

18. Material for this section has been drawn from Geiger and Hallacher 2005 and www .ed.psu.edu/cshe/nano/index.htm. See this report and Web site for additional detail and documentation.

19. Most recently, in 2005, $200,000 of Ben Franklin funding was provided to the Pittsburgh Technology Council to support a new Nanomaterials Commercialization Center involving Carnegie Mellon University, the University of Pittsburgh, Penn State, and industry in the southwestern Pennsylvania region.

20. Pennsylvania has a notoriously dysfunctional legislature, which showed its true colors in 2005 by illegally voting itself a substantial pay raise in a midnight session. This large, full-time body devotes itself largely to minutia. Significant initiatives originate with the governor, and some must run the legislative gauntlet.

21. Also in 2001, the DCED supported the Pittsburgh Digital Greenhouse with annual funding of $3.5 million per year. In 2005 this became the Technology Collaborative. With Penn State, Carnegie Mellon, and Pittsburgh universities as partners, and numerous small and large firms as members, the Technology Collaborative focuses on advanced electronics, cybersecurity, and robotics. With links to other membership organizations, particularly in robotics, the Collaborative is a giant network. It also is a conduit for research funds aimed at technology commercialization.

22. Another upstream Pennsylvania program is "Starter Kits" to support the recruitment of faculty, under the KIZ Program.

23. www.newpa.com/. The details of the JSLF have undergone many permutations as its chances for legislative approval have grown increasingly remote.

24. The conclusion of a study by the American Association of Medical Colleges in *Chronicle of Higher Education,* September 6, 2007.

25. www.newpa.com/newsDetail.aspx?id=816.

26. New York's nanotechnology policy has emphasized cooperation with IBM, in particular (Geiger and Sá 2008, chapter 3).

27. The Seagate Research Center located in Pittsburgh to work closely with the data storage systems center at Carnegie Mellon (see Geiger and Hallacher 2005).

REFERENCES

America's Best Graduate Schools: 2007 Edition. (2006) Washington, DC: U.S. News & World Report.

ANGLE Technology Group. (2004) Commonwealth of Pennsylvania Nanotechnology Strategy, August 10.

Armstrong, T. (n.d.) Nanomics: The Economics of Nanotechnology and the Pennsylvania Initiative for Nanotechnology. DCED, [2008].

Bagley, R. (2007) Pennsylvania Initiative for Nanotechnology (PIN). Presentation at Materials Day: 2007. Pennsylvania State University, April 10–11.

Battelle Memorial Institute. (2008) *Technology, Talent, and Capital: State Bioscience Initiatives 2008.* Columbus, OH: Battelle, www.bio.org/local/battelle2008/State_Bioscience_Initiatives_2008.pdf.

Battelle Memorial Institute, Technology Partnership Practice. (2007) *Positioning Ben Franklin Technology Partners for the Next Decade: Entrepreneurial Development, Risk Capital, and Technology Commercialization.* Columbus, OH: Battelle.

Braunerhjelm, P. and M. Feldman (eds.). (2006) *Cluster Genesis: Technology-Based Industrial Development.* Oxford: Oxford University Press.

Brint, S. (2007) How Do Public Research Universities Compete? In R. L. Geiger, C. Colbeck, R. L. Williams, and C. Anderson (eds.), *The Future of Public Research Universities,* 91–118. Rotterdam: SensePublishers.

Carnegie Commission on Science, Technology, and Government. (1992) *Science, Technology, and the States in America's Third Century.* New York: Carnegie Commission.

Chubin, D. E. and E. J. Hackett. (1990) *Peerless Science: Peer Review and U.S. Science Policy.* Albany, NY: SUNY Press.

Coburn, C. (ed.). (1995) *Partnerships: A Compendium of State and Federal Cooperative Technology Programs.* Columbus, OH: Battelle.

Cole, S. (1992) *Making Science: Between Nature and Society.* Cambridge, MA: Harvard University Press.

Department of Community and Economic Development (DCED). (2005a) Pennsylvania Initiative for Nanotechnology (PIN). White paper. DCED, Harrisburg, PA.

———. (2005b) Pennsylvania Techformation 2005. DCED, Harrisburg, PA.

Douglass, J. A (2006) Universities and the Entrepreneurial State: Politics and Policy and a New Wave of State-Based Economic Initiatives. Research and Occasional Paper Series No. CSHE 14.06. Center for Studies in Higher Education, University of California, Berkeley.

Feller, I. (1992a) American State Governments as Models for National Science Policy. *Journal of Policy Analysis and Management* 11 (2): 288–309.

———. (1992b) The Impacts of State Technology Programmes on American Research Universities. In T. G. Whiston and R. L. Geiger (eds.), *Research in Higher Education: The United Kingdom and the United States,* 64–88. Buckingham, UK: Open University Press.

Geiger, R. L. (2004a) *Knowledge and Money: Research Universities and the Paradox of the Marketplace.* Stanford, CA: Stanford University Press.

———. (2004b) *Research and Relevant Knowledge: American Research Universities since World War II.* New Brunswick, NJ: Transaction. Originally published in 1993.

———. (2004c) *To Advance Knowledge: The Growth of American Research Universities, 1900–1940.* New Brunswick, NJ: Transaction. Originally published in 1986.

———. (2007) Expert and Elite: The Incongruous Missions of Public Research Universities. In R. L. Geiger, C. Colbeck, R. L. Williams, and C. Anderson (eds.), *Future of Public Research Universities,* 15–34. Rotterdam: SensePublishers.

———. (2008) The Riddle of the Valley. *Minerva* 46 (1): 127–132.

Geiger, R. L. and P. M. Hallacher. (2005) Nanotechnology and the States: Public Policy, University Research, and Economic Development in Pennsylvania. Report to the National Science Foundation on Grant #0403783, www.ed.psu.edu/cshe/nano/index.htm.

Geiger, R. L. and C. Sá. (2005) Beyond Technology Transfer: U.S. State Policies to Harness University Research for Economic Development. *Minerva* 43 (1): 1–21.

———. (2008) *Tapping the Riches of Science: Universities and the Promise of Economic Growth.* Cambridge, MA: Harvard University Press.

Hallacher, P. M. (2005) *Why Policy Issue Networks Matter: The Advanced Technology Program and the Manufacturing Extension Partnership.* Lanham, MD: Rowman and Littlefield.

Heller, D. E. (2006) State Support for Public Higher Education in Pennsylvania. In R. G. Ehrenberg (ed.), *What's Happening to Public Higher Education?* 207–228. Westport, CT: Praeger.

Information Technology and Innovation Foundation. (2007) *2007 State New Economy Index.* Washington, DC: ITIF.

Jaffe, A. B. (1989) Real Effects of Academic Research. *American Economic Review* 79:957–970.

Jankowski, J. E. (1999) *What Is the State Government Role in R&D Enterprise?* Arlington, VA: National Science Foundation, Division of Science Resource Studies.

Kirchhoff, B. A., I. Hasan, C. Armington, and S. Newbert. (n.d.) *The Influence of R&D Expenditures on New Firm Formation and Economic Growth.* Maplewood, NJ: BJK Associates.

Leslie, S. W. (2001) Regional Disadvantage: Replicating Silicon Valley in New York's Capital Region. *Technology and Culture* 42:236–264.

Lester, R. K. (2005) Universities, Innovation, and the Competitiveness of Local Economies: A Summary Report from the Local Innovation Systems Project—Phase I. MIT Industrial Performance Center Working Paper 05-010, December 13.

National Nanotechnology Workshop. (2003) Regional, State, and Local Initiatives in Nanotechnology. Report of the National Nanotechnology Initiative Workshop, Nanoscale Science, Engineering, and Technology Subcommittee, Washington, DC, September 30-October 1.

Nexus Associates Inc. (2003) *A Continuing Record of Achievement: The Economic Impact of the Ben Franklin Technology Partners.* Harrisburg: Ben Franklin Technology Partners.

O'Mara, M. P. (2005) *Cities of Knowledge.* Princeton, NJ: Princeton University Press.

Pennsylvania Department of Health. (2007) 2006–2007 Annual C.U.R.E. Report. Pennsylvania Department of Health, Harrisburg, PA.

Plosila, W. H. (2004) State Science- and Technology-Based Economic Development Policy: History, Trends and Developments, and Future Directions. *Economic Development Quarterly* 18 (2): 113–126.

Porter, M. (1998) *On Competition.* Cambridge, MA: Harvard Business School Press.

Prabhu, R. (2006) Knowledge Creation and Technology Transfer in Nanotechnology. PhD Diss., Pennsylvania State University.

Rendell Gives Funding for PSU Science Building. (2005) *State College (PA) Centre Daily Times,* May 19.

Resnick, D. P. and D. S. Scott (eds.). (2004) *The Innovative University.* Pittsburgh: Carnegie Mellon University Press.

Richardson, S. (2002) State Higher Education Database, 1870–1965: An Introduction. *History of Higher Education Annual* 22:1009–1031.

Romer, P. M. (1990) Endogenous Technological Change. *Journal of Political Economy* 98 (S5): S71-S102.

Rosenberg, N. and R. R. Nelson. (1994) American Universities and Technical Advance in Industry. *Research Policy* 23:323–348.

Sá, C. M., R. L. Geiger, and P. M. Hallacher. (2008) Universities and State Policy Formation: Rationalizing a Nanotechnology Strategy in Pennsylvania. *Review of Policy Research* 25 (1): 3–20.

Salter, A. J. and B. R. Martin. (2001) The Economic Benefits of Publicly Funded Basic Research: A Critical Review. *Research Policy* 30:509–532.

Tassey, G. (2001) R&D Policy Models and Data Needs. In M. P. Feldman and A. N. Link (eds.), *Innovation Policy in the Knowledge-Based Economy,* 37–71. Boston: Kluwer Academic.

Tripp Umbach & Associates. (2004) *Pennsylvania State University Economic Impact Statement, 2003.* University Park, PA: Tripp Umbach & Associates.

Zucker, L. G. and M. R. Darby. (2005) Socio-Economic Impacts of Nanoscale Science: Initial Results and NanoBank. NBER Working Paper 11181. NBER, Cambridge, MA.

CHAPTER TWELVE

California

WILLIAM M. ZUMETA

California is a vast state, stretching some eight hundred miles from north to south, with a land area similar to that of Great Britain. It has high mountains, vast deserts, a stunning coastline, and great, highly productive farmlands. It is home to more than 36 million people, about one-eighth of the US population, and the number continues to grow rapidly. The demographics of this population have undergone remarkably rapid change in recent years. Whites now represent less than half the working-age population (ages 25–64), down from 71% in 1980, and their percentage is projected to be below 40% by 2020 (fig. 12.1). By that year adults of Hispanic origin are expected to constitute about 38% of the California workforce (up from 16% in 1980) and Asian Americans 17% (up from 6% in 1980).

California's economy is a dynamo. Its gross state product, similar to the GDP of France, would make it the fifth-largest in the world, were it a nation. In recent decades, the economy has become increasingly technology-based, although agriculture remains very important. The Silicon Valley south of San Francisco, the city of San Francisco itself and its more immediate environs, the vast Los Angeles metropolitan area, and the San Diego area are meccas for firms in the electronics, communications, information technology, biotechnology and other life sciences, and most recently, nanotechnology industries. The technology-based aerospace industry continues to be important in the Los Angeles area as well.

In the modern economy it is hardly surprising that these key industries are all knowledge-based and research-intensive. They have natural links to universities, since many of their employees are highly educated. They recruit from the universities, and many of them seek direct connections to and even financially support university research. Indeed, many of Silicon Valley's early companies were spin-offs

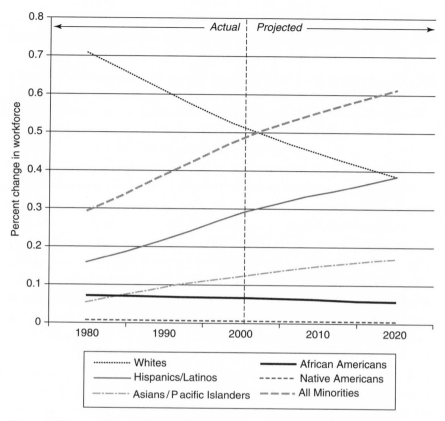

Fig. 12.1. Demographic trends in California's workforce, 1980–2020. *Sources:* National Center for Public Policy and Higher Education, 2005. Data is from US Census Bureau, 5% Public Use Microdata Samples (based on the 1980, 1990, and 2000 censuses) and US Population Projections (based on the 2000 census). *Notes:* Population projections are based on historical rates of change for immigration, birth, and death. Projections for Native Americans are based on the 1990 census. The census category "other races" is not included.

from Stanford University research, and this pattern continues today there and in other locales within the ten-campus University of California (UC) as well as in such academic powerhouses as the California Institute of Technology (Caltech) and the University of Southern California (USC).[1] The prosperous agricultural sector in California is also closely tied to university research and extension services, as has long been the case.

These two major trends—rapid population growth and demographic change and the shift in the economy toward knowledge- and technology-based industries—are key to comprehending the state's future; their intersection or lack thereof is another essential element. In short, the state will be faced with a workforce crisis before long if it is not able to do a far better job than at present of educating the burgeoning population groups of color and preparing them for the jobs its economy is creating (National Center for Public Policy and Higher Education 2005; Campaign for College Opportunity 2006). But efforts to do this and to provide for the elements essential to the health of the economically crucial research mission in California's universities must work through the state's political and policymaking processes. Here there are serious shortcomings and major challenges to be explored.

In this chapter I will analyze these issues with a primary focus on the research mission, defined broadly to include the fiscal health and prospects of the state's public research university system; the University of California and the status and future of its graduate programs; and issues surrounding the explicitly state-funded research programs in that university and outside it. I will proceed by first summarizing a bit of the relevant history of state policy toward public higher education in California, including the continuing importance of the 1960 Master Plan for Higher Education. I will look next at the current strains and pressures facing the state that affect its support of higher education and the University of California in particular, as well as some relevant policies of the state and the university itself. Then comes a brief look at the emergence of California's high-tech economy and its connections to academe, followed by an analysis of specific policies toward financial support, faculty salaries, and especially graduate education and prospects for future developments in these.

Next, a major part of the chapter is devoted to a description and analysis of the political economy of the large volume and wide range of explicitly state-initiated and state-supported research programs. In the language of Roger Geiger (this volume, chapter 11), these are largely "upstream" policies designed to support broadly defined research areas of interest to the state rather than "downstream" policies focused primarily on local technological applications. Most but not all of the state research programs are based at the University of California, including the unique new state entity called the California Institute for Regenerative Medicine (CIRM).

Finally, I offer a concluding assessment of this odd assortment of historical legacies, current policies, and policymaking mechanisms, giving attention to the factors underlying the challenges and prospects ahead for the state, the University of California, and its research mission. A key conclusion pertinent to the themes of this volume is that, large and resource-rich as it is, even California—a single state within a large nation—cannot efficiently pursue research policies in the same way a nation can. Fundamentally, the state's major research policies should be to maintain the quality and vitality of the University of California and the productive climate of interaction between UC (as well as other universities in the state) and research-intensive industries and employers. Secondarily, I also seek to point out how the state could take steps to improve its capacity to perform better in selected areas in which state-level policymaking could be most effective.

The Historical Legacy and Current Realities

California has a proud history of commitment to broad access and excellence in higher education. Granted constitutionally autonomous status from its founding,[2] the University of California was a leader in public higher education from early on. Although based in Berkeley (near San Francisco), it began developing outposts around the huge state early in the twentieth century. These eventually grew into ten campuses, including eight well-developed "general campuses" offering a wide range of undergraduate, graduate, and research programs; a specialized health sciences campus in San Francisco; and a new general campus recently launched in the agricultural Central Valley (at Merced). No other US state has anything approaching this large array of research institutions. The central administration of the university and especially the systemwide Academic Senate maintain and generally enforce common standards for faculty appointments and promotions and basic student qualifications throughout the system.

The standards are high, and the results have been impressive. In the most recent professionally based rankings of graduate program quality published by the National Research Council in 1995,[3] more than half of UC's 229 doctoral programs in the rated fields were ranked among the top twenty in the United States.[4] An aggregation of these discipline-based rankings found UC Berkeley placing higher than any other US university, public or private, and both UC San Diego and UCLA were in the top twelve.[5]

Table 12.1 shows more recent data pertinent to UC's academic and research quality: the rankings of the UC campuses in obtaining federal research and development funds, the vast majority of which are competitively awarded. In fiscal year 2005, three UC campuses ranked among the top dozen US universities by this measure, and the recent trends in UC's standing in these rankings over time are generally positive.[6]

A key juncture in California higher education history is represented by the 1960 adoption by the state legislature of the Donahoe Higher Education Act, California's famed Master Plan for Higher Education. In response to rapid in-migration from other states and the coming graduation from high school of a "tidal wave" of postwar baby boom children, a political free-for-all among state institutions seeking

TABLE 12.1.
Federal R&D spending and national ranking of UC campuses, 1997, 2000, 2005

	1997		2000		2005	
	$ (1,000s)	Rank	$ (1,000s)	Rank	$ (1,000s)	Rank
UC San Diego	238,569	6	326,037	5	463,946	8
UC Los Angeles	215,937	10	274,162	12	469,889	6
UC San Francisco	206,749	12	248,878	14	438,988	12
UC Berkeley	159,275	22	208,338	21	290,960	26
UC Davis	104,943	38	141,740	35	240,003	38
UC Irvine	64,293	74	88,274	67	161,524	56
UC Santa Barbara	64,915	73	80,754	69	103,955	88
UC Santa Cruz	24,005	132	25,959	135	62,301	121
UC Riverside	24,006	131	21,085	147	52,919	129
UC total	1,102,692		1,415,227		2,284,485	

Source: National Science Foundation/Division of Science Resources Statistics, Survey of Research and Development Expenditures at Universities and Colleges, FY2004 and FY2005.

Note: UC Total for 2005 includes $519,000 appropriated to the University of California Office of the President.

authorization to serve them began. UC and the State Board of Education worked together to develop a plan for the orderly expansion of public higher education that became the Master Plan. This plan codified a state commitment to provide free access to high-quality public higher education to all who were deemed able to benefit from it.[7]

In order to make these commitments manageable, the plan formally assigned the University of California, the State Colleges (now California State University [CSU]), and the California Community Colleges responsibility for educating different segments of the population. The University of California received responsibility for the most academically qualified undergraduates; for doctoral education; for professional education in the most prestigious fields, such as law and medicine; and for academic research. The other segments of the statewide system were to handle the large bulk of undergraduate demand: the community colleges were to be accessible to all, and the state colleges were to provide master's degree programs in certain fields.[8] Funding to achieve excellence in each segment's mission would presumably be allocated accordingly by the state.

During the 1960s California's economy surged, and the state indeed funded a vast expansion of the public higher education system. During the late 1950s and the 1960s, the University of California established three entirely new general campuses (Irvine, San Diego,[9] and Santa Cruz) and made three more much smaller operations into full general campuses (Davis, Riverside, and Santa Barbara). This was a remarkably ambitious effort to expand research-based public higher education throughout a large part of the state and to extend it to a substantial share of the population. It largely succeeded, although the abrupt end of the baby boom and major political and economic changes[10] in the late 1960s and early 1970s ended free tuition at UC and CSU and led to much slower growth of the new campuses than had been originally envisioned, especially in graduate programs. State policy, together with great economic prosperity, had permitted university leaders to greatly expand in a short period the base of the University of California, which the Master Plan called the primary state agency for academic research. During the 1970s and 1980s they were able to build substantially upon this base, albeit much more slowly, and they did it while maintaining high academic standards in both teaching and research.

RECENT STRAINS

The rapid demographic change in California's population that has already been described has had a number of indirect effects on the University of California. The resulting increase in the state's dependent population, together with the legacy of Proposition 13 and its aftermath, has strained the state government's finances (see note 10). The public schools have been largely overwhelmed by the numbers of students and the complexities of teaching large numbers from low-income and different language backgrounds (Schrag 2006). Both resources per student and student achievement have slipped. Relatively small proportions of Latino students (as well as African Americans, whose achievement has long lagged) are competitive for admission to the University of California, and the problem carries over to the graduate level. The clear need for better early preparation of these students has bolstered other forces at work to shift relative policy focus, including within the University of California, over the past decade and more toward undergraduate and even K-12 education at the expense of graduate study.

Yet, increasingly vital linkages between elementary and secondary (or K-12) education and the higher education segments remain weak, and they are not much better across the several postsecondary components of the state's educational system. As a result, California's high school graduation rates are low compared to the rest of the country and stagnant at best (National Center for Public Policy and Higher Education 2006). Most who do graduate are not fully prepared for college and require remedial classes. The vaunted community colleges enroll many students but graduate and transfer to universities distressingly few of them (Shulock and Moore 2007). The University of California and CSU systems evidently continue to do a very good job with the students they accept,[11] but without much more state money than has been made available in recent years, they are not enthusiastic about taking on a great many more. And no one at the state level has taken a statewide view of these problems and made system-level integration and performance, together with more funding, a political priority. An interested and determined governor could conceivably do this, but no governor has shown such an inclination for decades.

Like the rest of the United States, California experienced three deep recessions in recent years, in the early 1990s, again about ten years later, and a third beginning at the end of 2007. The first two hit harder and the effects lasted longer in California than in most other states. (The full dimensions of the most recent recession are unclear at this writing.) In the downturn of the early 2000s, the state took drastic measures to keep afloat financially, including deep budget cuts and long-term borrowing to balance its annual operating budget, the latter a highly unusual practice in US state government finance. The service on this debt is a continuing obligation, and an unfortunate precedent has been set. According to the legislature's fiscal analysis agency, the state faces a structural gap between projected expenditures and revenues of $4–5 billion per year through at least 2009–10 and a smaller deficit thereafter (California Legislative Analyst's Office 2006).[12] Competition for state operating budget funds is thus quite fierce. Entities like the University of California are poorly positioned structurally in this competition, because the growth of other major state expenditure items is protected by statutory formulas (e.g., education spending for K-12 and community colleges under voter-enacted Proposition 98) or is driven by federal and judicial requirements to fully fund caseload growth (in indigent health care and prisons) that do not apply to university enrollments.[13]

Figure 12.2 shows the dramatic impact of the earlier recessions on state support and overall financing of the University of California and makes clear that the period since 1990 has not been favorable fiscally speaking. In FY1990, before California felt the impact of the US recession that began later in the same year, UC received about $24,500 per full-time-equivalent student (FTES) from the state (in 2006 dollars). That figure dipped to $18,133 by 1995 and did not come close to the early 1990s levels again until 2001 ($22,916), the height of the "dot.com" economic boom. The subsequent downswing led to steady and cumulatively drastic cuts (a reduction of 38% from 2001 to 2005), so that state support in FY2007 was less than 60% of the 1990 level in real per-student terms.

As figure 12.2 illustrates, to mitigate the impacts of these swings, university policymakers, with the state's encouragement, turned primarily to tuition (called "fees" in California). Fee revenue per student more than doubled in real terms between 1991 and 1995 but after that declined somewhat as state support recovered and then surged. But between 2003 and 2006 there were sharp increases again as the state budget plunged, a total of 58% in fee revenue growth over just these three years. Thus, over the fifteen years from 1990 to 2006, fee revenues as a percentage of the

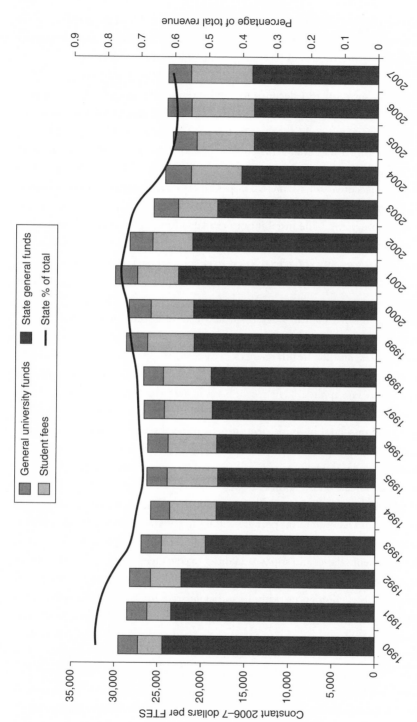

Fig. 12.2. UC state, university, and fee revenues per full-time equivalent student (FTES), FY1990–2007. *Source:* California Postsecondary Education Commission, *Fiscal Profiles 2006*, display 71. *Note:* Revenues for FY2006 and FY2007 are estimates. Total revenue excludes lottery funds.

total (state plus fee revenues) grew from just 9.2% to 29.3%. This certainly raises the possibility that the university is fast approaching the political limits of tuition increases as a source to sustain revenue growth, at least for a while.[14]

The fee increases clearly helped to stabilize the university's finances, but they did not fully stave off real declines in total revenue during the state's economic downturns, particularly in the early 2000s. The sum of state and fee revenues fell from $30,200 per student in 2001 to a nadir of $23,635 in 2004 in constant dollars (a 21.7% reduction). Total revenue per student in 2007 was 18.8% lower than the 1990 figure. It should also be noted here that about one-third of the increased fee revenues generated in recent years have not been spent on instruction or other core academic needs, but rather have been recycled into need-based student financial aid under the UC Grants program.[15] The instability in available funding, almost as much as the level, is clearly a major challenge for institutional leadership seeking to maintain the university's basic quality and effectiveness in all its areas of endeavor. The pattern of drastic increases in charges during hard times also impacts many students and their families adversely.

Although the fee increases described above have affected undergraduates the most, in recent years UC's heavily oversubscribed professional schools, particularly in law, business, and the health sciences, have also increased their charges substantially. In the arts and sciences, a standard, much lower graduate fee rate linked to the undergraduate rate applies.[16] The recent hikes in graduate fees affect some students, but most graduate students in the arts and sciences have fees covered for at least part of their time in school by initial recruitment offers, by external fellowships, or as part of their compensation for service in teaching or research. Still, there are impacts on graduate students when they are not covered by one of these sources of support, and there are effects on the sources as well.

In particular, graduate fees have been rising to levels higher than agencies providing research and fellowship support are willing to pay in full, thus putting unfavorable competitive pressure on UC researchers or forcing them to consider cuts in graduate student salaries and stipends in order to cover fees. Budgets for teaching assistants (TAs) face similar dilemmas. If TA fees are fully covered from these sources, budgets for instruction go up without sufficient state funds to cover the increases. UC faculty and administrative officials are quite worried about the impacts of these pressures on the campuses' ability to compete for the best graduate students, especially in a state with very high living costs.

California's future fiscal outlook is not particularly positive. As mentioned, the state carried a kind of consumer debt forward from the early-century recession as a claim against future revenues. Its FY2007 operating budget was balanced by one-time revenues, and a substantial structural gap between recurring expenditures and revenues at current tax rates was projected at that time through at least 2010 (California Legislative Analyst's Office 2006). The economic crisis of 2008–9 rapidly brought to a head the structural problems in California's finances, with the projected gap between expenditures and revenues reaching a staggering $42 billion in February 2009. Since the state has little political appetite for permanent tax increases—indeed quite the reverse—higher education support seems likely to face continuing strong competitive pressures at best.

On the other hand, California's political culture has become highly populist in orientation, and hence, unpredictable. After the current crisis has passed, a voter initiative could conceivably be mounted—with the support of a governor who chose to make higher education a political issue, probably by linking it closely to economic competitiveness—to guarantee this sector a share of the state budget, as has been done for many other functions.[17] Almost certainly, though, such a guarantee would include the other public higher education segments and would require a commitment to additional undergraduate enrollments and moderation in student fees from UC as a quid pro quo.

In this fiscal context, certain policies of the higher education segments themselves may work to divert some of what state resources are available from support of UC's nine well-established campuses. First, the CSU system has long sought to lift the Master Plan's restrictions on its ability to offer doctoral degree programs on its own.[18] For just as long, the University of California has opposed this, fearing an expansion of CSU's mission that would divert scarce resources for advanced education. CSU finally succeeded in getting legislative authority to offer EdD degrees in 2005 and now seeks to broaden this permission to other types of professional doctorates. PhD programs could be next, and the ultimate result would probably be the fiscal dilution UC fears.

Another questionable policy decision was initiated by the University of California itself. For some years there has been concern inside and outside the system that the university was neglecting an important part of California where there was no UC campus: the Central Valley, the four-hundred-mile-long agricultural corridor where population has been growing rapidly. Not coincidentally, many Latinos reside in this area, and the university felt it could serve them better with a campus there. Hence, UC began planning in earnest for a new campus at Merced in the late 1990s. The campus, though far from complete, opened its doors in 2005 with the blessing of state policymakers. Building a new general campus with programs at all levels is a very expensive undertaking, and enrollments have so far fallen well below projections (Ashton 2006). Ultimately the university's strategy to go where the students are may be quite sound, but one wonders about the priority of undertaking such a large and continuing financial commitment in the present fiscal context.

In the long run, if for no other reason than simply to sustain its political and financial support from the state, the University of California surely needs to find ways to ensure that its student bodies are much more reflective of the changing ethnic complexion of the state. Reflecting very mixed feelings among the citizenry about this emotional issue, in 1995 the Regents of UC (the system's governing board) voted to end "affirmative action" policies that had permitted vigorous recruitment and acceptance of qualified students of color at both the graduate and undergraduate levels.[19] Soon thereafter the voters of the state passed a similar ballot initiative (Proposition 209) that applied to all its public colleges and universities. At UC, enrollments of minority students (except Asian or Pacific Islander Americans) at both levels declined for a few years after this prohibition took effect. The university made great efforts to find legal ways to reach out to underrepresented groups of students—even to help improve their prior educational preparation—and to evaluate applicants more "holistically," that is, not relying so heavily on prior academic performance and test scores alone. After a few years, enrollments of most minorities again began to increase, but the gains among African Americans at both the undergraduate and graduate levels have been very modest; their numbers remain below pre-1998 levels. Native American undergraduates were 28% fewer in 2005 than in 1998.[20]

UC has made serious efforts to increase minority representation in its student body, but it seems clear that the self- and then state-imposed inability to take race or ethnic origin into account in admissions has taken a toll on progress among Latinos and other seriously underrepresented groups, such as African Americans and Native Americans. Comparisons of enrollments of each of these groups as a percentage of all enrollments by level shows that Latinos have gained only 2.1 percentage points at the undergraduate level since 1991 and just 0.4 points at the graduate level in spite of this group's rapid population growth, while African Americans have actually lost ground at both levels. Only Asian Americans and the small "other" ethnicity category have made strong gains.[21] Clearly, more will have to be done to increase the rate of progress in the Latino category in particular if the university is to retain full political viability over time as California's population changes and this group moves toward majority status.

California's Economy and Its Links to Academe

California is a state oriented toward research, science, and engineering. Table 12.2 provides some statistics from the National Science Foundation that support this

TABLE 12.2.
Science and engineering profile: California

Characteristic	CA	US	Rank
Doctoral scientists, 2003	76,410	566,330	1
Doctoral engineers, 2003	22,650	118,540	1
S&E doctorates awarded, 2004	3,499	26,275	1
Engineering (%)	22	22	n/a
Life sciences (%)	22	27	n/a
Social sciences (%)	17	16	n/a
S&E and health postdoctorates in doctorate-granting institutions, 2003	7,693	46,807	1
S&E and health graduate students in doctorate-granting institutions, 2003	51,989	507,247	1
Population, 2004 (thousands)	35,894	297,550	1
Civilian labor force, 2004 (thousands)	17,552	148,769	1
Personal income per capita, 2004 ($)	35,172	33,041	12
Federal spending			
Total expenditures, 2003 ($M)	219,706	2,024,246	1
R&D obligations, 2003 ($M)	17,410	91,359	1
Total R&D performance, 2003 ($M)	59,664	277,577	1
Industry R&D, 2003 ($M)	47,142	198,244	1
Academic R&D, 2003 ($M)	5,363	40,055	1
Life sciences (%)	58	59	n/a
Engineering (%)	13	15	n/a
Physical sciences (%)	11	8	n/a
Number of SBIR awards, 1999–2004	6,476	31,847	1
Utility patents issued to state residents, 2004	19,488	84,268	1
Gross state/national product, 2004 ($ billions)	1,551	11,744	1

Source: National Science Foundation/Division of Science Resources Statistics, Science and Engineering State Profiles: 2003–4.

Notes: SBIR = small business innovation research; S&E = science and engineering. Rankings and totals are based on data for the fifty states, the District of Columbia, and Puerto Rico. Reliability of estimates of industry R&D and of doctoral scientists and engineers varies by state, because sample allocation was not based on geography. Rankings do not take into account the margin of error of estimates from sample surveys. Data on doctoral scientists and engineers include only recipients of doctoral degrees from US institutions in S&E and health fields. The field percentages represent the largest three fields within the state.

claim. The state ranks first in the United States in most gross measures of scientific and R&D (research and development) activities. Even allowing for its large population (about one-eighth of the US total) and economy (13% of US GDP), nearly all of the indicators in this compilation show California to be well ahead of national norms. Thus, it is not surprising that California's economy has long been linked to academe and public policies relating to it.

THE STATE, THE UNIVERSITY, AND CALIFORNIA AGRICULTURE

For much of the first century after statehood, agriculture was California's leading industry, and it continues to be very important today, owing largely to the state's favorable climate and soils. In 1867, Governor Frederick Low helped bring together the supporters of the small, private College of California with the state officials responsible for the federal land grants under the 1862 Agricultural College Land Grant Act (the Morrill Act) to form what became, in 1868, the University of California (Scheuring 1995). The university has long had a close relationship to the state's farmers and, more recently, its agribusinesses, through research and development, degree-oriented education—both supported substantially by the state[22]—and Cooperative Extension (originally the Agricultural Extension Service), which provides a variety of outreach services and is jointly supported by federal, state, and local governments. While not unique to California, this extension model, taking university research findings directly to the users and maintaining close touch with their problems and needs, is generally considered a great success story in public science and technology policy (Rasmussen 1989). Among the notable successes of UC agricultural research was critical support of the now huge California wine industry through early research on grape diseases and the development of disease-free plants at UC Davis. Also, beginning in the 1940s, scientists at Berkeley developed very productive strains of strawberries that now represent the vast majority of those planted commercially. As of 1995, California strawberries accounted for about 80% of US production (Farrell 1995).

In recent decades, in response to trends in environmental and social awareness and criticism that it was too focused on the needs of its agribusiness clientele, UC's agricultural research and applied activities have shifted in significant measure away from pure production efficiency and toward development of sustainable practices and studies of the social impacts of technology as well as the problems of small-scale farmers and low-income farm workers (Scheuring 1995; Walker 2004). Yet, stringencies in state and federal general funding have also cut in the other direction by pushing faculty to look more to private funding and governmental project grants, many of which emphasize production and efficiency issues.

CALIFORNIA'S HIGH-TECH ECONOMY AND ITS TIES TO ACADEME

The roots of California's high-tech economy go back at least to early-twentieth-century aviation, when the likes of Glenn Curtis and Glenn Martin, Donald Douglas, the Lockheed brothers, John Northrop, and T. Claude Ryan were all pioneering in this field in Southern California—Douglas with the aid of aeronautics research at Caltech. The Los Angeles area in particular has become a world leader in what is now "aerospace" R&D and manufacturing.[23] Southern California's entertainment industry has also been a technology leader since the earliest days of both radio and television engineering, fields in which early pioneers had backing from members of the Stanford University community (Starr 2005).

Stanford played a leading role in the origins of the uniquely vibrant innovative complex of technology-oriented firms and spin-offs known as Silicon Valley. As early as 1909, Stanford was involved financially in the creation of a significant small company, eventually named Federal Telegraph Corporation, led by one of its recent engineering graduates (Sturgeon 2000).[24] A Stanford professor, Frederick Terman, worked on vacuum tubes and supported, with about $100 in university resources, the work of a graduate named Russell Varian, whose discoveries ultimately led to the radar used in World War II. Terman also taught and advised two young engineers named William Hewlett and David Packard. After the war, he became Stanford's dean of engineering and later its provost and senior vice president. Key policies of his were to bring promising inventors into contact with Stanford faculty, the creation of the Stanford Industrial Park in 1951 (the first tenant was Varian and Associates, from whom Stanford earned more than $2.5 million in licensing fees on patented inventions from its $100 investment), and the establishment of the Stanford Research Institute. Terman's fertile mind in both engineering and institutional design played a significant role in the creation of the hotbed of technology-oriented entrepreneurialism around Stanford and environs that is Silicon Valley. Stanford had an indirect role in the development of silicon-based semiconductor and microprocessing technology developed by new companies like Fairchild Semiconductor and later Intel (Starr 2005). Thus, though a private university played an important part, state policy had little to do with the emergence of Silicon Valley.

Meanwhile, UC Berkeley contributed to the development of usable atomic energy under the leadership of physicists Ernest Lawrence and J. Robert Oppenheimer but was not much involved in its commercialization. The San Diego area was an early player in biotechnology, and this had much to do with the presence of the scientific prowess of the UC campus there. The rapid development of this campus in the 1960s, at a very high level of quality from the outset and with a focus on the life and other sciences, was a deliberate public policy step by the state and university leadership (Kerr 2001).[25] Biotechnology companies have since sprung up as well around the university's Bay Area campuses (UC San Francisco and UC Berkeley) and in the sprawling Los Angeles metropolitan area, where several UC campuses are located. Early in the twenty-first century California boasted no less than 40% of the nation's research and manufacturing in biotechnology (Starr 2005).

It is no accident that research- and technology-oriented industries tend to locate in proximity to leading research universities. Firms in these industries seek such proximity in order to facilitate access to relevant academic research and, even more importantly, to university researchers and students. Firms may employ faculty members as consultants and occasionally lure one away to lead a corporate lab or project, but the real prize is often access to students and new graduates with fresh ideas and entrepreneurial zeal. As in the early days in Silicon Valley, in crucial fields like computer software and biotechnology, small firms begun by academics and/or recent graduates have played a key role in the development of new products and processes and thereby have affected regional economic development. Thus, beginning in the 1980s, state policymakers around the country have taken a renewed interest in trying to seed these kinds of developments in various direct ways, with mixed results (see Geiger this volume, chapter 11). Here, as Geiger and other analysts suggest is important in understanding long-term impact, I first take a broad view of the role of state policy in supporting the research base.

It is important to remember, though, that California is increasingly a bifurcated society. Its high-tech economy is becoming increasingly dependent upon immigrants from Asian countries, some of whom probably will not stay as opportunities improve in their homelands. Critically, most of the large and fast-growing Latino population group sees only a very distant connection between elite universities like UC and their own immediate needs. The university and the state need to break out of the Proposition 209 straitjacket to begin building deeper bonds with this group and prepare them for the contemporary economy and society. Yet, as Schrag's (2006) analysis suggests, whites will likely continue to resist competition for prized university spaces from both Asians and eventually Latinos, especially when anything that could be interpreted as ethnic preference is involved. Unless successfully addressed, these demographic cleavages will eventually erode support for the university and thus for a key engine of research-based growth in the state's economy.

State Policies of Direct Importance to Academic Research Capacity
THE HISTORICAL LEGACY

Very likely, the state's earlier support of the development of the University of California's eight general research university campuses and the health sciences campus (UC San Francisco) at a high level of academic quality is the most important policy underlying the state's academic research capacity. As suggested above, the key policy challenge now is how the state can afford to continue to sustain this base adequately while also supporting the development of a tenth campus, UC Merced, without resorting to tuition escalation. Such an escalation would be risky to both its commitment to broad citizen access to higher education of the highest quality and to its political standing in an increasingly diverse state.

Another core state policy with a long history is the autonomy of the University of California that was enshrined in the state constitution when the university was founded (Stadtman 1970). Many observers believe that the legal independence of UC from the state and the tradition of providing most of the institution's annual state appropriation in a single or a few budget line items (albeit with many specific expectations and understandings underlying) have played a key role in supporting the university's high academic quality standards (Douglass 2000; Glenny and Dalglish 1973; Trow 1993). These basic academic autonomy norms appear to remain strong—a good thing in view of the nimbleness required in an era of heightened global academic competition. This is all the more true in a polity characterized by tendencies toward political meddling and overregulation that are clearly manifested in state budgetary and other relations with the other segments of public higher education.

More broadly, this autonomy has been interpreted to permit the university substantially more control over its finance, purchasing, contracting, and personnel policies than is typical of American public universities. Of particular importance are two major finance policies: the university's appropriation from the state and its tuition-setting authority. As suggested, the University of California's state appropriation is provided with very few line items directing how the money is to be spent, although there are negotiations with state officials that create understandings about this, and the institution is subject to state audits. The university's Board of Regents retains the authority to set tuition (fee) rates, even for state resident students. Although there are discussions with state officials about fee rates in the context of negotiations about state appropriations, the university's legal control of fee-setting gives it considerable leverage, because elected officials generally do not wish to see fees rise too fast, as usually occurs when state support is considered inadequate. While this tactical game applies most prominently to undergraduate fees, graduate and professional fees are also part of the negotiations.

FACULTY SALARIES

The most basic resource for developing and maintaining academic quality and research capacity is, of course, a quality faculty. The state and the University of California developed such quality in large part by being willing to pay premium salaries to attract and retain top faculty (Kerr 2001). This aspiration is nicely illustrated by the official institutional comparison group that UC and the state use to calibrate where average UC salaries should be. Unlike most public research universities, the University of California has an official peer group that includes four of the top private universities (Harvard, Yale, MIT, and Stanford) in addition to several strong public universities (Illinois, Michigan, SUNY-Buffalo, and Virginia).[26]

The peer group is indeed strong, but the state's recent performance in actually funding the university's faculty salaries to this level is another matter. Table 12.3 shows the actual percentage salary increases provided in each year from 1990–91 through 2005–6 and the percentages that would have been needed to attain parity with the average faculty salaries in the peer group.[27] After a much better performance in the 1980s, in only four years of the most recent sixteen was the actual increase equal or nearly so to the parity level: in 1990–91 and in three years during the state's economic boom in the late 1990s and very early 2000s. In all the other years, the peer parity target was substantially higher than the amount the state provided for faculty salaries. After 2000–2001, the gap widened considerably so that, as of 2006–7, a 14.5% increase would have been needed to reach parity with the average of the eight official peer universities' salaries. It is also worth noting that UC faculty received no salary increase at all for three consecutive years in the early 1990s and did little better during the economically difficult years from 2001–2 to 2004–5. These trends are cause for serious concern about the university's ability to continue to attract and retain faculty equal to the best in the country.

GRADUATE ENROLLMENTS

Strong graduate programs and abundant graduate students are crucial to the success of university research programs. For many years, though, the University of California was unable to expand graduate enrollments as it wished because of financial constraints imposed by the state and pressures to expand undergraduate enrollments as the higher priority. During the halcyon days of the university's expansion, the state paid a premium in its appropriation for each additional graduate student enrolled, that is, a substantially larger amount than for undergraduates. Although this premium was eliminated by the state budget authorities in the 1970s, the UC administration, with its constitutional autonomy and independent funding sources, was able to continue it at a reduced level for considerably longer. Eventually, in the 1990s, the premium payment for graduate enrollments was eliminated entirely. That, plus the state's fiscal travails already recounted, made it virtually impossible to expand graduate programs substantially.[28]

Figure 12.3 illustrates the stagnancy in systemwide graduate enrollments during

TABLE 12.3.
UC salary increases compared to parity with official
peer group, 1991–92 to 2005–6, in percentages

	Peer Parity Figure	Actual Salary Increase
1991–92	3.5	0.0
1992–93	6.7	0.0
1993–94	6.5	0.0
1994–95	12.6	3.0
1995–96	10.4	3.0
1996–97	10.3	5.0
1997–98	6.7	5.0
1998–99	4.6	4.5
1999–2000	2.9	2.9
2000–2001	3.0	3.0
2001–2	3.9	0.5
2002–3	6.9	0.5
2003–4	9.2	0.0
2004–5	9.3	0.0
2005–6	13.9	2.0

Source: California Postsecondary Education Commission,
Faculty Salaries at California's Public Universities,
2006–7, 3.

the 1990s while undergraduate numbers grew by 16%, all in the middle and latter parts of the decade (1994–2000). From 2000 to 2005 however, graduate enrollments started climbing for the first time in decades, increasing nearly 20% over these five years, compared to 12.7% for undergraduates. Much of this increase was the product of two initiatives: one to double graduate enrollments in engineering and computer science, inspired by Governor Pete Wilson with considerable corporate encouragement; and the second, also largely externally driven, to sharply increase the university's output of teachers and administrators for the public schools.[29] In many other fields there was little change even in the recent period. UC administrators think that the fiscal and political climate is such that new graduate programs and substantial enrollment increases need strong external interest to generate the political and financial support to be viable. This usually means that specific fields, often those of clear economic relevance, must be emphasized.[30]

In addition to some targeted state support and reallocations from other parts of the budget, the recent gains in graduate enrollment have been financed in part by increased fee (tuition) revenue from undergraduates as well as from graduate students, particularly graduate professional students. Faculty and administrators generally feel that the limits of this strategy have been reached. The necessary cross-

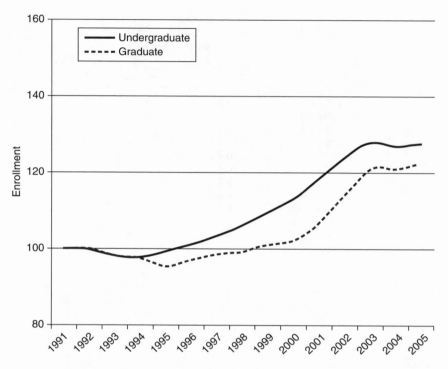

Fig. 12.3. Trend in UC undergraduate and graduate enrollments, 1991–2005 (indexed to 1991 = 100). *Source:* California Postsecondary Education Commission On-line Data System (accessed October 3, 2006).

subsidies among fields are unpopular internally, and those from undergraduates to graduate programs could become politically problematic. Also, grants and instructional budgets must be charged for much of the higher fees, which these budgets must cover for graduate student research and teaching assistants. Finally, graduate students who are unsupported must pay the higher fees, which usually means higher loan debt. This will eventually make recruiting the best students difficult.

Scope and Range of State-Supported Research in California

Having provided essential background on California's historical policies with regard to higher education and UC, including its recent challenges that have led to flagging support of the core human infrastructure underlying the research mission, I now turn to an examination of the state's policies toward research. While nearly all of these involve the University of California in some way, a recent development is that not all of the state's academically oriented research is now overseen by UC. The entire picture is a fascinating one and in many ways typically Californian.

First, it is important to note again that the Master Plan designated the University of California as the primary academic research agency of the state. This status has been taken quite seriously over the years, so that the university has been assigned a number of research-related tasks—only partially compensated financially—that might best be described as technical assistance to state agencies. Two prominent examples are assisting with various aspects of the research program of the state environmental protection agency and participating in the regular review of the impacts of proposed changes in health benefits in the state's public assistance and social services programs (Auriti 2006). This role appears to be more extensive than that typical of public research universities in the United States. The university and the state have gone as far as to set up a California Policy Research Center that performs over $1 million per year of research and policy analysis for the state and gets considerable attention from the UC vice provost for research, who sits on its board. Within the center are several ongoing applied research programs on such topics as access to health care for low-income populations, the Welfare Policy Research Project, and the UC Latino Policy Institute. The center also maintains a technical assistance capability for responding to policymaker requests in a wide range of fields and supports an annual grants competition for UC faculty interested in researching topics of importance to the state.

As described earlier, the University of California is also the state's land grant university under the federal Morrill Act and thus has special responsibilities and resources for agricultural research and extension services. While many other states also have land grant universities with such responsibilities, the function is quite important in California given the continuing importance of its agricultural sector. Moreover, in California these activities are not limited to a single campus as is usually the case elsewhere.[31]

According to UC figures, the university received about $189 million in state support for research[32] in FY2006, of which $173 million was for direct research costs. The total peaked at $201 million in FY2002, but in the subsequent state budget reductions, it was cut back to a low of $186 million in FY2005 before recovering a bit.[33] Using 2004 for comparison purposes (the latest year for which federal data were available at the time of writing), UC's state research support represented almost 8% of its federal plus state R&D funding.

A very wide range of research programs receive state support across the vast ten-campus University of California system. Of course, mission-oriented state agencies, such as the state department of agriculture and the California Environmental

Protection Agency, support specific research studies. But much of the state money goes to support substantial, ongoing research programs. A number of these were prompted by the legislature or by ballot initiative, as is explained below.

THE SPECIAL RESEARCH PROGRAMS

One important and long-standing set of state research programs is dubbed the Special Research Programs. Included are three specific state-supported programs on HIV/AIDS (established in the early 1980s), tobacco-related disease (established in 1988), and breast cancer (approved in 1993). The HIV/AIDS program was begun in the early years of what became the AIDS pandemic, on the initiative of UC San Francisco and UCLA medical school faculty who recognized that research was needed to understand and combat the new scourge. California's effort largely predated research support on this topic by the National Institutes of Health, the nation's major medical research support agency. The faculty involved in the program then worked with the powerful Speaker of the Assembly,[34] who was from San Francisco, to get legislation passed authorizing and funding a research program at the university. Aware that the problem was beyond solution by faculty at two campuses alone, and to ensure that the highest-quality research was funded, the university initiated an external competition, with peer-review processes, to fund AIDS research at California universities and institutes. They also used the state support as seed funding to begin developing proposals for larger-scale federal support. The state's support has now continued for a quarter century and, unlike the case of the other two Special Research Programs, provides core institutional support for centers as well as project grants.

Whereas the HIV/AIDS research program is funded through the state's General Fund budget, much of the funding for the tobacco-related disease and breast cancer programs comes from statutorily designated shares of state cigarette-excise-tax revenues. The tobacco disease program was created as part of a ballot initiative in 1988, and the breast cancer program was established five years later by the legislature as a result of advocacy by breast cancer activists and a key legislative champion. The same powerful Assembly speaker who championed the HIV/AIDS program became a supporter and pushed through the legislature a further increase in the cigarette tax to fund the breast cancer program. Its first grants were made in 1994.

Collectively, the three Special Research Programs received some $37 million in state funds in 2006–7.[35] At the peak in 1997–98, their total state support was more than $80 million (fig. 12.4). Over their lifetime the Special Research Programs have received more than $700 million in state support and have awarded close to three thousand grants (Gruder 2007).[36] As the graph shows, however, the support for the

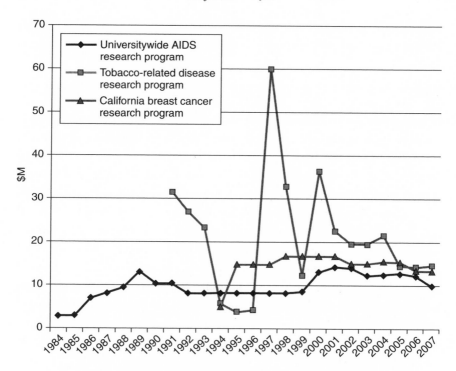

Fig. 12.4. Annual Special Research Programs (SRP) appropriations, FY1984–2007. *Source:* University of California, Office of the President. *Note:* Data in this figure have not been adjusted for inflation.

tobacco-related disease program has been subject to dramatic swings, and all three programs have experienced erosion of the real value of their state support over time. The funding problems of the tobacco and breast cancer programs are largely attributable to the erosion of cigarette tax revenue as antismoking programs in California have proven remarkably effective.[37] A proposed tripling of the state cigarette tax (from $.87 to $2.60 per pack) was on the November 2006 general election ballot; if it had passed, it would have provided a large infusion of funds to these two research programs and created yet another program.[38] The measure failed by a 52% to 48% margin.

The HIV/AIDS program's budgetary ups and downs and general slow decline in real terms are a function of its having to compete for state general funds in the face of much larger, often formula-driven budget competitors and the vicissitudes of the state's economy. It never had a designated revenue stream like the two excise-tax-supported programs. The SRP's executive director reports that the programs are all actively involved in efforts to either broaden their revenue sources or restructure their strategies to ensure continuing impact with declining constant dollar budgets (Gruder 2007).

INDUSTRY-UNIVERSITY COOPERATIVE RESEARCH PROGRAMS (IUCRP)

In 1996, driven largely by an initiative of (then) UC president Richard Atkinson,[39] a new program was created with both state ($5 million in the first year) and UC ($3 million) support.[40] It is called the Industry-University Cooperative Research Program, or now more commonly "UC Discovery Grants." The idea is to leverage corporate support in five targeted research areas by offering the state and UC funds as a match for corporate grants to UC researchers working closely with industrial partners. The purpose of the grants is less seed funding than support for projects with discrete targeted outcomes that may range across the spectrum from basic research to "proof of concept" as a prelude to new product or process development.[41] The targeted areas are biotechnology, electronics manufacturing technology, digital media, network communications, and information technology for the life sciences. In addition, a much earlier state-corporate research matching program in micro-electronics (MICRO), dating back to the early 1980s, was brought loosely under the IUCRP umbrella.

MICRO receives about $5 million in state funding per year. State funding for the other programs was quickly ramped up from the initial $5 million in 1996, when only the biotechnology program existed, to a peak of $17 million in 1999 for all the programs. The university's annual contribution, having served its initial purpose of helping to bring in state funding, remained at its original $3 million level. The IUCRP suffered cutbacks in its state support in the economic downturn of the early 2000s, so that currently it receives about $14.7 million from the state. Corporate support overall has more than matched the state and UC contributions. According-ing to the IUCRP leadership, all the programs continue to attract many quality grant applications from UC-industry partners, though the IT-in-the-life-sciences program—the one most recently developed—has had some trouble attracting the expected level of interest.

CALIFORNIA INSTITUTES FOR SCIENCE AND INNOVATION

Still another notable state venture into research at UC is the program called California Institutes for Science and Innovation, or "Cal ISIs." This effort began officially in 2000 as an initiative of Governor Gray Davis, with important support from influential corporate leaders and at least one UC regent acting independently. The state's economy was riding very high at the time, and the governor was intrigued by the idea of developing at the university large-scale interdisciplinary research programs relevant to the state's economy. The result was a proposal for substantial state support for three such multicampus institutes in specific fields identified by the governor and the key supporters. The university insisted on a broader competition, with campuses or combinations of them submitting proposals that then went through both external and internal peer-review processes. Initially three institutes were approved and funded, and a fourth followed soon thereafter. The four California Institutes for Science and Innovation now operating are

- California Institute for Quantitative Biological Research; the lead campus is San Francisco, with collaboration from Berkeley and Santa Cruz.
- California Nanosystems Institute; the lead campus is Los Angeles, with collaboration from Santa Barbara.
- California Institute for Telecommunications and Information Technology; the lead campus is San Diego, with collaboration from Irvine.
- Center for Information Technology Research in the Interest of Society; the lead campus is Berkeley, with collaboration from Davis, Merced, and Santa Cruz.

State funding for the four institutes totaled $400 million over four years, which went mostly toward buildings and other capital assets. Acquisition of federal and private gifts and grants was a requirement for receiving the state funds, and the institutes have built substantial physical and human infrastructure and supported considerable research. However, as the main institute buildings neared completion, concerns arose on the affected campuses and in the systemwide Academic Senate about the institutes' ongoing operating funding, since the state has provided only about $5 million annually for this (Coleman 2007).[42] Somewhat surprisingly in light of the history of the institutes as a Davis administration initiative,[43] the university persuaded Governor Arnold Schwarzenegger to seek $19.8 million in his FY2008 budget proposal for core operating support for the four institutes. The university hopes eventually to see the state's core support increase to around $35 million per year, but the 2008 budget passed by the legislature again provided only about $5 million (UC Office of the President 2007).

OTHER STATE-FUNDED RESEARCH PROGRAMS

Although it is difficult to compile a complete inventory, there is a substantial list of other state-funded research programs within the UC system, including several often politically controversial labor studies centers, occupational health research centers, the Ernest Gallo Clinic and Research Center at UC San Francisco, and the

MIND Institute at UC Davis. The last is illustrative of the way new state research programs are often initiated. It is a multimillion-dollar institute created in 1999 for research on neurodevelopmental disorders, largely through the efforts of affluent and influential parents of autistic children.[44] They had help from campus leaders and a key supportive legislator (Associated Press 1999). For a few years there was also a state-supported research program at UC San Diego on the medicinal uses of marijuana, but this politically controversial effort has fallen out of favor and is no longer funded by the state.

In addition to research programs run through the University of California, the state has in recent years created other mechanisms for administering certain research programs that utilize UC researchers to carry out some of their work. In these cases researchers at other California universities and nonprofit institutions are also eligible for grants, which is true in only some of the UC-run programs. One example was a program created with state general funds in the booming 1990s to research "gender-related cancers" other than breast cancer. This program was established as a result of lobbying from advocates for research on particular types of cancer (e.g., ovarian cancer but not cervical cancer). Prostate cancer was included before passage, in part to broaden political support. Governor Pete Wilson, who had to sign the legislation to make it law, directed that the program be administered by the State Department of Health Services rather than the University of California. The program ran for several years with UC researchers receiving grants from it, but its funding was eliminated during the state budget retrenchment in the first years of the present century. It was slated for a rebirth had the ballot initiative to raise the tobacco excise tax, Proposition 86, been passed by the electorate in November 2006.

THE CALIFORNIA INSTITUTE FOR REGENERATIVE MEDICINE

The most notable case of a state research program operated outside the auspices of UC, however, is that of CIRM, the California Institute for Regenerative Medicine. This institute is an autonomous state agency created by a November 2004 ballot initiative, the campaign for which was spearheaded by Robert Klein II, a wealthy real estate developer who is for personal and family reasons a strong supporter of stem cell research. He had the support of a small group of very wealthy and influential friends, including corporate leaders. Since the Bush administration had largely stymied federal support for this potentially pathbreaking medical work, with typical California hubris, Klein and his supporters felt that the huge state should take the matter into its own hands.[45]

In media-frenzied California, it is now not at all unusual for wealthy, media-savvy individuals with an issue that can appeal to the public to propose ballot initiatives, pay most of the costs to get them on the ballot, advertise widely, and sometimes get them approved by the voters. The stem cell research initiative was a tall order; the topic was controversial, and the bill to the state's taxpayers would be $3 billion over ten years, funded by bonds to be serviced from the state's general funds. Yet, it was approved by a substantial margin (59% to 41%). Significantly, the initiative's drafters did not propose that the institute be part of the University of California, but rather chose to set up a new public entity, CIRM, in order to maintain more control. Klein himself is chairman of the board and is widely reported to "effectively manage the agency" (Hamilton 2006), although a respected neuroscientist served as president and chief scientific officer until early 2007.[46]

The emotional politics surrounding stem cell research in the United States largely hamstrung the fledgling institute for more than two years. Both anti-stem-cell-research advocates and fiscal conservatives challenged CIRM in court and prevented it from utilizing any state funds beyond an initial $3 million loan (Hamilton 2006; Schwarzenegger Orders 2006).[47] Yet, during this period the institute managed to hire a staff of nineteen, to occupy a headquarters building in San Francisco, and in late 2006 to make $12.1 million in training grants to 169 young researchers[48] at sixteen California institutions, the majority at nine different UC campuses. All this was made possible by multimillion-dollar private gifts raised by founder Klein and his friends. Later, they raised larger sums, some $32 million, from philanthropists by issuing, with the state treasurer's approval, "bond anticipation notes" to be repaid to the lenders from the state's bond proceeds once the bonds were sold and the funds released (Bole 2006).

Anticipating successful resolution of opposition lawsuits, Governor Schwarzenegger ordered the state's finance officials to provide the institute with a $150 million loan early in 2007 in anticipation of release of the bond funds (Hamilton 2006). CIRM officials initiated their first research grant competition in fall 2006 and awarded $45 million in grants in February 2007 (Stem Cell Research 2007). Shortly thereafter, CIRM awarded another $75.7 million in research grants, and $48.5 million in capital funding was awarded in June 2007 (Somers 2007a). A further $220 million in capital grants was expected to follow shortly after the sale of the first set of bonds authorized by the initiative and now freed up by the courts (Somers 2007b).[49] So, the flow of new state funds for research in this field has begun, and it will be very substantial.

In addition to the lawsuits—claiming, among other things, that the line of stem cells CIRM proposed to acquire had patents too broad to be legal and that its operations as an independent state agency were illegal—CIRM's critics dogged its processes for developing intellectual property policies and ethical guidelines for the research. State legislators and agencies have become involved because many feel that revenue from intellectual property—such as new pharmaceuticals, stem cell lines, and processes that could be used in R&D or manufacturing—resulting from research supported by the state should be owned by it, at least to the point of recouping its investment (Somers 2007c).

CIRM officials feel that such claims by the state would be unrealistic in the marketplace and that research and corporate partners would be unlikely to participate on this basis. The University of California, which has played a role in helping CIRM establish policies, generally supports CIRM's position.[50] One knowledgeable UC official even expressed concern that some state officials and the press seem to believe that revenues from CIRM's future intellectual property should play a substantial role in paying off the Proposition 71 bonds that finance the institute. The insiders think that this is an unrealistic expectation, given past experience with unpredictable and uneven intellectual property revenue streams. Another politically potent expectation is that any new drugs or treatments resulting from CIRM's research would be "affordable" to Californians (Somers 2007c), also a demand that could deter potential partners.

CIRM's intellectual property policies, adopted in 2006, reflect a compromise position in which the institute is free to grant intellectual property rights to its grantees (as is the case with federally funded research grants to universities) but retains rights and responsibilities to the public interest, such as "march in" rights when a licensee is slow to develop useful products with the CIRM-funded intellectual property (IP), requirements for publication and sharing of biomedical materials, and requirements that nonprofit grantees use their share of IP revenues for scientific research or education. Also, the IP policies require that exclusive licenses to CIRM-funded intellectual property be granted "only to organizations with plans to provide access to resultant therapies and diagnostics for uninsured California patients. In addition, such licensees will agree to provide to patients whose therapies and diagnostics will be purchased in California by public funds the therapies and diagnostics at a cost not to exceed the federal Medicaid price" (CIRM 2006a, 17).

In short, the CIRM venture is highly politicized. Yet, the University of California, like the leading private universities in California, is clearly anxious to participate. UC San Francisco and UC Irvine as well as private Stanford University and the University of Southern California (USC) have already made large capital and hiring commitments in order to be prepared for the time when the spigot of CIRM grants opens fully (Hamilton 2006).

Strategy and Politics in State-Supported Research

The state seems to have no overall plan or strategy in research and technology development and does not appear to be capable of or much interested in developing one. The variety of programs described above, which developed in a more-or-less ad hoc fashion and accreted over time with some signs of overlap and no real coordination or apparent overarching strategy, are suggestive evidence of this. When one looks inside California government, there is little to be found in the way of institutional structures supportive of strategic or even coordinated policymaking in this area.[51] The legislature has no science and technology policy committees or specialized staff resources that might facilitate taking a broad, strategic view. Facing strict term limits, California legislators have little time to learn about complex science or technology and research policy issues and have generally weak political incentives to specialize in this area.

Business, a logical ally, tends to be fragmented in California and surprisingly weak in influence in the capital. Rather, to make a mark in a few years before being forced "up or out" by term limits, legislators are often better advised to focus on fields like health or education that are more easily understood and can produce regular tangible benefits for constituents. These more established policy arenas also provide more valuable political resources, such as a potential committee chairmanship, access to a well-informed staff cadre, and more-or-less guaranteed media coverage.

A similar gap in appropriate institutional structures appears in the executive branch. In the budget crisis of the early 2000s, the state eliminated its Technology, Trade, and Commerce Agency, which had included, for a few years, a Division of Science, Technology, and Innovation.[52] Only a small remnant of this unit remains at the state Department of Business, Transportation, and Housing. The only other science- and technology-oriented agency is the California Commission on Science and Technology, a small, state-chartered though privately funded entity with an advisory role but limited capacity and visibility.[53] The governor's office itself has never had a formal science and technology strategy or analysis capability, depending instead largely on input from stakeholders.

Governors have only rarely shown an interest in broad strategic thinking about science and technology or research issues. The current governor, Arnold Schwarzenegger, has shown some interest, especially once it became clear that the state's economy had fully recovered from the recession-induced budget straits that dogged his predecessor. Governor Schwarzenegger has advocated strongly for stem cell research and has been supportive of CIRM. He has recently shown considerable interest in environmental issues, alternative energy sources, and nanotechnology. His signature initiative is an ambitious greenhouse-gas-reduction program, enacted into law as Assembly Bill 32 in 2006. He has asked the University of California to work with the California Energy Commission and the State Air Resources Board to develop targets and benchmarks for emissions reductions (CCST 2007b).

In 2007, Governor Schwarzenegger packaged as a "$95 Million Research and Technology Initiative" his support of $19.8 million in increased state operations support for the California Institutes for Science and Technology; $30 million in lease-revenue-bond funding for a new energy-nanotechnology building for Lawrence Berkeley Laboratory's Helios project (a solar energy technology project); $40 million in similar bond funding for Berkeley's Energy Biosciences Institute, which helped the campus win a $500 million commitment over ten years for this from the energy giant British Petroleum; and a $5 million commitment of matching funds for the University of California's bid for federal support to build the world's first Petascale computer, a project expected to cost $200 million in total.[54] It appears that this was less an executive initiative in the classic sense than a way of packaging the governor's responses to the University of California's requests for modest support of its research funding priorities in what was then a relatively favorable budget climate.[55]

A few years earlier Schwarzenegger's predecessor, Gray Davis, had access to a large budget surplus and became interested—with no small help from individuals connected to the University of California—in the California Institutes for Science and Innovation (Cal ISI) initiative, which could be considered a strategic research initiative. But, the vicissitudes of the California economy and politics being what they are, nearly all the money and the governor himself were gone within a few years. It is notable, though, that Governor Schwarzenegger supported this initiative in his FY2008 budget proposal rather than allowing it to wither, as often happens when gubernatorial administrations change. In the end, the increase in appropriations was not supported by the legislature, however.

The impact of ups and downs in the economy on state budgets is one important factor that makes it harder for states than for the federal government to develop consistent science policies: states face much stronger norms of budget balance than does the national government, often necessitating deep cuts in "nonessential" state programs during downturns. The political vicissitudes, at least in their intensity, are more specific to the California environment. Governors in California can be recalled by the voters (although Davis's recall was the first one in the state's history), and the state's politics are generally more polarized and mercurial than is true elsewhere.[56]

Indeed, one aspect of the state's politics makes coherent S&T (science and technology) and related policymaking arguably nearly impossible: the ascendancy of the initiative process. California voters can make their own policies at the ballot box, trumping any carefully laid plans of the governor, the legislature, or the University of California. The number of such ballot initiatives has grown dramatically in recent decades, and many of them bear little or no relation to the programs of leading elected officials. Examples directly affecting academic research go back at least to the cigarette tax initiatives of the 1980s and early 1990s that provided the (not very satisfactory) funding source for UC's tobacco-related disease and breast cancer research programs. More recent examples include both CIRM itself and the failed 2006 initiative to raise more revenue for targeted research from higher cigarette taxes.

Whether the initiative route is used or not, the policymaking vacuum created by the lack of legislative or executive strategic capacity in S&T and research policy sets the stage for ad hoc and often politically driven policymaking processes to hold sway. In the majority of cases it seems that the state-supported research programs at UC were created more by the impetus of individual legislators and citizen advocacy groups, often with the help of individual faculty members operating outside the purview of established university review and priority-setting processes, than as part of either state or UC strategic thinking.

Strategy, Politics, and Challenges within UC's State-Supported Research Programs

The University of California is populated by smart people who very much want to do serious science and institution-building. In dealing with the state in cases where the latter offers—by ballot, gubernatorial, or legislative initiative—a new research program not necessarily high on UC's priority list, the university's administration seems to make an effort to shape the terms of the program to make it compatible with academic norms and values.[57] The early history of the HIV/AIDS research program already recounted is one such example. Another is President Atkinson's insistence on an open competition for the California Institutes for Science and Innovation, complete with several stages of peer review of proposals, rather than simply accepting the ideas for the institutes of Governor Davis and the project's early supporters. Most recently the university has sought to influence the governance arrangements and basic policies of CIRM to help ensure appropriate scientific and ethical review procedures in this controversial area of research as well as what it regards as realistic policies regarding intellectual property. In at least one important case, UC itself took the initiative when President Atkinson developed the basic idea for the Industry-University Cooperative Research Program and, with industry help, persuaded the state to help fund it.

Although sometimes the parameters of state-funded programs are not subject to much negotiation, as in the case of initiative-created programs and some others that have aspects of a crusade in their enactment (e.g., the breast cancer research program, the MIND Institute), the university tries to guide even these as best it can to produce quality science and useful results for the state. Where the enabling legislation and the politics of oversight permit, university program managers generally seek to devise—or guide oversight boards toward—strategies that emphasize gaps in the federal research portfolio in the field or topics of special interest in light of the state's population groups, industry mix, or natural systems. This is not made easier, however, by mandates, such as one for the breast cancer program: to find a *cure* for this affliction.[58] Some programs explicitly seek to broaden their impact by emphasizing seed grants (or institutional support for research centers, in the case of the HIV/AIDS program or the Institutes for Science and Innovation) that can be used to support competitive proposals for much larger federal funding. Others, such as the IUCRP and the ISIs, seek to maximize their impact by requiring corporate funding matches and active partnerships.

Serious challenges for research managers are created, however, by California's mercurial, polarized, and increasingly populist politics and the intense glare created by its media, not to mention its fiscal rollercoaster. New mandates for research programs can emerge unpredictably from the political process, and so can sharp budget cuts or even program eliminations that have little to do with the quality or productivity of the research. Thus, program managers must spend a good deal of effort on managing publicity and relations with key stakeholders and on keeping tabs on external politics. The state-supported research programs are thus subject to a high degree of political accountability but remarkably little *performance* accountability.[59] Not only is it hard to find any sign of serious performance oversight from Sacramento—which might not be desirable anyway in regard to academic research, if the oversight was not expertly done—but the university itself seems to have done little to evaluate these programs through its normal periodic academic review processes.[60] The IUCRP is not subject to the five-year academic reviews normally required of UC's multicampus research programs.[61]

It may well be that, in some cases, this lack of oversight makes it possible for good R&D managers, working with a well-constructed structure of advisory boards, peer-review panels, and high-quality researchers, to simply do their work efficiently with UC's strong internal academic quality standards serving as the primary form of accountability. Yet, it is clear that in some cases the political pressures have permeated rather deeply into these structures. A case in point is the breast cancer research program. Created by ballot initiative, the statute specifies some of the research priorities, including the mandate to seek a cure, and disease advocates (along with scientists) serve not only on the advisory body that determines broad priorities, but also on the panels that review individual research proposals.[62] They may bring not only unrealistic expectations, but also nonscientific penchants about causes and remedies that they want to see researched. Although they are supposed to defer to scientific reviewers on questions of pure scientific merit in proposals, the odd mix of scientists and passionate advocates with political and media connections on the boards and panels is a considerable challenge to manage, while ensuring that the most significant and sound science is pursued.

Yet, the breast cancer program has faced fewer budget challenges than either the HIV/AIDS or the tobacco program, in part because the advocates are so committed to the program and are seen as a formidable bloc. The tobacco program has been less able to ward off attacks on its budget. The HIV/AIDS research program has experienced attempted raids on its budget and successful skewing of its research priorities by pressures to allocate large resources to the testing of drugs for treating patients and the investigation of a particular, high-cost treatment approach (organ transplantation) that was not widely accepted as a high scientific priority.

The California Institutes for Science and Innovation faced the usual types of questions about skewing of academic research priorities to meet the interests of industry, but the issues became more acute as the institutes built substantial infrastructures that needed continual financing, while lacking much ongoing, earmarked state operating support. If the efforts by the university and the governor had been successful in securing this support, tensions would likely be considerably alleviated. Faculty concerns about these institutes have been manifest in the Academic Senate, and it will be illuminating to see how the Senate-initiated program reviews of them, now begun, turn out.

The UC Discovery Grants Program (IUCRP) faces a different kind of internal management problem. California, however large, is a state, not a nation. Under some pressure to see that its long-stagnant state dollars are spent mostly on grants, the program uses only peer reviewers from within the University of California, who serve without compensation. Since three rounds of grants are made each year within fairly narrowly specified fields, it is hard to vary the peer-review panels sufficiently. This schedule also makes for some potentially serious compromises with the assumption of anonymity at the heart of the peer-review system, an assumption that is much more plausible in a more broadly based (i.e., national) system. Thus, reviewers are provided access to the applicant's vitae and grant track record.[63] It is also probably possible in many cases for applicants to surmise who their reviewers are likely to be. Over time, even without any explicit collusion, a mutual "back scratching" pattern of application approvals could easily develop within this closed system. In addition, individual judgments by one or two researchers about the merits of another academic or of a particular research direction or approach may come to play too great a role. These potentially serious problems, though not currently discernible, may be difficult to avoid given the constraints the program faces.

The State Role in Technology Transfer

The University of California, which performs some $3 billion worth of externally funded R&D each year, has a substantial technology transfer operation, as would be expected. The university grossed about $110 million in revenue from patent-related activities in 2005–6, of which more than $52 million was distributed as discretionary revenue to the campuses (University of California 2007). The balance covers legal and administrative costs associated with technology transfer and is also used to reward inventors and their departments by established formulas. These are fairly modest amounts in relation to UC's total research budget, but all significant discretionary revenue is especially valuable when budgets are tight. According to technology transfer officials, the university's goals in its technology transfer policies are not primarily financial, but rather are intended to ensure that university research benefits the public and fosters the advancement of nonprofit research generally, for example, by permitting other researchers to use data, methods, cell lines, software, and the like on reasonable terms.

Until fairly recently, the state of California was not a large player in UC research and did not have much to say about technology transfer policies, which were largely seen as being between UC, the federal government, and corporations.[64] Recently, as state research support has grown, there has been much more state interest in intellectual property and related issues and the associated potential revenue streams. There is no overarching state policy, however. Different supporting agencies ask for different things and are not necessarily consistent over time. The legislature has become involved at times, but there is still no consistent policy. The Davis administration authorized a commission to study what the state's policies should be. This commission reported shortly after Governor Davis was removed from office (California Business, Transportation, and Housing Agency 2003), and Governor Schwarzenegger has evidently not paid much attention. The legislature subsequently requested a report from the California Council on Science and Technology, which sought to "lay the groundwork for an informed discussion on building a comprehensive set of state policies governing the creation and administration of IP developed with state support" (CCST 2006, 1). These reports generally make sensible suggestions about standardizing state intellectual property policies. The problem seems to be that there is no body competent to receive them or motivated to act on them at the state level.

As a result of the debates surrounding CIRM's intellectual property policies, UC officials fear that the state may become increasingly aggressive in asserting owner-ship rights to inventions and copyrights derived from research it has funded.[65] The state seems to be moving toward seeking a guarantee of "recoupment" of its costs on projects that generate significant revenues. The university is concerned about this trend, for it feels it needs this revenue to help fund the costs of technology transfer efforts—which, as noted, are designed more to achieve public diffusion goals than to make money—on the many projects with public benefits from such transfer that do not generate revenues.[66] State agencies also sometimes seek to assert ownership of data generated in state-funded research projects. The university has resisted these attempts vigorously in order to be able to utilize such data in future research—either its own or that of other nonprofits—but results of negotiations with state agencies have varied as to precise terms.

Of particular concern to the university are some of the technology-transfer-related issues recently debated in the context of getting CIRM up and running. Although the ballot Proposition 71, which created CIRM, does not state that rev-enues from its intellectual property must be dedicated to helping service the state general obligation bonds that support it, such an expectation has grown in political and media circles. The University of California, in providing comments to CIRM and its advisers regarding intellectual property policies for the institute, sought to damp down this expectation, fearing that it would create unrealistic expectations, undesirable incentives, and difficulties in securing private-sector participation. The university may also be concerned that it might find itself subject to similar pressures in regard to its own state-supported research. Further, political efforts to ensure that pharmaceuticals and therapies created from CIRM research are judged affordable to Californians could, if regulations are too tightly drawn, also deter corporations seeking to license CIRM technologies. Finally, although CIRM is planning to assert "march in" rights (i.e., to declare licensees in noncompliance with terms of their license agreements)[67] similar to those in federal law, UC is concerned that CIRM may come under political pressure to interpret these more aggressively than federal agencies have, thus creating further hesitancy in the minds of potential licensees.

These concerns may apply specifically to CIRM at the moment, but given the scope and visibility of that enterprise, its policies in this fluid field could easily influence those of the state and even the federal government.

Assessment of the State-Supported Research Programs in California

California is a large, varied, and complex state with particular research needs for which it turns to its world-renowned public university for help. The state's Master Plan for Higher Education in 1960 declared the University of California the state's primary agency for academic research, and this designation has been taken seriously. While some of the research on broad human afflictions like breast cancer and tobacco-related disease that the state pays for might, in an ideal political economy, be financed at the federal level, there is little doubt that the considerable additional state investment adds social value, even if not exclusively for Californians. In this huge and traditionally optimistic state, Californians often have the hubris to think that they can cure cancer or AIDS or develop powerful new therapies from human stem cells on their own—and indeed perhaps they will.[68] In reality, most of the state research programs give particular attention to state needs and to the fact that they can often best serve by strategically complementing much larger federal efforts in their fields.

The fact that the passions and priorities of the public can be made known directly at the ballot box in California has almost certainly generated more state tax money for research than would have been forthcoming through more conventional political processes, especially in light of the limited capacity of the state's governmental institutions in this sphere. In the unique case of stem cell research, where ideological disputation has stymied the federal effort, California voters have, in the view of many, performed a valuable function by stepping into the breach. The sheer size of their contribution may well move the field forward significantly and also help retain leading researchers in the country, until the federal government is ready to assert its customary leadership role.

On the negative side of the ledger regarding the state's role in supporting research, there are a number of points to make. Feller (1997), Geiger and Sá (2005), and Geiger (this volume, chapter 11) have pointed out some of the limitations of states as instruments of science and technology policies, and California's efforts seem to suffer from all of these in addition to unique challenges of its own. Most fundamental perhaps is that California, like other states, lacks "honest broker" institutions comparable to the National Science Foundation or the National Institutes of Health at the federal level, with strong norms of scientifically based selection of both priorities and projects within broad research fields. After long and generally very successful experience, these norms are widely accepted by national political leaders, sufficiently to create a strong counterweight to the inevitable political pressures that tend to distort priorities and push toward excessive, politically driven geographic spread of grants at the expense of peer-review processes.[69] The presence of respected oversight agencies with such norms also makes it more difficult to de-emphasize or terminate sound research programs when administrations change than is the case when research programs are less insulated from politics. The University of California seems to do its best to perform this broker role but cannot be fully effective when it is a research performer as well as a broker and is so dependent upon the goodwill of state politicians for its core institutional financial support.

A second major challenge for state research and science policies is the instability in state budgets, which, combined with structural rigidities that work against "discretionary" research programs in budgetary competition, can often lead to abrupt cuts or even elimination of programs for reasons unrelated to their value or productivity. While the fluctuations of capitalist economies are at the root of budget instability, state-supported programs suffer more from ups and downs than federally supported ones simply because there are laws and usually strong norms against deficit spending in state finance that do not apply at the federal level. California's revenue structure, with its heavy dependence on volatile capital gains taxes, is particularly prone to dramatic fluctuations. Political polarization and legal limits on the length of service of elected officials (just six years in the Assembly and eight years in the state Senate) in California add to instability and generally make it more difficult to sustain a stable environment for complex programs with long-term horizons, such as those that support research.

Feller (1997) and Geiger (this volume, chapter 11) both suggest that states are much less well equipped than are national authorities to demand accountability from science and technology programs, since at the state level the programs are driven largely by the priorities of the politically influential. This is certainly the case in California, where the pressures on research programs for political accountability are all too clear but the constituency for substantive accountability is remarkably weak.[70] Even the University of California has done relatively little to formally review the performance of state-supported research programs to date, and there seems to be little or no pressure from Sacramento to change this.

Unlike at least a few other states, California seems particularly unable to generate broad strategic—as opposed to ad hoc and politically driven—thinking about state research policy. Surely high-technology-oriented California could benefit from such thinking. While occasionally governors have moved in this direction and could do so again—indeed, Governor Schwarzenegger has shown some such inclinations—there is little guarantee that a governor's plan, however well conceived, would long survive his or her term of office. And there are no permanent, competent institutional structures in either the executive or the legislative branch to help develop or sustain any such hypothetical state R&D strategy. Interestingly, the initiative process could conceivably help with continuity in the sense that an initiative can provide a permanent, statutory basis and revenue source for a program as well as a sense of political mandate that may survive quite a while. In general, though, the politically saturated initiative process would seem an unlikely vehicle to enact a comprehensive research or science policy strategy for the state.

Finally, as the discussion of the challenges facing peer review in the UC Discovery Grants program suggests, even a state as large as California faces difficulties in sustaining sufficiently broadly based peer-review processes, which are fundamental to effective research programs. The independent California Institute for Regenerative Medicine plans to work around such problems by utilization of national (and even international) advisory groups and peer-review panels. This may prove expensive to this now well-funded agency but is surely well worth the expense. However, such a strategy may not be practical for many state research programs for which the funding is small and for which there are powerful incentives to spend most of it on the research itself, narrowly defined.

Conclusion

California's policies regarding research and the academic infrastructure that largely supports it are clearly a mixed bag. They certainly reflect both the history of the state and its current fiscal and political climate. The latter is both volatile and problematic for the academic enterprise and the state's research initiatives.

In terms of core financial support for the University of California, the historical legacy has created an unusually strong base, but recent trends have been corroding the edges and probably beyond. The immediate future is problematic, as the state's general fiscal policies are probably unsustainable; a new and particularly deep recession is now under way; and the university may be near the political limits of large tuition increases and cross-subsidization, particularly of graduate education by undergraduate tuition revenue. Although recent, widely publicized accountability concerns related to compensation of its executives have not helped, the more serious long-term threats to the university, in light of this sober fiscal environment, may be the financial dilution threatened by the need to develop the new Merced campus and the aspirations of the much larger California State University system to initiate doctoral education. Most importantly, it can be persuasively argued that the state's highest priority needs in education policy are at the K-12 and undergraduate levels, where the performance of the burgeoning populations of color is seriously lagging. If not corrected, this weakness will ultimately impact the university's graduate programs and research capacity. In this rapidly changing demographic environment, UC's medium- and long-term capacity to serve the state is further hindered by the limitations created by the voters in Proposition 209. This ballot initiative severely hampers UC's ability to diversify its student bodies, both undergraduate and graduate, in the direction of the ethnic composition of the state's young population.

Given this context, it is not clear that UC's graduate programs will be able to get much additional priority, even in the face of real concerns about their competitiveness in funding graduate students. Programs that can best compete for emphasis and funding will likely be those with demonstrably close connections to state workforce and economic needs, which may mean that those in science- and technology-oriented fields will do tolerably well. If so, at least the research mission in those fields would be supported, but seriously uneven health and development across graduate fields within the institution may result.

The research programs that the state explicitly funds are, overall, impressive in range and level of state support. Many undoubtedly produce significant results in scientific terms and in progress toward certain state objectives, although evaluation mechanisms are very weak outside of traditional scholarly assessments. Besides lack of evaluation, the main problem with these programs is their politicization.[71] Although the politically driven birthing processes of most of the state research programs may produce more resources for research (at least for a while) than would otherwise be forthcoming, they also create serious challenges for ensuring strategic prioritization, sensible allocations within programs, and even on occasion appropriate peer review of individual project proposals.

For all its problems, the University of California remains a great and very broad and deep system of public research universities, infused by strong quality norms and academic values. The taxpayers—not limited to residents of California—doubtless get their money's worth from the vast bulk of the state-funded research at the university. There are many ways in which the results could be made still more impressive, but progress is mainly dependent upon more effective state political leadership and institutional reform—especially building of state science and technology policymaking capacity—on research and related matters. It is far from clear that these will be forthcoming anytime soon.

NOTES

An earlier draft of this chapter was presented at the Conference on the Public Interest and the Academic Research Enterprise at Seville, Spain, November 11, 2006. I wish to acknowledge excellent and very diligent research assistance from Deborah Frankle.

1. Stanford, Caltech, and USC are all private institutions, while the multicampus University of California is public.

2. The roots of this state institution's remarkable academic prowess are often attributed to its constitutionally guaranteed autonomy from the state government (Glenny and Dalglish 1973; Pelfrey 2004; Stadtman 1970; Trow 1993), which largely exempts it from direct state controls over personnel and contracting, allows the university's Board of Regents to set tuition, and provides its operating appropriation with few direct fiscal controls. Of course, there is some pragmatic negotiation with the state authorities over how much money is needed and how it will be spent, and tuition decisions are influenced by how much state funding is expected.

3. Goldberger, Maher, and Flattau 1995. A new national quality assessment study is now under way.

4. Cited in Pelfrey (2004, 77).

5. Cited in Pelfrey (2004, 77). All the other universities in the top dozen were private.

6. The newest campus, Merced, did not exist in the years shown in table 12.1. It should be noted that Berkeley, Riverside, Santa Barbara, and Santa Cruz are all hampered in these rankings by their lack of a medical school. Federal R&D support from the largest federal granting agency, the National Institutes of Health, increased much more than support in other scientific fields over most of the period shown.

7. Good accounts of developments during this period are provided by Stadtman (1970), Kerr (2001), and Pelfrey (2004).

8. A student aid program was also established to support state resident undergraduates attending private colleges and universities. These "Cal Grants" served to divert a substantial number of students from the public systems at considerable savings to the state. California now has well over one hundred private colleges and universities, including a number of very strong institutions.

9. San Diego had a long history as a UC research installation but had only begun to educate a few graduate students in the sciences when the decision to develop a full general campus (at a new site) was made (Kerr 2001).

10. Key events were student unrest on the UC campuses, the election of Ronald Reagan as governor, and the recession of the early 1970s with its aftermath of slow growth and high inflation. By 1978 the state experienced a "tax revolt" and passage of the property-tax-limiting Proposition 13 ballot initiative, which has hamstrung its finances ever since. See Schrag

2006, especially chapter 2, which is aptly titled, "Dysfunction, Disinvestment, Disenchantment."

11. The undergraduate completion rates in these two systems are among the best in the country for public institutions of their type.

12. Projections in early 2009 put the state's budget deficit in the tens of billions.

13. Other states are in a broadly similar predicament, but California's voters have created more than the typical number of constraints.

14. This seems particularly likely in a state with a stronger-than-usual policy commitment to access and in particular to low tuition (fees). In FY2007, the governor used improved state revenues to "buy out" scheduled 8% undergraduate fee increases for UC and the CSU system (Gledhill 2005).

15. There have been some variations in this percentage over time, but for most years since 1995, policy has been for the university to use one-third of incremental fee revenue for need-based financial aid. Overall, about 27% of tuition and fee revenue is spent on such aid (Alcocer 2006).

16. In 2006–7, the general rate for California resident graduate students was $6,162 for the academic year, compared to $5,409 for undergraduates. Non-state-resident graduate students nominally are charged much more, but what they actually pay varies greatly by field.

17. This would be dubious fiscal policy, for the state's General Fund budget is already excessively constrained by such "earmarked" funding streams.

18. The Master Plan permitted joint doctorates between California State University and private universities. Only a few such programs exist, however.

19. The ban took effect with the fall 1998 entry cohort.

20. Enrollment figures came from the California Postsecondary Education Commission On-line Data System, www.cpec.ca.gov/OnLineData/OnLineData.asp.

21. The large percentage increase in the "other" category seems to reflect rapidly changing views about race and ethnicity in America with increasing immigration and intermarriage.

22. Federal and private funds are also important in supporting research in the agricultural sciences.

23. Lockheed Corporation also has substantial R&D and manufacturing operations in the Silicon Valley area of northern California.

24. The company developed some of the technology for wireless telephone and telegraph services on the West Coast of the United States and became an important Navy contractor before and during World War I. Stanford's High Voltage Laboratory and several of its professors played important roles, and in turn the lab received donations from the company (Sturgeon 2000, 21–22), an early example of a mutually productive relationship between academe and corporate R&D.

25. To be sure, the state and the regents acted in response to aggressive prompting from business and scientific leaders in San Diego led by Roger Revelle (Kerr 2001; Starr 2005).

26. The State University of New York–Buffalo is the only one in the group that seems

somewhat anomalous. It ranks far below the others in receipt of competitive federal research and training funds.

27. The parity levels are estimates necessarily made before some of the schools have announced their annual salary increases, so there is some inaccuracy.

28. California's historical ability to import highly educated people could conceivably have played a role in the state's unwillingness to finance expansion of costly graduate programs. (In general, the state-dominated US system of financing public research and graduate universities with a national reach may suffer from tendencies toward underinvestment by state patrons who feel they cannot capture all the benefits of their investments.) At present, there is more concern that historical patterns of in-migration of the highly educated may be stifled by a combination of federal immigration restrictions and the state's high urban cost of living.

29. Other fields with notable increases over this period were health sciences and professions, business, and both the physical and the life sciences (California Postsecondary Education Commission On-Line Data System, accessed October 18, 2006).

30. Faculty (i.e., the Academic Senate) are more wary, though, preferring that the university set its own academic priorities in this matter as in others.

31. Although the center of UC's agricultural research and education activities long ago migrated from the Berkeley campus to UC Davis (originally the site of the University Farm), Berkeley continues to house programs in the agricultural sciences. Additionally, there remains considerable activity at the Riverside campus in southern California, originally the site of UC's Citrus Experiment Station.

32. As is customary, this does not count the less targeted but in all much larger state support for research that is built into the workload expectations (i.e., division of time between research and teaching) of state-supported faculty.

33. The total will likely rise substantially now that the California Institute for Regenerative Medicine (CIRM) is fully operational and releasing its grant funds (see below).

34. The Assembly is the lower house of the bicameral state legislature.

35. It should be noted that researchers at other California universities and nonprofit institutions receive many of the grants under these UC-run programs. As of 2007, the share of SRP grants held by non-UC institutions was just under 50% (Gruder 2007).

36. Moreover, since one explicit purpose of many of the grants is to "seed" projects that will attract larger-scale federal or private funding, the total expended on the projects is likely considerably larger.

37. The tobacco-related disease program has also been adversely affected by the state's diversion of a portion of the funds earmarked for research to the state Department of Health Services to support the department's cancer registry database (Gruder 2007).

38. The executive director of the Special Research Programs (Gruder 2006, 2007) reported that the tobacco program would have particularly benefited from additional funding as its support has declined more while proposals received have increased sharply in recent years. He asserted that the quality of proposals is such that the program could readily fund at least twice the current number of grants to support scientific projects that peer reviews of applications indicate are rated "outstanding" or "excellent" using NIH criteria.

39. Atkinson had formerly been director of the National Science Foundation and also chancellor of the UC San Diego campus, a university with a strong science and engineering emphasis and close ties with industry.

40. History and data in this paragraph and the next came from an interview with Julie Stein, acting director, IUCRP (Stein 2006).

41. In terms of Geiger's upstream-downstream distinction, the IUCRP is probably the one California research policy that strays somewhat from the general upstream policy thrust, but it would still best be classified as midrange rather than downstream in its primary orientation.

42. The university's Office of the President also provides about $5 million from internal funds.

43. Governor Gray Davis, a Democrat, was recalled (turned out of office) by voter referendum in 2003; at that time Republican movie star and former world champion body builder Arnold Schwarzenegger was elected governor. Evidently, the university considered the institutes' operating costs item a high priority in its budget request, and Schwarzenegger was able to package this support as part of a Governor's Research and Innovation Initiative in his FY2008 budget proposal (see http://gov.ca.gov/index.php?/press-release/5004).

44. The 2000–2001 state budget included $34 million for capital and other costs of the institute.

45. The states of Connecticut, Illinois, Maryland, New Jersey, and New York have launched stem cell research efforts of their own, though on a smaller scale (Fischer 2007; Hamilton 2006; Schwarzenegger Orders 2006).

46. Dr. Zach Hall resigned in April 2007. One reason he cited was the ill will created by disagreements within CIRM's complex governance bodies over the extent of public involvement in processes for oversight of capital projects (Somers 2007d).

47. CIRM's two-year courtroom odyssey likely came to an end in May 2007 when the California Supreme Court declined to review a lower court decision upholding its constitutionality (Somers 2007b). Under the US legal system, however, creative legal assaults on different grounds cannot be ruled out.

48. Fifty-four of the researchers supported were graduate students (CIRM 2006b).

49. The state treasurer announced the sale of the first $250 million in bonds in early October 2007 (California state treasurer 2007). However, after the legal triumph, there remained disagreement within the CIRM governance structure about how much should be spent on capital grants to universities and how such grants should be distributed (Brainard 2007). In December 2007, CIRM was forced to disqualify ten grant applications because letters of support for them were written by deans who also serve as CIRM board members; it was judged that an unacceptable conflict of interest was involved. Another case of intervention by a board member in a grant application has provoked audits by the Fair Political Practices Commission and the state auditor (California Rejects 2007).

50. Streitz (2006). Streitz suggested that a state policy of outright ownership would deter corporate interest in partnering with CIRM or licensing intellectual property resulting from its research, just as such efforts at the federal level had done when they were attempted.

51. The discussion on institutional structures over the next several paragraphs draws on Barbour 2005. The general finding of lack of capacity for science policymaking is consistent with Geiger's (this volume, chapter 11) characterization of state incapacity in this area.

52. Barbour (2005, 22) reports that during its brief life, this unit had a mandate to "track, support, inform, and provide coherence to state S&T policy" and "guided initiatives in biomass, next-generation Internet, rural e-commerce, high-tech manufacturing, and aerospace, among others."

53. CCST depends on membership dues provided by major scientific institutions and corporations in the state, together with foundation grants solicited for some individual projects.

54. CCST 2007a; Governor Schwarzenegger's State of the State 2006. UC San Diego was also a finalist for the BP Energy Biosciences Institute and would have received the $40 million in bond funding had it been successful. UC's Lawrence Berkeley and Lawrence Livermore Laboratories and the San Diego campus were all involved in the Petascale computer bid.

55. Geiger (this volume, chapter 11) suggests that such gubernatorial behavior in S&T policy is typical.

56. See Schrag 2006, chapter 2, for an insightful analysis of the reasons for this. He ties much of it to the rapidity of massive demographic and economic changes described earlier.

57. On occasion the university has even turned down state offers of support for research programs it did not feel were a good fit, but such action would be unlikely once a program is legislatively enacted. The University of California is, after all, the primary state agency for academic research, according to the Master Plan, and is also dependent to a considerable degree on legislative goodwill for its general financial support.

58. An interviewee told of a conversation at the outset of this program in which the legislative champion asked a university vice president whether he judged that the cure would take three years or four. The program spends about $15 million per year (see fig. 12.4), a very small amount in comparison to the sum of federal and private efforts.

59. This is not to say that the programs themselves do not make efforts to measure their performance. The IUCRP maintains the Economic Research Unit that surveys companies, researchers, and students about the impacts of its grants and publishes the results.

60. A recent exception is the process for academic reviews of the Institutes for Science and Innovation, which was initiated only after considerable agitation by the Academic Senate.

61. There is some measure of internal accountability, however, in that the IUCRP steering committee includes members of the university's committees on research and budget (Stein 2007).

62. Such advocates must be from outside California, but their local counterparts are invited to observe the panel meetings.

63. Discovery Grant applicants must also have a liaison and at least a one-for-one dollar match pledged from a California firm.

64. An exception was in agricultural research, which, as explained earlier, has a long history of state involvement.

65. It should be noted that this thrust runs generally counter to the post-Bayh-Dole-Act policies of the federal government.

66. The state has long declined to pay for any of the costs of the technology transfer function.

67. Examples of such terms might include requirements for due diligence by the licensee in utilizing the technology to create therapeutic products; mandates to provide Californians with "affordable" prices for such products; and requirements for payment of specified percentages of license revenues in royalties to the state.

68. This, of course, could only be true in a limited sense, for even a "breakthrough" advance would owe much to the published science from all over the world that has gone before.

69. The growth of non-peer-reviewed earmarks in the federal academic and science budgets in recent years is, to be sure, a worrisome trend. Notably, though, the Democratic congressional majority elected in November 2006 largely rejected such earmarks in its first major appropriations legislation.

70. This seems all the more paradoxical in light of the professionalism of the state governmental staff and the high level of sophistication of the state's intelligentsia generally.

71. Of course, these two shortcomings are related. The political stakes undermine incentives for objective evaluation.

REFERENCES

Alcocer, D. (2006) E-mail correspondence with Deborah Frankle, October 25. Alocer is a member of student support staff, University of California.

Ashton, A. (2006) UC Merced Lowers Bar on Student Enrollment; High Hopes for Growth Don't Survive the First Year. *Modesto Bee,* August 16, B1.

Associated Press. (1999) Four Fathers with Autistic Sons Raise $6 Million to Establish New Treatment Institute at UC Davis. *Ascribe,* March 23.

Auriti, E. (2006) Interview by author, September 14. Auriti is executive director for research, policy, and legislation in the office of the vice provost for research, University of California.

Barbour, H. (2005) The View from California. *Issues in Science and Technology* 21 (3): 21–24.

Bole, K. (2006) California's Stem Cell Fund Gets off the Ground. *San Francisco Business Times,* May 29.

Brainard, J. (2007) California Stem-Cell Researchers Ponder Next Steps after Court Victory. *Chronicle of Higher Education,* June 1.

California Business, Transportation, and Housing Agency. (2003) *Recommendations Improving the Effectiveness of the University of California Technology Transfer Process and Enhancing Technology Commercialization.* Sacramento: California Business, Transportation, and Housing Agency.

California Council on Science and Technology (CCST). (2006) Policy Framework for Intellectual Property Derived from State-Funded Research. Final Report to the California Legislature and Governor of California, CCST, Sacramento.

————. (2007a) BP Awards $500 Million Energy Research Program to UC Berkeley, Lawrence Berkeley National Laboratory. CCST, Sacramento, www.ccst.us/calinews/2007/20070201helios.php, February 1.

————. (2007b) Research, Education, and Energy Central to Schwarzenegger's Plans for 2007. CCST, Sacramento, www.ccst.us/calinews/2007/20070110sots.php, January 10.

California Institute for Regenerative Medicine (CIRM). (2006a) *Intellectual Property Policy for Non-Profit Organizations.* San Francisco: CIRM.

————. (2006b) Stem Cell Institute Awards First Scientific Grants. News release, April 10.

California Legislative Analyst's Office. (2006) *California's Fiscal Outlook: LAO Projections 2006-07 through 2011–12.* Sacramento: LAO.

California Rejects Stem-Cell Applications over Conflicts of Interest. (2007) *Chronicle of Higher Education Today's News,* December 14.

California state treasurer. (2007) Treasurer Lockyer Announces Results of First Stem Cell Research Bond Sale. News release, October 4.

Campaign for College Opportunity. (2006) *Keeping California's Edge: The Growing Demand for Highly Educated Workers.* Oakland, CA: Campaign for College Opportunity.

Coleman, L. (2007) Telephone interview by author, June 5. Coleman is vice provost for research, University of California.

Douglass, J. A. (2000) *The California Idea and American Higher Education: 1850 to the 1960 Master Plan.* Stanford, CA: Stanford University Press.

Farrell, K. R. (1995) Foreword to A. F. Scheuring, *Science and Service: A History of the Land-Grant University and Agriculture in California.* Oakland: ANR Publications, University of California.

Feller, I. (1997) Federal and State Government Roles in Science and Technology. *Economic Development Quarterly* 11 (4): 283–295.

Fischer, K. (2007) New York State Approves $600-Million for Stem-Cell Research. *Chronicle of Higher Education,* April 1, A23.

Geiger, R. and C. Sá. (2005) Beyond Technology Transfer: US State Policies to Harness University Research for Economic Development. *Minerva* 43:1–21.

Gledhill, L. (2005) Governor to Cancel Tuition Increases: Extra Costs Would be Covered by State's Improved Revenue. *San Francisco Chronicle,* December 29, A1.

Glenny, L. A. and T. K. Dalglish. (1973) *Public Universities, State Agencies, and the Law: Constitutional Autonomy in Decline.* Berkeley: Center for Research and Development in Higher Education, University of California, Berkeley.

Goldberger, M. L., B. A. Maher, and P. E. Flattau. (1995) *Research Doctorate Programs in the United States: Continuity and Change.* Washington, DC: National Academy Press.

Governor Schwarzenegger's State of the State: Research and Innovation. (2006) http://gov.ca.gov/sots/research_innovation.html, released December 27.

Gruder, C. L. (2006) Interview by author, September 14. Gruder is executive director, Special Research Programs, University of California.

————. (2007) Telephone interview by author, May 22.

Hamilton, D. (2006) Donors Sustain Stem Cell Effort in California amid Funding Battle. *Wall Street Journal,* August 16, A1.

Kerr, C. (2001) *The Gold and the Blue: A Personal Memoir of the University of California, 1949–1967.* Vol.1, *Academic Triumphs.* Berkeley: University of California Press.

National Center for Public Policy and Higher Education. (2005) Policy Alert: Income of US Workforce Projected to Decline If Education Doesn't Improve; and Policy Alert Supplement: Projected Drop in Income for California Most Severe in US, National Center for Public Policy and Higher Education, San Jose, CA.

———. (2006) *Measuring Up 2006: The National Report Card on Higher Education.* San Jose, CA: National Center for Public Policy and Higher Education.

Pelfrey, P. A. (2004) *A Brief History of the University of California.* 2nd ed. Berkeley: Regents of the University of California.

Rasmussen, W. (1989) *Taking the University to the People: Seventy-five Years of Cooperative Extension.* Ames: Iowa State University Press.

Scheuring, A. F. (1995) *Science and Service: A History of the Land Grant University and Agriculture in California.* Oakland: ANR Publications, University of California.

Schrag, P. (2006) *California: America's High-Stakes Experiment.* Berkeley: University of California Press.

Schwarzenegger Orders $150M Loan for State's Stem-Cell Research Institute. (2006). *Seattle Times,* July 21.

Shulock, N. and C. Moore. (2007) *Rules of the Game: How State Policy Impedes Completion in the California Community Colleges.* Sacramento: Institute for Higher Education Leadership and Policy, California State University.

Somers, T. (2007a) State Gives More Grants to Stem Cell Researchers. *San Diego Union-Tribune,* March 17, A1.

———. (2007b) Stem Cell Institute Bonds Can Be Issued. *San Diego Union-Tribune,* May 17, A1.

———. (2007c) Stem Cell Institute Opposes State Bill. *San Diego Union-Tribune,* April 11.

———. (2007d) Stem Cell Institute to Gather Public Input. *San Diego Union-Tribune,* May 3.

Stadtman, V. A. (1970) *The University of California, 1868–1968.* New York: McGraw-Hill.

Starr, K. (2005) *California: A History.* New York: Modern Library.

Stein, J. (2006) Interview by author, September 14. Stein is acting director, University-Industry Cooperative Research Program, University of California.

———. (2007) Telephone interview by author, August 13.

Stem Cell Research Gets $45 Million in California. (2007) *Albany (NY) Times Union,* February 17.

Streitz, W. (2006) Interview by author, September 15. Streitz is director, Office of Technology Transfer Policy, Analysis, and Campus Services, University of California.

Sturgeon, T. J. (2000) How Silicon Valley Came to Be. In M. Kenney (ed.), *Understanding Silicon Valley: The Anatomy of an Entrepreneurial Region,* 15–47. Stanford, CA: Stanford University Press.

Trow, M. A. (1993) Federalism in American Higher Education. In A. Levine (ed.), *Higher Learning in America, 1980–2000,* 39–66. Baltimore: Johns Hopkins University Press.

University of California. (2007) University of California Technology Transfer Annual Report 2006. UC Office of the President, Office of Technology Transfer, Oakland, CA.

University of California, Office of the President. (2007) Governor Signs 2007-08 Budget. News release, August 24.

Walker, R. A. (2004) *The Conquest of Food: 150 Years of Agribusiness in California.* New York: New Press.

CHAPTER THIRTEEN

Conclusion

DAVID D. DILL AND FRANS A. VAN VUGHT

Economists have argued for several decades that international forces have altered the basis of economic development. Markets have become increasingly interconnected. Goods, services, capital, labor, and knowledge move around the world in response to the best conditions. Natural resources are no longer the dominant factor in economic growth. In many developed nations there is now an observable trend toward international outsourcing of increasing numbers of traditional industries as well as routine service activities.

Globalization leads to increasing national specialization. This process of specialization, which is amplified by scale and learning effects, creates a reallocation of production processes between countries and forces nations to look for their comparative advantages in the international arena. Given this situation, national governments try to identify and develop their specific strengths. They try to increase their location attractiveness for business firms; they try to attract mobile production factors; they develop their sociocultural profiles; and they try to increase their innovation capacity.

Many nations now promote innovation as a principal means for economic growth. In particular, Western industrialized nations seek to identify their comparative advantages in the production of knowledge-intensive goods and services. To better compete in a globalized economy, these countries focus increasingly on knowledge, creativity, and innovation. In this context the Academic Research Enterprises (AREs) are becoming crucial objects of national policy. They form a large part of the environment of the knowledge economy and therefore are increasingly addressed by national innovation policies.

The adoption of national innovation policies in the developed nations is altering the framework conditions that have traditionally guided the ARE. The case studies collected in this volume indicate that the various governmental actors have similar motives for developing and implementing their national innovation policies. They all refer to the growth and importance of the "knowledge society," in which knowledge has become *the* crucial production factor and in which the creation, transfer, and application of knowledge are of prime importance for further social and economic development. They also speak of the processes of globalization and increasing international competition, in which the capacity to make use of new and applicable knowledge may lead to important strategic benefits. The creation, dissemination, and application of knowledge are regarded as core factors for the international competitiveness of regions, nations, and even whole continents. Therefore, these factors have become key strategic areas for policies at subnational, national, and supranational levels (World Bank 2007).

As a consequence, over the past decades all of the governments in our study have become more assertive in their efforts to influence the behavior of the ARE. The ARE, which is primarily funded by public sources, is perceived by policymakers to be one of their remaining means of influencing international competitiveness. Our collected cases suggest that a multilevel system is emerging in which the different tiers of government, with diverse amounts of authority and competence, interact in an attempt to find appropriate policies for steering the ARE.

In this final chapter we look at policies inspired by the National Innovation Systems (NIS) perspective in the collected case studies and suggest some possible lessons for the design of future policy instruments. Our focus, like that of the chapter authors, is on the three dimensions of the ARE—academic research, doctoral training, and knowledge transfer. We examine the linkages among the various actors and organizations, and we analyze the national framework conditions influencing behavior and beliefs in the university sector. After examining the general characteristics of the policy strategies implemented by the various governments, we examine the specific policy instruments employed and their impacts on the ARE as revealed in the reports of our selected case studies. Next, we explore two issues that emerged in the cases: the challenges of designing appropriate university autonomy and of governing in a multilevel system. We conclude by suggesting a new strategy of "mutual learning" that is likely to be particularly effective for improving the performance of the ARE in the new globally competitive environment of higher education.

Policy Strategies

In the present-day international context, policymakers are seeking to redesign their systems of innovation and research and adapt them to the new demands of globalization and competitiveness. For this they employ certain policy strategies, that is, processes in which policy instruments are related to policy objectives in the attempt to realize these objectives. Generally speaking, as our collected case studies show, the policy strategies consist of some combination of the basic notions of market coordination and central governmental planning.

The coordinative capacity of the market mechanism is well known. In a free market with perfect competition, prices carry the information on the basis of which decisions are made with respect to demand and supply. However, actual markets often fail to match the model of the perfectly competitive free market. In reality, one has to allow for transaction costs, scale effects, actors that are less than perfectly informed, production factors that are less than perfectly mobile, and nonhomogeneous goods. In addition, high barriers to entry to a market may give existing organizations monopoly power, or competition may take place by means of other mechanisms than prices (e.g., quality or reputation). In short, the perfectly competitive free-market mechanism seldom is a realistic option for policymakers (Teixeira et al. 2004; Weimer and Vining 2005).

But central governmental planning clearly also has its drawbacks. Central governmental planning is an approach to public-sector steering in which the knowledge of the object of steering is assumed to be firm; the control over this object is presumed to be complete; and the decision-making process regarding the object is completely centralized. In reality, governmental actors are unable to form comprehensive and accurate assessments of policy problems and to select and design completely effective strategies. Furthermore, governments are unable to monitor and totally control the activities of other societal actors involved in a policy field and therefore run the risk of noncompliance, inefficiency, and nepotism (Lindblom 1959; Van Vught 1989).

A "third way" thus has to be found. And this is what, in our case studies, the policy actors at the various levels of government appear to be seeking. These third ways appear to be specific combinations of the two basic notions of the free market, on the one hand, and of central planning, on the other. They are policy strategies that possess a set of policy characteristics: features that result from the relative emphasis on market coordination and central planning and create the specific appearance of these policies.

PRIORITIZATION STRATEGIES

The innovation policies analyzed in this volume can be divided into two broad categories. The first and largest category includes policies that can be described as *prioritization strategies.* These policies show characteristics like foresight analyses in the science and technology sectors, priority allocation and concentration of resources, and quality assessments of research outputs. They reflect continuation of the notions of central planning.

For example, in Australia both the commonwealth and the state governments have engaged in research priority setting, emphasizing areas of science that will enhance economic competitiveness. In Canada the governments have defined and funded Centres of Excellence in areas deemed strategic to the country's prosperity. In Finland TEKES explicitly funds university research programs in technology fields that are assumed to be priorities of the Finnish policy of industrial development. In the Netherlands the national Innovation Platform has selected a limited set of "national key areas" in which both fundamental research and knowledge transfer should be increased. The Foresight Assessments begun in the United Kingdom in the early 1990s were one of the earliest prioritization strategies in research funding. In the United States, the president's National Science and Technology Council has recently defined interagency research programs in areas of strategic importance to the national economy, and several states are now identifying specific technical fields and funding academic research in them with the expectation of stimulating economic growth.

These prioritization strategies also include national efforts to assess the quality of research outputs of the ARE. The Research Assessment Exercises (RAE) have been a major driver of the significant changes in UK university behavior. Similar, if less ambitious, efforts to link general university funding for research to government-determined output measures are also being experimented with in Australia, as part of the Institutional Grants Scheme, in Finland with performance-based contracts, and in the Netherlands with the "Smart Mix" program.

COMPETITION STRATEGIES

The other category of innovation policies targets market forces. These *competition strategies* show policy characteristics such as emphasizing competitive allocation of research-related resources, encouraging entrepreneurial university behavior, deregulating the university sector, and encouraging multiple sources of funding for the ARE. Thus, these strategies reflect a greater reliance on market coordination. The preeminent example of this strategy is the US federal science policy, with its emphasis on a national market composed of rivalrous private and state-supported universities, its limited federal control, and its competitive allocation of funding through a set of overlapping research agencies. But all of the countries examined in our cases are now experimenting with competition strategies, for example, by allocating less money for research via institutional block grants or general university funds and providing more resources via research councils and competitive grant schemes.

Aspects of the competition strategy can be found in more or less intensive forms in various OECD (Organisation for Economic Co-operation and Development) countries. For example, Australia, Canada, Finland, Germany, Japan, and the Netherlands have adopted a competitive approach to strengthening research doctoral training, either through competitive national fellowships to support PhD students or through competitive grants for the development of selected graduate or research schools, or both. Australia is also utilizing competitive funding for the allocation of university research facilities, Canada and Finland for the allocation of well-funded faculty chairs, and Germany for funds designed to identify and support university "excellence." The United Kingdom is attempting to further diversify the funding base of its universities by offering competitive "third-stream" funding to promote greater knowledge transfer between universities and industry. Similarly, Canada and several of the US states competitively award matching funds for research facilities and research projects as a means of inducing private industry to participate in and financially support university research.

THE STATE SUPERVISING MODEL

Although the prioritization and competition strategies that have developed as part of governmental innovation policies can be clearly distinguished, neither is a clear-cut specimen of the respective notions of market coordination or central planning. Rather, the two strategies are both examples of the "third way" mentioned previously. The two strategies in this sense can be interpreted as manifestations of the "state supervising policy model" (Van Vught 1989). This model is a combination of market coordination, which emphasizes decentralized decision-making by providers and clients; framework setting; and supervision by government. In the general policy model of state supervision, the influence by governmental actors is limited. Governments do not intrude into the detailed decisions and operations of other actors. Rather, a certain level of autonomy of these actors is respected and their self-regulating capacities are acknowledged. Governments in this policy model see themselves as the providers of the regulatory, financial, and communicative frameworks within which other actors can operate and as the supervisors of these frameworks.

However, as our cases show, the setting and supervision of governmental policy frameworks in this model can nevertheless have major impacts on the behavior of other actors. By introducing certain general quality assessment instruments or financial allocation mechanisms into their national policy frameworks, governments are able to strongly steer the ARE without introducing detailed regulation. The differences between the prioritization and competition strategies previously mentioned reflect the levels of impact governmental policy frameworks have on the national AREs. The policy characteristics of the prioritization strategy clearly show a higher level of guidance and restriction of the ARE than does the competition strategy. In the sections that follow, we explore these issues more systematically by examining the specific policy instruments described in our case studies and their impacts upon the ARE.

Policy Instruments

In the literature on policy sciences, the concept of policy instruments has a notable history. Classical authors in the field (Dahl and Lindblom 1953; Etzioni 1968; Mitnick 1980) have underlined the importance of a clear understanding of the various types of policy instruments. Others have pointed at the need to study the instruments' impacts and effectiveness (Ingram and Mann 1980; Mazmanian and Sabatier 1981). Several categorizations of policy instruments have been suggested in the literature (Bardach 1979; Elmore 1987; Hood 1983; Schneider and Ingram 1990; Weimer and Vining 2005), and we will follow a generally accepted categorization by distinguishing between legal, financial, and information policy instruments (Van Vught 1994). In the following discussion we describe the generic policy instruments that are being applied to the ARE in our case studies.

LEGAL INSTRUMENTS

Legal policy instruments are the strong tools used by governmental actors to assert their formal authority. Legal instruments are intended to command and to forbid, to commend and to permit. The adoption of national innovation policies in the leading OECD nations, however, has transformed the traditional use of laws and regulations to influence the ARE. The expressed goals of improving the quality and productivity of academic research and research doctoral training, particularly in science and technology, and better linking academic knowledge to social and economic development have motivated national governments to rely less on direct regulation of inputs, processes, and outputs and more on shaping the institutions that influence the behavior of universities. The detailed laws and rules by which many governments traditionally directly regulated publicly supported universities are being systematically reduced (i.e., universities are being deregulated) in order to provide more space for entrepreneurial action in service to society. For example, in the place of line-item budgets and civil service laws, many of the governments in our case studies are adopting new framework conditions for the ARE that are less detailed. However, these new framework conditions do not necessarily imply less governmental steering. In fact, they appear to be more effective as mechanisms of "remote control."

This shift from direct regulation to "steering at a distance" is illustrated in Finland and Japan. In these countries, the formal designation of universities as publicly supported corporations provides greater latitude for institutions to respond to emerging market opportunities; it also awards them greater discretion in their financial, personnel, and research activities. Because of the new policy focus on framework conditions rather than direct regulation, universities in the European Union, particularly, are now gaining much of the institutional autonomy, management flexibility, and contractual freedom that permitted Penn State University and the University of California to achieve international prominence as state-supported research universities in the United States, as explained by Geiger (chapter 11) and Zumeta (chapter 12).

Ironically, this shift in framework conditions is also detectable in the United Kingdom and the other Westminster countries (e.g., Australia and Canada), where publicly supported universities traditionally possessed, in comparative terms, remarkable professional discretion in their internal affairs (a situation once characterized as "the private management of public money"). Henkel and Kogan (chapter 9) stress that the adoption of the Higher Education Reform Act of 1988 and the Further and Higher Education Act of 1992 substantially altered the UK government's relationship to the universities; there was a shift from government's subsidizing independent professional institutions to government's making a formal exchange for public service. As is discussed below, this alteration is also reflected in the increasing adoption of contractual agreements between government bodies and universities, illustrated in the UK and Finland cases.

A further example of the new focus on framework conditions rather than direct regulation is the emphasis on reassigning intellectual property rights (IPR) in the university sector, lifting the restrictions imposed by government ownership of publicly funded research, and providing greater latitude and incentive for universities and their academic staffs to interact with private industry. First adopted in the United States with the Bayh-Dole Act, IPR-related reforms have been introduced in Japan, Germany, and the United Kingdom.

FINANCIAL INSTRUMENTS

The second category of policy instruments is financial instruments. Financial instruments reflect "the power of treasure," the influence of signing checks (Hood 1983). Crucial instruments in this category include contracts, bounties, and transfers (i.e., subsidies). Financial instruments may be used directly—as when governments introduce "quasi-markets" in which universities, as sellers, must compete with each other for research contracts from a monopsonistic government buyer—or indirectly through intermediaries, as when the government subsidizes doctoral students or private industries to purchase education or research, respectively, from higher education institutions.

Under the growing influence of national innovation policies, financial instruments have become an increasingly dominant means of steering the ARE. All our cases show a strong reliance by governmental actors on financial instruments, especially, as previously suggested, instruments that help implement the competition strategy.

With the prominent exception of US federal policy for the ARE, which utilizes competitive funding for nearly all academic research, research training, and related infrastructure grants, most countries in our study have utilized a dual funding model to support the ARE (Jongbloed and Vossensteyn 2001). General university funding (GUF) instruments have traditionally allocated funds incrementally on the basis of past history, formulas linked to input elements, and/or negotiation. Grants from research councils, in contrast, have traditionally been allocated competitively. However, research council grants have been determined not by formulas or past history, but by peer review, qualitative judgments based upon *expected* performance, and the capacity of applicants. Consistent with recent trends in OECD nations (Geuna 2001), several of our case studies reveal the continuation of dual funding policies, but with an effort by governments to simulate market conditions by allocating less research funding via GUF and larger proportions competitively through research councils.

A second new element in the financial policy instruments observed in our case studies is the association of GUF with output measures of "performance." In the case of the RAE in the United Kingdom, core institutional funding for research is fully determined by a formula based upon peer ratings of research quality at the subject level as well as indicators of publication productivity and grants received. In our other country cases, the linkage between GUF and performance is less pronounced. In these countries, peer reviews as an indicator for performance funding are not employed; instead, specific indicators of research output are used to assess eligibility for some portion of GUF. The design of these performance-based funding instruments in our collected case studies suggests that there is an emerging international consensus on the output measures most useful for assessing the effectiveness of the ARE. These include research publications and citations, research doctoral graduates, and competitive grants received. As Henkel and Kogan emphasize (chapter 9), even in the United Kingdom future versions of the RAE will likely rely to a much greater extent on similar metrics rather than on costly peer evaluations.

A third new element is the allocation of direct government support through competitive tenders, with specific objectives and limited budgets, on the basis of past performance and the "excellence" of the proposal (Geuna 2001). For example, competitive funding for research doctoral support is often linked to PhD completion rates as well to various measures of the research strength of the relevant departments or programs (e.g., existence of a Center of Excellence). In addition, our cases reveal that competitive funding instruments are being used to allocate the following:

- distinguished faculty chairs,
- grants for research infrastructure,
- research centers of excellence,
- graduate or research schools, and
- funds to achieve institutional "world-class" status.

In both the United Kingdom and Japan, competitively allocated infrastructure awards are also being used as incentives for institutional mergers that could help foster greater critical mass for research. The German competitively awarded Excellence grants have similarly encouraged mergers between universities and public research institutions.

In contrast, matching-fund instruments, which also are often competitively allocated, are being used by supranational, national, and subnational governments. These are most often employed to better link universities and private industry, for example, in cooperative research programs or in the development of cooperative research facilities. These matching-grant instruments encourage the creative pull of market forces in a way that the similar UK third-stream funding initiative, awarded on a competitive basis directly to universities, may not. One innovative financial instrument, the Canadian Foundation for Innovation—an institution capitalized by major grants from the national governments, thus assuring its independence—competitively allocates research infrastructure awards of up to 40% of a project, contingent upon matching funds from local governments and/or the private sector.

A financial policy instrument distinctive to research funding is indirect cost reimbursement. The United States has been unique in its heavy reliance on negotiated indirect costs as a substitute for GUF or institutional block grants. However, a number of countries have expressed concern that, if not fully costed, competitive grant systems may encourage universities to cross-subsidize research projects with funds intended for other purposes, such as teaching, in an effort to maintain the visibility and vitality of their research programs. These potentially dysfunctional effects have encouraged nations such as Canada and the United Kingdom to commit to the eventual full-costing of competitively awarded government research.

Finally, in recognition of the distributional impacts of competitive research funding, several countries—notably Australia, the United Kingdom, and the United States—are experimenting with financial policy instruments designed to enhance research capacity and strengthen the productivity of universities that have been less successful in attracting competitively awarded research funds. In a period when research support for higher education is increasingly motivated by its potential economic impact, compensatory research funding policies are often a reaction to political pressures for a more even distribution of national research funding. The EPSCOR program in the United States, for example, was an initial response by the National Science Foundation to congressional concerns that threatened the existing merit-based competitive research funding system.

INFORMATION INSTRUMENTS

Information instruments, the third category of policy instruments, allow governmental actors to send out messages and to provide responses. The instruments of information are used to communicate with other actors, to "launch" certain initiatives, to ask for reactions, to report on certain conditions, to facilitate decisions, and to assure quality and accountability.

With the emergence of national innovation policies, information instruments are playing an increasingly important role in the ARE. They are applied in evaluation and review processes as well as to help identify priority areas and centers of excellence. Several examples of the use of information instruments to identify research excellence are found in our case studies. The most comprehensive such approach is the German Excellence Initiative, designed to identify and strengthen a selected number of universities with "world class" potential. The more typical form of this instrument is to identify, through a competitive, merit-based process, national centers of excellence in strategically important research fields. This instrument is being applied in Canada with the Networks of Centres of Excellence program, in Japan with the Centers of Excellence initiative, in Finland by the Centers of Excellence program conducted by the Academy of Finland, and in the United States by the

Engineering Research Centers program of the National Science Foundation. In each of these cases, the policy not only designates particular research programs or centers as eligible for special financial support, but also signals to the world that the selected center has achieved or soon will achieve international standing in research.

In addition, information instruments are being used to assess and strengthen the quality of existing subjects and academic programs. This is a critical role of the RAE in the United Kingdom as well, but because of its linkage to institutional research support noted above, the RAE is most often perceived as an instrument for allocating financial resources and concentrating research funding. In Australia a similar instrument, the Research Quality Framework, was aborted, but a new research quality–assurance process, using metrics, is under consideration. In a number of our case studies, information instruments similar in design to the RAE are being applied to the university sector, but they are not as directly connected to research funding. For example, the German Science Council is experimenting with research evaluations in selected science fields; the US National Research Council publishes systematic evaluations and rankings of research doctoral programs; and the Netherlands Royal Academy of Sciences provides recognition for distinguished research schools. However, the goal of these instruments is clearly longer-term improvement rather than immediate accountability.

A related evaluation instrument focuses not on academic fields or programs, but on each university's capacity for managing research. Australia, the Netherlands, and the United Kingdom have all developed information-based policy instruments requiring the review of university research programs and/or research management plans. The design of this instrument is well illustrated by its emergence in the Netherlands. Initially, as Jongbloed (chapter 8) notes, the Association of Universities in the Netherlands (VSNU) conducted research assessments, similar in focus to those described above, by international panels of all subject fields in the university sector. In 2003, however, the focus of the assessments shifted to each university's management of research. Universities in the Netherlands are now required to arrange their own external assessment of their research program, following a standard evaluation protocol. Some accountability is provided by the publication of the research assessment, but again, and unlike the RAE, the assessments are not used to inform government research funding.

An interesting and innovative approach to the use of information instruments is seen in the Open Method of Coordination (OMC), as it has been developed in the European Union's innovation policy. The OMC uses agreed-upon indicators for benchmarking the policy performance of the EU member states. By doing so, peer pressure and "naming and shaming" mechanisms are triggered, which appear to result in stronger commitments to the joint European policy objectives and greater efforts to reach them. As we review the legal, financial, and information policy instruments implemented in our collected case studies, the increasing relevance and effectiveness of information-based policies for steering the ARE stand out. Henkel and Kogan (chapter 9) argue in their analysis of the RAE in the United Kingdom that information policy instruments that are applied consistently over time, that are grounded in the core academic process of peer review, and that focus on improving research outcomes relevant to the reputations of academic staff and institutions are highly consistent with academic values and have therefore inspired a significant degree of related self-regulation by universities. Such instruments also help provide an influential signal to potential customers of university-based research. As we suggest in our concluding section, well-designed, information-based policy instruments may therefore offer greater potential for improving the contribution of universities to social and economic development than either legal or financial instruments (Majone 1997).

Impacts on the Academic Research Enterprise

As noted, the new and rapidly growing focus on national innovation in many countries has led to substantial alterations in the framework conditions of the ARE. What are the impacts of these changes on the ARE? What are the effects with regard to the performance of universities? How do the various policies influence universities' strategies and operations? Do differences in the use of policy instruments create different sorts of impact? In the following paragraphs we explore such issues that may be relevant for the present dynamics and the future perspectives of the Academic Research Enterprise.

In exploring the impacts of the discussed policy instruments, it is obvious that some caution is appropriate. Although we purposely selected a sample of the leading OECD nations in order to provide a more valid basis for comparison, the complexity of the forces influencing the behavior of the ARE in each country makes the discovery of associations between particular instruments and particular effects challenging. Nonetheless, as outlined above, we gain some confidence from the common calls for change in the ARE motivated by the shared perspective on national innovation and from the similarity of policy strategies and even policy instruments being implemented in our focal countries as well as from the comparable observations about both positive impacts and problems reported in our collected case studies. In this section we try to clarify the nature of these reported impacts.

IMPACTS ON RESEARCH PRIORITIES

The adoption of the national innovation perspective has encouraged supranational, national, and subnational governments to be more assertive in defining particular topics or areas of research through Foresight processes, specified research programs in fields like nanotechnology, the designation of national centers of excellence, and other related instruments. The impact of this approach is not yet clear, although the similarity in the applied fields defined across countries, and particularly among subnational governments, is striking and raises the question whether these decisions reflect best scientific insight or some degree of policy mimicry.

At the supranational and national levels, where scientific agencies are apt to be most knowledgeable and where peer-review processes are common, there is likely less need for concern about this apparent redundancy. At the subnational level, as both Geiger (chapter 11) and Zumeta (chapter 12) describe in the cases of California and Pennsylvania, governments may lack the institutions to act as honest brokers in the designation of fields for research. Reviewing the new and growing investment of the US states in applied areas of academic research, Geiger questions the value of these largely "downstream" strategies, which may be defined by regional government bureaucrats with little academic input. Instead, he calls for a greater focus on more upstream, peer-reviewed, and basic research, appropriate to the economic needs of a particular region. Similarly, a recent analysis of the "European paradox" (Dosi, Llerena, and Labini 2006)—the acknowledged weak relationship between the quality of academic research and economic development in many European countries—criticizes the overly applied focus of the EU Framework Programmes and recommends a greater emphasis on competitively awarded basic research support at the supranational level, now the major focus of the newly initiated European Research Council.

IMPACTS ON UNIVERSITY RESEARCH

The policy strategies employed by national governments and their related use of policy instruments also appear to have had direct effects on the behavior of universities, thereby producing discernible changes in overall national AREs. The combination of information and financial policy instruments is leading universities to develop more specific institutional strategies in the three basic segments of their research mission: research, research doctoral training, and knowledge transfer. International forces and the market competition introduced by these new policy instruments have also led to major reforms in the organization of publicly supported universities. Universities in all of our case-study countries are now being encouraged by government to adopt a more corporate type of organization, with a stronger central administration, better ties to external stakeholders, and greater independence in the management of their internal affairs—a form well illustrated by Clark's (1998) concept of the "entrepreneurial university."

The national emphasis on competitive financial instruments for the ARE has also affected the internal research allocations of universities. The typical reaction of individual universities to the national research policies is to increase the quality and size of their successful research fields and hence to focus and concentrate their academic efforts in certain specialized areas. The outcomes of these institutional specialization and concentration processes, of course, differ according to the conditions of the various institutions. Previous academic performance, the affiliation of top-level researchers, and in particular the financial resources of a university are of crucial importance in developing an institutional research profile. But the general effect appears to be a trend within universities toward "focus and mass," toward specialization and concentration.

The new policies also appear to be making universities in nearly all of our case countries more productive in their output of publications and research doctoral graduates as well as in their patenting and licensing activities. In Australia and the United Kingdom, this improvement has also occurred in universities newly designated after the abolition of the binary line, but the recent evidence from the United Kingdom suggests that any closing of the performance gap between the old and the new universities brought about by these new policies has now slowed if not ended (Crespi and Geuna 2004). This analysis suggests that the adoption of performance-based research funding for the ARE creates a one-time shock to the overall system, which initially motivates increased research productivity in all universities eligible for the funding but over time is most likely to lead to an increased concentration of research in those institutions with richer resources, larger numbers of internationally recognized academic staff, and established reputations (Soo 2008).

Marked improvements in the organization and management of university research activities and programs were also reported in most of our cases. It is likely that this improvement in university research programs is due not only to the related policy instruments reviewed above, but also to the general reductions in funding for publicly supported universities that have occurred in conjunction with the massification and expansion of higher education in most countries (Williams 2004). As a consequence, universities in some of our case-study countries have necessarily become more highly motivated to pursue alternative sources of revenue for their research programs and therefore have been required to develop the research centers and internal research management processes necessary to survive in this competitive market.

A reported negative impact of these new instruments is the diminishment of research support in particular fields, often in unanticipated ways. Historically, the social sciences and humanities have received substantially lower levels of research support in all of our case countries than have the basic sciences, medical sciences, and engineering. The current concern with national innovation and economic development, as well as the use of the new policy instruments of academic research, further disadvantages research in the "softer fields." Although research support for the social sciences and humanities has been stronger in countries other than the United States that subscribed to a dual-funding model for universities, these fields have suffered there, too, as a result of the general shift of funding to research councils and the increasing national emphases on the applied sciences and technology. In recognition of this problem, both Canada and the United Kingdom have recently established research councils in the humanities and social sciences.

Less obviously, the case studies suggest that the strong emphasis on research programs in the applied sciences and technology along with performance-based funding can also result in inadequate support for research in some basic science subjects, such as chemistry, physics, and mathematics, which serve as the critical foundation for many technical and applied fields (Cohen, Nelson, and Walsh 2002). In the United Kingdom, the concentration of research funding brought about by the RAE has led many universities to reduce or eliminate basic science departments that do not receive the highest rating. In the United States, despite a recent initiative by the National Science Foundation to increase funding for the basic sciences, shifts in research priorities by the large, mission-oriented agencies like the Department of Defense and NASA (the National Aeronautics and Space Administration), which fund significant amounts of academic basic research, may still result in reduced funding in foundational science fields. These concerns suggest that the more competitive and dynamic environment of the ARE, which these new policy instruments helped create, may now require national governments to take more active steps to define particular subjects as in the national interest and to assure that these fields receive adequate support for research and doctoral education.

Finally, the increased emphasis on research performance and productivity reflected in our case studies is frequently accompanied by expressed concerns about the possible effects of these policies on the nature of the research conducted. Several of our case studies note that the increased incentives for applied research and knowledge transfer may reduce the amount of basic research and over the longer run actually retard or diminish innovation by reducing the number of significant discoveries in fundamental knowledge. However, the studies on this issue in both the United Kingdom and the United States, where performance-based and competitive research policies, respectively, are arguably most advanced, does not yet indicate a reduction in the proportion of basic research being conducted (Dill, chapter 10; Henkel and Kogan, chapter 9). As noted in the US case, the emphasis on competitive research funding also requires the investment of time by researchers in developing and revising grant proposals. Unless this increase in effort is accompanied by commensurate increases in the amount of funds awarded, which is not always the case, the overall productivity of the ARE may be negatively affected.

IMPACTS ON RESEARCH DOCTORAL TRAINING

Research doctoral training, particularly in the sciences and technology, is an important component of the national innovation perspective that has motivated the development of new policies designed to enhance PhD education in our focal countries. This effort has been largely successful, as our country case studies illustrate, increasing both the numbers of PhDs and the program completion rates, primarily through policies focused on financial support for doctoral students. Outside North America, policy instruments providing incentives for restructuring doctoral education have also been effective. Policies encouraging the adoption of more structured, "taught" research doctoral programs appear to have been a valuable and influential innovation. Initially, in order to make structured research training more practical and economically viable, many European universities developed collaborative, cross-institutional doctoral training in selected fields. The Finnish National Graduate School, the National Research Schools in the Netherlands, and the joint doctorate networks in the European Union are examples of the effort by universities and government policy to combine research specialization with sufficient critical mass to make taught doctoral programs more feasible. But there is some evidence from the Netherlands and Finnish cases that single university-based graduate schools, which permit better institutional control over the design, development, and assurance of quality in PhD programs, may ultimately be found superior to decentralized research doctoral networks or schools. There is also some evidence from the United States that well-designed research doctoral rankings may be particularly influential instruments for improving PhD programs in the increasingly competitive global market for doctoral training.

While the necessary critical mass for high-quality research doctoral programs remains controversial, the case studies do suggest a policy emphasis on concentrating research doctoral education in universities with proven, high-quality programs of research. In several countries financial support for research doctoral programs is increasingly linked to indicators of research quality, such as numbers of publications, competitive grants received, existence of centers of research excellence, and assessments of subject and/or research doctoral program quality. In contrast, while the US universities that win the largest number of competitively awarded research grants thereby have greater resources to support research doctoral students, policies formally linking financial support for PhD programs to indicators of research quality have been less common in the United States. Perhaps because of this, a significant amount of recent growth in US PhD production has occurred in universities possessing lower-ranked doctoral programs.

All of the countries studied are also seeking to make research doctoral education more globally competitive and are therefore attempting to attract more international students as a means of strengthening their programs. In both the United Kingdom and the United States, a large proportion of recent PhDs have been foreign nationals. The United States especially has historically pursued a strategy of recruiting the "best and the brightest" from around the world, and recent proposals advocate an aggressive continuation of this policy to safeguard the American position as a leader in technical innovation (Galama and Hosek 2008). However, as Dill notes in chapter 10, such a policy is basically a zero-sum game. It is not a viable policy for all nations; and, given the global changes in higher education systems, it may be a questionable long-term policy for the United States. The evidence in the US case suggests that heavy reliance on an international recruitment strategy, particularly a focus on able students from developing nations, restrains salaries for all PhD graduates in a country and thereby suppresses domestic demand for research doctorates. The number and attractiveness of research doctoral and postdoctoral fellowships and scholarships available for domestic students, therefore, may be a critically important variable in designing effective national policies for innovation.

IMPACTS ON KNOWLEDGE TRANSFER

As our case studies illustrate, a major impact of the national innovation policies is that knowledge transfer has become an accepted and valued element of the general mission of most universities. Despite initial reluctance and even controversy in some institutions, most of our cases comment on the significant changes in university culture that have occurred over recent decades; a much more entrepreneurial and utilitarian orientation to both university education and research programs has developed. Universities now increasingly focus on their potential role as regional partners in innovation "clusters"; they develop programs with business and industry; they open up technology transfer offices; they offer consultancy and training activities in order to assist entrepreneurs in making use of new knowledge; and some even adopt an innovative entrepreneurial character as an institutional identity. In Europe a group of "entrepreneurial universities" have organized themselves into a cooperative network, the European Consortium of Innovative Universities.

As with publications and doctoral students, our case studies report increases in knowledge transfer activity, as indicated by the numbers of patents, licenses, and industrial start-ups, although the economic benefit to society of this substantial growth in activity is less clear. A much-debated topic in the context of knowledge transfer is policies on intellectual property rights (IPR). The original changes in the IPR legislation in the United States—the Bayh-Dole Act—were motivated by a desire to speed knowledge to market; patent and licensing rights were reallocated to universities through new laws designed to increase university incentives for knowledge transfer. The policy was never expected to create a major new source of funding for the ARE. However, with the growing competition for academic research monies in the United States and around the world, universities in all of our case countries are more aggressively seeking research revenues from other sources and, in many instances, have interpreted new IPR legislation as an exhortation to "cash in" on their research outcomes. The evidence from our collected cases is that the majority of universities in the OECD countries are at best breaking even, and many are suffering net losses from their investments in technology transfer offices and affiliated activities. While many universities see their technology transfer expenses as a necessary investment that they expect to bear significant fruit over time, Geiger's (2007) research in the United States suggests that over the longer term, the institutions that do reap some financial benefit from patenting and licensing are the most highly ranked and best-known research universities. But even in these institutions, there tends to be a natural limit to the amount of such revenue that can be earned.

One unintended impact of public policies that emphasize IPR as a means of stimulating academic knowledge transfer is their influence upon the core processes of academic science. Because of increased incentives for universities to patent and license their discoveries as a means of raising revenues, some theoretical results and research tools that have traditionally been freely available to other scholars and researchers are now being restricted. This constriction of open science may in fact lessen the economically beneficial "spillovers" that are a primary rationale for the public support of basic academic research. Policy instruments intended to provide incentives for knowledge transfer, therefore, have to be designed with particular care to maintain the benefits of open science.

Research on sources of innovation in industry raises additional questions regarding the effect of national knowledge transfer policies on the "hard" artifacts of academic research (Cohen, Nelson, and Walsh 2002). Patents and licenses are influential on innovation and profits in a relatively small number of industries and technical fields, biotech being the most prominent example. This reality helps explain the natural ceiling on patenting and licensing revenues that Geiger (2007) discovered in leading US universities. More influential for most industries are the "softer" knowledge transfer processes, such as publications, meetings, the use of consultants, and the hiring of new PhD graduates, whose added expertise is a primary means of transferring academic knowledge to industry (Agarwal and Henderson 2002; Cohen, Nelson, and Walsh 2002). As Geiger (2007) notes, public policies that highlight the "hard" outputs of academic research are, therefore, likely to undersupport knowledge transfer that is beneficial to society. In the policies implemented by the European Commission and by some of the European countries studied in this volume, the emphasis on patenting and licensing appears to be more limited than in the United States. Instead, the knowledge transfer focus is largely on the exchange of people, the increased production of research doctorates, and the stimulation of start-up firms. This European approach to knowledge transfer is "softer" than the US focus on licensing and patents but, as a first comparative study shows, not necessarily less effective. Despite less effort in terms of invention disclosures and patent applications, the EU countries execute more licenses and create more start-up firms (but have fewer patents granted) than the United States (Van Vught 2007).

IMPACTS ON INSTITUTIONAL DIVERSITY

Reviewing the policy impacts discussed in our cases raises one consistent question: is there is an overall diversification effect at the level of the *system* of the ARE as a result of the various reactions by universities to their altered framework conditions? The introduction of market forces and greater competition into higher education should, according to economic theory, lead not only to greater productivity in research outputs, but also to greater allocative efficiency for society as universities are required to respond more effectively to the needs of their various research patrons.

Our case studies offer some interesting insights into this issue. The cases suggest that there has been greater diversity historically in research and research doctoral training performance among the publicly supported universities in our focal countries than has often been acknowledged. Because of the distinctive national policies of the United States, the US ARE has long been considered a system with substantial diversity in quality, with highly ranked academic research concentrated in a minority of its universities. About one-third of the nation's universities conduct more than two-thirds of federal academic R&D in addition to graduating over two-thirds of research doctorates. In contrast, the national policies of other countries in our sample were designed to achieve a common "gold standard" of performance, or what Enders (chapter 7) has called an "egalitarian homogeneity," among publicly supported universities. But with the possible exception of the universities in the Netherlands, all of our country cases reveal that research and PhD training have been concentrated in a relatively small part of the university sector.[1] The general impact of the new policy instruments is to concentrate academic research in a smaller number of institutions as well as in universities in economically advantaged regions.

In Canada, according to Fisher and Rubenson (chapter 3), the key development in the ARE over the past fifteen years has been the emergence of a clear and separable strata of research universities. In Finland the government has made a public commitment to concentrate research and PhD training in a small number of comprehensive universities. As Van Vught argues (chapter 5), a perhaps unintended effect of the EU Framework Programmes is a growing stratification between those universities that are successful in receiving Framework Programme funding and those that are not, thereby producing an elite category of European research universities. In other countries national innovation policies have clearly been designed to create a group of "world class universities." The RAEs in the United Kingdom and the Excellence Initiative in Germany are obvious examples.

Although the evidence of increasing research concentration is apparent, we conclude that the new policy instruments are not completely successful in encouraging a diversity of university roles and missions. These policies certainly stimulate universities to engage in international competition, but they provide insufficient incentives for the development of true system diversity. While global market forces as well as government-designed prioritizing and competition strategies have been effective in helping differentiate a class of international research universities, the existing policy instruments appear inadequate for steering the majority of a country's universities into constructive roles as part of a national ARE. Academic autonomy—although different in different systems, as discussed below—is such that scholarly norms and values have become major drivers of institutional homogeneity. The forces of academic professionalism and the eagerness to increase individual and institutional academic reputations impels all universities in the new, more competitive environment to imitate the leading research universities rather than to diversify their missions and profiles.

All universities try to recruit and employ the best scientists, that is, those scholars with the highest recognition and rewards, the highest citation impact scores, and the largest numbers of publications. In order to be able to do so, they need to increase their research expenditures (since the research context attracts scholars), creating a continuous need for extra resources. Given their wish to increase their reputation, universities also try to attract the most talented PhD students. They use selection procedures to find them, but they also offer grants and other facilities in order to attract them, again leading to a continuing need for additional resources. The major dynamic driving all universities is therefore an increasingly costly "reputation race" (Brewer, Gates, and Goldman 2002; Van Vught 2008), in which universities are constantly trying to display their best possible academic performance and which prompts a permanent hunger for financial revenues. In this sense Bowen's famous law of higher education still holds: "in quest of excellence, reputation and influence . . . each institution raises all the money it can . . . [and] spends all it raises" (Bowen 1980, 20).

The result of these forces is that the new policy instruments for the ARE have not yet engendered the allocative efficiency for society that they were expected to achieve. In our concluding section we suggest a strategy for addressing this problem, but first we explore two related policy issues that emerged in our review of the case studies—university autonomy and multilevel governance.

University Autonomy and Regulation

While the legal, financial, and information policy instruments applied in our selected cases have had a major impact on the behavior and performance of the universities, the framework conditions for each of our countries are unique, a product of each nation's particular history. They are what economists call "path dependent" (Nelson and Winter 1982) and hence also potentially susceptible to "lock-in," when stakeholders resist the significant reforms necessary to maintain a productive and efficient ARE. Our case studies reinforce two key points discovered in recent research on university autonomy. First, there is substantial variation in university autonomy across nations as well across states and provinces within federal systems (Aghion, Boustan, et al. 2005; Aghion, Dewatripont, et al. 2007; Martins et al. 2007). Second, institutional control over budgeting and academic personnel is an important contributor to academic research performance (Aghion, Boustan, et al. 2005; Aghion, Dewatripont, et al. 2007). Zumeta (chapter 12) refers to "the nimbleness required [of universities] in an era of heightened global academic competition." Institutions such as the US private universities that operate under framework conditions permitting great independence and managerial flexibility may, all other things being equal, possess a comparative advantage in the international academic reputation race. A crucial policy challenge, of course, is to define the level of institutional autonomy and flexibility publicly supported universities require to operate effectively in the current competitive international environment.

From our cases we deduce the following characteristics that universities seek in their efforts to become vital players in the growing international competition for research reputation (see Dill 2001):

- lump-sum funding,
- autonomy in purchasing and contracting,
- ownership of facilities,
- autonomy in faculty hiring and wage setting,
- self-accrediting status, and
- authority to set and retain tuition and fees.

Several of these characteristics, particularly those regarding flexible funding and authority over contracting and facilities, are relatively uncontroversial and consistent with the general trend in most developed countries toward "state-supported" rather than completely state-financed universities. In their new context, universities require the managerial authority to efficiently administer revenues from different sources. However, in many developed countries, regulations regarding academic personnel, academic programs, and fees are contentious.

First, with regard to personnel, in some of our case-study countries, civil service regulations and state-determined salary schedules govern the employment and work conditions of academic and research staff as well as research doctoral students. These national regulations, not in all cases designed originally for university personnel, may be too inflexible for universities engaged in fast-moving, frontier research. Autonomy over hiring and wage setting is essential to effectively compete in the international market for the best academic and research staff and research doctoral students (Aghion, Dewatripont, et al. 2007). Furthermore, state restrictions on personnel often have the unintended effect of making universities more inbred. The hiring by a university of its own doctoral graduates has been shown to have a negative effect on both the productivity and impact of academic research (Horta, Veloso, and Grediaga 2007). In addition, an inbred academic staff is less likely to import into its university new research and technology from the international system of science—one of the influential ways universities contribute to national systems of innovation (Mowery and Sampat 2004).

Second, many of the world's leading universities are essentially "self-accrediting institutions." Thus they gain the comparative advantage of being able to quickly and easily introduce new and innovative research doctoral programs. It is clear that the academic integrity of these universities' internal quality assurance systems can be secured by external institutional accreditation and/or academic audit procedures as well as by their ambitions to maintain a high academic reputation. But the universities' ability to lead in innovative academic programming is not slowed by bureaucratic processes.

Finally, our case studies reinforce empirical research indicating that institutional autonomy to decide on the sources and structure of funding (e.g., the capacity to independently set and retain tuition and fees) is a necessary institutional prerogative with significant potential influence for strengthening academic research performance (Aghion, Dewatripont, et al. 2007; Martins et al. 2007). Distinguished private universities in the United States have long enjoyed a comparative advantage in fee-setting. While their international research reputation would not be possible without access to the generous federal research funds competitively available in the United States, the ability to independently set and retain tuition fees provides a substantial and totally flexible source of funds that these universities have skillfully exploited to improve their research activities. Over the past two decades, this ability to independently control tuition fees has been a primary reason for the increasing gap in reputation between distinguished private universities in the United States and both US and non-US publicly supported universities.[2]

Our national cases suggest that this autonomy has also proven crucial for publicly supported universities. Both the University of California system and Pennsylvania State University in the United States were able to weather substantial state reductions in per-student financial support primarily because they possessed the independent authority to set and retain tuition fees. Utilizing this authority, they have been able to offset state cuts in appropriations and to strengthen their research reputations over time. Publicly supported universities in the United States and, for instance, in Germany that lack this independent authority to raise fees have been forced to absorb increasing numbers of students without commensurate financial support. As a consequence, these universities have experienced difficulties in maintaining their international research reputations.

While important arguments exist regarding the relationship between university access and tuition fees (Johnes 2004), universities possessing an international research reputation attract students from around the globe in part because these institutions provide substantial private benefits to their students. Permitting such universities to charge those students who can afford to pay market-appropriate tuition and fees is a rational means of sustaining an ARE in a competitive, international market.

The nations discussed in this volume show substantial differences with regard to the level of autonomy afforded their universities. Particularly in Japan and some

European countries, the legal policy instruments used to steer the AREs create serious limitations for needed university autonomy. As a result, the universities in these countries do not have sufficient managerial authority to compete effectively in the international reputation race. In the competition with US private universities, some state-supported US universities as well as universities in our selected OECD countries are confronted with comparative disadvantages. They lack a sufficient level of flexibility or "nimbleness" to act effectively in the international competition.

The universities that are successful in the international rivalry for research reputation are operating in essentially a different category of competition. This special position is largely a consequence of regulatory frameworks that allow these universities to use their autonomy for competing internationally. In those countries where the regulatory frameworks imposed by governments are still fairly restrictive, the universities largely lack this ability. In this sense, different framework policies do cause substantial differences in the outcomes of the international academic reputation race.

Of course, the danger of the less restrictive regulatory approach outlined above is that, if inappropriately designed, it could further undermine institutional diversity by encouraging all universities to participate in the costly reputation race to become world-class institutions. For this reason the type of autonomy we are recommending would need to be assigned, as currently is done in Japan, selectively. What is needed, in our view, is a regulatory framework that recognizes the existing differences in market contexts, in which the institutional autonomy necessary for competing internationally is awarded not by institutional category or title, as is now too often the case, but rather by university performance. Only those universities that already have developed a capacity for productive research with international impact, for high-quality research doctoral education, and for attracting significant numbers of competitive research grants would be awarded the type of autonomy suggested here. As in international soccer leagues, access to this regulatory framework would be permeable, based upon current performance. New institutions might develop the capacity over time to become eligible for this additional autonomy.

New Forms of Governance

Our cases suggest that a multilevel system of governance is emerging in which the policy strategies of supranational, national, and subnational sectors of government must be more systematically related in order to increase the performance of the ARE. Depending on the policy roles they seek to play and the policy instruments they apply, these governmental actors, to a large extent, shape the contexts in which universities and other research organizations operate.

The spatial principle of public goods (Oates 1999) offers a potentially useful guideline to the assignment of governmental responsibilities in the vertical governance structure related to the ARE. Research and research doctoral education, for example, are jointly produced by national and subnational governments and, in the EU countries, involve the supranational government as well. Even in apparently unitary systems, such as in Finland and Japan, local and regional governments are increasingly influential in the knowledge-transfer activities associated with the ARE. As suggested in the majority of our case studies, there is a significant policy debate concerning the appropriate roles and authority of these different government levels with regard the ARE.

The spatial principle suggests that "the provision of public services should be located at the lowest level of government encompassing, in a spatial sense, the relevant costs and benefits" (Oates 1999, 1122). As we suggest in the following sections, this spatial perspective is potentially useful for determining the appropriate level for the provision of public goods in both federal and unitary systems of higher education. For example, from this perspective, published basic research and research doctoral students are highly mobile public goods, and the relevant social benefits they generate may not be returned to the region or state in which they are produced. Consequently, as suggested in a number of our case studies, basic research and research doctoral education may be insufficiently supported by subnational governments. This principle therefore provides some justification for the observed centralization of financial support for basic research and research doctoral education at the national level (and, in the European Union, at the supranational level).

Consistent with this perspective, national governments in both unitary and federal countries are placing an increased emphasis on specification of academic research priorities and the financing of academic research as a means of better steering their AREs. One new aspect of the changing governance systems of the ARE is the elaboration and restructuring of national governmental bodies to better coordinate science and innovation policies. Our case studies show that the new national innovation policies—focusing on the knowledge flows of discovery, dissemination, and application—create pressures for more coordination across ministries, funding councils, or agencies as well as across governmental policy fields that historically have functioned quite separately. The performance of the ARE is now understood to be influenced by a wider array of policy issues, including immigration policies, regulations regarding Intellectual Property Rights, and labor regulations affecting the mobility of researchers and students.

Finland, for example, one of the earliest adopters of a national innovation policy, in 1987 established the Science and Technology Policy Council, chaired by the prime minister, to advise the government and the ministries on issues related to science and technology, scientific research, and researcher training. At about the same time, Canada created the Ministry of State for Science and Technology and the National Advisory Board for Science and Technology, composed of university, industry, labor, and government representatives, to advise the prime minister. In the Netherlands the National Innovation Platform has been formed to better coordinate the work of the ministries related to knowledge, innovation, and R&D policy. In the United Kingdom Prime Minister Gordon Brown instituted the Department of Innovation, Universities and Skills to integrate innovation, research, and higher education and also created the Funders Forum to better coordinate the research allocations of regional development agencies, charities, business, the Higher Education Funding Councils, and the research councils. The emergence of these new government coordinating mechanisms is a further symptom of the obsolescence of traditional governance structures and the need to reconceptualize the appropriate framework conditions for the ARE.

Historically, decentralized state-level governments, such as those in the United States and Germany, have supported university research in the fields of engineering and agriculture as a means of aiding the development of local business and industry. These policies have often yielded highly productive regional "clusters," where closely related and geographically proximate networks of university researchers and related businesses provided a strong base for regional development (Braunerhjelm and Feldman 2006).

Our cases suggest that in the globally competitive economy, many subnational governments in federal systems, such as those of Australia, Canada, Germany, and the United States, are now seeking to utilize their publicly supported universities as a means of local economic development. However, as described above, subnational governments may lack the institutional structures and the political will to invest effectively in university research. As in Pennsylvania, political pressures from more proximate stakeholders may influence subnational governments to invest in science policies that support immediate job creation in business, or "downstream" research projects closely related to industry, rather than building links between fundamental "upstream" university research and the economic needs of the region. Or, for similar reasons, subnational governments may engage in "scatteration" (Geiger, chapter 11), allocating research funds among all their universities rather than concentrating research funds on the most worthy research projects through competitive awards and merit-based peer review. Even when subnational governments do invest in upstream research programs that aim to create knowledge in fields such as biotechnology, nanotechnology, and materials sciences, they may, as in California, permit users to be too closely involved in defining the nature of research rather than allowing a more independent scientific analysis of relevant research options. Finally, both of the US state cases suggest that the more immediate political perspectives of subnational governments may lessen incentives to conduct truly objective and independent evaluations of the effectiveness of regional knowledge-transfer policies and university research investments. Therefore, while the identification, support, and dissemination of regionally relevant university research remain an appropriate and necessary role for subnational governments, an increasingly important policy role for national governments is to develop the capacity of the lower-level governments to perform this role effectively.

Consistent with the previously mentioned spatial principle of public goods, subnational governments also appear especially motivated to support academic programs in professional and/or vocational fields, because the graduates of those programs will likely remain in the region. Subnational governments, as noted in the California case, are therefore unlikely to provide adequate support for research doctoral programs, whose highly mobile graduates follow the best job opportunities, wherever in the world they are found (Aghion, Dewatripont, et al. 2007).

"ONE SIZE FITS ALL" KNOWLEDGE TRANSFER

Conversely, while subnational governments may pursue inefficient university knowledge-transfer policies, "one size fits all" national government policies can steer universities away from the type of knowledge transfer that fosters regional economic development. Comparative research involving several of our focal countries (Finland, Japan, the United Kingdom, and the United States) revealed that the knowledge-transfer processes—patenting, licensing, and new business formation—favored by national innovation policies were often not the most important contributors to local and regional development (Lester 2007). While some "global" universities produce technology artifacts that are transferable worldwide, effective knowledge transfer for most universities is a more local process and depends upon the nature of industrial development occurring in the regional economy. Universities do contribute to the creation of new businesses, but much more commonly they help to upgrade mature industries, support the diversification of existing businesses into new fields, and assist in the transplantation of industries. In these roles traditional publications, the provision of skilled S&T graduates for the regional economy, and technical problem-solving with local business and industry through consulting and contract research are much more significant channels for influencing technical innovation than patents and licenses (Agarwal and Henderson 2002; Cohen, Nelson, and Walsh 2002). Universities also play a crucial role by providing a "public space" in which, through meetings, research conferences, and industrial liaison programs, local business practitioners can discuss the future direction of technologies, markets, and regional industrial development.

This contribution to regional development is potentially a role all universities with scientific and/or technical faculties, not just "world class" institutions, can perform. National policies encouraging this type of local and regional focus would therefore also promote the development of socially beneficial diversity in the ARE. Such policies should provide incentives for universities to focus less on their possibly wasteful investments in conventional technology transfer and more on developing a strategy for encouraging innovation in their region, as well illustrated by policies adopted by the national governments in Finland and Japan (Kitagawa and Woolgar 2008; Nilsson 2006). This approach would encourage universities to systematically assess the circumstances and development of local industry, the research strengths of the institution, and the most appropriate channels for aligning the university's capabilities with the needs of the local economy (Lester 2007). The Finnish National Centres of Expertise Programme provides one well-regarded national model for developing universities as nodal points in regional networks of innovation by helping them better integrate their research expertise with local industry and business along the lines suggested here (OECD 2007).

In a multilevel system of governance, the various sectors of government each have their own authorities and competences but are also mutually interdependent. Generally speaking, the "higher-level actors" condition, but cannot completely control, the "lower-level actors," who therefore have the authority to develop and implement their own policies (Van Vught, chapter 5). Nevertheless, in all systems, the *joint* impacts of the interactive governments on the ARE are becoming stronger and therefore need to be more carefully coordinated, a point we address in our concluding section.

A New Innovation Policy Strategy: Mutual Learning

The national innovation policies adopted by the leading OECD nations have positively affected the productivity of the Academic Research Enterprise in most countries and have encouraged a more entrepreneurial culture within universities, particularly in the development of active processes of knowledge transfer. At the same time, our discussion of the impacts of these policies reveals several limitations. The apparent positive relationship between the adoption of competition strategies and academic research performance may not be linear, and the actual impact of the observed increased research outputs on technical innovation and economic development has yet to be fully established. Furthermore, the new policies may be encouraging all institutions to engage in a costly race for world-class reputations, a race that relatively few can win and that diminishes the diversity in research missions most beneficial to society. These problems are particularly visible in the US ARE, arguably the country with the most competitive and market-oriented system of universities, where recent econometric analyses have raised serious questions as to whether the existing framework conditions of the ARE promote allocative efficiency in academic research, PhD training, and knowledge transfer (Dill, chapter 10).

We would suggest that the previously identified weaknesses of current public policies appear to be symptoms of market and government failures associated with inadequate information on the performance of both universities and related public policies. In the competitive and multilevel political environment now shaping the ARE, what is needed, in our view, is a new policy perspective, the *mutual learning strategy*.[3] Such a strategy would focus less on the identification and prioritization of promising technology fields (i.e., the prioritization strategy) or on stimulating competition between ARE institutions (i.e., the competition strategy) and more on the provision of information to enhance university research performance and policy learning.

Mutual learning consists of three elements: a continuous search for new, better policies, a process of trial and error, and the gaining of experience and results under real-world conditions. Mutual learning, in this sense, is the "deliberate attempt to adjust the goals and techniques of policy in response to past experience and new information" (Hall 1993, 278). It implies the search for more effective policies through the application of existing policies. It combines application with analysis and thus focuses on learning.

A mutual learning strategy underscores the necessity of providing valid, publicly accessible information on the performance of the ARE. Mutual learning can take place only if the access to knowledge is a public good, open to all participants in the process, and if no specific ownership of information exists. The mutual learning strategy is therefore clearly related to the concept of "open innovation" (Chesbrough 2003) and the Open Source approaches to software and information, in which ownership and protection of information are seen as restricting the circulation of knowledge and the consequent social benefits for society. A mutual learning policy strategy, therefore, would stress the importance of open data environments and of public provision to stakeholders of information about research performance and the effectiveness of public policies, in order to stimulate learning and dissemination processes.

INFORMATION AND MUTUAL LEARNING

A traditional role of government is to provide information in strategically important policy areas to help the public evaluate socially beneficial behavior. However, the increased economic value of academic research, PhD graduates, and university reputation has motivated development of a worldwide industry of publications designed to provide information on university rankings and research doctoral program quality, usually at a profit, by nongovernmental agencies. The *U.S. News and World Report* pioneered the publication of university quality rankings for students in 1983. But more recent rankings, such as the Shanghai Jiao Tong University rankings (commenced in 2003), the *Times Higher Education Supplement* rankings (commenced in 2004), and PhD rankings by the commercial firm Academic Analytics in the United States (commenced in 2005) have focused more explicitly on institutional research performance, research doctoral quality rankings, and worldwide university reputation. These rankings provide extra stimuli for universities and governments to clamber up the global ladder of university reputation. The measures employed in these league tables represent the private interests of those who design them, and the validity and reliability of their indicators of research performance are highly debatable (Dill and Soo 2005). It is likely, therefore, that the information they provide contributes to the types of market and government failures in the ARE identified in our collected case studies. In the new worldwide competitive market that confronts the ARE, there is a need for more valid "signals" of research performance—information-oriented public policy instruments designed to assure a more efficient rivalry among universities as they vie to better serve society.

Similarly, we suggest that the policy strategy of mutual learning is best understood as a continuous search by multiple sets of actors for the improvement of existing policies. The focus is on trying to find interesting results and "good practices" gained from the implementation of policies in various real-world contexts. In this sense policies should be seen as "social experiments" that are simultaneously tried out by different groups of actors, allowing comparisons and mutual learning. The strategy of mutual learning is a policy strategy of parallel social experimentation with the objective of finding results that can be diffused to other environments.

The case studies in this volume offer some interesting leads for the application of a mutual learning strategy to the design of public policy for the ARE. The various instruments for identifying effective research performance by universities, such as the research centers of excellence programs adopted in a number of our focal countries and the research doctoral evaluations subsidized by the US federal government, all provide useful signals to the clients and supporters of universities that would likely improve the allocative efficiency of the ARE in the competitive global environment. The research evaluations in the Netherlands, Jongbloed suggests (chapter 8), also proved extremely influential in improving research performance in the university system. Because this latter instrument supports the development of a qualitative peer-review evaluation process, it promotes useful ongoing learning by the universities themselves about means of improving their programs of research. As such, it does not appear to suffer from the mentioned limitations of metric-based evaluation systems or of performance-based subject evaluations, such as the much more costly RAE process in the United Kingdom. Finally, the Open Method of Coordination (OMC), as it is being applied in the context of the Lisbon Strategy of the European Union, and the various experiences with the multilevel governance structures in federal political systems provide us with some further elements to elaborate the mutual learning strategy.

THE OPEN METHOD OF COORDINATION

The Open Method of Coordination in Europe offers a creative new form of information-based policy instruments. It is based on the assumption that coordination of national policies can be achieved without the transfer of legal competences or financial resources to the European level. It works through the setting of common goals; translating these into national policies; defining explicit, related performance indicators; and measuring and comparing the performance of these policies. With regard to national innovation, performance measurement takes place by using standardized indicators for benchmarking processes and progress monitoring as well as by means of peer reviews of the outcomes (Bruno, Jacquot, and Mandin 2006; European Commission 2000; Gornitzka 2007).

The OMC clearly is an arrangement that promotes policy learning among the EU member states. Its basic idea is to create, in a two-level structure of jurisdictions, systemically organized mutual-learning processes. At the level of the European Union, the member states evaluate their various policy performances according to the joint objectives set and the indicators agreed upon. In the variety of experiences, "good practices" are identified and their diffusion is supported. The coordination of the process is largely in the hands of the European Commission, which analyses the progress reports of the member states, identifies good practices, suggests recommendations for each member state, and drafts an overall report that must be approved by the European Council (the heads of state or government of the member states and the president of the European Commission). Though the European Commission cannot make mandatory recommendations, it nevertheless plays a crucial role in organizing the process by suggesting common goals, collecting and analyzing information, and drafting recommendations. The OMC stimulates the member states to experiment with different policies, evaluate their outcomes, and then identify good practices. It is a process of mutual learning, coordinated at the level of the European Commission but with substantial flexibility and openness for the national governments.

A criticism of the OMC policy is that it is too "top-down" and may, ironically, lessen mutual learning by limiting the process of competitive policy experimentation among states (Kerber and Eckardt 2007). This argument relies on the principle of laboratory federalism (Oates 1972), which suggests that in the provision of complex public goods, such as the university's contribution to national innovation, progress in public policy is more likely to occur from experiments by subnational governments than by central government dictate. The basic idea of laboratory federalism is that decentralized governments offer the advantage of better knowledge of local and regional circumstances and hence are better able to efficiently provide public goods (Tanzi 1995). In addition—and of interest for our discussion—the theory suggests that "learning by doing" experiments among decentralized governments and among universities themselves will produce more effective policy innovations that will then be identified and adopted by other institutions, states, or regions. In the parallel processes of competitive experimentation, superior policies are selected and diffused through imitation. Mutual learning is therefore the expected outcome of the competition among policies.

However, the empirical research on laboratory federalism suggests that the generation and diffusion of policy innovations among subnational governments may be contingent upon factors that need to be considered in the design of the framework conditions for the ARE (Musgrave 1997; Tanzi 1995; Volden 2006). For example, insufficient information on the costs and benefits of ARE policy innovations may limit the identification of best practice policies among subnational governments, and, as suggested in the German case, path-dependent traditions within states and provinces may cause both the generation and the diffusion of superior policies to be more the exception than the rule. Subnational government bureaucracies may also be less competent on issues of science policy than national agencies or more subject to manipulation by local political forces, both suggested in the US state cases. For these reasons, and in the absence of appropriate incentives and supporting institutions, generic decentralization strategies may not yield the expected social benefits in the performance of the ARE.

The EU experience with the OMC can be usefully compared with the lack of comparable information-oriented instruments to promote mutual learning among the US states. The National Science Foundation provides extensive data on science and technology in the US system, and federal science agencies subsidize the research doctoral rankings conducted by the National Academies of Science. But the federal government has not formally supported the provision of systematic comparative data, similar to that contained in the European Innovation Scoreboard, on the innovation performance of the fifty states (European Commission 2008) or provided comparative data on the publication performance of US universities such as that included in the *European Report on Science and Technology Indicators* (European Commission 2003). Nor has it provided related indicators or incentives for policy learning that would help guide the rapidly increasing investments in academic science and technology by the many states. As a consequence, the regulatory framework of publicly supported universities among the fifty states continues to vary widely, with one state providing to its ARE the autonomy that has been discovered to be associated with high levels of research performance and neighboring states pursuing highly restrictive and inefficient regulatory policies (Aghion, Dewatripont, et al. 2007). Similarly, as indicated in the California and Pennsylvania case studies, the states are failing to implement policy strategies for academic research that would help improve the performance of the overall US ARE. As our cases suggest, this general problem of regulatory "lock-in" and lack of policy learning is also exhibited within other federal systems of higher education, such as Australia's, Canada's, and Germany's.

In summary, the policy strategy of mutual learning provides a potentially valuable and important supplement to the policy strategies of prioritization and competition, the two strategies that are so far still dominant in national innovation policies. The mutual learning strategy assumes a minimal level of policy heterogeneity and therefore is particularly appropriate for multi-echelon political systems, like federal states and the European Union. But mutual learning may be applicable in unitary nation states as well, such as Finland, with its emphasis on regional diversification. Finally, the heterogeneity of policy contexts also offers a new and interesting means of addressing the issue of university autonomy in different AREs and the inequalities regarding global academic competition that result from these differences. In diversifying their policy contexts in order to stimulate policy learning, national governments may create different conditions for different categories of universities and hence allow at least some of these institutions to really compete on the international platform of academic reputation. National governments that take global competition processes seriously and accept the fact that the capacity to create, disseminate, and apply knowledge is of crucial importance in these processes may, in this sense, find important extra strategic advantages in developing their ability to learn.

Conclusion

We end our global study of national innovation policies and the academic research enterprise with the conclusion that public policies designed to strengthen academic research and its contributions to economic development need to focus on promoting mutual learning among universities, their various patrons, and policymakers in the different strata of multilevel governance. For this to occur, governments need to invest in information-based policy instruments that provide to the many stakeholders of the universities valid and reliable information on academic research performance as well as comparably objective information on the social costs and benefits of public policies intended to enhance academic research, improve research doctoral training, and boost knowledge transfer.

Universities by their very nature are learning organizations dedicated to gathering, comprehending, and conveying information and knowledge to the larger society (Dill 1999). They are also institutions that appear particularly attentive and responsive to valid information about their own academic performance, quality, and reputation. Therefore, in the now highly competitive global environment, appropriately crafted information-based policies that promote the activity of mutual learning are likely to be particularly powerful instruments for shaping university behavior and improving the benefits to society of academic research.

NOTES

1. The following observations were noted in the case analyses:

- Of the 283 US universities engaged in research and doctoral education, 99 (66 public, 33 private) perform 72% of federal academic R&D and grant 69% of research doctorates (Dill, chapter 10).
- In Canada 80% of the Can$10 billion competitively allocated to universities by the national government since 1997 has been awarded to 15 universities (Fisher and Rubenson, chapter 3).
- In Germany 30 of 100 universities account for 78% of German Research Foundation project funds (Enders, chapter 7).

- In the European Union the larger, older universities receive a disproportionate amount of the competitively awarded Framework Programme research funds (Van Vught, chapter 5).
- In the United Kingdom a clear variation existed even among the pre-1992 universities in research resources, reputations, and outputs. By 2000 the top 15 of the now 108 universities accounted for 50% of research income and the bottom 50 for less than 10% (Henkel and Kogan, chapter 9).

In addition, in Australia 6 of 38 universities produce 70% of publications and consume 65% of R&D resources (Bourke and Butler 1998).

2. It is reported that a Nobel Laureate economist at Stanford University once argued at a faculty meeting that it made no economic sense for the university to continually set its tuition charges below what the market would truly bear. His administrative colleagues agreed with his economic analysis but felt then (and continue to feel) that maintaining Stanford's public reputation required that the institution's tuition fees remain comparable with its distinguished private university competitors.

3. This mutual learning strategy has a strong foundation in political science (Braybrooke and Lindblom 1963; Lindblom 1959, 1965, 1977), but there are also significant relationships with the "learning approaches" in the planning literature and policy sciences (Dunn 1971; Friedmann 1973; Jantsch 1975; Michael 1973). Mutual learning is also seen as a strategy superior to rational "constructivist" planning in the field of evolutionary economics because it takes the experiences gained under real-world conditions into account (Nooteboom 2000).

REFERENCES

Agarwal, A. and R. Henderson. (2002) Putting Patents in Context: Exploring Knowledge Transfer from MIT. *Management Science* 48 (1): 44–60.

Aghion, P., L. Boustan, C. Hoxby, and J. Vandenbussche. (2005) Exploiting States' Mistakes to Identify the Causal Impact of Higher Education on Growth. Working Paper, Harvard University Department of Economics, Cambridge, MA.

Aghion, P., M. Dewatripont, C. Hoxby, A. Mas-Colell, and A. Sapir. (2007) Why Reform Europe's Universities? Bruegel Policy Brief, 2007/04, Bruegel, Brussels.

Bardach, E. (1979) *The Implementation Game.* Cambridge, MA: MIT Press.

Bourke, P. and L. Butler. (1998) The Concentration of Research in Australian Universities: Six Measures of Activity and Impact. Higher Education Series, Report No. 32, DEETYA, Canberra.

Bowen, H. R. (1980) *The Costs of Higher Education.* San Francisco: Jossey-Bass.

Braunerhjelm, P. and M. Feldman (eds.). (2006) *Cluster Genesis: Technologically-Based Industrial Development.* Oxford: Oxford University Press.

Braybrooke, D. and C. E. Lindblom. (1963) *A Strategy of Decision: Policy Evaluation as a Social Process.* New York: Free Press.

Brewer, D. J., S. M. Gates, and C. A. Goldman. (2002) *In Pursuit of Prestige: Strategy and Competition in U.S. Higher Education.* New Brunswick, NJ: Transaction Press.

Bruno, I., S. Jacquot, and L. Mandin. (2006) Europeanization through Its Instrumentation: Benchmarking, Mainstreaming, and the Open Method of Coordination . . . Toolbox or Pandora's Box? *Journal of European Public Policy* 13 (4): 519–536.

Chesbrough, H. W. (2003) *Open Innovation: The New Imperative for Creating and Profiting from Technology.* Boston: Harvard Business School Press.

Clark, B. R. (1998) *The Entrepreneurial University.* Oxford: Pergamon Press.

Cohen, W. M., R. R. Nelson, and J. P. Walsh. (2002) Links and Impacts: The Influence of Public Research on Industrial R&D. *Management Science* 48 (1): 1–23.

Crespi, G. and A. Geuna. (2004) The Productivity of Science. Science and Technology Policy Research Unit (SPRU), University of Sussex, UK, www.sussex.ac.uk/spru/documents/crespiost2.pdf.

Dahl, R. A. and C. E. Lindblom. ([1953]1976) *Politics, Economics, and Welfare.* Chicago: University of Chicago Press.

Dill, D. D. (1999) Academic Accountability and University Adaptation: The Architecture of an Academic Learning Organization. *Higher Education* 38 (2): 127–154.

———. (2001), The Regulation of Public Research Universities: Changes in Academic Competition and Implications for University Autonomy and Accountability. *Higher Education Policy* 14 (1): 21–35.

Dill, D. D. and M. Soo. (2005) Academic Quality, League Tables, and Public Policy: A Cross-National Analysis of University Ranking Systems. *Higher Education* 49 (4): 495–533.

Dosi, G., P. Llerena, and M. S. Labini. (2006) The Relationships between Science, Technologies, and Their Industrial Exploitation: An Illustration through the Myths and Realities of the So-Called "European Paradox." *Research Policy* 35:1450–1464.

Dunn, E. S., Jr. (1971) *Economic and Social Development: A Process of Social Learning.* Baltimore: Johns Hopkins Press.

Elmore, R. F. (1987) Instruments and Strategy in Public Policy. *Review of Policy Research* 7 (1): 174–186.

Etzioni, A. (1968) *The Active Society.* New York: Free Press.

European Commission. (2000) Development of an Open Method of Co-ordination for Benchmarking National Research Policies: Objectives, Methodology, and Indicators. Working Document, SEC(2000) 1842, EC, Brussels.

———. (2003) *Third European Report on Science and Technology Indicators 2003: Towards a Knowledge-Based Economy.* Luxembourg: EC.

———. (2008) *European Innovation Scoreboard 2007: Comparative Analysis of Innovation Performance.* Luxembourg: EC.

Friedmann, J. (1973) *Retracking America: A Theory of Transactive Planning.* New York: Anchor Press and Doubleday.

Galama, T. and J. Hosek. (2008) *U.S. Competitiveness in Science and Technology.* Santa Monica, CA: RAND.

Geiger, R. L. (2007) Technology Transfer Offices and the Commercialization of Innovation in the United States. Paper presented at the CHER Conference, Dublin, Ireland, August 30- September 1.

Geuna, A. (2001) The Changing Rationale for European University Research Funding: Are There Negative Unintended Consequences? *Journal of Economic Issues* 35 (3): 607–632.

Gornitzka, Å. (2007) The Open Method of Coordination in European Research and Education Policy: New Political Space in the Making? Paper presented at the CHER Conference, Dublin, Ireland, August 30- September 1.

Hall, P. A. (1993) Policy Paradigms, Social Learning, and the State: The Case of Economic Policymaking in Britain. *Comparative Politics* 25:275–296.

Hood, C. (1983). *The Tools of Government.* London: Macmillan.

Horta, H., F. M. Veloso, and R. Grediaga. (2007) Navel Gazing: Academic Inbreeding and Scientific Productivity. Paper presented at the CHER Conference, Dublin, Ireland. August 30–September 1.

Ingram, H. and D. Mann. (1980) *Why Policies Succeed or Fail.* Beverly Hills, CA: Sage.

Jantsch, E. (1975) *Design for Evolution: Human Systems in Transaction.* New York: Braziller.

Johnes, G. (2004) The Evaluation of Welfare under Alternative Models of Higher Education Finance. In Teixeira et al. 2004, 87–112.

Jongbloed, B. and H. Vossensteyn. (2001) Keeping up Performances: An International Survey of Performance-Based Funding in Higher Education. *Journal of Higher Education Policy and Management* 23 (2): 127–145.

Kerber, W. and M. Eckardt. (2007) Policy Learning in Europe: The Open Method of Coordination and Laboratory Federalism. *Journal of European Public Policy* 14 (2): 227–247.

Kitagawa, F. and L. Woolgar. (2008) Regionalization of Innovation Policies and New University-Industry Links in Japan: Policy Review and New Trends. *Prometheus* 26 (1): 55–67.

Lester, R. K. (2007) Universities, Innovation, and the Competitiveness of Local Economies: An Overview. In R. K. Lester and M. Sotarauta (eds.), *Innovation, Universities, and the Competitiveness of Regions,* 9–30. Helsinki: TEKES.

Lindblom, C. E. (1959) The Science of Muddling Through. *Public Administration Review* 19:79–88.

———. (1965) *The Intelligence of Democracy.* New York: Free Press.

———. (1977) *Politics and Markets.* New York: Basic Books.

Majone, G. (1997) The New European Agencies: Regulation by Information. *Journal of European Public Policy* 4 (2): 262–275.

Martins, J. O., R. Boarini, H. Strauss, C. de la Maisonneuve, and C. Saadi. (2007) The Policy Determinants of Investment in Tertiary Education. Economics Department Working Papers, No. 576, OECD, Paris, www.oecd.org/eco.

Mazmanian, D. A. and P. A. Sabatier. (1981) *Effective Policy Implementation.* Toronto: Lexington Books.

Michael, D. N. (1973) *On Learning to Plan—and Planning to Learn.* San Francisco: Jossey-Bass.

Mitnick, B. M. (1980) *The Political Economy of Regulation: Creating, Designing, and Removing Regulatory Reforms.* New York: Columbia University Press.

Mowery, D. C. and B. N. Sampat. (2004) Universities in National Innovation Systems. In J. Fagerberg, D. C. Mowery, and R. R. Nelson (eds.), *Oxford Handbook of Innovation,* 209–239. Oxford: Oxford University Press.

Musgrave, R. A. (1997) Devolution, Grants, and Fiscal Competition. *Journal of Economic Perspectives* 11 (4): 65–72.

Nelson, R. R. and S. G. Winter. (1982) *An Evolutionary Theory of Economic Change.* Cambridge, MA: Harvard University Press.

Nilsson, J.-E. (2006) *The Role of Universities in Regional Innovation Systems: A Nordic Perspective.* Copenhagen: Copenhagen Business School.

Nooteboom, B. (2000) *Learning and Innovation in Organization and Economics.* Oxford: Oxford University Press.

Oates, W. E. (1972) *Fiscal Federalism.* New York: Harcourt, Brace, Jovanovich.

———. (1999) An Essay on Fiscal Federalism. *Journal of Economic Literature* 37 (3): 1120–1149.

Organisation for Economic Co-operation and Development (OECD). (2007) *Higher Education and Regions: Globally Competitive, Locally Engaged.* Paris: OECD.

Schneider, A. and H. Ingram. (1990) Behavioral Assumptions of Policy Tools. *Journal of Politics* 52 (2): 510–530.

Soo, M. (2008) The Effect of Market-Based Policies on Academic Research Performance: Evidence from Australia, 1992–2004. PhD diss., University of North Carolina–Chapel Hill.

Tanzi, V. (1995) Fiscal Federalism and Decentralization: A Review of Some Efficiency and Macroeconomic Aspects. In M. Bruno and B. Pleskovic (eds.), *Annual World Bank Conference on Development Economics, 1995,* 295–316. Washington, DC: World Bank.

Teixeira, P., B. Jongbloed, D. D. Dill, and A. Amaral (eds.). (2004) *Markets in Higher Education: Rhetoric or Reality?* Dordrecht, Netherlands: Kluwer.

Van Vught, F. A. (ed.). (1989) *Governmental Strategies and Innovation in Higher Education.* London: Jessica Kingsley.

———. (1994) Policy Models and Policy Instruments in Higher Education. In J. C. Smart (ed.), *Higher Education: Handbook of Theory and Research,* 10:88–126. New York: Agathon Press.

———. (2007) Knowledge Transfer in the European Union. Paper presented at the CHER Conference, Dublin, Ireland, August 30-September 17.

———. (2008) Mission Diversity and Reputation in Higher Education. *Higher Education Policy* 21 (2): 151–174.

Volden, C. (2006) States as Policy Laboratories: Emulating Success in the Children's Health Insurance Program. *American Journal of Political Science* 50 (2): 294–312.

Weimer, D. and A. R. Vining. (2005) *Policy Analysis: Concepts and Practice.* Upper Saddle River, NJ: Pearson Prentice Hall.

Williams, G. (2004) The Higher Education Market in the United Kingdom. In Teixeira et al. 2004, 241–269.

World Bank. (2007), *Global Economic Prospects: Managing the Next Wave of Globalisation,* Washington, DC: World Bank.

AKIRA ARIMOTO is Director and Professor at the Research Institute for Higher Education of Hijiyama University and Professor Emeritus of Hiroshima University in Japan.

DAVID D. DILL is Professor Emeritus of Public Policy at the University of North Carolina–Chapel Hill in the United States.

JÜRGEN ENDERS is Professor and Director of the Center for Higher Education Policy Studies (CHEPS) at the University of Twente in the Netherlands.

DONALD FISHER is Professor of Sociology in Educational Studies and Co-Director of the Center for Policy Studies in Higher Education and Training (CHET) at the University of British Columbia in Canada.

ROGER L. GEIGER is Distinguished Professor of Education and Senior Scientist in the Center for the Study of Higher Education at Pennsylvania State University in the United States.

LEO GOEDEGEBUURE is Academic Programs Director of the LH Martin Institute for Higher Education Leadership and Management at the University of Melbourne and Associate Professor in the Centre for Higher Education Management & Policy (CHEMP) at the University of New England in Australia.

MARY HENKEL is Professor Associate of Politics and History at Brunel University in the United Kingdom.

SEPPO HÖLTTÄ is Professor of Higher Education Administration and Finance in the Department of Management Studies at the University of Tampere in Finland.

BEN JONGBLOED is Senior Research Associate in the Center for Higher Education Policy Studies (CHEPS) at the University of Twente in the Netherlands.

MAURICE KOGAN (April 10, 1930–January 6, 2007) was Director of the Center for the Evaluation of Public Policy and Practice and Professor Emeritus of Government and Public Administration at Brunel University in the United Kingdom.

V. LYNN MEEK is Professor and Foundation Director of the LH Martin Institute for Higher Education Leadership and Management at the University of Melbourne in Australia.

KJELL RUBENSON is Professor of Adult Education in Educational Studies and Co-Director of the Center for Policy Studies in Higher Education and Training (CHET) at the University of British Columbia in Canada.

JEANNET VAN DER LEE is Senior Research Fellow and Research Project Officer in the Centre for Higher Education Management & Policy (CHEMP) at the University of New England in Australia.

FRANS A. VAN VUGHT is Professor of Comparative Higher Education Policy Studies, Center for Higher Education Policy Studies (CHEPS), and former President of the University of Twente in the Netherlands.

WILLIAM M. ZUMETA is Professor of Public Affairs and Education Policy at the University of Washington and Senior Fellow at the National Center for Public Policy and Higher Education in the United States.

Page numbers in *italics* indicate figures and tables.